**Soil Biology**

Volume 59

Series Editor

Ajit Varma, Amity Institute of Microbial Technology,
Amity University, Noida, Uttar Pradesh, India

The Springer series *Soil Biology* publishes topical volumes in the fields of microbiology, environmental sciences, plant sciences, biotechnology, biochemistry, microbial ecology, mycology and agricultural sciences.

Special emphasis is placed on methodological chapters or volumes. This includes coverage of new molecular techniques relevant to soil biology research or to monitoring and assessing soil quality as well as advanced biotechnological applications. Leading international authorities with a background in academia, industry or government will contribute to the series as authors or editors.

Key Topics: microbial-plant interactions; microbial communities; root symbiosis, mycorrhiza; rhizosphere environment; soil fauna, e.g. termites etc.; biochemical processes, soil proteins, enzymes, nucleic acids; degradation of biomaterials, decomposition, nutrient cycles; soil genesis, mineralization; bioremediation of contaminated sites; biotechnological applications of soil microorganisms.

More information about this series at http://www.springer.com/series/5138

Bhoopander Giri • Ajit Varma
Editors

# Soil Health

Springer

*Editors*
Bhoopander Giri
Department of Botany,
Swami Shraddhanand College
University of Delhi
Delhi, India

Ajit Varma
Amity Institute of Microbial Technology
Amity University
Noida, Uttar Pradesh, India

ISSN 1613-3382 ISSN 2196-4831 (electronic)
Soil Biology
ISBN 978-3-030-44363-4 ISBN 978-3-030-44364-1 (eBook)
https://doi.org/10.1007/978-3-030-44364-1



This Springer imprint is published by the registered company Springer Nature Switzerland AG.
The registered company address is: Gewerbestrasse 11, 6330 Cham, Switzerland

# Foreword

The principles of functionality and sustainability of the ecosystem are incorporated into the definition of soil health, also known as soil quality. The disturbances, natural or anthropic, affect the functions of the soil and, therefore, their health. The evaluation of this is not easy, since it is dynamic and can change in a short period of time. These changes depend on the specific characteristics of the soil, the environmental conditions, the use, and the management practices.

In recent decades, to achieve high agricultural production, intensive agricultural practices have been developed and applied; these practices, types of irrigation, and, especially, chemical fertilizers and pesticides have had very negative environmental consequences. In the same way, large disturbances of natural or anthropic origin have progressively diminished the world's forest area, degrading large areas of soil.

This volume addresses the current state of knowledge about soil health; the properties of the soil, especially the biological component, are related to its sanitary state. In some chapters, some parameters of these components are analyzed as indicators of soil health; in other chapters, the potential use or application of the biological component to restore or improve soil conditions and the production of sustainable crops is addressed. In addition, the agricultural practices called "green manures" and some novelties in biological components or fertilizer materials are discussed as possible strategies to restore or improve soil health and productivity in a sustainable way. In short, this book addresses agri-environmental sustainability from the point of view of soils. We are sure that this book will help us face such important challenges for the future as global food security and climate change.

I thank the editors for this invitation, and I am sure that this book *Soil Health* will be a great resource for knowledge about the thematic area of soil health and a reference for academics, researchers, and students.

I convey my best wishes to Springer Nature and the editors and collaborators who are experts in the field of this remarkable book.

ETSI Montes, Forestal y del Medio Natural, Universidad Politecnica de Madrid, Madrid, Spain

José Alfonso Domínguez-Núñez

# Preface

Soil health can be defined as the capacity of a soil to function within ecosystem frontiers, to sustain biological productivity, to maintain environmental quality, and to promote plant, animal, and human health. Soil fertility is a feature of soil health showing the capacity of a soil to provide nutrients for plant's growth and development. Soil health is established through the interactions of physical, chemical, and biological properties including soil texture, measurement of percent of the sand, silt, and clay, arrangement of individual soil particles like sand, silt, and clay into aggregates (soil structure), nutrient-carrying capacity and pH, the group of soil organisms like bacteria, fungi, and actinomycetes, etc. A healthy soil provides adequate levels of macro- and micronutrients to plants and holds adequate population of soil microorganisms. Soil representing good tilth or structure often found to resist soil erosion and compaction and thereby degradation provides adequate aeration and water to plants, promotes good root growth, and maintains worthy biotic habitat and diverse populations of beneficial organisms and low populations of pests and pathogens. Indeed, soil health plays a central role in the economic and social development of a country. The production of food, fodder, renewable energy, and several other essential commodities vital for sustaining human, animal, and plant life depends on soil health. In the past few decades, human population has increased rapidly, stimulating a tremendous increase in the intensification of agriculture to meet the ever-increasing demand for food. Soils are now showing symptoms of exhaustion and stagnating or declining crop yields. Inadequate and imbalanced nutrient management and desertion of organic manures is gradually triggering nutrient deficiencies in crop fields. Overextracting of mineral nutrients, neglect of organic fertilizers, insignificant nutrient replenishment, excessive use of chemical fertilizers and pesticides, and irrigation with poor quality water/saline water are found to be the major causes of festering crop productivity. Since the prospects for further increase in the area under cultivation are diminishing, much of the desired increase in food grain production needs to be achieved by enhancing crop productivity per unit area through holistic and sustainable approaches.

The volume *Soil Health* comprises 19 provocative chapters written by the experts of this field, covers latest research, and provides up-to-date knowledge of different aspects of soil health, factors influencing soil health, consequences of degradation of soil health on sustainable agriculture, and solutions to improve and maintain soil health so as to achieve the goal of higher productivity and sustainability without any damage to the soil system and the environment. We believe that with the opulence of information on different aspects of soil health and its sustainability, this extensive volume is a valuable resource for researchers, academicians, and students in the broad field of botany, ecology, microbiology, and agriculture.

We are highly delighted and thankful to all our contributing authors for their endless support and outstanding cooperation to write altruistically these authoritative and valuable chapters. We extend our sincere thanks to all our colleagues who helped us in the preparation and compilation of this generous volume. We also thank Springer officials, especially William F Curtis, Eric Schmitt, Sabine Schwarz, Paul Roos, Nathalie Berg, and Anand Ventakachalam for their generous support and efforts to accomplish this wide volume. We especially thank our families for consistent support and encouragement.

Delhi, India Bhoopander Giri
Noida, Uttar Pradesh, India Ajit Varma

# Contents

# Contributors

**K. A. Adegbite** Landmark University, Omu-Aran, Kwara, Nigeria

**Kehinde Abodunde Adegbite** Crop and Soil Science Department, Landmark University, Omu-Aran, Kwara, Nigeria

**A. O. Adekiya** Landmark University, Omu-Aran, Kwara, Nigeria

**Aruna Olasekan Adekiya** Crop and Soil Science Department, Landmark University, Omu-Aran, Kwara, Nigeria

**Imtiaz Ahmad** Department of Entomology, Pennsylvania State University, University Park, PA, USA

**E. T. Alori** Landmark University, Omu-Aran, Kwara, Nigeria

**Elizabeth Temitope Alori** Crop and Soil Science Department, Landmark University, Omu-Aran, Kwara, Nigeria

**Una Andersone-Ozola** Faculty of Biology, Department of Plant Physiology, University of Latvia, Riga, Latvia

**K. S. Anil Kumar** ICAR–National Bureau of Soil Survey and Land Use Planning, Regional Centre, Bangalore, India

**Manju Balhara** Department of Bioscience, Aarhus University, Ny Munkegade, Denmark

**Pooja Baweja** Department of Botany, University of Delhi, New Delhi, India

**N. Biswakarma** ICAR–Indian Agricultural Research Institute, New Delhi, India

**Ranjit Chatterjee** Department of Vegetable and Spice Crops, Uttar Banga Krishi Viswavidyalaya (UBKV), Pundibari, West Bengal, India

**Thomas W. Culliney** United States Department of Agriculture, Animal and Plant Health Inspection Service, Plant Protection and Quarantine, Center for Plant Health Science and Technology Plant Epidemiology and Risk Analysis Laboratory, Raleigh, NC, USA

**K. Das** ICAR–Central Research Institute for Jute and Allied Fibres, Barrackpore, India

**Ankita Debnath** Department of Vegetable and Spice Crops, Uttar Banga Krishi Viswavidyalaya (UBKV), Pundibari, West Bengal, India

**Riddha Dey** Department of Botany, Institute of Science, Banaras Hindu University, Varanasi, Uttar Pradesh, India

**S. W. Ejue** Landmark University, Omu-Aran, Kwara, Nigeria

**P. C. Ghasal** ICAR–Indian Institute of Farming Systems Research, Meerut, India

**Manju M. Gupta** Sri Aurobindo College, University of Delhi, Delhi, India

**Sandeep Kumar Gupta** Department of Botany, Institute of Science, Banaras Hindu University, Varanasi, Uttar Pradesh, India

**Nguyen Hue** University of Hawaii, Honolulu, HI, USA

**Gederts Ievinsh** Faculty of Biology, Department of Plant Physiology, University of Latvia, Riga, Latvia

**Anu Kalia** Electron Microscopy and Nanoscience Laboratory, Department of Soil Science, College of Agriculture, Punjab Agricultural University, Ludhiana, Punjab, India

**Andis Karlsons** Laboratory of Plant Mineral Nutrition, Institute of Biology, University of Latvia, Salaspils, Latvia

**K. S. Karthika** ICAR–National Bureau of Soil Survey and Land Use Planning, Regional Centre, Bangalore, India

**Harleen Kaur** Department of Microbiology, College of Basic Sciences and Humanities, Punjab Agricultural University, Ludhiana, Punjab, India

**Inderdeep Kaur** Sri Guru Tegh Bahadur Khalsa College, University of Delhi, Delhi, India

**Sukhjinder Kaur** Department of Soil Science, College of Agriculture, Punjab Agricultural University, Ludhiana, Punjab, India

**Gaurav Kumar** Department of Environmental Studies, PGDAV College, University of Delhi, New Delhi, India

**A. Kumar** ICAR–Indian Institute of Maize Research, Ludhiana, Punjab, India

**Sanjeev Kumar** Department of Genetics and Plant Breeding, School of Agriculture Lovely Professional University, Jalandhar, India

**Savindra Kumar** Department of Botany, Zakir Husain Delhi College, University of Delhi, New Delhi, India

**Zaiga Landorfa-Svalbe** Faculty of Biology, Department of Plant Physiology, University of Latvia, Riga, Latvia

**Subhalaxmi Mishra** Department of Vegetable Science, OUAT, Bhubaneswar, Odisha, India

**Neemisha** Department of Soil Science, Punjab Agricultural University, Ludhiana, Punjab, India

**Anita Osvalde** Laboratory of Plant Mineral Nutrition, Institute of Biology, University of Latvia, Salaspils, Latvia

**Apostolos Papadopoulos** Crop Intellect, Riseholme College, Lincoln University, Lincoln, UK

**V. Pooniya** ICAR–Indian Agricultural Research Institute, New Delhi, India

**Jay Kishor Prasad** Department of Botany, Institute of Science, Banaras Hindu University, Varanasi, Uttar Pradesh, India

**Maddur Puttaswamy Raghavendra** Maharani's Science College for Women, Mysuru, Karnataka, India

**Richa Raghuwanshi** Department of Botany, Mahila Mahavidyalaya, Banaras Hindu University, Varanasi, Uttar Pradesh, India

**Pratibha Rawat** Department of Plant Physiology, College of Basic Sciences & Humanities, G. B. Pant University of Agriculture & Technology, Pantnagar, Uttarakhand, India

**Samiksha** Department of Biotechnology, School of Biosciences, Lovely Professional University, Jalandhar, India

**Aralakuppe Narayana Santhoshkannada** Maharaja's Government Pre-University College, Mysuru, Karnataka, India

**Deepti Shankhdhar** Department of Plant Physiology, College of Basic Sciences & Humanities, G. B. Pant University of Agriculture & Technology, Pantnagar, Uttarakhand, India

**S. C. Shankhdhar** Department of Plant Physiology, College of Basic Sciences & Humanities, G. B. Pant University of Agriculture & Technology, Pantnagar, Uttarakhand, India

**Sat Pal Sharma** Department of Vegetable Science, College of Horticulture, Punjab Agricultural University, Ludhiana, Punjab, India

**Premasis Sukul** Department of Soil Science, School of Agriculture, Lovely Professional University, Jalandhar, India

**Olayanju TMA** Landmark University, Omu-Aran, Kwara, Nigeria

**Sania Zaib** Faculty of Biological Sciences, Department of Biochemistry, Quaid-i-Azam University, Islamabad, Pakistan

**R. Zhiipao** ICAR–Indian Agricultural Research Institute, New Delhi, India

# Chapter 1
# Soil Health in India: Past History and Future Perspective

**Priyanka Srivastava, Manju Balhara, and Bhoopander Giri**

**Abstract** Soil is a very important and sensitive resource of any country as the crop yield of a nation on which the whole population depends is directly linked to the soil health. The dependence of a nation on others for meeting the demand of its population for food and other crop products is a matter of great concern. The Green Revolution in India intensified agricultural productivity to meet urgent public needs and for the commercialization of crop products. Indeed, Green Revolution helped in achieving goal up to a certain level with the use of high input of chemicals in the form of fertilizers, pesticides, fungicides, insecticides, nematicides and weedicides along with intense irrigation practices. After Green Revolution, the decline in crop yield in spite of fertilizer application reveals the loss of soil fertility. Toxic chemicals in soil affected the life of beneficial soil organisms, which indeed are responsible for maintaining soil fertility. Further, these chemicals polluted groundwater, air and adversely affected human and animal health. Hence, restoration of soil health and environment is an urgent need. Avoidance of chemical fertilizers and use of natural fertilizers like biofertilizers, vermicompost, farm yard and green manure, and biopesticides can be a sustainable approach in achieving the crop productivity along with nourishing the soil and environment. Present chapter discusses about the effects of Green Revolution on soil health in India and suggests for consideration of techniques with eco-friendly approaches to heal soil loss and to manage soil fertility for sustainable agriculture.

**Keywords** Green Revolution · Soil health · Sustainable agriculture · Productivity · Restoration

P. Srivastava
Department of Botany, Sri Guru Tegh Bahadur Khalsa College, University of Delhi, Delhi, India

M. Balhara
Department of Bioscience, Aarhus University, Aarhus, Denmark

B. Giri (✉)
Department of Botany, Swami Shraddhanand College, University of Delhi, Delhi, India



B. Giri, A. Varma (eds.), *Soil Health*, Soil Biology 59,
https://doi.org/10.1007/978-3-030-44364-1_1

## 1.1 Introduction

Soil is a vital natural resource that must be kept in harmony with the environment for agroecosystem's sustainability and services to the benefit of the mankind. By virtue of geologic erosion, some of the world's most productive soils have been developed, for instance, in India the Indo-Gangetic Plains and in China the Nile Delta and Loess Plateau. Unfortunately, accelerated erosion due to anthropogenic activities has fostered drastic effects on the soil ecosystem and its services. Indeed, soil ecosystem is the foundation which nourishes and sustains the life of plants, animals and human beings on this planet. Soil is the home for diverse microorganisms where these microbes flourish and maintain nutrients. Soil nutrients support plant community and the basis of agriculture on which the livelihood dependents and for optimal continuation of soil functions, and the health of soil is a deciding phenomenon for performance outputs.

Indeed, a healthy soil is imperative for worthy plant growth and yield; the way of its utilization influences physical, biological and chemical properties and thereby the plant growth and production. The physical properties of soil (like structure, depth, available water, texture, colour, consistency, porosity, density and water flow capacity) play vital role in maintaining soil health. The components of soil such as sand, silt and clay form aggregates, and the association of these aggregates forms peds—the larger units of soil structure. Soil structure, which itself gets affected by soil water content, largely affects aeration, water movement, heat conduction and resistance to erosion and also the growth of plant root. Soil profile exhibits several horizons (Fig. 1.1) viz. O-horizon, which comprises of humus and loose and partly decayed organic material at the ground surface; A horizon or topsoil (2″ to 10″ deep from ground surface) is generally rich in mineral and organic matter, and is dark coloured;

**Fig. 1.1** Soil Profile

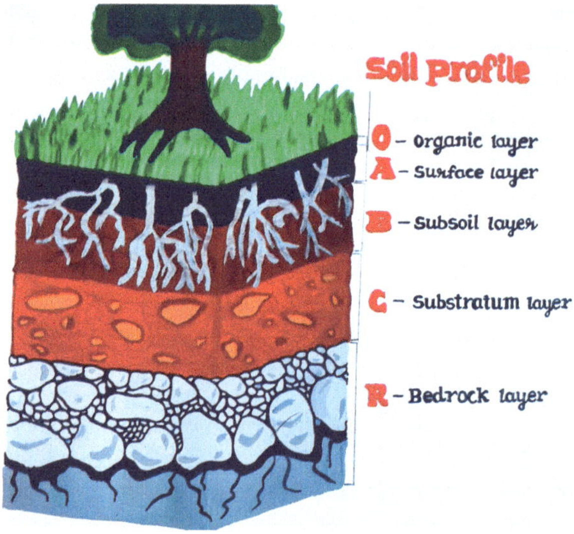

E-horizon is found in acidic soils, between A and B horizons and also known as zone of leaching or zone of eluviation. Metals, nutrients and dissolved organic acids from A horizon are transported and removed through this zone and accumulate in B horizon. B horizon or subsoil (10″ to 30″ in depth) is therefore also known as zone of accumulation. It may contain soluble minerals such as calcite in arid climates and often possesses roots and biological activity, hence considered as an effective soil depth; C horizon (30″ to 48″ in depth) is an area of least weathered portion of soil profile; R horizon (bedrock) contains a mass of rock such as granite, basalt, quartzite, limestone or sandstone that forms the parent material for some soils provided that the bedrock is close enough to the surface for weathering.

Moisture storage capacity of a soil is of prime concern as it influences soil formation, structure, stability and erosion and thereby altering plant growth. Soil colour depends on the content of organic matter, drainage conditions and the degree of oxidation taking palce in the soil however, having no active role in predicting the soil characteristics. Whilst, soil texture largely affects soil behaviour. Soil consistency is important in predicting cultivation problems and can be measured at three moisture conditions viz. air-dry, moist and wet. Moreover, soil porosity is indispensable for oxygen supply to organisms decomposing organic matter and for the movement and storage of water and dissolved nutrients. The biological properties of soil rely on the microbial and faunal activities in soil and is indispensable for a healthy or good quality soil. These soil organisms include earthworms, nematodes, protozoa, actinomycetes, fungi, bacteria and different arthropods. These organisms decompose organic matter and thus help in mineralization, making minerals available for plant uptake. Microorganisms release certain organic substances, which help in maintaining soil structure. Earthworms are indeed vital for bioturbation in the soil. Soil microorganisms such as mycorrhizal fungi play a significant role in making nutrient available to plant's uptake, while rhizobacteria participate in nutrient cycling by the process of mineralization, nitrification, nitrogen fixation, denitrification, enriching soil with mineral nutrients. Some microbes recycle nutrient like carbon by consuming material accumulated at the soil surface or within the rhizosphere and thereby produce a bulk of carbon annually and produce a by-product called humus. Since humus is less palatable to the microbes, it is decomposed sluggishly and takes a long time to be completely decomposed. Gradually, part of shallow decomposed humus starts moving downward as a clay–humus complex. Due to lesser oxygen availability in the lower parts of soil, it becomes more difficult for microbes to further decompose humus. Eventually, due to various processes that blending soil, this humus moves back up to where there is more oxygen and then the microbes eventually destroy the humus and release some more carbon dioxide.

Soil health refers to the quality and capacity of soil for sustenance of the dependent living community that essentially depends on soil's biological, chemical and physical environment. The characteristics of a soil could influence with changes in the concentration of soil organic matter and the nutrient status (Fig. 1.2). Considering today's scenario, the concept of soil health is modified and explained as 'the continued capacity of a specific kind of soil to function as a vital living system within natural or managed ecosystem boundaries, to sustain animal and plant productivity,

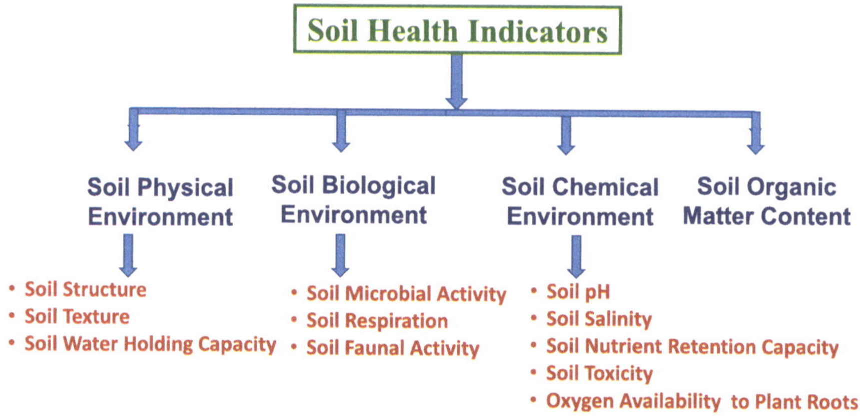


**Fig. 1.2** Measures of Soil Health

to maintain or enhance the quality of air and water environment, and to support human health and habitation' (Doran and Zeiss 2000). Soil health is not only confined to the enhanced crop productivity while a balance between many soil functions, environmental protection and in maintaining the health of plants and animals (Doran 2002). In India, agricultural crop productivity level is very low and unsatisfactory in comparison to the per hectare (ha) crop yield of developed countries. India after being the major producer of various crops like rice, wheat and pulses rank lower than other countries in terms of crop productivity (yield per hectare). The production capacity of Indian agricultural soils before and after Green Revolution, impact of cumulative use of fertilizers and pesticides, over-exploitation of agricultural soils and groundwater and future requisitions for improvisation and maintenance of soil health for sustainable crop production is discussed as well in this chapter.

## 1.2 Indian Agriculture Before Green Revolution

Due to poor irrigation facilities, fertilizers availability, crop varieties and other agricultural practices in India, the agricultural production before the period of Green Revolution was very low and in fact insufficient to fulfil the growing demand of rapidly increasing population. Major famines at that time had worsen the situation and the country was in a very bad phase in terms of per capita food availability. During the Independence phase (around 1946), the per capita food availability was only 417 gm per day, which was very low for the survival and sustenance of population. At the same time, rural indebtedness tremendously increased, converting farmers to landless labourers (NCAR 1976). Further, partitioning of country created severe shortage of food crops as well as of commercial crops (NCAR 1976), which needed an urgent management to overcome the situation. After Independence, urgent steps were taken by the agricultural scientists and policy makers, and between 1947 and 1950 the net irrigated area of India increased from 18.9 to 20.2 million hectare and further introduction of irrigation facilities the irrigated area continued to grow. During the 1950s and the early 1960s, the crop production was sufficient to fulfil the requirement of growing population but it was insufficient for the commercialization of crop products. For modernization and industrialization, India needed high agricultural output to mitigate the demand of market and for the agriculture-dependent industries. Despite of higher agricultural production, the output was not enough to touch the industrial sector. In fact, poor rural people were unable to purchase the daily required food for their survival (Dandekar and Rath 1971). In 1959, an initiative was undertaken to start the Intensive Agricultural District Programme (IADP), which was expanded to Intensive Agricultural Area Programme (IAAP) in 1964 with the aim to concentrate supplies and efforts to 'intensive cultivation areas' to achieve productivity. Both IADP and IAAP did not introduced any of the new technologies, while remain dependent on irrigation and on high use of fertilizers. The productivity under IADP was not up to the expectation of agrarian community because only marginal increase in crop yield was observed. In 12 rice-growing districts, the yield increased from 12.4 quintals to only 13.3 quintals while for wheat from 10.2 quintals to 13.5 quintals with added cost (Desai 1969). Hence, advanced agricultural technologies were needed to overcome the situation and prerequisite goal but no such techniques were available to the farmers. Introduction of 'Miracle seeds' in Mexico by Norman Borlaug in the 1960s and its success in field production provided a technological tool. The seeds were genetically modified for enhanced production and were capable of performing under high chemically fertilized soils. The technology was highly expensive as the seeds required very protected conditions like chemical pesticides, chemical fertilizers and proper irrigation practices. The maintenance cost of the technique was out of the reach of Indian cultivators. However, in India monsoon failure during 1965 and 1966 and threat of major famine led to prediction of about one million starvation deaths in Bihar alone and a huge damage (NCAR 1976). The situation and demand of the time proceeded towards the acceptance of the expensive technology. In 1965–1967 the technology was introduced with new strategy to fulfil the urgent needs. The technology was launched in the areas which were already adopted by IADP and IAAP and received great success which further recognized as the Green Revolution in Indian Agriculture.

## 1.3 Green Revolution in India

The Green Revolution was initiated in India in the 1960s with the goal to feed millions of people undergoing malnutrition by increasing the crop production throughout the country. The Rockefeller Foundation was concerned with increasing the crop productivity using new techniques in developing countries to fulfil the food demand. Norman Borlaug provided new technique in the form of High-Yielding

Variety (HYV) seeds, which were 'stocky, disease resistant, fast growing and highly responsive to fertilizers' (Hardin 2008). The technique was adopted and quickly spread all over the developing counties, including India with great success (Hardin 2008). Indian government with the help of Indian geneticists Dr. MS Swaminathan launched Green Revolution for which he is also known as the father of Green Revolution in India. The period of Green Revolution lasted from 1967 to 1978 with great achievements. At first stage, the Green Revolution was focused on states like Punjab and Tamil Nadu with proper irrigation facility. The technique was focused on food grain crops like wheat and rice and not on commercial and cash crops. Chemical fertilizers were available and used to increase the crop productivity. Similarly, pesticides and weedicides were frequently used to minimize any loss or damage to the growing crops. Further introduction of more agricultural tools like drills, harvesters, tractors etc. assisted commercial farming with more production. The wheat yields tremendously increased with this amazing technology and from only 11 million tonnes in 1960 increased to 55 million tonnes in 1990 (Chakravarti 1973; Sebby 2010). The yield of wheat increased from 850 kg/hectare to 2281 kg/hectare by 1990 (Sebby 2010) and productivity was enough for food supply to growing population and for the commercialization and stocking for future requirements. Wheat yield increased by 270% while rice yield grew by 145% over the past half century in India (Tripathi 2019).

In Punjab, during the period of 1968–1980, the rice yield increased tremendously and growth rate was around 9%, which later became stagnant (1981–1990) and even dropped down to 1.13% during 1995–1996 (Singh 2016). Similarly, wheat yield started declining from the mid-1950s may be due to decrease in its genetic potential and to monoculture cropping pattern which otherwise attracts pest and diseases (Andow 1983). Now the productivity of potato, cotton and sugarcane in the state had also become stagnant. Even high dose of fertilizer application was not helpful in overcoming the situation. The initial increase in crop productivity further declined due to imbalanced nutrient ratio and deficiency of some micronutrients along with the disturbance in soil structure (Singh 2008). Farm mechanization, especially intense use of tractors, had damaged the physico-chemical properties of soil and disturbed the biological activities that resulted in decreased crop yield (Kumar and Singh 2010). Preferred cultivation of some crops had shifted interest of farmers from cultivating native pulses like moong, gram, tuar etc. along with many oilseed crops such as mustard, sesame etc., which was making the country dependent for importing the produce of such crops at high cost, and at the same time crops grown on the chemical-rich soils were deprived of nutrition and loaded with toxins (Shiva 2015).

## 1.4 Soil Health Past Green Revolution: Consequences on Soil Fertility

The Green Revolution besides providing surplus food to the country with intensified agriculture over the years degraded the fragile agro-ecosystems in India (Rahman 2015). The revolutionary success of the techniques used during the period of the Green Revolution was based on modern methods. Under traditional methods soil maintains the nutrient status by recovering itself after any stressed condition (Das 1999). The land was used efficiently to grow more types of crops in the same field (Rosset 2000). While in modern method during the Green Revolution monocropping was favoured where only one type of crop was grown and the soil cannot replenish its vanished-off nutrients, therefore destroying soil fertility. However, with mixed cropping the recharge of nutrients is indeed possible due to differential requirement of different crops.

High-yielding seeds of the Green Revolution involved input of high amount of chemical fertilizers to increase nutrients status of soil and along the high amount of pesticides to protect crops from pathogens. These chemicals severely harmed indigenous species and resulted in the displacement of thousands of native species (Siddiq 1994). The basic ecosystem of the agricultural field underwent a severe change and disturbance (Shiva 1993). Addition of chemical fertilizers to the soil damaged biological activities of the soil organisms. Soil microbial processes like nutrient immobilization and mineralization, which are the common sources of soil nutrients, got affected and resulting in the depletion with decreased soil fertility and nutrient status (Li et al. 2014; Damodaran et al. 2013). To mitigate the requirement of new seeds farmers continued to supply increasing quantity of fertilizers, which further continued to decrease the important nutrients from the soil (like nitrogen, iron, phosphorus and manganese) (Zwerling 2009). Organic carbon in soil was around 0.5% in the 1960s which declined to 0.2% in the 1990s (Singh 2016). Poor fertilizer use policy of the government and inappropriate use of pesticides and fertilizers by the farmers, specially of small farmers, further worsened the situation. The farmer's spirit for obtaining more production and having reduced risk of crop failure subjected to the intensive use of chemicals in the agricultural fields (Vasavi 2009). Nitrogen-rich fertilizer like urea had been applied in a high ratio than that of the recommended dosages that resulted in generating soil toxicity as well as reduced soil and plant health (Saidur 2015).

Clearing of trees from farm fields also affected the fertility of soil as the soil became unprotected, loose and more prone to wind or water erosion. Production of crops such as wheat and rice was targeted during the phase of Green Revolution. These crops need high water and sufficient irrigation system for an ideal yield, which is difficult and even much costlier for the farmers, particularly those belonging to the arid and semi-arid regions. The crops like millets easily grow in the arid and semi-arid conditions due to their low water requirement; however, the unavailability of high-yielding seeds of these crops motivated farmers to grow wheat and rice. To meet the water requirement of these crops, water was extracted from the groundwater reserves with the help of pumps/tube wells, depleting it to a critical level. The groundwater continued to sink between 0.3 and 1.0 meter per year during 1993–2003 in certain areas specially in Punjab and Haryana, which persistently reached to a level down to 28 m by 2006, which necessitated more efforts and high cost extraction technologies (Zwerling 2009). Diminishing water resources and soil toxicity increased pollution of underground water (Vaidyanathan 2000). Salinity of soil increased with negative impact on the soil health. Long-term presence of chemical residues from fertilizers and pesticides continued to degrade the soil environment and heath, and also the health of people and animals. Study and survey during the last 45 years indicated the presence of organochloride, organophosphate, nitrates in the soil (Saidur 2015). Such chemicals have entered our food chain, for instance, synthetic pyrethroid and carbamates are present in the feed of cattle grown on highly fertilized soils, also found in milk and dairy products, hence affecting human health (Saidur 2015).

Due to overuse and exploitation of resources, the land of Indian states where Green Revolution took place became very sensitive. Decreased fertility of soil not only affected the crop yield, a major community dependent on agriculture later faced a bad phase of economic loss. Suicides by cultivators are common as they never came out of debt they had taken for application of the expensive techniques to their farm field. Immediate steps are needed with the sustainable traditional approach to get the health of Indian soil back along with the higher capability of productivity. Practices which are not solely focused on productivity, while with the consideration of environment health also should be prioritized.

## 1.5 Regaining Soil Health: A Sustainable Approach

### *1.5.1 Increase Inputs of Organic Matter*

Soil organic matter (SOM) is a reservoir of many important plant nutrients. Its availability in the soil is necessary for maintaining soil health. Addition of crop residues, compost, green manure and animal-derived material provide organic matter as well as certain soil nutrients, which generally occur in the immobile state in most of the soils, and hence they are not readily available to the plants (Chenu et al. 2015). By the process of decomposition, mineralization and nutrient cycling these nutrients become available in the soil and then easily acquired by the plant (Coleman et al. 2004). SOM maintains the cation exchange capacity (CEC) of the soil to hold the positively charged nutrients, helps in nutrient leaching, increases water-holding capacity and soil aggregation. The addition of organic matter in the form of mulch prepared using weeds, crop residues increases crop production and also protects soil from erosion (Roose and Barthes 2001). Crop residues influence the rate of mineralization and decomposition thereby releasing nitrogen in the soil (Jarvis et al. 1996; Rahn et al. 1992) and the residues with low nitrogen helps in nitrogen immobilization hence preventing leaching of nitrogen (Jenkinson 1985). To improve the level of soil organic matter, compost manure may be prepared using green waste of parks, gardens and households that could help in maintaining soil fertility as it effectively increases soil microbial activity. It is also found to be effective in reducing the incidence of plant diseases (Abawi and Widmer 2000). Frederiksson et al. (1997) observed that animal manures not only increase soil nutrition but also the quality and yield of cereals. Hence maintaining an ample amount of soil organic matter is crucial for managing soil health and its fertility.

### *1.5.2 Crop Rotation and Cover Crops*

An adequate level of soil nutrients is important for maintaining soil health and its fertility. Crop rotation moderates soil nutrients through growing diverse crops with different nutrient inputs and requirements (Stockdale et al. 2001). As we know that water-intensive crops such as rice and sugarcane lower the water table, crop rotation in the same field with less water requiring nutrient-rich crops like millets could recuperate the water table and maintain nutrient status of the soil. The practice is also beneficial for management of soil-borne plant pathogens as rotating crops interfere with the life cycle of a pathogen, therefore preventing the growth of pathogen (Knudsen et al. 1995). Use of cover crops reduces erosion of soil. Crops with tap root increase aeration in soil by creating macropores in compact soils and also improve infiltration, whereas the crop with fibrous root-like grasses supports aggregation and largely supports soil health. Cover crops help in enhancing soil fertility by adding organic matter to the soil through their biomass. Instantaneously, these crops also help in reducing the leaching of important soil nutrients like nitrate. The roots of legumes are colonized by nitrogen-fixing bacteria, and hence using legumes as cover crops could further enhance the nitrogen content of soil through their nitrogen-fixation capability (Stockdale et al. 2001). About 90% of terrestrial plants are colonized by mycorrhizal fungi, which facilitate host plant for better acquisition of soil nutrients and to cope up with environmental stresses (Giri 2017). Cover crops also influence mycorrhizal fungi and thereby modulate the composition of rhizosphere with favoured fungal composition to maintain soil processing and health.

### *1.5.3 Tillage Practices*

Any physical disturbance to the soil affects plant root and also leads to the modification of soil biota. Soil compaction makes roots difficult to grow easily and the decreased water infiltration and drainage further affect the plant growth. Tillage has many adverse effects on the soil productivity especially by altering biological properties of soil. Some sensitive microbes severely get affected and impact soil's biological diversity. The extraradical hyphal network of arbuscular mycorrhizal fungi may be disturbed due to frequent tillage practices, resulting in the poor supply of soil nutrients to their host plants. Tillage exposes the soil to air, which reduces the soil moisture content and disturbs the living place of many soil microorganisms, which are indeed important for managing soil health. Tillage affects soil coverage and makes the soil more prone to erosion. Non-inversion tillage is preferred over the traditional inversion tillage as the former tillage practice causes less disturbance in soil and to the soil organisms. Farming practices in which tillage is a must for good crop production, the increased input of organic matter through compost, vermicompost, green manure or other means are required to overcome this loss.

### *1.5.4 Nutrient and pH Management*

Soil pH plays an important role in maintaining the diversity of soil organisms and the availability of several important micronutrients. In acidic soils, nutrients like Fe, B, Cu, Zn, Mn, Ni are readily available and the plants requiring these nutrients can grow well in these soils. While, alkaline soils are rich in Mo, K, S, Ca, P and Mg and are good for the crops with nitrogen-fixation ability. Legume crop like alfalfa grows well in soil pertaining pH around 6.2. Most of the soil microorganisms prefer pH around 7.0. In general, the microbial activity decreases at lower pH (5.0 or less) as has been reported in the case of microbes like nitrifying bacteria (Rousk et al. 2009; Munns 1986). Hence, it is quite important to maintain soil pH according to the crop requirement and for both soil health management and crop productivity (Kidd and Proctor 2001). The soil pH can be maintained with the help of certain amendments to soil, such as $FeSO_4$ and $Al_2(SO_4)_3$ can be used to lower the pH of the alkaline soils (Brown et al. 2008). Alkaline soil with high concentration of $CaCO_3$ can also be treated with sulphur amendments to lower soil pH (Mitchell and Huluka 2008). Acidic soil (low pH) can be treated with lime to increase its pH. In addition, lime reduces soil Al and Mn toxicity. Amendments like $CaCO_3$, $Ca(OH)_2$ and CaO can be used to increase the activity of sensitive nitrogen-fixing microbes. Supplementing soil with $Ca(NO_3)_2$ usually increases its pH, if leaching does not take place (Zhang et al. 2016). Whereas $(NH_4)_3PO_4$ and urea slowly induce soil acidification as well as promote Al and Mn toxicity (Fageria et al. 2010).

Amending soils with SOM is of great significance as it buffers soil pH by releasing anions and cations on its decomposition. Interestingly, in acidic soils SOM binds $H^+$ at negatively charged site whereas in basic soil releases $H^+$ ions. It is pertinent to understand that during initial stage of decomposition of SOM, there is an increase in soil pH. Soil pH further increases due to mineralization by soil microbes; however, long-term microbial decomposition and leaching of nitrate may decrease the soil pH (McCauley et al. 2017). Overall, the effect of SOM on soil pH depends on quality and quantity of organic matter along with the rate of decomposition and uptake or loss of decomposition products (Wang et al. 2009).

### 1.5.5 *Biofertilizers*

A sustainable approach is required to maintain soil health and fertility without affecting the natural health of ecosystem. We know that chemical fertilizers are degrading soil ecosystem and imposing many more nasty effects on human, animal and environment (Youssef and Eissa 2014). Therefore, to achieve sustainable agriculture, the use of biofertilizers could be a good choice over chemical fertilizers. Biofertilizers contain living microorganisms like actinomycetes, bacteria, fungi and blue-green algae (BGA) along with the sterilized carrier material such as manure, saw dust, earthworm cast (Khosro and Yousef 2012). Carrier material helps in easy handling, long-term storage of the microbial inoculum and increases the water ration capacity and effectiveness of microbial inoculum (Ritika and Uptal 2014). Biofertilizers' application to soil positively alters the rhizosphere dynamics by inviting favourable microorganisms inhabiting this area. These microorganisms provide required macro- and micronutrients to the plant root and help in nutrient management for sustainable agriculture (Adesemoye and Kloepper 2009). Biofertilizers release nutrients slowly to the soil and plant with continuous supply for longer period, hence helping in maintaining nutrient status of soil (Itelima et al. 2018).

Biofertilizers comprise various category of soil microbes, which indeed help in maintaining soil and plant health. The different categories of biofertilizers with specified microorganisms include: (1) Nitrogen-fixing biofertilizers comprise nitrogen-fixing microbes and provide nitrogen to the soil and plants. These nitrogen fixers are grouped as a) free-living microbes viz., *Azotobacter*, *Clostridium*, *Bejerinkia*, *Klebsiella*, *Nostoc* and symbiotic microbes viz., *Rhizobium*, *Anabaena*, *Frankia*, *Azospirillum*. They convert atmospheric nitrogen into available organic form of nitrogen; (2) Phosphate solubilizing biofertilizers enrich soils with phosphorus (P). In soils, P is present in insoluble form of phosphate, therefore not readily available for plant uptake. Bacteria such as *Bacillus subtilis*, *Bacillus circulans*, *Phosphaticum* and *Pseudomonas putida* are capable enough to solubilize the insoluble phosphate, so these microbes may be used as biofertilizers for P-requiring crops. Some fungi like *Penicillium spp.* and *Aspergillus spp.* due to their phosphate solubilizing ability, are also used as phosphate solubilizing biofertilizers; (3) Phosphate mobilizing biofertilizers include various categories of mycorrhizal fungi such as arbuscular mycorrhiza (*Glomus spp.*, *Sclerocystis spp.*, *Acaulospora spp.* etc.) and ectomycorrhiza (*Amanita spp.*, *Boletus spp.*, *Laccaria spp.*, involved in mobilizing the phosphate from the soil to plant roots (Chang and Yang 2009); (4) Potassium (K) is generally present in the form of mineral silicates and is beyond the approach of plant acquisition, hence potassium solubilizing biofertilizers can be applied for helping the plant in uptake of such kind of elements from the soil. The microorganisms like *Bacillus spp.* and *Aspergillus spp.* can solubilize silicates, releasing the K from the metal, mobilize it, and make available to plant's acquisition; (5) Sulphur oxidizing biofertilizers are composed of microbes like *Thiobacillus spp.* and used as sulphur oxidizers as these microbes oxidize sulphur into sulphates, which are nicely utilized by the plants; (6) Plant growth-promoting biofertilizer are composed of plant growth-promoting rhizobacteria, which play a vital role in plant growth and development. They promote the synthesis and concentration of plant growth hormones and enhance plant growth viz. *Azospirillum* secretes auxins, gibberellins and ethylene; *Rhizobium* and *Bacillus* synthesize indole acetic acid; *Pseudomonas* promote root growth and influence the decomposition of organic matter for better nutrient availability (Bhattacharyya and Jha 2012).

Biofertilizers play an important role in surviving plants under environmental stresses. Arbuscular mycorrhizal fungi (AMF) alone or in combination with nitrogen-fixing bacteria have been found to improve plant productivity under salt and drought stress conditions (Evelin et al. 2009; Ansari et al. 2013; Hussain et al. 2002). AMF increases the photosynthetic efficiency and anti-oxidative response of plants under drought stress condition (Ruiz-Sanchez et al. 2010). Inoculation of seed with rhizobacteria *Pseudomonas spp.* show improvement in seed germination and growth under water stress condition (Backman and Sikora 2008). Biofertilizers also provide defence to the plant against various pathogens. PGPRs such as *Bacillus subtilis* provide resistance against various fungal and viral diseases of banana, tomato, pepper, cucumber, etc. (Backman and Sikora 2008). Therefore, using biofertilizers for plant disease management could be a better approach for both sustainable crop production and soil environment (Ritika and Uptal 2014).

### *1.5.6 Vermicompost*

To improve the quality and fertility of soils, vermicompost can be a good option as an organic manure, which is produced by the earthworms through the process of vermicomposting. Earthworms, after feeding on organic matter such as plant residues and other biological wastes, produce humus-like substance—the vermicast (worm casting), which is the compost of worms (Chanda et al. 2011). After passing through the digestive tract of worms, the vermicast becomes clean, odourless and rich in nutrients like nitrogen (2–3%), phosphorus (1.55–2.25%) and potassium (1.85–2.25%) along with many micronutrients (Lazcano and Domínguez 2011). The method of vermicomposting is a very efficient technology to recycle organic waste into very nutritious compost that indeed could improve soil health and productivity of the crop plants (Azarmi et al. 2008; Guerrero 2010; Yadav and Garg 2011; Hema and Rajkumar 2012). Vermicompost also improves soil microbial activity and properties, such as porosity and water infiltration rate of soil. It increases soil oxygen availability and maintains the temperature of soil and helps in achieving high yield of crop (Arora et al. 2011). The earthworms like, *Lampitomauritii*, *Eiseniafoetida*, *Perionyx excavates* are commonly used for vermicomposting (Reddy and Pattnaik 2009; Subbulakshmi and Thirunee, 2015) and produce eco-friendly organic fertilizers in easily available form to crop plants (Parthasarthi and Ranganathan 2002). Vermicompost application in crops like rice and lentil has shown higher yield in comparison to chemical fertilizers (Karmakar et al. 2013;

Sreevidya et al. 2016). Not only crop yield, vermicompost significantly influences the chemical and physical properties of soil which were poorly deteriorated due to the imbalanced and unjustified use of chemical fertilizers. Further, vermicompost also protects plants from pathogens and diseases (Rostami et al. 2014), hence it is persuasive to state that vermicomposting is a fast and very effective process for the production of organic manure and to manage soil health and ecosystem productivity.

### *1.5.7 Organic Pesticides in Relation to Soil Health*

Certain plants such as marigold, pepper, garlic exhibit ability to produce insect-repelling substances; therefore, planting and growing such insect-repelling plants for pesticidal use is an organic and natural way to avoid applications of chemical pesticides in agricultural fields. Similarly, herbal sprays are organic sprays prepared from plant and animal extracts, exhibiting insect-repellent quality due to its odour, volatile ingredients or presence of certain compounds. The oils of plants like neem, citrus, citronella, eucalyptus could be applied in the form of foliar sprays to keep away insects from crop plants (Phasomkusolsil and Soonwera 2010). Due to their biodegradable nature, such insect repellents are eco-friendly and safe; they neither harm plant nor the soil and environment, and indeed are very effective against insects, pests and fungus. Planting mixed crops in the same field could be a companion planting where one crop protects the other against those pathogens or insects that have tendency to be repelled by them naturally (Peralta et al. 2018). For example, tomatoes can naturally repel diamond-backed moth larvae. If tomatoes are grown with cabbage, it protects the cabbage against these larvae. Similarly, basil is insect repellent and has larvicidal activity against mosquitoes and flies (Mahmoud et al. 2017; Hassan et al. 2015). Spray made up of mild soap such as castile soap with water is an efficient insecticide to control flies, beetles and other insects. The mixture can be directly sprayed on the infected surface of plant without contaminating the soil. Salt spray made with epsom is an effective natural pesticide (a magnesium-rich salt) that not only repel the pests though provide nutrients to the plants. The application of such organic, biodegradable and natural pesticides would keep away the soil and groundwater from the contamination of toxic chemicals that not only gravely affect the soil health and its sustainability but also the human health and environment.

### *1.5.8 Soil Health Card Scheme in India*

Soil is the base of agriculture, hence protecting and managing its health is of utmost significance for obtaining good quality crop produce as well as for sustainable agriculture. In India, due to over-exploitation and intensive use of chemicals (fertilizers and pesticides) to obtain maximum crop yield for meeting the growing demand

**Table 1.1** Consumption of NPK in India (in million tonnes) (modified from Reddy 2017)

| Year | N | P | K | Total |
|---|---|---|---|---|
| 1956 | 0.1 | 0.0 | 0.0 | 0.1 |
| 1981 | 3.7 | 1.2 | 0.6 | 5.5 |
| 1991 | 8.0 | 3.2 | 1.3 | 12.5 |
| 2001 | 10.9 | 4.2 | 1.6 | 16.7 |
| 2011 | 16.6 | 8.0 | 3.5 | 28.1 |
| 2015 | 16.9 | 6.1 | 2.5 | 26.6 |

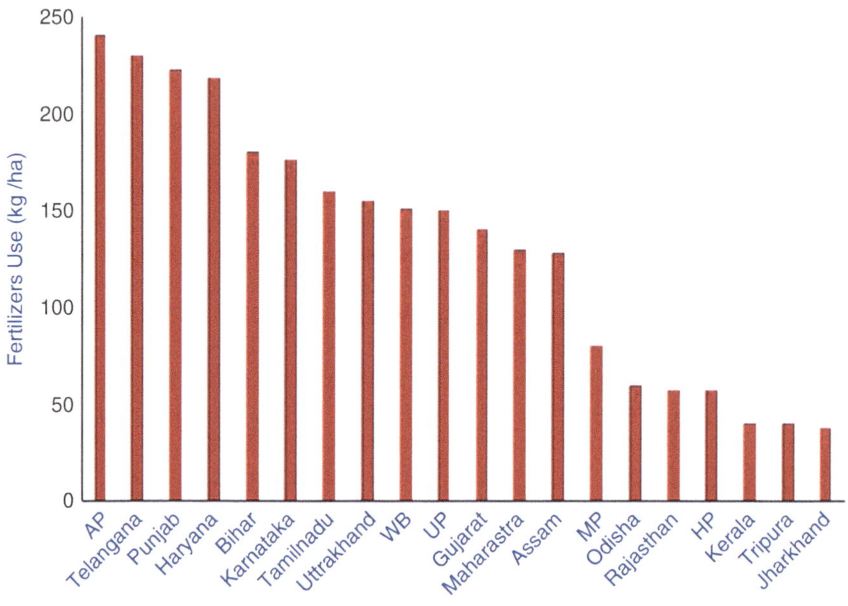


**Fig. 1.3** Insights of the consumption of chemical fertilizers in some States of India

of increasing population for food has largely degraded the life of agricultural soils. India is an agriculture-dominant country. Agriculture share about 4% of Indian GDP. The current consumption of NPK ratio in India is 6.7:2.4:1 against an ideal ratio of 4:2:1 (Reddy 2017). Estimates indicate that in 1956, the consumption of nitrogen in India was just 0.1 million tonnes, which has increased to 17.0 million tonnes in 2015 and the consumption of phosphorus and potassium has increased from 0.0 to 6.1 to 2.5 million tonnes, respectively (Table 1.1). However, the consumption of NPK in terms of fertilizer use per acre varies among the states of India (Fig. 1.3). It is noteworthy to state that India is spending about one lakh crore rupees every year for providing fertilizer subsidy. The subsidy amount is about Rs. 6500 per hectare of net cropped area and about Rs. 7000 per farmer (Reddy 2017). This has led to the excessive and unbalanced use of fertilizers, particularly NPK (Reddy 2017). As a consequence, the amount of food grain produced/kg of fertilizer applied has reduced from 13 to 4 kg since 1970 to 2010, respectively (Reddy 2017).

While the production capacity of agricultural soils is decreasing due to the deficiency of various macro- and micronutrients, certain nutrients like nitrogen and phosphorus are being excessively used by the farmers. In fact, farmers are not fully aware of various physical, chemical and microbial activities of their agricultural fields; therefore, farmers are unfortunately not able to apply chemical fertilizers in a balanced and required quantity. Indeed, they are deliberated to grow at least two or more crops a year, instead of one without proper soil health management. This unplanned use of soil is aggravating the problem of nutrient deficiencies and largely changing chemical composition of soils. Moreover, soils are facing the problem of depletion of organic carbon content throughout the country, which is rendering soils more vulnerable to erosion. Increasing dosage of fertilizers and pesticides is eventually dropping the number of beneficial soil organisms and affecting the productivity and health of soil. In fact, the soils with reduced nutrients composition are resulting in the poor nutritional value of the food crops. To deal with the problem of soil health management and ensuing the United Nations resolution, Ministry of Agriculture, Government of India has introduced soil health card scheme (SHCS) on 5 December 2015 to facilitate agrarian community for better understanding of soil health status of their agricultural fields and integrated nutrient management. The main objective of SHCS is to generate awareness among farmers for the judicious use of fertilizers, biofertilizers, organic fertilizers to manage soil health and for better understanding of cropping pattern, cost reduction, farm profitability and sustainability.

## 1.6 Conclusion

The goal of high crop productivity for growing population and commercial requirements has increased our dependence on various chemicals. In the form of fertilizers, pesticides, insecticides etc., these chemicals have entered in the environment and eventually in the food chain. These chemicals are life-threatening and gradually destructing the health of flora, fauna and of soil ecosystem. They have contaminated our food and now continuing to affect the food chain and biodiversity. We have lost a huge area of fertile land all over the world, due to addition of these chemicals and over-exploitation of the soil; therefore, the sustenance of soil fertility and restoration of health of degraded and stressed soils is of great concern. To sort out the problem, replacing chemical fertilizers with organic matter, FYM, vermicompost, green manure and biofertilizers can be the eco-friendly approaches. Biofertilizers have potential of enhancing productivity in a sustainable way along with increasing the soil health. Modifications in agricultural practices where more crops can be grown in the same field and sustainable approach to maintain and replenish the nutrient in the soil is the demand of time. Application of cost-effective natural technologies should be preferred which not only provide nutrient to the soil but improves its chemical and physical properties. More approaches and strategies are required to make our soil much more productive but not at the cost of environment health, while it needs to be in the environment-friendly manner.

**Acknowledgement** Authors sincerely acknowledge the support and cooperation provided by Ms. Radha Rana.

## References

Abawi G, Widmer T (2000) Impact of soil health management practices on soil borne pathogens, nematodes and root diseases of vegetable crops. Appl Soil Ecol 15:37–47

Adesemoye AO, Kloepper JW (2009) Plant-microbes interactions in enhanced fertilizer use efficiency. Appl Microbiol Biotechnol 85(1):1–12

Andow D (1983) The extent of monoculture and its effects on insect pest populations with particular reference to wheat and cotton. Agric Ecosyst Environ 9(1):25–35

Ansari MW, Trivedi DK, Sahoo RK et al (2013) A critical review on fungi mediated plant responses with special emphasis to *Piriformospora indica* on improved production and protection of crops. Plant Physiol Biochem 70:403–410

Arora VK, Singh CB, Sidhu AS, Thind SS (2011) Irrigation, tillage and mulching effects on soybean yield and water productivity in relation to soil texture. Agric Water Manag 98 (4):563–568

Azarmi R, Giglou MT, Taleshmikail RD (2008) Influence of vermicompost on soil chemical and physical properties in tomato (*Lycopersicum esculentum*) field. Afr J Biotechnol 7 (14):2397–2401

Backman PA, Sikora RA (2008) An emerging tool for biological control. Biol Control 46(1):1–3

Bhattacharyya PN, Jha DK (2012) Plant growth-promoting rhizobacteria (PGPR): emergence in agriculture. World J Microbiol Biotechnol 28(4):1327–1350

Brown TT, Koenig RT, Huggins DR, Harsh JB, Rossi RE (2008) Lime effects on soil acidity, crop yield, and aluminum chemistry in direct-seeded cropping systems. Soil Sci Soc Am J 72:634–640

Chakravarti AK (1973) Green revolution in India. Ann Am Assoc Geogr 63(3):319–330

Chanda GC, Bhunia G, Chakraborty SK (2011) The effect of vermicompost and other fertilizers on cultivation of tomato plants. J Hortic For 3:42–45

Chang CH, Yang SS (2009) Thermotolerant phosphate solubilizing microbes for multifunctional bio-fertilizer preparation. Bioresour Technol 100(4):1648–1658

Chenu C, Rumpel C, Lehmann J (2015) Methods for studying soil organic matter: nature, dynamics, spatial accessibility, and interactions with minerals. In: Paul EA (ed) Soil microbiology, ecology and biochemistry, 4th edn. Academic Press, Amsterdam, pp 383–419

Coleman DC, Crossley DAJ, Hendrix PF (2004) Fundamentals of soil ecology. Elsevier Academic Press, Boston, MA

Damodaran T, Sah V, Rai RB, Sharma DK, Mishra VK, Jha SK, Kannan R (2013) Isolation of salt tolerant endophytic and rhizospheric bacteria by natural selection and screening for promising plant growth-promoting rhizobacteria (PGPR) and growth vigour in tomato under sodic environment. Afr J Microbiol Res 7(44):5082–5089

Dandekar VM and Rath N (1971) Poverty in India. Economic and Political Weekly, June 2 & June 9.

Das BM (1999) Shallow foundations: bearing capacity and settlement. CRC Press, Boca Raton, FL

Desai DK (1969) Intensive Agricultural District Programme. Economic and Political Weekly. June 28: A-83-89

Doran JW (2002) Soil health and global sustainability: translating science into practice. Agric Ecosyst Environ 88(2):119–127

Doran JW, Zeiss MR (2000) Soil health and sustainability: managing the biotic component of soil quality. Appl Soil Ecol 15:3–11

Evelin H, Kapoor R, Giri B (2009) Arbuscular mycorrhizal fungi in alleviation of salt stress: a review. Ann Bot 104:1263–1280

Fageria N, Santos A, Moraes M (2010) Influence of urea and ammonium sulfate on soil acidity indices in lowland rice production. Commun Soil Sci Plant Anal 41:1565–1575

Frederiksson H, Salomonsson L, Salomonsson AC (1997) Wheat cultivated with organic fertilizers and urea. Acta Agr Scand B-S P 47:35–42

Giri B (2017) Mycorrhizal dependency and growth response of *Gliricidia sepium (Jacq.) Kunth ex Walp.* under saline condition. Plant Sci Today 4(4):154–160

Guerrero RD (2010) Vermicompost production and its use for crop production in the Philippines. Int J Global Environ Issues 10(3/4):378–383

Hardin LS (2008) Meetings that changed the world: Bellagio 1969: the green revolution. Nature 455:470–471

Hassan MI, Hammad KM, Saeed SM (2015) Repellent effect of *Ocimum basilicum* and *Glycyrrhiza glabra* extracts against the mosquito vector, *Culex pipiens* (Diptera: Culicidae). J Egypt Soc Parasitol 45(2):241–248

Hema S, Rajkumar N (2012) An assessment of vermicomposting technology for disposal of vegetable waste along with industrial effluents. J Environ Sci Comp Sci Engineer & Technol 1(1):5–8

Hussain N, Mujeeb F, Tahir M et al (2002) Effectiveness of rhizobium under salinity stress. Asian J Plant Sci 4:124–129

Itelima JU, Bang WJ, Onyimba IA et al (2018) A review: biofertilizer; a key player in enhancing soil fertility and crop productivity. J Microbiol Biotechnol Rep 2(1):22–28

Jarvis SC, Stockdale EA, Shepherd MA, Powlson DS (1996) Nitrogen mineralization in temperate agricultural soils: processes and measurement. Adv Agron 57:187–235

Jenkinson DS (1985) How straw incorporation affects the nitrogen cycle. In: Hardcastle J (ed) Straw, soils and science. AFRC, London, pp 14–15

Karmakar S, Brahmachari K, Gangopadhyay A (2013) Studies on agricultural waste management through preparation and utilization of organic manures for maintaining soil quality. Afr J Agric Res 8:6351–6358

Khosro M, Yousef S (2012) Bacterial bio-fertilizers for sustainable crop production: a review. APRN J Agric Biol Sci 7(5):237–308

Kidd PS, Proctor J (2001) Why plants grow poorly on very acidic soils: are ecologists missing the obvious? J Exp Bot 52:791–799

Knudsen IMB, Debosz K, Hockenhull J, Jensen DF, Elmholt S (1995) Suppressiveness of organically and conventionally managed soils towards brown foot rot of barley. Appl Soil Ecol 12:61–72

Kumar S, Singh P (2010) Determinants of stagnation in productivity of important crops in Punjab. Agro Economic Research Centre, Punjab Agricultural University, Ludhiana, India

Lazcano C, Domínguez J (2011) The use of vermicompost in sustainable agriculture: impact on plant growth and soil fertility. In: Mohammad M (ed) Soil nutrients. Nova Science, New York

Li X, Rui J, Xiong J, Li J, He Z, Zhou J, Yannarell AC, Mackie RI (2014) Functional potential of soil microbial communities in the maize rhizosphere. PLoS One 9(11):e112609

Mahmoud EMAH, Bashir HHN, Assad OHY (2017) Effect of basil (*Ocimum basilicum*) leaves powder and Ethanolic-extract on the 3rd larval instar of *Anopheles arabiensis* (Patton, 1905) (Culicidae: Diptera). Int J Mosquito Res 4(2):52–56

McCauley A, Jones C and Olson-Rutz K (2017) Soil pH and organic matter. Nutrient Management Module No. 8. http://landresources.montana.edu/nm/documents/NM8.pdf

Mitchell C and Huluka G (2008) Lowering soil pH; Alabama cooperative extension system, Agronomy and soil series, S-04-08, pp 1–5

Munns DN (1986) Acid soil tolerance in legumes and rhizobia. Adv Plant Nutr 2:63–91

NCAR (National Commission on Agriculture, Delhi). (1976) for Evaluation of Intensive Agricultural District Programme (IADP) 1: 411

Parthasarthi K, Ranganathan LS (2002) Supplementation of presumed vermicast with NPK enhances growth and yield in leguminous crops (*Vigna mungo* and *Arachis hypogaea*). J Curr Sci 2:35–41

Peralta AL, Sun Y, McDaniel, Lennon JT (2018) Crop rotational diversity increases disease suppressive capacity of soil microbiomes. Ecosphere 9(5): e02235. https://doi.org/10.1002/ecs2.2235

Phasomkusolsil S, Soonwera M (2010) Insect repellent activity of madicinal plant oils against *Aedes aegypti* (Linn.), *Anopheles minimus* (Theobald) and *Culex quinquefasciatus* say based on protection time and biting rate. SE Asian J Trop Med Public Health 41:831–840

Rahman S (2015) Green revolution in India: environmental degradation and impact on livestock. Asian J Water Environ and Pollut 12(1):75–80

Rahn CR, Vaidyanthan LV, Paterson CD (1992) Nitrogen residues from brassica crops. Asp Appl Biol 30:263–270

Reddy AA (2017) Impact study of soil health card scheme. National Institute of Agricultural Extension Management (MANAGE), Hyderabad, p 210

Reddy MV, Pattnaik S (2009) Vermi-composting of municipal (organic) solid waste and its implications. In: Singh SM (ed) Earthworm ecology and environment. International Book Distributing, Lucknow, pp 119–113

Ritika B, Uptal D (2014) Bio-fertilizer a way towards organic agriculture: a review. Acad J 8 (24):2332–2342

Roose E, Barthes B (2001) Organic matter management for soil conservation and productivity restoration in Africa: a contribution from francophone research. Nutr Cycl Agroecosyst 61:159–170

Rosset P (2000) Lessons from the green revolution. Food First: Institute for Food and Development Policy, 8 April 2000 and Institute for Food and Development Policy, 14 October 2009

Rostami M, Olia M, Arabi M (2014) Evaluation of the effects of earthworm Eisenia fetida-based products on the pathogenicity of root-knot nematode (*Meloidogyne javanica*) infecting cucumber. Int J Recycl Org Waste Agricult 3:58

Rousk J, Brookes PC, Bååth E (2009) Contrasting soil pH effects on fungal and bacterial growth suggest functional redundancy in carbon mineralization. Appl Environ Microbiol 75:1589–1596

Ruiz-Sanchez M, Aroca R, Monoz Y et al (2010) The arbuscular mycorrhiza symbiosis enhances the photosynthetic efficiency and the antioxidative response of rice plants subjected to drought stress. J Plant Physiol 167:862–869

Saidur R (2015) Green revolution in India: environmental degradation and impact on livestock. Asian J Water Environ Pollut 12(1):75–80

Sebby K (2010) The green revolution of the 1960's and its impact on small farmers in India. Environmental studies undergraduate student theses 10, University of Nebraska, Lincoln. https://digitalcommons.unl.edu/envstudtheses/10

Shiva V (1993) The violence of the green revolution: third world agriculture, ecology and politics, 2nd edn. Zed Books, London

Shiva V (2015) The system that drives farmers into a debt trap creates malnutrition. The solution lies in shifting from a toxic, high-cost system to a nutritious, low-cost, sustainable food production model. In: Nothing green in green revolution. India Today, Published on 24 August 2015

Siddiq A (1994) Sustainability of the Indus Basin: impact of tertiary salinity/Sodicity on wheat productivity, damage assessment and future public policy. Ph.D. dissertation, Dept. of Agricultural Economics, University of Illinois, Urbana-Champaign

Singh MV (2008) Micronutrient deficiencies in crops and soils in India. In: Alloway BJ (ed) Micronutrient deficiencies in global crop production. Springer, Dordrecht, pp 93–125

Singh RB (2016) Grey side of the green revolution. In 67th ICID Foundation Day Seminar- Second Green Revolution: Role of Irrigation and Drainage, 24 June 2016

Sreevidya M, Gopalakrishnan S, Kudapa H, Varshney RK (2016) Exploring plant growth-promotion actinomycetes from vermicompost and rhizosphere soil for yield enhancement in chickpea. Braz J Microbiol 47(1):85–95

Stockdale EA, Lampkin NH, Hovi M, Keatinge R, Lennartsson EKM, MacDonald DW, Padel S, Tattersall FH, Wolfe MS, Watson CA (2001) Agronomic and environmental implications of organic farming systems. Adv Agron 70:261–327

Subbulakshmi G, Thirunee R (2015) Commerce agricultural sciences promotion of organic farming using vermicomposting Technology in Tamil Nadu, India. Department of Chemistry, GRT Institute of Engineering and Technology, Tiruttani, Thiruvallur, Tamil Nadu, pp 26–27

Tripathi AK (2019) India faces the risk of a decline in growth of crop yields. In: The Hindu-business line, Published on 2nd October 2019

Vaidyanathan A (2000) India's Agricultural Development Policy. Econ Polit Wkly 35 (20):1735–1741

Vasavi AR (2009) Suicides and the making of India's agrarian distress. S Afr Rev Sociol 40 (1):94–108

Wang N, Li J, Xu R (2009) Use of agricultural by-products to study the pH effects in an acid tea garden soil. Soil Use Manag 25:128–132

Yadav A, Garg VK (2011) Recycling of organic wastes by employing Eiseniafetida. Bioresour Technol 102:2874–2880

Youssef MMA, Eissa MFM (2014) Biofertilizers and their role in management of plant parasitic nematodes. A review. E3 J Biotechnol Pharm Res 5(1):1–6

Zhang Y, Zhang S, Wang R, Cai J, Zhang Y, Li H, Huang S, Jiang Y (2016) Impacts of fertilization practices on pH and the pH buffering capacity of calcareous soil. Soil Sci Plant Nutr 62 (5–6):432–439

Zwerling D (2009). 'Green revolution' trapping India's Farmers in debt. Retrieved from http://www.npr.org/templates/story/story.php?storyId=102944731&ps=rs

# Chapter 2
# Biochar for Maintaining Soil Health

**Nguyen Hue**

**Abstract** Soil, like human, must have good health to function well. This condition can only be achieved when its biological, chemical, and physical aspects are in their optimal capacity and in balance. Such requirements could be partially met by using biochar as a soil amendment. More specifically, biochar is a solid material, high (>50%) in organic carbon (C) content, and is produced by heating biomass, such as wood scraps, crop residues, or animal wastes, in an environment of elevated temperature (350–900 °C) with low or no oxygen. First, with properties such as large surface areas, numerous pores, and variable pore size, biochar can improve soil physical attributes, including aggregate stability, water holding capacity, root penetration, and reduced erosion. Second, given commonly alkaline pH value, abundant reactive surface functional groups, relatively high CEC, ash content, and labile C (5–10% of total fixed C), biochar can enrich soil fertility and soil organic matter. Lastly, with those desirable characteristics, application of biochar to soils, especially highly weathered, nutrient-poor, and/or acidic soils, has been proven to enhance soil microbial abundance, diversity, and activity. On the other hand, since climate and time also affect biochar properties and performance via oxidation by water and oxygen, long-term field experiments with different soil types and biochars should be actively conducted, keeping in mind that interactions between biochar and soil are inevitable, complex, and difficult to predict. Thus, biochar properties should be clearly characterized, and standardized, including pyrolysis process (e.g., fast, slow, highest treatment temperature, residence time); also feedstock (e.g., wood, crop residues, animal manure) should be clearly identified and publicized. Such clear specifications would help strengthen the use of biochar as a soil amendment to maintain and enhance soil health.

**Keywords** Biochar · Pyrolysis · Feedstock · Highest temperature treatment · Soil quality

N. Hue (✉)
University of Hawaii, Honolulu, HI, USA
e-mail: nvhue@hawaii.edu



B. Giri, A. Varma (eds.), *Soil Health*, Soil Biology 59,
https://doi.org/10.1007/978-3-030-44364-1_2

## 2.1 Soil Health: Definition and Measurable Indicators

Along with air and water, soil sustains life on earth (Biswas et al. 2019). Soils are materially (e.g., different minerals, organic fractions) and characteristically (e.g., different pH, surface area and charge) varied and must have the proper balance in physical, chemical, and biological conditions to optimally provide its many essential ecosystem services, which can range from supplying nutrients for good plant growth, sustaining productivity of agriculture and forestry, to preserving water and environmental quality. Thus, the definition and measurement of soil health depend on the services sought. According to Doran and Parkin (1994), soil health (the authors used the term "soil quality," which includes both dynamic and static soil properties) is defined as "the capacity of a soil to function, within ecosystem and land use boundaries, to sustain productivity, maintain environmental quality, and promote plant and animal health." Recently, the definition of soil health has been slightly rephrased as "the continued capacity of the soil to function as a vital living ecosystem that sustains plants, animals, and humans" as adapted by the Cornell University Comprehensive Assessment of Soil Health (Cornell-CASH 2016) and the US Natural Resources Conservation Services (USDA-NRCS 2017). Thus, soil health cannot be assessed by measuring only crop yield, number of earthworms, soil available nutrient levels, or any single outcome (Dick 2018). It requires a combination of many indicators. Informative and practical indicators of soil health need to (1) be easy to measure; (2) be responsive to changes in management and climate; (3) include biological, chemical, and physical properties of the soil; (4) represent the soil function of interest; and (5) be accessible to, affordable and interpretable by users.

Some recommended indicators of soil health are (USDA-NRCS 2017):

1. Soil organic matter (SOM). This parameter would reflect the soil capacity to affect nutrient supply and retention for the needs of both plants and microbiota. SOM also affects many physical properties, such as aggregate stability, water holding capacity and infiltration, etc. SOM is usually measured as total organic carbon (C), oxidizable C (Cornell-CASH 2016; Weil et al. 2003; Culman et al. 2012), or $CO_2$ respiration (USDA-NRCS 2017).
2. Soil structural stability. This indicator would affect soil erosion, water and air movement as well as root penetration. Wet-sieving method for aggregate stability, bulk density measurement, water retention curve, infiltrometry, and penetrometry are common techniques to assess soil structural stability (Grossman and Reinsch 2002; Dane and Hopmans 2002).
3. Bioavailable nitrogen (N) and other nutrients. Along with C, N is essential to the growth and function of both plants and microbes. So, total organic N and bioavailable N are key indicators of soil health. Nitrogen mineralization rate and quantity based on certain incubation/extraction methods are often used to assess soil health nutrients (Haney et al. 2018; Hurisso et al. 2018). Extractable and presumably bioavailable nutrients, such as phosphorus (P), potassium (K),

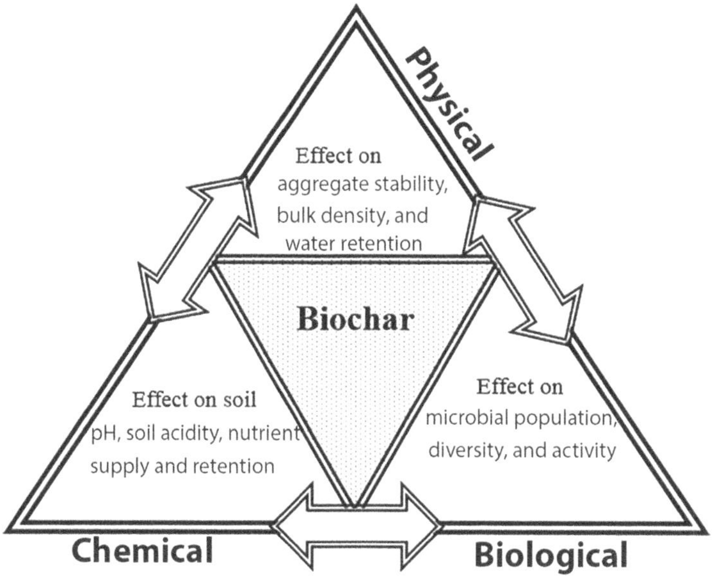


**Fig. 2.1** Potential effects of biochar on soil's biological, chemical, and physical properties (Modified from O'Toole and Rasse 2017)

calcium (Ca), magnesium (Mg), and many other micronutrients, can be routinely measured by soil testing laboratories (Hue et al. 2000).

4. Microbial population, diversity, and enzyme activity. Metagenomics has been used to identify microbes that cannot be cultured in the laboratory (Streit and Daniel 2017). Some commercial laboratories have used phospholipid fatty acid analysis to provide microbial community structural information (Buyer and Sasser 2012). Enzymes, such as β-glucosidase, N-acetyl- β-D-glucosaminidase, have been used to measure microbial activity (Deng and Popova 2011; Lammirato et al. 2011).

Since organic C is essential to soil health, and biochar contains large quantity of C (>50%), a fraction (5–10%) of which is labile and reactive (Lorenz and Lal 2018), biochar role in maintaining soil health deserves a close evaluation as proposed in Fig. 2.1.

## 2.2 Biochar

### *2.2.1 History, Definition, and Production*

In the 1960s, the late Dutch soil scientist, Wim Sombroek (1934–2003) discovered "dark soils" called Terra Preta in the Amazon basin of Brazil (Sombroek et al. 2002;

Harder 2006; Marris 2006). These soils were found to contain burned wood, crop C residue, and bone from animals (Sombroek et al. 2002). Some archeologists surmised that these fertile black soils helped sustain a relatively large population of the local Indians whose land mostly consisted of nutrient-poor Oxisols and Ultisols in the Tropics (Lehmann and Rondon 2006; Steiner et al. 2007). In fact, the role of biochar (it was known as charcoal before this millennium) in soil quality/health was acknowledged long before the twentieth century as pointed out by Spokas and Novak (2015).

According to the International Biochar Initiative (IBI), biochar is defined as a solid material obtained from the thermochemical conversion (i.e., heating or pyrolysis) of biomass (e.g., wood, crop residue, manure, biosolids, etc.) in an oxygen limited environment (IBI 2012).

Early pyrolysis produced charcoal for energy or metallurgical uses, and has evolved from amateur homemade devices, small-scale commercial pits and mound kilns to modern fast reactors (Brown et al. 2015; Oaks 2018; Cox 2019). Simple and inexpensive charcoal kilns consisted of pits or mounds (Fig. 2.2 a and b). These kilns usually used wood as feedstock and soil or brick as insulator. The process may take several days or even weeks, and the finished charcoal is rather low ($< 25\%$) in yield and quality (Oaks 2018).

On the other hand, fast pyrolysis reactors are characterized by high mass and heat transfer rates, which can be several hundred degrees (°C) per second (Boateng et al. 2015). Antal and Gronli (2003) at the Hawaii Natural Energy Institute (University of Hawaii at Manoa) produced biochar by the flash ignition in a reactor containing a packed bed of biomass at elevated pressure. Their process took less than 30 min and could yield over 90% fixed C, which can be improved further by higher pressure and the removal of the released gases. The high flow rates of gas and short residence time of biochar produced in the fast pyrolysis process would yield biochars with properties quite different from those produced by slow pyrolysis as discussed in the following sections.

Between the two options, there have been some biochar making inventions, such as the top-lit up-draft gasifier (Cox 2019) and Kon-tiki open-air conical kiln (Oaks 2018), that are mobile and could be homemade (Figs. 2.3 a & b). These devices are designed for wood-based feedstock and produced in batch mode small quantities of biochar suitable for individual or family uses (e.g., home gardening or conducting research).

### 2.2.2 *Biochar Structural Properties*

Biochar properties, from physical and structural strength to chemical and composition, depend on both the feedstock and pyrolysis process used (Chia et al. 2015; Cong et al. 2017). Feedstock for biochar can range from forest products, crop residues, to animal and municipal wastes. In general, the original biomass structure strongly influences the final biochar structure. For example, biochar pore structure

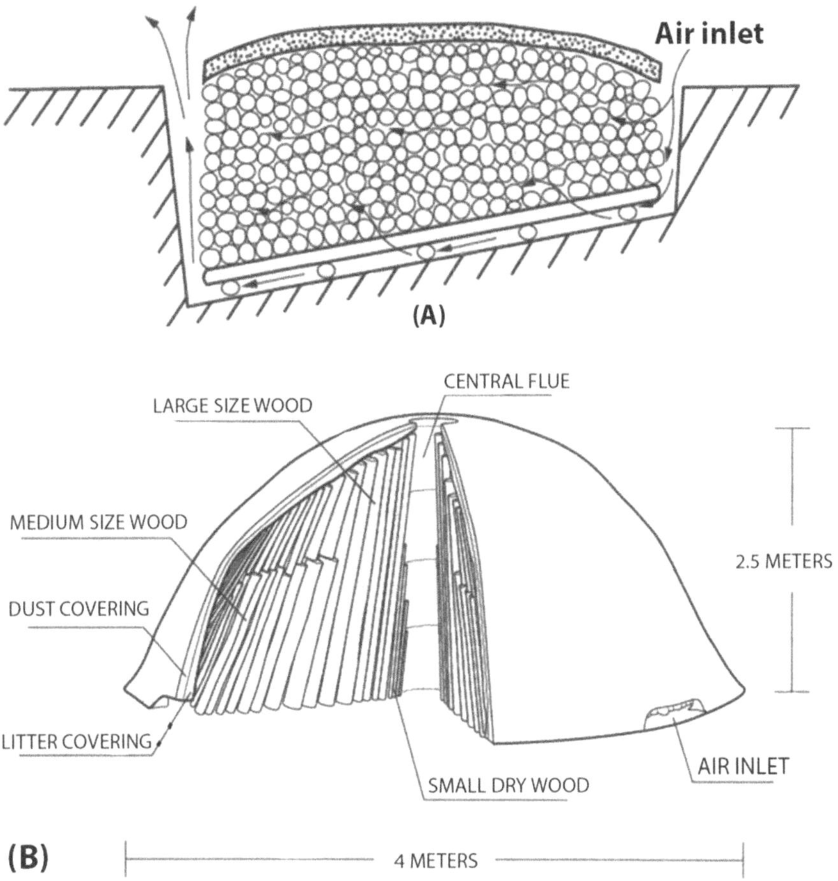


**Fig 2.2** Pit kiln (**a**), Mound kiln (**b**) (adapted from FAO 1987)

closely resembles the cellular structure of its wood-based feedstock (Fuertes et al. 2010). As an example, Fig. 2.4 shows the scanning electron microscope images of shape and size of pores from six different wood-based biochar (Berek and Hue 2016). In a review paper, Chen et al. (2019) mentioned that the volume due to small pores of a rice husk biochar was 2.1 $cm^3/g$, which was 12.3 times larger than that of a biochar produced from sludge (biosolids) as measured with the nuclear magnet technology (Fig. 2.4).

The highest treatment temperature (HTT) during pyrolysis and the residence time significantly affect biochar structure (Kim et al. 2012; Ronsse et al. 2013). As the HTT increases, the aromatic C structure, the nano-pore size, and the total surface area of the biochar increase (Chia et al. 2015). However, when HTT exceeds 700–750 °C, some microporous structures of biochar may break down, reducing its surface area (Huang et al. 2014). Consequently, the specific surface area of biochar generally ranges from 1.5 to over 500 $m^2/g$ (li et al. 2018, Liu et al. 2019),

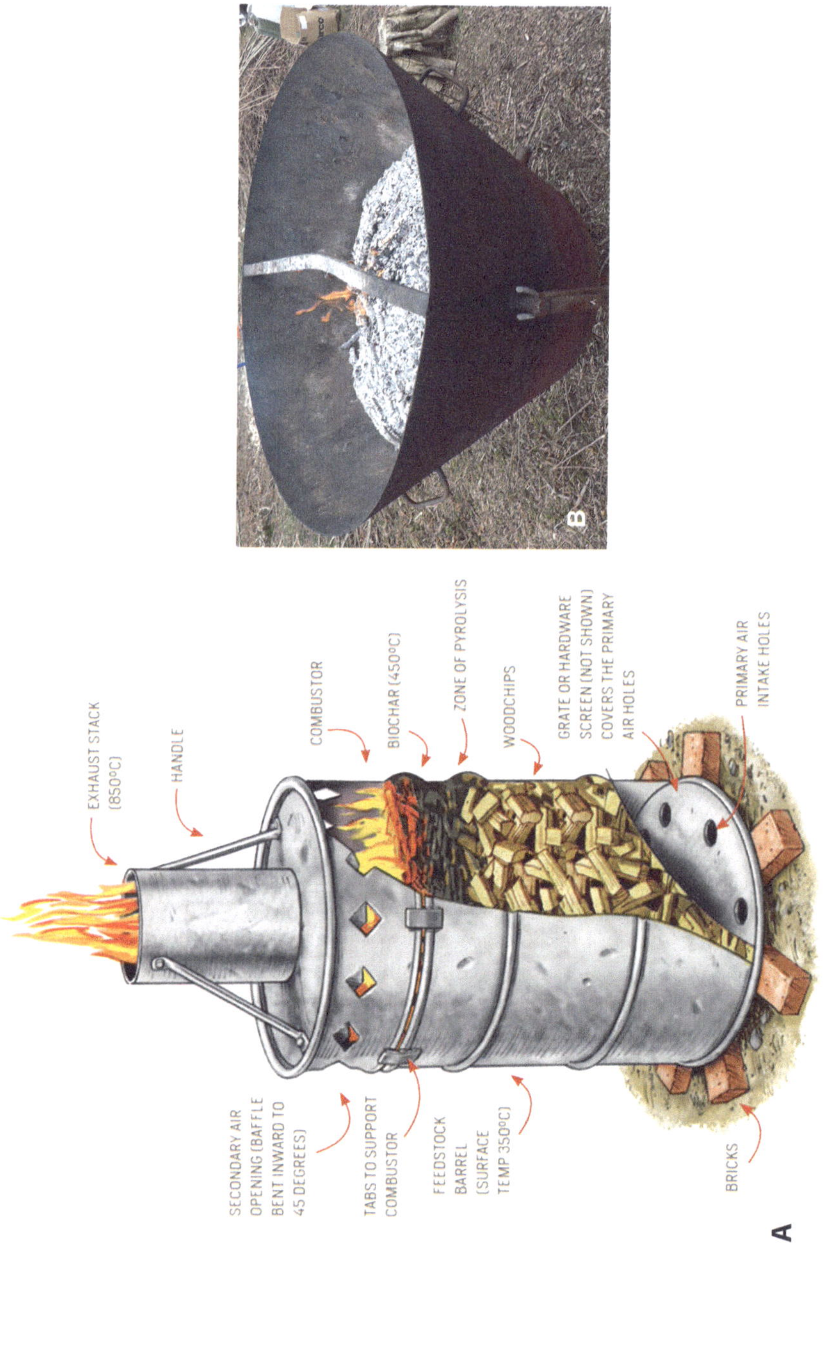


**Fig. 2.3** Top-lit up-draft gasifier (**a**, left) and Kon-tiki kiln (**b**, right) (Adapted from Cox 2019 and Oaks 2018, respectively)

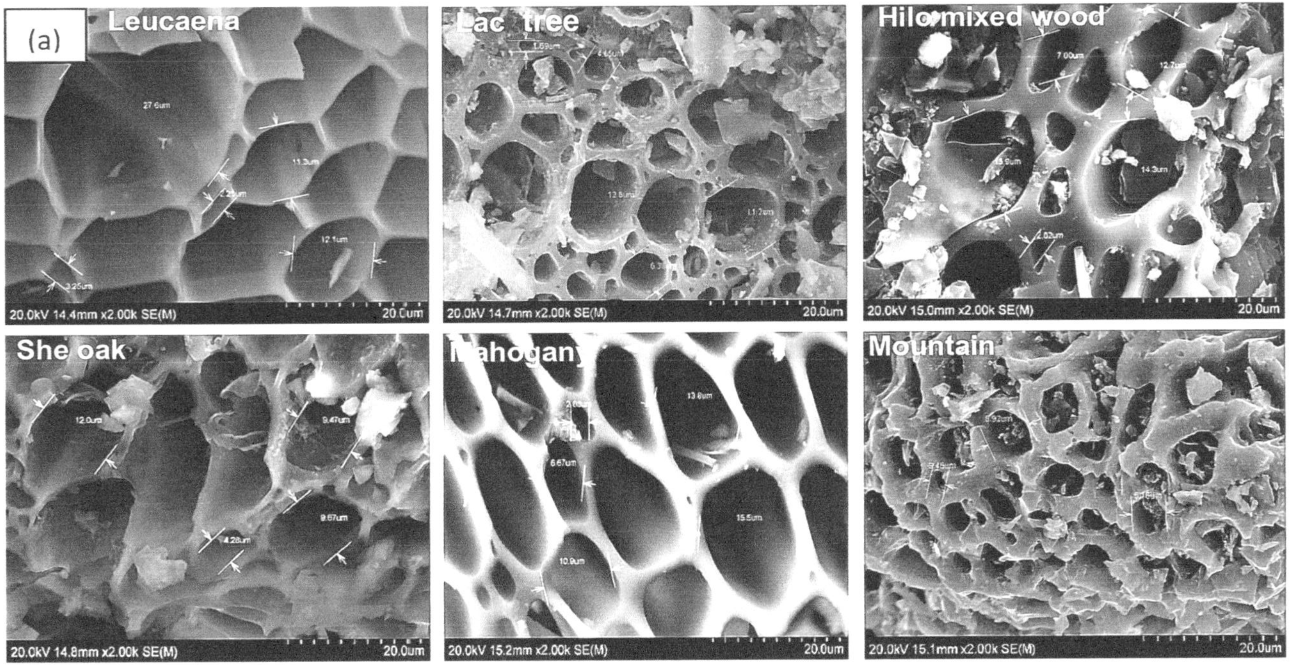


**Fig. 2.4** Scanning electron microscope images of the surface structure and porosity of six different wood-based biochars (adapted from Berek and Hue 2016)

and reaches a maximum and then declines as HTT increases further. At lower HTT and with fast pyrolysis, tars and volatile products (bio-oils) from the thermal decomposition of biomass may block micropores and reduce surface areas. As the temperature increases, the same substances volatile and escape from the pores, yielding more volume and larger surface areas. Moreover, it should be noted that having numerous micropores, biochar, especially made at high HTT, could trap significant amount of water and nutrients, such as nitrate (Prendergast-Miller et al. 2011).

### 2.2.3 Biochar Nutrients

Total nutrient concentrations in biochars are strongly influenced by the feedstocks as illustrated in Table 2.1. In general, plant-based biochars contain more C but less N than manure-based biochars. The values closely reflect the C and N content in the corresponding feedstocks. That is because manure usually has lower C and higher N (as proteins) than plants, especially wood. Also, increasing pyrolysis temperature increases the total nutrient concentration in biochar as shown in Table 2.2. It is not surprising that total (fixed) C averages above 60% in plant-based biochar and around 40% for manure-based biochars. More specifically, the biochar C is mostly aromatic, which is formed in an irregular stack of condensed rings when the HTT exceeds 400 °C (Kleber et al. 2015; Chia et al. 2015). Nitrogen is mainly present on the surface of biochar as C-N heterocyclic structure and the bio-availability of this N is very low (Chen et al. 2019; Deenik et al. 2010). Similarly, biochar P is not readily available. According to Ippolito et al. (2015), available P ranges from 0.4 to 34% of total P in biochar. In contrast, the authors also reported that between 55 and 65% of

**Table 2.1** Average total nutrient concentrations (dry weight basis) of biochars from various feedstocks (modified from Table 7.1 of Ippolito et al. 2015)

| Source | C | N | P | K | Ca | Mg |
|---|---|---|---|---|---|---|
| | ←------ | | | % | | ------ |
| Corn | 58.8 | 1.06 | 0.23 | 1.90 | 0.86 | 0.71 |
| Rice straw/husk | 43.6 | 1.40 | 0.12 | 0.07 | – | – |
| Peanut shell | 75.3 | 1.83 | 0.21 | 1.10 | 0.33 | 0.15 |
| Bagasse | 78.6 | 0.87 | 0.07 | 0.22 | 0.73 | 0.18 |
| Coconut coir | 73.8 | 0.88 | | | | |
| Hardwood | 74.4 | 0.72 | 0.11 | 0.95 | 1.01 | 0.95 |
| Softwood | 74.6 | 0.79 | 0.07 | 1.69 | 2.07 | 1.80 |
| Food waste | 44.4 | 3.28 | 0.66 | 1.92 | 5.18 | 0.49 |
| Poultry/manure/litter | 35.3 | 2.15 | 3.31 | 6.02 | 10.3 | 1.22 |
| Swine manure | 44.9 | 2.79 | 6.08 | 2.34 | 4.80 | 2.90 |
| Cattle manure | 48.5 | 1.90 | 0.92 | 4.06 | 2.88 | 0.99 |
| Biosolids/sludge | 23.8 | 1.22 | 4.24 | | | |

**Table 2.2** Average total nutrient concentrations (dry weight basis) of biochars based on pyrolysis temperature and pyrolysis type (modified from Table 7.2 of Ippolito et al. 2015)

| Source | C | N | P | K | Ca | Mg |
|---|---|---|---|---|---|---|
| **Pyrolysis Temp.** | ←---------------------------% --------------------------- | | | | | |
| <300 °C | 53.6 | 1.25 | 1.14 | 0.49 | 0.11 | |
| 300–399 °C | 57.1 | 1.99 | 1.37 | 2.11 | 3.91 | 0.71 |
| 400–499 °C | 62.1 | 1.29 | 1.30 | 1.77 | 5.24 | 0.51 |
| 500–599 °C | 63.2 | 1.15 | 1.18 | 1.49 | 4.99 | 0.69 |
| 600–699 °C | 62.4 | 0.94 | 1.14 | 1.49 | 5.56 | 0.67 |
| 700–799 °C | 63.7 | 1.50 | 4.29 | 5.40 | 4.68 | 1.88 |
| > 800 °C | 63.2 | 0.84 | 2.54 | 7.72 | 7.84 | 7.26 |
| **Pyrolysis type** | | | | | | |
| Fast | 56.2 | 0.74 | 1.48 | 5.32 | 6.05 | 6.06 |
| Slow | 60.2 | 1.44 | 1.54 | 2.08 | 4.78 | 0.87 |

**Table 2.3** Average pH and cation exchange capacity (CEC) of biochars from various feedstocks (modified from Table 7.5 of Ippolito et al. 2015)

| Source | pH | CEC ($cmol_c$/kg) |
|---|---|---|
| Corn | 9.27 | 60.7 |
| Rice straw/husk | 9.17 | 21.2 |
| Peanut shell | 8.52 | – |
| Bagasse | 7.59 | 11.5 |
| Hardwoods | 7.94 | 13.8 |
| Softwoods | 7.48 | 14.5 |
| Food waste | 9.09 | 8.1 |
| Poultry manure/litter | 9.80 | 53.8 |
| Swine manure | 9.37 | – |
| Cattle manure | 8.99 | – |
| Biosolids/sludge | 6.90 | 2.36 |

the K, Ca, and Mg available from biochars can be related to their total concentration. Most of biochar K is water soluble and readily available, especially when produced from slow pyrolysis (Cantrell et al. 2012; Berek et al. 2018). Liu et al. (2019) mixed 36 different biochars with water (1:75 mass ratio) and measured several macronutrients (nitrate, phosphate, ammonium, K, Ca, chloride, etc.) and micronutrients (copper, iron, manganese, etc.) in the extract after 2 days of incubation. The authors found elevated concentrations of these nutrients, especially in manure-based biochars. Biochar Ca and Mg, in most situations, are probably present in carbonate, phosphate, and/or oxide forms (Berek and Hue 2016) (Table 2.2).

### *2.2.4 Biochar pH and Cation Exchange Capacity (CEC)*

Most biochars are alkaline, and manure-based biochars often have higher pH than wood-based biochars (Table 2.3). Such alkalinity is probably caused by the presence

**Table 2.4** Average pH and cation exchange capacity (CEC) of biochars based on pyrolysis temperature and pyrolysis type (modified from Table 7.6 of Ippolito et al. 2015)

| Source | pH | CEC ($cmol_c/kg$) |
|---|---|---|
| **Pyrolysis temperature** | | |
| <300 °C | 5.01 | 32.7 |
| 300–399 °C | 7.60 | 37.1 |
| 400–499 °C | 8.10 | 19.1 |
| 500–599 °C | 8.71 | 28.3 |
| 600–699 °C | 9.00 | 12.6 |
| 700–799 °C | 9.83 | 3.9 |
| > 800 °C | 10.80 | 4.4 |
| **Pyrolysis type** | | |
| Fast | 8.38 | 2.9 |
| Slow | 8.50 | 25.0 |

of alkali salts such as KOH, NaOH, $CaCO_3$, and $MgCO_3$ formed during pyrolysis (Berek and Hue 2016; Cao and Harris 2010). In fact, increasing pyrolysis temperature decomposes acidic functional groups, such as carboxylic COOH, phenolic OH, and lactonic O, forming alkali bases and making biochar more basic (Yuan et al. 2011; Table 2.4).

As Table 2.4 specifically demonstrates, biochar pH increases from 7.6 at the 300–399 °C pyrolysis temperature range to 8.7 at 500–599 °C and 9.8 at 700–799 °C range. Since biochars are mostly basic, they could be used as liming materials, and their calcium carbonate equivalent (CCE) can range from 6 to 30% as reported by Hue and co-workers (Berek and Hue 2016; Ahmad et al. 2018).

Like soil organic matter, biochar can carry pH-dependent (variable) charge, most often negative charge, giving rise to cation exchange capacity (CEC). Biochar CEC is generated mainly by oxygen containing functional groups, such as carboxylate, phenolate, or lactonate, on the biochar surface (Fig. 2.5).

This partially explains why wood-based biochars often have higher CEC than manure-based biochars (Table 2.3). That is because cellulose and lignin in wood contain much more functional groups than manure even when pyrolyzed. Furthermore, as biochar ages, and is exposed to oxygen and water, more functional groups on the surface can be generated through oxidation, thus increased CEC is attained (Clough and Condron 2010).

## 2.3 Biochar's Impacts on Soil Health

Although the effects of biochar on soil health depend on many factors, including biochar properties (mainly, particle and pore size, porosity, surface area, surface functional groups), soil properties (e.g., texture, pH, C content), and their complex interactions, it has been observed so far that biochar provides significant impacts, mostly beneficial, on soil health, crop production, and the environment (Laird et al. 2010; Lorenz and Lal 2018; Biswas et al. 2019; Chen et al. 2019; Wu et al. 2019).

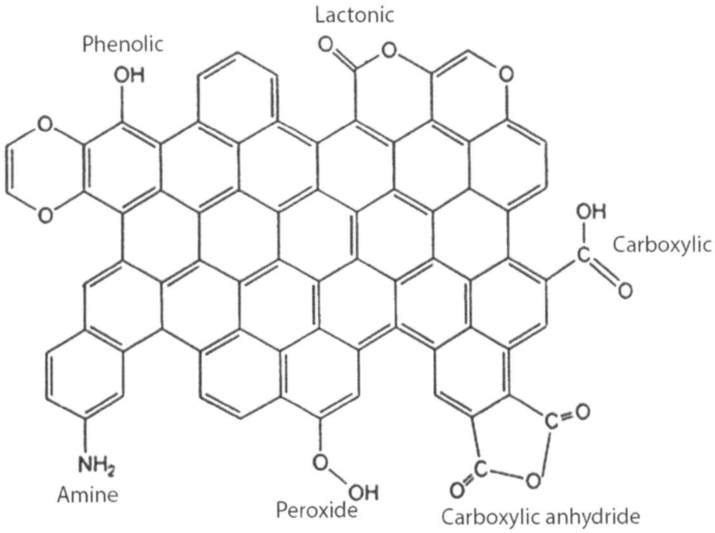


**Fig. 2.5** Functional groups commonly found on biochar surfaces

Such benefits are most pronounced when biochar is applied to soils with low fertility and acidic as those in the humid Tropics (Lehmann and Rondon 2006; Steiner et al. 2007; Jeffery et al. 2011; Berek et al. 2018).

### 2.3.1 *Biochar and Soil Physical Properties*

Despite the fact that most C in biochar is condensed aromatic and recalcitrant, a small fraction of it (< 10%) is labile and bioavailable, particularly if biochar is made at low HTT and fast pyrolysis (Bruun et al. 2012; Maestrini et al. 2014; Mukherjee and Zimmerman 2013; Meng et al. 2019; Wang et al. 2017). As an example, Steiner et al. (2007) reported that 4–8% of biochar C was lost during four cropping cycles in a field trial in Manaus, Brazil (humid tropical conditions). In fact, C from biochar plays a key role in soil aggregate stability based on mean weight diameter measurement (Liu et al. 2014; Wang et al. 2017). More specifically, Liu et al. (2014) applied 40 ton/ha of a wheat straw biochar (pyrolyzed at 350–550 °C) to a red soil (Ultisol) of Southern China, and reported that the soil water stable aggregate (>0.25 mm) was enhanced by 28% over the control. Furthermore, soil organic C, total N and C:N ratio were also significantly increased in the >2 mm, 2–0.5 mm, and < 0.25 mm aggregate fractions of the biochar treatment. Similarly, Wang et al. (2017) showed a remarkable improvement in aggregation of a fine texture (silty loam) soil (Yolo series from California) with 217% and 126% average increases in mean weight diameter when incubated for 60 weeks with a softwood biochar (pyrolyzed at 600–700 °C with algal digestate) and a walnut shell biochar gasified at 900 °C,

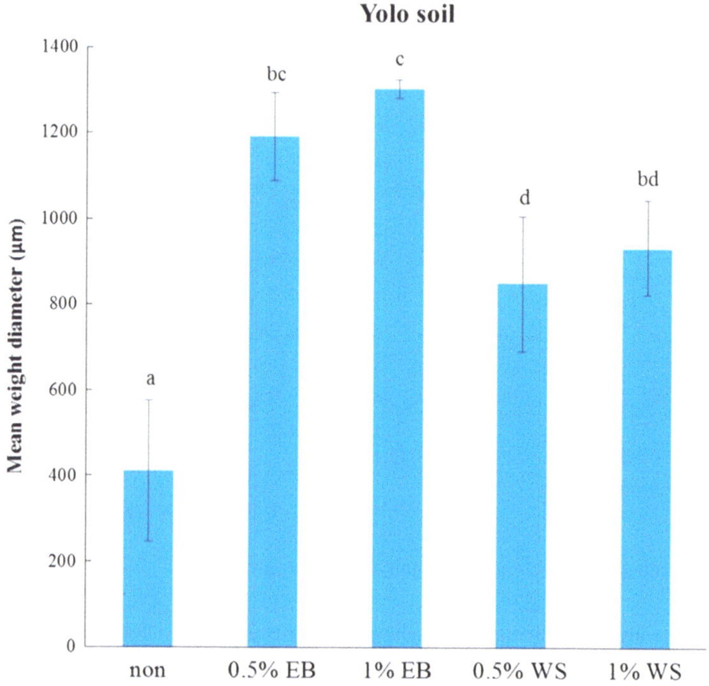


**Fig. 2.6** Soil aggregate stability (mean weight diameter) after 60 weeks of incubation in the Yolo silty loam soil with and without the addition of two biochar types (EB, softwood biochar; WS, walnut shell biochar) at 0.5 and 1.0% (W:W) application rates. The error bars represent standard errors and bars with different letters indicate statistically significant ($P < 0.05$) differences (adapted from Wang et al. 2017)

respectively (Fig. 2.6). The authors suggested that biochars enhanced the proportion of C stored within the soil macroaggregates and strengthened aggregate stability.

The bulk density of most biochars ranges from 0.20 to 1.0 g/cm$^3$, depending on the feedstock, with an average of 0.5 g/cm$^3$ (Chia et al. 2015; Laird and Novak 2017). Thus, adding biochar at common rates of 0.5–5.0% to mineral soils having an average bulk density of 1.2 g/cm$^3$ will reduce the overall bulk density of the amended soil significantly (Laird et al. 2010; Obia et al. 2016; Verheijen et al. 2019). For example, Case et al. (2012) reported that soil bulk density (in field moist condition) decreased from 0.95 g/cm$^3$ to 0.89, 0.87, and 0.84 g/cm$^3$ with the application of 0, 2, 5, and 10% of a hardwood biochar (HTT = 400 °C, 24-h residence), respectively.

As shown in Fig. 2.4 (Berek and Hue 2016), biochars have high porosity, which was caused by the pyrolytic emission of structural water and the decomposition into gases of feedstock tissues (e.g., cellulose, lignin, proteins). With numerous and variable pores, biochars help reduce the bulk density and increase the water holding capacity of the amended soil (Duong et al. 2017; Obia et al. 2016; Fisher et al. 2019).

For example, Duong et al. (2017) showed that 1% biochars made from rice husk or coffee husk (HTT = 550 °C) increased the water holding capacity of a sandy gray soil of Vietnam by 26–33%. Cautions should be taken, however, because the effect of biochar on water retention could vary significantly, depending on particle size of biochar, quantity applied, as well as the soil texture (Fisher et al. 2019; Masiella et al. 2015). That is because biochar disrupts the soil matrix by changing the pore size distribution: It promotes larger pores in fine textured soils (loam and clay), but makes pore space narrower in coarse textured sandy soils.

## 2.3.2 *Biochar and Soil Chemical Properties*

### 2.3.2.1 Biochar's Impact on Soil pH and Acidity

Given the alkaline pH of most biochars, incorporating biochar into acid soils can increase soil pH up to 73% with an average increase of 28% (Mukherjee and Lal 2017). As an example, Xu et al. (2012) reported an increase over 2 pH units, from 5.0 to >7.0, when 5% peanut shell biochar (pyrolyzed at 350 °C) was applied to four acid soils (Oxisols and Ultisols) of Southern China. The authors also showed a significant increase in pH buffering capacity, defined as the slope of the linear response line of soil pH as a function of acid/base additions in the pH range of 4.0 to 7.0, of these biochars amended soils (Table 2.5).

Biochar alkalinity can come from four sources (Fidel et al. 2017): (1) Surface functional groups (as conjugates bases such as carboxylate, phenolate), (2) Soluble organic compounds (also conjugate bases of weak organic acids), (3) Carbonates (salts of bicarbonate and carbonate), and (4) Other inorganic alkalis (oxides, hydroxides, sulfates, phosphates) as illustrated in Fig. 2.7.

Total alkalinity, however, strongly correlates with total base cations (Na + K + Ca + Mg) extracted with 0.05 *M* HCl for eight biochars produced under different temperatures (300, 500, and 600 °C) and pyrolysis conditions of slow, fast, and gasification as shown in Fig. 2.8.

Alkalinity of biochar and its liming potential are often expressed as calcium carbonate equivalent (CCE). A CCE range of 5.0–30.0% is common for many wood-based biochars (Laird et al. 2010; Berek and Hue 2016). It is worth noting that a biochar with 8% CCE applied at 2% (w:w) to an acid Ultisol of Hawaii lowered exchangeable aluminum (Al) from 2 $cmol_c$/kg to virtually zero, thus completely eliminating Al toxicity in this soil (Berek and Hue 2016; Fig. 2.9).

In fact, Al and to a lesser extent, manganese (Mn) in acid soils can also be complexed and detoxified by reactive functional groups on the biochar surfaces. Most of these groups contain oxygen, such as carboxyl, carbonyl, and hydroxyl and closely related to pyrolysis conditions: their number and density usually decrease as the HTT increases (Gul et al. 2015; Zhao et al. 2017). In contrast, upon aging and exposed to oxygen and water, biochar can develop more of these reactive functional groups (Mukherjee et al. 2014). Consequently, heavy metals, such as lead (Pb),

**Table 2.5** Effect of two biochars incorporated on properties and pH buffering capacity of soils (CSBC = canola straw biochar, PSBC = Peanut shell biochar; adapted from Xu et al. 2012)

| Soil and location | Depth (cm) | Treatment | pH | Organic matter (g/kg) | CEC (cmol/kg) | pH buffering capacity (mmol/kg/pH) |
|---|---|---|---|---|---|---|
| Ultisol from Liuzhou, Guanxi | 60–120 | Control | 5.38 | 4.4 | 5.15 | 20.8 |
| | | 1% CSBC | 6.72 | 15.5 | 5.90 | 22.3 |
| | | 5% CSBC | 7.46 | 23.0 | 6.17 | 27.3 |
| | | 3% PSBC | 6.83 | 27.7 | 8.26 | 30.5 |
| | | 5% PSBC | 7.35 | 41.2 | 9.28 | 36.1 |
| Oxisol from Chengmai, Hainan | 60–130 | Control | 5.05 | 8.4 | 5.97 | 20.1 |
| | | 3% CSBC | 6.68 | 19.1 | 6.12 | 23.0 |
| | | 5% CSBC | 7.29 | 26.3 | 7.14 | 27.0 |
| | | 3% PSBC | 6.85 | 31.0 | 8.01 | 29.4 |
| | | 5% PSBC | 7.20 | 44.4 | 9.03 | 38.6 |
| Ultisol from Kunlun, Hainan | 50–110 | Control | 5.00 | 10.9 | 5.30 | 15.5 |
| | | 3% CSBC | 6.70 | 21.4 | 6.53 | 18.4 |
| | | 5% CSBC | 7.47 | 26.8 | 7.04 | 23.6 |
| | | 3% PSBC | 7.04 | 32.9 | 7.80 | 25.7 |
| | | 5% PSBC | 7.45 | 46.2 | 9.69 | 34.6 |

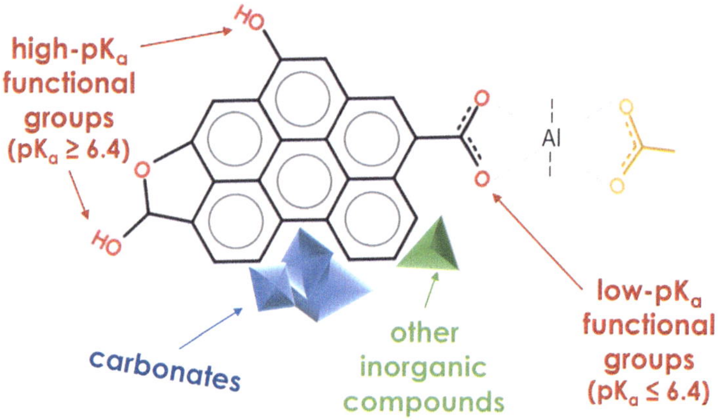


**Fig 2.7** Four likely sources of alkalinity in biochars (modified from Fidel et al. 2017)

cadmium (Cd), are readily sorbed and detoxified by similar mechanisms (e.g., complexation, cation exchange, and precipitation) as they do for Al and Mn (Beesley et al. 2015; Li et al. 2017).

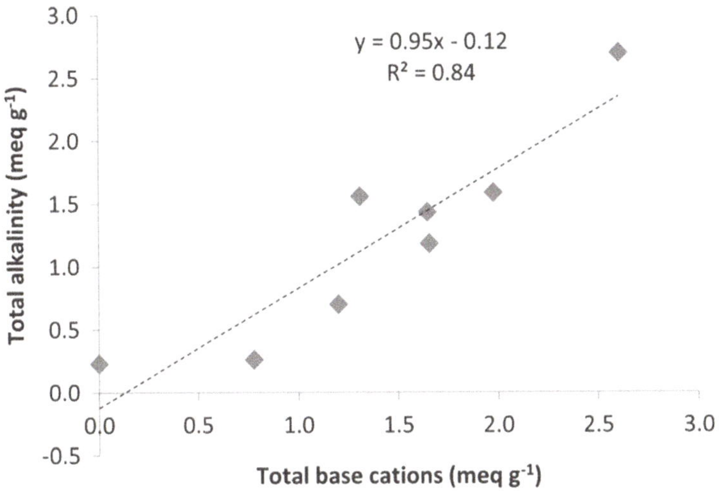


**Fig. 2.8** Relationship between total soluble base cations (sum of meq $g^{-1}$ for 0.05 M HCL extractable Na+ K+ Ca+ Mg) and total alkalinity (meq $g^{-1}$) for all eight untreated biochars. Each data point represents the mean of three measurements. Total alkalinity was determined by titration of 0.05 MHCL extracts. (Taken from Fidel et al. 2017)

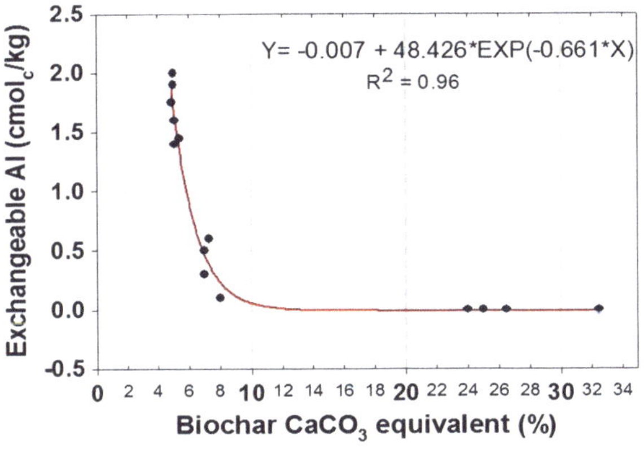


**Fig. 2.9** Exchangeable Al as a function of $CaCO_3$ equivalent of wood-based biochars applied to an acid Ultisol of Hawaii (adapted from Berek and Hue 2016)

#### 2.3.2.2 Biochar's Impact on CEC, Nutrient Retention and Supply

Given that the $pK_a$s of these oxygen containing groups range from 2 to 7 or perhaps 9, which are not much different from common soil pHs (pH 3–9), negative charges and CEC will be markedly increased when biochar is aged and is mixed with soil. Many studies (Silber et al. 2010; Laird et al. 2010; Mukherjee et al. 2014; Martinsen et al. 2014) have shown that soil CEC may increase up to 30% on average.

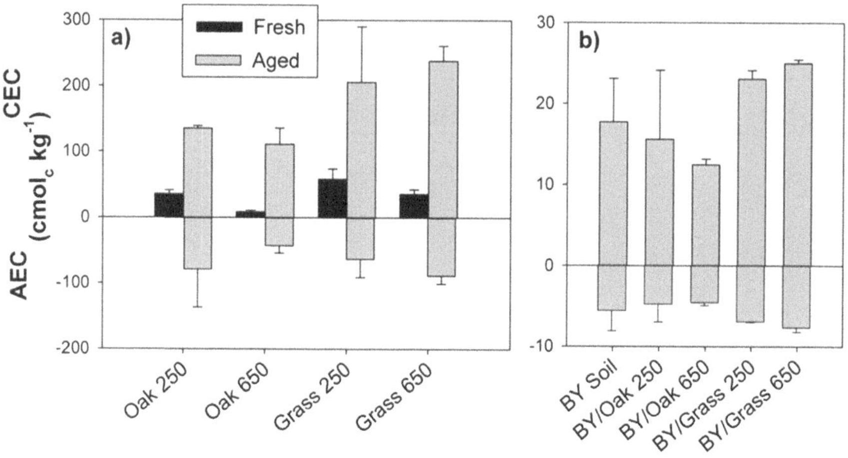


**Fig 2.10** Cation and anion exchange capacities (CEC and AEC, respectively) measured at pH 6–7 on (**a**) fresh and "aged" oak and grass biochars produced at 250 and 650 °C, and (**b**) aged BY soil and BY soil–biochar mixtures (Adapted from Mukherjee et al. 2014)

However, conflicting evidence also exists (Mukherjee and Lal 2017). For example, Mukherjee et al. (2014) reported that after aging for 15 months, biochars made by pyrolysis of wood (oak and pine) and grass at 250, 400, and 650 °C exhibited fivefold increases in CEC. When added to soil, the CEC of the biochar amended soil (a forest Spodosol from Florida) increased with the grass biochar but decreased with the oakwood biochar (Fig. 2.10). Thus, the effect of biochar on soil CEC has not clearly determined, perhaps due to the interactions between biochar and soil.

Having many negatively charged functional groups on the surface and increased CEC of the amended soil, biochar can effectively retain nutrient cations, such as $NH_4^+$, $K^+$, $Ca^{2+}$, and $Mg^{2+}$ (Laird et al. 2010; Wang et al. 2015). Mehlich III extractable P, K, Mg, and Ca were increased significantly when a hardwood (oak and hickory) biochar was applied at 0.5–2.0% to a Mollisol of Iowa (Laird et al. 2010). Ammonium in wastewater (concentration range 2–20 mg/L) was removed by 66% by a filter made of peanut hull biochar (Saleh et al. 2012). Nutrient entrapment caused by porous structure, and high water holding capacity has been suggested as a responsible mechanism for anions, such as nitrate and arsenate, retention (Ippolito et al. 2015).

Besides being an efficient adsorbent, biochar itself contains nutrients (Table 2.1). Depending on feedstock and pyrolysis process, and also on individual nutrient, nutrient availability may be immediate or gradual. For example, biochars derived from animal manure or grass and pyrolyzed at lower temperature release more nutrients than those made from woody biomass at higher HTT (Mukherjee and Zimmerman 2013). Also as discussed previously, over 50% of total K in biochar is water soluble and readily bioavailable. Thus, biochar can be a good source of K for crop uptake, especially in organic farming (Martinsen et al. 2014; Butnan et al. 2015; Berek et al. 2018).

On organic N mineralization, biochar can have positive, neutral, or negative effects (Prommer et al. 2014; Maestrini et al. 2014). For example, an increase of 7% in N mineralization was obtained when 5% of a wheat straw biochar (slow pyrolysis at 525 °C) was mixed with a sandy loam soil (Spodosol), while a 43% reduction was resulted from the application of the same feedstock but fast pyrolyzed biochar after 65 days of incubation (Bruun et al. 2012). Similarly, the direct contribution of N from biochar has a mixed result, particularly in terms of plant responses (Gul and Whalen 2016; Hood-Nowotny et al. 2018). Since the C/N ratio in many biochars is much greater than the 25–30 range, which deems optimal for N mineralization, N deficiency in crops due to N immobilization in biochar amended soils may occur, at least in the short term (Deenik et al. 2010; Cely et al. 2014).

## *2.3.3 Biochar and Soil Biological Properties*

As discussed previously, biochar can change soil physical properties via its large surface areas and numerous and size-variable pores; it can modify soil chemical properties via its alkaline pH, CCE, considerable CEC, high ionic strength (expressed as electrical conductivity), along with some labile organic C. These changes, in turn, affect the growth, composition, and activity of soil biota. Figure 2.11 summarizes such probable causes and expected effects of biochar on soil biological properties.

### 2.3.3.1 Biochar as a Potential Habitat and Growth Promoter for Soil Biota

The porous structure of biochar, its large internal surface area, and its high capacity to retain water provide favorable habitats for soil biota (Quilliam et al. 2013; Jaafar et al. 2015). Bacteria (size 0.3–3 mm) and hyphae (<16 μm) of different fungi can colonize biochar macropores (sizes of 2 mm–2 μm are common), and avoid predators, such as mites and nematodes (Ezawa et al. 2002; Jaafar et al. 2015; Ogawa and Okimori 2010). SEM images from Palansooriya et al. (2019) clearly show fungal hyphae grown on the surface of a peanut shell biochar (Fig. 2.12).

In addition, water is essential to all living organisms, and its presence in biochar pores would enhance the microbial habitability (Batista et al. 2018). Such habitat may also help some microbes that are less competitive in the "hostile" environment of the unamended soils become established (Ogawa and Okimori 2010; Wong et al. 2017).

Depending on the pyrolysis conditions (HTT and residence time) and the feedstock from which biochar derived, a significant quantity of labile C can be added to soil when biochar is applied. For example, flash carbonizing and low HTT leave residual bio-oils and other condensed volatile compounds on the biochar surfaces (Deenik et al. 2010). Such C materials can serve as substrates (energy sources) for

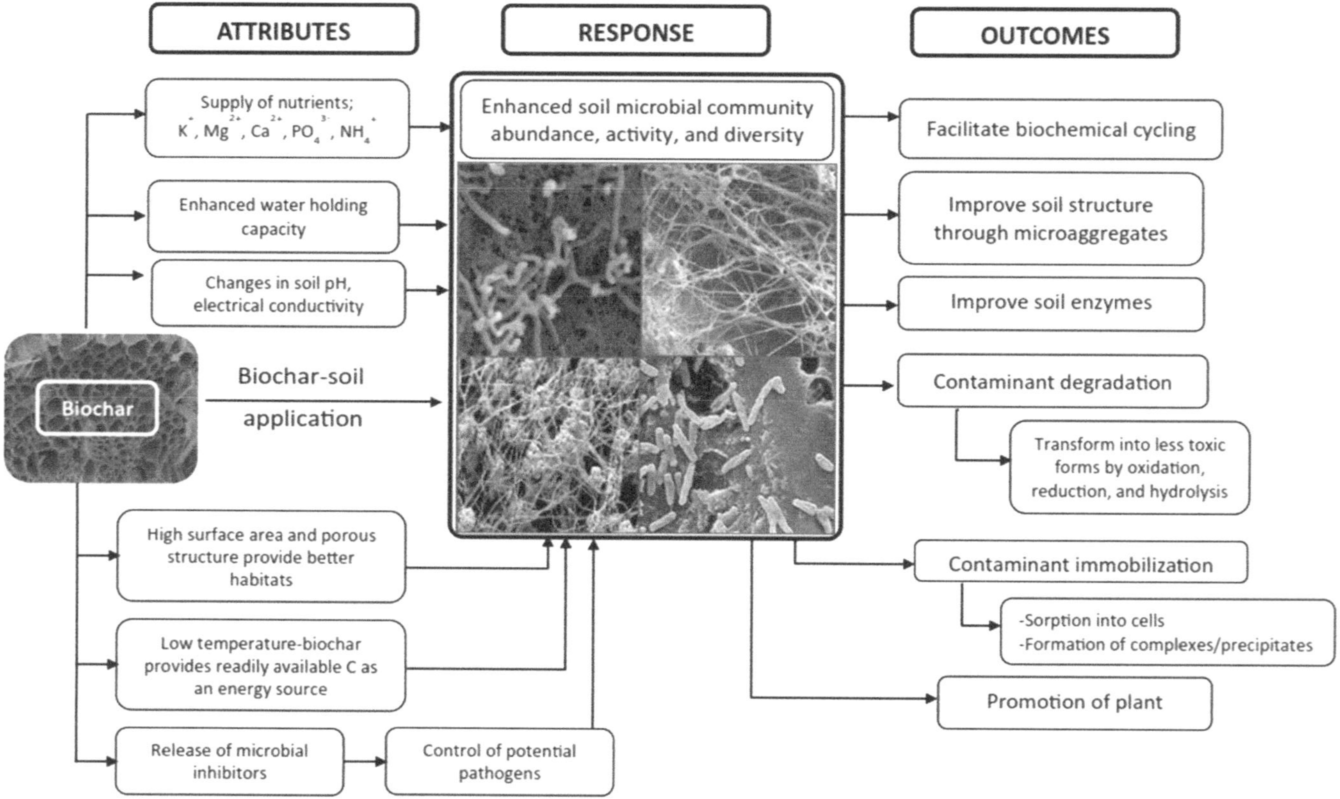


**Fig. 2.11** Schematic diagram showing the effect of biochar application on soil microorganisms and microbial responses (adapted from Palansooriya et al. 2019)

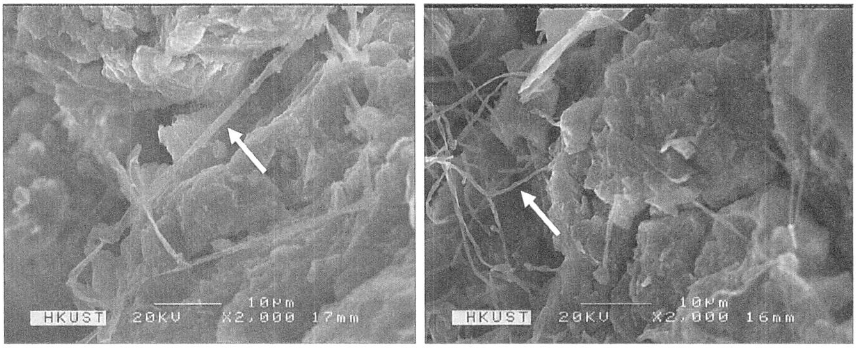


**Fig 2.12** SEM images of fungal hyphae (pointed by white arrow) grown on peanut shell biochar (pyrolyzed at 500 °C) (adapted from Palansooriya et al. 2019)

microbe growth and metabolism, or even be toxic to certain microbial pathogens (Graber et al. 2014). Maestrini et al. (2014) used $^{13}$C-labelled ryegrass biochar to estimate microbial uses of biochar C in a forest soil. The authors found that 4.3% of biochar $^{13}$C was mineralized after a 5-month incubation, of which 0.45% was in microbial biomass. The labile/mineralized C contribution from biochar may interact with native soil organic matter (SOM), causing a priming effect, which can speed up or slow down the mineralization of native SOM (Whitman et al. 2015).

Besides C, biochar pH and ash content likely alter pH and nutrient status of the amended soil as earlier discussed. Since most bacteria thrive at near neutral pH, whereas fungi favor acidic or alkaline media, biochar application will strongly influence the bacteria to fungi ratio, microbial communities (Chen et al. 2013; Wong et al. 2019). Microbial feeders and their predators may also change as a result (Thies et al. 2015).

#### 2.3.3.2 Biochar's Effects on Soil Enzyme Activities and Microbial Community Structures

There are strong interactions between biochar and extracellular enzymes (Bhaduri et al. 2016). These enzymes are needed to degrade substrates, particularly C- and/or N-containing materials (e.g., cellulose, proteins) for their food. Biochar will affect the activity of these enzymes in various ways, depending on the relative location (folding conformation) of the active sites of the enzyme and the reactive functional groups on biochar surfaces, pH, and concentration of ionic species in the surrounding environment (Thies et al. 2015). Increased enzyme activity of dehydrogenase, β-glucosidase, and urease in a red soil (an Ustult) of China was recorded when amended with an oakwood or bamboo biochar at 0.5, 1.0, and 2.0% after 372 days of incubation (Demisie et al. 2014). Such increases have often attributed to the labile C from biochars (Demisie et al. 2014; Bhaduri et al. 2016; Gasco et al. 2016).

Using $^{13}C$-labelled phospholipid fatty acid (PLFA) analysis to study microbial community and their main C food source in a very acidic soil (pH 3.7) amended with Miscanthus biochar (HTT 350 and 700 °C), Luo et al. (2018) showed that all microbial groups (Gram-positive, Gram-negative bacteria, actinobacteria, and fungi) were more abundant in the biochar treated soil after 14 months of incubation. These microbes used the C from the 350 °C biochar, but not the 700 °C biochar as substrate. Similar findings were reported by Gomez et al. (2014) in a study using four soils from the Midwest, USA, which received a fast pyrolysis biochar at rates of 0, 1, 5, 10, and 20%. The authors concluded that biochar stimulated microbial activity and growth. More specifically, biochar addition proportionally increased microbial abundance in all four soils, and altered the community composition, most strikingly at the 20% rate, toward a more Gram− bacteria, relative to Gram+ and fungi. Also, biochar can serve as a C substrate for microbial activity.

Changes in microbial community can be further studied using more modern molecular techniques, such as 16S rRNA and 18S rRNA gene, which are characterized with terminal restriction fragment length polymorphism (T-RFLP) combined with clone library analysis, denaturing gradient gel electrophoresis (DGGE) and quantitative real-time PCR assay (qPCR) as reported by Chen et al. (2013). The authors found that gene copy numbers of bacterial 16S rRNA was increased by 28% and 64% and that of fungal 18S rRNA decreased by 35% and 46% under biochar applications of 20 and 40 ton/ha, respectively, over the control in a rice paddy of China.

Based on these abovementioned studies, it is likely that biochar can change microbial community structure, but the effect varies with biochar type, soil type, climate, and time. A fraction of C from biochar could also be used as a food source for microbial growth, and could shift microbial distribution from one group to another (Palansooriya et al. 2019).

## 2.4 Concluding Remarks

Biochar's use as a soil amendment can significantly maintain and benefit soil health. First, with properties, such as large surface areas, numerous pores and variable pore size, biochar can improve soil physical attributes, including aggregate stability, water holding capacity, root penetration, and reduced erosion. Second, given commonly high pH value, abundant reactive functional groups, relatively high CEC, ash content, and labile C, biochar can enrich soil fertility and soil organic matter. Lastly, with those desirable characteristics, application of biochar to soils, especially highly weathered, nutrient-poor, and acidic soils, has been proven to enhance soil microbial abundance, diversity, and activity. On the other hand, since climate and time also affect biochar properties and performance via oxidation by water and oxygen, long-term field experiments with different soil types and biochars should be actively conducted, keeping in mind that interactions between biochar and soil are inevitable, complex, and difficult to predict. Thus, biochar properties should be clearly characterized, and standardized, including pyrolysis process (e.g., fast, slow, highest treatment temperature, residence time); also feedstock (e.g., wood, crop residues, animal manure) should be clearly identified and publicized (Spokas and Novak 2015).

Finally, the cost of biochar production should be reduced substantially (currently, commercial biochar costs over US$1000/ton in Hawaii per the author's knowledge). The task could be accomplished either through developing new and more efficient pyrolysis processes or changing policy that would monetize the value of carbon sequestration in soils where biochar has demonstratively served as an effective amendment for maintaining and enhancing soil quality.

**Acknowledgments** This work was partially supported by a grant (No. SW16-021) from the USDA-Western Region Sustainable Agricutlure Research and Education (W-SARE) given to the author from 2016 to 2019.

## References

Ahmad AA, Berek AK, Radovich TJK, Hue NV (2018) Biochar as a soil amendment and nutrient regulator. ASHS Conf. Prog, July 30–August 3, 2018, Washington DC, Am Soc Hort Sci Abst, pp 227

Antal MJ, Gronli M (2003) The art, and technology of charcoal production. Industrial Eng Chem Res 42:1619–1640

Batista E, Shultz J, Matos T, Fornari M (2018) Effect of surface and porosity of biochar on water holding capacity aiming indirectly at preservation of the Amazon biome. Sci Report 2018:10677. Pages 9

Beesley L, Moreno-Jimenez E, Fellet G, Melo L, Sizmur T (2015) Biochar and heavy metals. In: Lehmann J, Joseph S (eds) Biochar environmental management. Routledge, New York, pp 563–594

Berek AK, Hue NV (2016) Characterization of biochars and their use as an amendment to acid soils. Soil Sci 181:412–426

Berek AK, Hue NV, Radovich TJK, Ahmad AA (2018) Biochars improve nutrient phyto-availability of Hawaii's highly weathered soils. Agron 8:203–221

Bhaduri D, Saha A, Desai D, Meena HN (2016) Restoration of carbon and microbial activity in salt-induced soil by application of peanut shell biochar during short-term incubation study. Chemosphere 148:86–98

Biswas B, Nirola R, Biswas JK, Pereg L, Willett IR, Naidu R (2019) Environmental microbial health under changing climates: state, implication and initiatives for high-performance soils. In: Lal R, Francaviglia R (eds) Sustainable agriculture reviews 29. Springer, Cham, pp 1–32

Boateng AA, Garcia-Perez M, Masek O, Brown R, Campo BD (2015) Biochar production technology. In: Lehmann J, Joseph S (eds) Biochar environmental management. Routledge, New York, pp 63–87

Brown R, Campo BD, Boateng AA, Garcia-Perez M, Masek O (2015) Fundamentals of biochar production. In: Lehmann J, Joseph S (eds) Biochar environmental management. Routledge, New York, pp 39–61

Bruun EW, Ambus P, Egsgaard H, Hauggaard-Nielsen H (2012) Effects of slow and fast pyrolysis biochar on soil C and N turnover dynamics. Soil Biol Biochem 46:73–79

Butnan S, Deenik JL, Toomsan B, Antal MJ, Vityakon P (2015) Biochar characteristics and application rates affecting corn growth and properties of soils contrasting in texture and mineralogy. Geoderma 237–238:105–116

Buyer JS, Sasser M (2012) High throughput phospholipid fatty acid analysis of soils. Appl Soil Ecol 61:127–130

Cantrell KB, Hunt PG, Uchimiya M, Novak JM, Ro KS (2012) Impact of pyrolysis temperature and manure source on physicochemical characteristics of biochar. Bioresour Technol 107:417–428

Cao X, Harris W (2010) Properties of dairy-manure derived biochar pertinent to its potential use in remediation. Bioresour Technol 101:5222–5228

Case S, McNamara NP, Reay DS, Whitaker J (2012) The effect of biochar addition on $N_2O$ and $CO_2$ emissions from a sandy loam soil. The role of soil aeration. Soil Biol Biochem 51:125–134

Cely P, Tarquis AM, Paz-Ferreiro J, Mendez A, Gasco G (2014) Factors driving the carbon mineralization priming effect in a sandy loam soil amended with different types of biochar. Solid Earth 5:585–594

Chen JH, Liu XY, Zheng JF, Zhang XH, Zhang B, Lu HF, Chi ZZ, Pan GX, Li LQ, Zheng JF, Zhang XH, Wang JF, Yu XY (2013) Biochar soil amendment increased bacterial but decreased fungal gene abundance with shifts in community structure in a slightly acid rice paddy from Southwest China. Applied Soil Ecol 71:33–44

Chen W, Meng J, Han X, Lan Y, Zhang W (2019) Past, present, and future of biochar. Biochar 1:75–87

Chia CH, Downie A, Munroe P (2015) Characteristics of biochar: physical and structural properties. In: Lehmann J, Joseph S (eds) Biochar environmental management. Routledge, New York, pp 89–109

Clough TJ, Condron LM (2010) Biochar and the nitrogen cycle: introduction. J Environ Qual 39:1218–1223

Cong RF, Abbruzzini TF, de Andrade CA, Milori D, Cerri A (2017) Effect of pyrolysis temperature and feedstock type on agriculture properties and stability of biochars. Agric Sci 8:914–933

Cornell-CASH (2016) Comprehensive assessment of soil health. Cornell University, Ithaca, NY, p 134. http://soilhealth.cals.cornell.edu

Cox J (2019) Gardening with biochar. Storey, North Adams, MA, p 129

Culman SW, Snap SS, Freeman MA, Schipanski ME, Beniston J, Lal R, Drinkwater LE, Franzluebbers AJ, Clover JD, Grandy AS, Lee J, Six J, Maul JE, Mirsky SR, Spargo JT, Wander MM (2012) Permanganate oxidizable carbon reflects a processed soil fraction that is sensitive to management. Soil Sci Soc Am J 76:494–504

Dane JH, Hopmans JW (2002) Water retention and storage. In: Dane JH, Topp GC (eds), Methods of soil analysis, Part 4: Physical methods. Soil Science Society of America Book Series, Vol. 5. Madison WI pp 675–720

Deenik JL, McClellan M, Uehara G, Antal MJ, Campbell S (2010) Charcoal volatile matter content influences plant growth and soil nitrogen transformations. Soil Sci Soc Am J 74:1259–1270

Demisie W, Lu Z, Zhang M (2014) Effect of biochar on carbon fractions and enzyme activity of red soil. Catena 121:214–221

Deng S, Popova I (2011) Carbohydrate hydrolases. In: Dick RP (ed) Methods of soil enzymology. Soil Science Society of America, Madison, WI, pp 185–209

Dick R (2018) Soil health. CSA News 63:12–17. https://doi.org/10.2134/csa2018.63.1114

Doran JW, Parkin TB (1994) Defining and assessing soil quality. In: Doran JW et al. (eds), Defining soil quality for a sustainable environment. Soil Science Society of America Special publ No 35. Soil Science Society of America and American Society of Agronomy, Madison, WI. pp 3–21

Duong VT, Khanh NM, Nguyen NTH, Phi NN, Duc NT, Duong Hoa XO (2017) Impact of biochar on the water holding capacity and moisture of basalt and grey soil. J Sci Ho Chi Minh Open Univ 7:36–43

Ezawa T, Yamamoto K, Yoshida S (2002) Enhancement of the effectiveness of indigenous arbuscular mycorrhizal fungi by inorganic soil amendments. Soil Sci Plant Nutr 48:897–910

FAO (1987) Simple technologies for charcoal making, second printing. FAO forestry paper 41. FAO, United Nations, Rome Italy
Fidel RB, Laird DA, Thompson ML, Lawrinenko M (2017) Characterization and quantification of biochar alkalinity. Chemosphere 167:367–373
Fisher B, Manzoni S, Morillas L, Garcia M, Johnson M, Lyon S (2019) Improving agricultural water use efficiency with biochar- a synthesis of biochar effects on water storage and fluxes across scales. Sci Total Environ 657:853–862
Fuertes AB, Camps-Arbestain M, Sevilla M, Macia-Agullo JA, Fiol S, Lopez R, Smernik RJ, Aitkenhead WP, Arce F, Marcias F (2010) Chemical and structural properties of carbonaceous products obtained by pyrolysis and hydrothermal carbonization of corn stover. Soil Res 48:618–626
Gasco G, Paz-Ferreiro J, Cely P, Plaza C, Mendez A (2016) Influence of pig maure and its biochar on soil $CO_2$ emissions and soil enzymes. Ecol Eng 95:19–24
Gomez JD, Denef K, Stewart CE, Zheng J, Cotrufo MF (2014) Biochar addition rate influences soil microbial abundance and activity in temperate soils. Europ J Soil Sci 65:28–39
Graber ER, Frenke O, Jaiswal AK (2014) How may biochar influence severity of diseases caused soilborne pathogens? Carbon Manag 5:169–183
Grossman RB, Reinsch TG (2002) Bulk density and linear extensibility. In: Dane JH, Topp GC (eds), Soil Sci Soc Am Book Series 5, Madison WI pp 201–228
Gul S, Whalen JK (2016) Biochemical cycling of nitrogen and phosphorus in biochar-amended soils. Soil Biol Biochem 103:1–15
Gul S, Whalen JK, Thomas BW, Sachdeva V (2015) Physico-chemical properties and microbial responses in biochar-amendmed soils: mechansisms and future directions. Agric Ecosyst Environ 206:46–59
Haney RL, Haney EB, Smith DR, Harmel RD, White MJ (2018) The soil health tool – theory and initial broad-scale application. Appl Soil Ecol 125:162–168
Harder B (2006) Smoldered-earth policy created by ancient Amazonia natives, fertile, dark soils retain abundant carbon. Sci News 169:133
Hood-Nowotny R, Watzinger A, Wawra A, Soja G (2018) The impact of biochar incorporation on inorganic nitrogen fertilizer plant uptake; an opportunity for carbon sequestration in temperate agriculture. Geosci 8:420
Huang H, Wang YX, Tang JC, Zhu WY (2014) Properties of maize stalk biochar produced under different pyrolysis temperatures and its sorption capacity to naphthalene. Environ Sci 35:1884–1890
Hue NV, Uchida R, Ho MC (2000) Sampling and analysis of soils and plant tissues. In: Silva JA, Uchida RS (eds) Plant nutrient management in Hawaii's soils. College of Tropical Agriculture & Human Resources, University of Hawaii, Honolulu, HI, pp 23–26
Hurisso TT, Moebius-Clune DJ, Culman SW, Moebius-Clune BN, Thies JE, van Es HM (2018) Soil protein as a rapid soil health indicator of potentially available organic nitrogen. Agric Envrion Lett 3:180006
IBI (2012) Standardized product definition and product testing guidelines for biochar that is used in soil. International Biochar Initiative, Washington, DC
Ippolito JA, Spokas KA, Novak JM, Lentz RD, Cantrell KB (2015) Biochar elemental composition and factors influencing nutrient retention. In: Lehmann J, Joseph S (eds) Biochar environmental management. Routledge, New York, pp 130–163
Jaafar NM, Clode PL, Abbott LK (2015) Soil microbial responses to biochars varying in particle size, surface and pore properties. Pedosphere 25:770–780
Jeffery S, Verheijen FGA, Van Dervelde M, Bastos AC (2011) A quantitative review of the effects of biochar application to soils on crop productivity using meta-analysis. Agric Ecosyst Environ 144:175–187
Kim K, Kim J, Chao T, Choi JW (2012) Influence of pyrolysis temperature on physicochemical properties of biochar obtained from the fast pyrolysis of pitch pine (*Pinus rigida*). Bioresour Technol 118:158–162

Kleber M, Hockaday W, Nico PS (2015) Characteristics of biochar: macro-molecular properties. In: Lehmann J, Joseph S (eds) Biochar environmental management. Routledge, New York, pp 111–137

Laird DA, Fleming P, Davis DD, Horton R, Wang B, Karlen DL (2010) Impact of biochar amendments on the quality of a typical Midwestern agricultural soil. Geoderma 158:443–449

Laird DA, Novak JM (2017) Biochar and soil quality. In: Lal R (ed) Encyclopedia of soil science, 3rd edn. CRC Press, Boca Raton, FL, pp 189–192

Lammirato C, Miltner A, Kaestner M (2011) Effects of wood char and activated carbon on the hydrolysis of cellobiose by β-glucosidase from *Aspergillus niger*. Soil Biol Biochem 43:1936–1942

Lehmann J, Rondon M (2006) Bio-char soil management on highly weathered soils in the humid tropics. In: Uphoff N et al (eds) Biological approaches to sustainable soil systems. CRC Press, Boca Raton, FL, pp 517–530

Li H, Dong X, da Silva EB, de Oliveira LM, Chen Y, Ma LQ (2017) Mechanisms of metal sorption by biochars: biochar characteristics and modifications. Chemosphere 178:466–478

Li JM, Cao LR, Yuan Y, Wang RP, Wen YZ, Man JY (2018) Comparative study of microsystin-LR sorption onto biochars producewd from various plant- and animal-wastes at different pyrolysis temperatures: influencing mechanisms of biochar properties. Bioresour Technol 247:794–803

Liu P, Ptacek CJ, Blowes DW (2019) Release of nutrients and trace elements from wood-, agricultural residue- and manure-based biochars. Int J Environ Res 13:747–758. https://doi.org/10.1007/s41742-019-00209-5

Liu Z, Chen X, Jing Y, Li Q, Zhang J, Huang Q (2014) Effects of biochar amendment on rapeseed and sweet potato yields and water stable aggregate in upland red soil. Catena 123:45–51

Lorenz K, Lal R (2018) Biochar. In: Lorenz K, Lal R (eds) Carbon sequestration in agricultural ecosystems. Springer, Cham Switzerland, pp 301–350. https://doi.org/10.1007/978-3-319-92318-5_8

Luo Y, Dungait JA, Zhao X, Brookes PC, Durenkamp M, Li G, Lin Q (2018) Pyrolysis temperature during biochar production alters its subsequent utilization by microorganisms in an acid arable soil. Land Degrad Dev 29:2183–2188

Maestrini B, Hermann AM, Nannipieri P, Schmidt M, Abiven S (2014) Ryegrass-derived pyrogenic organic matter changes organic carbon and nitrogen mineralization in a temperate forest soil. Soil Biol Biochem 69:291–301

Marris E (2006) Black is the new green. Nature 442:624–626

Martinsen V, Mulder J, Shitumbanuma V, Sparrevik M, Borresen T (2014) Farmer-led maize biochar trials: effect on crop yield and soil nutrients under conservation farming. J Plant Nutr Soil Sci 177:681–695

Masiella CA, Dungan B, Brewer CE, Spokas KA, Novak JM, Liu Z, Sorenti G (2015) Biochar effects on soil hydrology. In: Lehmann J, Joseph S (eds) Biochar environmental management. Routledge, New York, pp 543–562

Meng J, He T, Sanganyado E, Lan Y, Zhang W, Han X, Chen W (2019) Development of the straw biochar returning concept in China. Biochar 1:139–149. https://doi.org/10.1007/s42773-019-00019-0

Mukherjee A, Zimmerman AR (2013) Organic carbon and nutrient release from a range of laboratory-produced biochars and biochar-soil mixtures. Geoderma 193–194:122–130

Mukherjee A, Lal R (2017) Biochar and soil characteristics. In: Lal R (ed) Encyclopedia of soil science, 3rd edn. CRC Press, Boca Raton, FL, pp 184–188

Mukherjee A, Zimmerman AR, Hamdan H, Cooper WT (2014) Physicochemical changes in pyrogenic organic matter (biochar) after 15 months of field aging. Solid Earth 5:693–704

O'Toole A, Rasse D (2017) Biochar: soil carbon and fertility. In: Lal R (ed) Encyclopedia of soil science, 3rd edn. CRC Press, Boca Raton, FL, pp 191–197

Oaks R (2018) Making charcoal and biochar. The Crowood Press, Wiltshire

Obia A, Mulder J, Martinsen V, Cornelissen G, Borresen T (2016) In situ effects of biochar on aggregation, water retention and porosity in light-textured tropical soils. Soil Tillage Res 155:35–44
Ogawa M, Okimori Y (2010) Pioneering works in biochar research, Japan. Austr J Soil Res 48:489–500
Palansooriya KN, Wong JTF, Hashimoto Y, Huang L, Rinklebe J, Chang SX, Bolan N, Wanh H, Ok YS (2019) Response of microbial communities to biochar-amended soils: a critical review. Biochar 1:3–22. https://doi.org/10.1007/s42773-019-00009-2
Prendergast-Miller MT, Duvall M, Shohi SP (2011) Localisation of nitrate in the rhizosphere of biochar-amended soils. Soil Biol Biochem 43:2243–2246
Prommer J, Wanek W, Hofhansl F, Trojan D, Offre P, Unch T, Schleper C, Sassmann S, Kitzler B, Soja G, Hood-Nowotny RC (2014) Biochar decelerates soil organic nitrogen cycling but stimulates soil nitrification in a temperate arable field trial. PLoS One 9:1–16
Quilliam RS, Glanville HC, Wade SC, Jones DL (2013) Life in the 'charosphere'- does biochar in agricultural soil provide a significant habitat for microorganisms? Soil Biol Biochem 65:287–293
Ronsse F, van Hecke S, Dickinson D, Prins W (2013) Production and characterization of slow pyrolysis biochar: influence of feedstock type and pyrolysis conditions. GCB Bioenergy 5:104–115
Saleh ME, Mahmoud AH, Rashad M (2012) Peanut biochar as a stable adsorbent for removing $NH_4$-N from wastewater: a preliminary study. Adv Envron Biol 6:2170–2176
Silber A, Levkovitch I, Graber ER (2010) pH-dependent mineral release and surface properties of corn straw biochar: agronomic implications. Environ Sci Techn 44:9318–9323
Sombroek W, Rodriques T, Cravo M, Jarbas TC, Woods W, Glaser B (2002) Terra Preta and Terra Mulata: pre-Columbian Amazon kitchen middens and agricultural fields, their sustainability and their replication. In: Proceedings 17th World Congr Soil Sci. Thailand paper no. 1935
Spokas K, Novak J (2015) Biochar: the field experience. In: Goreau TJ et al (eds) Geotherapy. CRC Press, Boca Raton, FL, pp 235–248
Steiner C, Teixeira WG, Lehmann J, Nehls T, de Macedo JLV, Blum WEH, Zech W (2007) Long term effects of manure, charcoal and mineral fertilization on crop production and fertility on a highly weathered central Amazonian upland soil. Plant Soil 291:275–290
Streit WR, Daniel R (2017) Metagenomics: methods and protocols. Humana Press, New York, p 302
Thies JE, Rillig MC, Graber ER (2015) Biochar effects on the abundance, activity and diversity of the soil biota. In: Lehmann J, Joseph S (eds) Biochar environmental management. Routledge, New York, pp 327–389
USDA-NRCS (2017) Soil health. https://www.nrcs.usda.gov/wps/portal/nrcs/main/soils/health/. Access 5 August 2019
Verheijen F, Zhuravel A, Silva FC, Amaro A, Ben-Hur A, Keizer JJ (2019) The influence of biochar particle size and concentration on bulk density and maximum water holding capacity of sandy vs sandy loam soil in a column experiment. Geoderma 347:194–202
Wang D, Fonte SJ, Parikh SJ, Six J, Scow KM (2017) Biochar additions can enhance soil structure and the physical stabilization of C in aggregates. Geoderma 303:110–117
Wang Z, Guo H, Shen F, Yang G, Zhang Y, Zeng Y, Wang L, Xiao H, Deng S (2015) Biochar produced from oak sawdust by Lanthanum (La)-involved pyrolysis for adsorption of ammonium ($NH_4^+$), nitrate ($NO_3^-$), and phosphate ($PO_4^{3-}$). Chemosphere 119:646–653
Weil RR, Islam KR, Stein MA, Gruver JB, Samson-Liebig SE (2003) Estimating carbon for soil quality assessment: a simplified method for laboratory and field use. Am J Alt Agric 18:3–17
Whitman T, Singh BP, Zimmerman AR (2015) Priming effects in biochar-amended soils: implications of biochar-soil organic matter interactions for carbon storage. In: Lehmann J, Joseph S (eds) Biochar environmental management. Routledge, New York, pp 455–487
Wong JTF, Chen X, Deng W, Chai Y, Ng CWW (2019) Effects of biochar on bacterial communities in a newly established landfill over topsoil. J Environ Manag 236:667–673

Wong JTF, Chen Z, Chen X, Ng CWW (2017) Soil-water retention behavior of compacted biochar-amended clay: a novel landfill final cover material. J Soils Sediments 17:590–598

Wu P, At-Ul-Karim ST, Singh BP, Wang H, Wu T, Liu C, Fang G, Zhou D, Wang Y, Chen W (2019) A scientometric review of biochar research in the past 20 years (1998–2018). Biochar 1:23–43

Xu RK, Zhao AZ, Yuan JH, Jiang Y (2012) pH buffering capacity of acid soils from tropical and subtropical regions of China as influenced by incorporation of crop straw biochars. J Soils Sediments 12:494–502

Yuan JH, Xu RK, Zhang H (2011) The forms of alkalis in the biochar produced from crop residues of different temperatures. Bioresour Technol 102:3488–3497

Zhao Y, Feng D, Zhang Z, Sun S, Zhang Y (2017) Physical-chemical properties of sawdust biochar. In: Simmons CA (ed) Biochar. Nova Science, New York, pp 1–45

# Chapter 3
# Plant Growth-Promoting Rhizobacteria: A Booster for Ameliorating Soil Health and Agriculture Production

**Pratibha Rawat, Deepti Shankhdhar, and S. C. Shankhdhar**

**Abstract** Soil is a powerful nonrenewable asset that embraces life on earth by furnishing nutrients to plant. Degradation of soil health due to indiscriminate use of chemical fertilizers and industrialization has become predominant environmental concern with high preeminence. In view of the present scenario, soil microbes are the most important candidates for improving soil fertility and health. The plant growth-promoting microbes are used for enhancing soil fertility under stressed and normal environment. Soil holds variety of microbial species such as fungi, bacteria, mosses and liverwort. The prevalence of microbes is an indicator of soil biological activities and regulates physical and chemical properties of soil. It enhances soil health and crop productivity by diverse mechanisms like biofortification of nutrients, bioremediation of soil, regulation of nutrient cycling, antibiosis, rhizosphere competence, secretion of enzymes, stimulation of systemic resistance in host plant, and production of metabolites, volatile compounds and antifungal toxins against pathogens. Interaction of plant and microorganisms results in plant growth promotion and disease control under fluctuating environment and enables sustainable agriculture without compromising ecosystem balance. Thus, the inclusive use of plant growth-promoting rhizobacteria promotes soil fertility that encourages sustainable agriculture production under extreme condition.

**Keywords** Soil health · Rhizosphere · Plant growth-promoting rhizobacteria · Biofertilizers

P. Rawat · D. Shankhdhar · S. C. Shankhdhar (✉)
Department of Plant Physiology, College of Basic Sciences and Humanities, G. B. Pant University of Agriculture and Technology, Pantnagar, Uttarakhand, India



B. Giri, A. Varma (eds.), *Soil Health*, Soil Biology 59,
https://doi.org/10.1007/978-3-030-44364-1_3

## 3.1 Introduction

Global technological advancement in both agricultural and industrial sector along with an outburst of global population has disturbed the demand and food supply ratio drastically in recent years. High input cost and low production were the major challenges of conventional agriculture system. To accomplish the production demand, contemporary farming practice rely on the use of chemical fertilizers that escalate the yield but beyond a threshold level only and also decline soil fertility and degrade the environment and consumer health (Singh 2015). Nevertheless, application of inorganic chemicals in cultivable land causes heavy metal accumulation in soil degrading its quality (Azzi et al. 2017). Global climate change is another factor that challenges agriculture productivity. Therefore, for an eco-friendly and sustainable agricultural growth an efficacious alternative is required. In this regard biofertilizers can be a potential cost-effective substitute for producing safe and quality food.

Biofertilizers are the beneficial microbial inoculants that reside in the rhizospheric region in the soil and promote the growth and development of plants by improving nutrient accessibility from soil to plants. These plant growth-promoting rhizobacteria (PGPR) produce growth hormones, siderophores, exopolysaccharides, enzymes like 1-aminocyclopropane-1-carboxylic acid (ACC) deaminase, antioxidants and volatile compounds that provide environmental stress resistance mechanisms in plants. They also act as antagonists of plant pathogens by producing antifungal and antibacterial compounds in soil (Khadeejath Rajeela et al. 2018). Such traits of eco-friendly microbes are suitable for sustainable agricultural practices as chemical-based farming system results in eutrophication, carbon footprint, and acidification of soil. Moreover, biotic and abiotic factors in the environment also contribute to decline in crop productivity. Therefore, for accomplishing the inevitable requisite for a reliable agriculture system that utilizes organic resources for improving plant health plant beneficial microbes are the best remedy for farmers and soil health.

## 3.2 Rhizosphere: An Ecological Niche for Soil Microbes

The region of soil near the root proximity is termed as rhizosphere. It represents the most biologically active region of the soil where microbiome and plants interact. A variety of compounds are secreted by roots of plants that give signals and allure the soil inhabiting diverse microbial communities. Root exudates are organic compounds excreted by plant roots and are rich in carbon sources that facilitate microbial growth. These secretions change the physiochemical attributes of soil like pH, nutrient content, energy resources that harmonize the soil microbial dynamics (Dakora and Phillips 2002). Environmental factors like light, soil temperature, oxygen availability, pH, nutrient acquisition, and inherent microbial community

**Table 3.1** Components of root exudates

| Constituents of root exudates | Types | References |
|---|---|---|
| Organic acids | Citric acid, oxalic acid, malic acid, succinic acid, acetic acid | Wang et al. (2019) |
| Amino acids | Histidine, serine, lysine, arginine, alanine, aspartic acid, glutamic acid, threonine, proline, tyrosine, methionine, tryptophan, isoleucine, phenylalanine | Madhukar et al. (2018) |
| Enzymes | Protease, invertase, acid phosphatase, alkaline phosphatase, urease | Wang et al. (2019) |
| Sugars | Glucose, rhamnose, mannose, arabinose, fructose, galactose, ribose, xylose, maltose, raffinose | Madhukar et al. (2018) |
| Vitamins | Thiamine, riboflavin, niacin, pantothenate, biotin | Dakora and Phillips (2002) |
| Inorganic ions | $H^+$, $OH^-$, $HCO_3^-$ | Dakora and Phillips (2002) |
| Gaseous molecules | $CO_2$, $H_2$ | Dakora and Phillips (2002) |

regulate the formulations of root secretions compared to plant species (Singh et al. 2004). Dominant molecules of root exudates comprise amino acids, organic acids, vitamins, sugars, and nucleobases as depicted in Table 3.1 (Keiluweit et al. 2015). In addition to flourishing the microbial dynamics in the rhizosphere, root exudates assist in symbiotic plant–microbe interaction and also impede the development of competing plant. For instance, root exudates of *Tagetespatula* L. containing thiophenes restricts the growth of competing *Abutilon theophrasti* L. plant surrounding the *Tagetes* plant (Weidenhamer et al. 2019).

### 3.2.1 Soil Health

Soil is a hotspot of microbial biodiversity and is the ultimate source of nutrients and nourishment for plants. Soil health in broad term refers to potential of soil to sustain all life forms and nurture their growth and development in the ecosystem (Doran and Safley 1997). Recycling and storage of nutrients, maintenance of water and soil quality, repression of pathogen, decomposition of soil organic matter, and detoxification of toxic chemicals are some attributes of healthy soil (Dubey et al. 2016) whereas degeneration of water and soil quality is a trait of poor soil health (Takoutsing et al. 2016). Soil microbes are the major contributors of rich soil health. The predominant soil microflora comprises bacteria, fungi, and actinomycetes. Out of these rhizobacteria forms symbiotic relations with plants and improve nutrient accessibility, water uptake, biogeochemical cycles in the ecosystem and provide resistance to biotic and abiotic stress. Soil aeration, moisture content, and porosity are also influenced by soil microbiota.

Climate fluctuations and anthropogenic activities have negative impact on diverse microbiome of soil. Shift in soil microbial dynamics has been observed owing to use of chemicals, monotype cultivation that turn down the soil enzymatic activity like phosphatases, proteases, dehydrogenases, and β-glucosidases that balances soil biological activities (Lazcano et al. 2013). It was evident from the research that application of chlorantraniliprole, an insecticide against leaf folder and yellow stem borer in paddy field, reduced the dehydrogenase and phosphatase activity in rhizospheric soil compared to control soil and also had adverse effect on soil microflora of treated soil indicating that soil deterioration via chemicals is a consequential warning for limited and vital soil asset which is the baseline of crop productivity (Sahu et al. 2019).

### *3.2.2 Plant Growth-Promoting Rhizobacteria: A Savior for Soil Health*

A healthy soil furnishes a diversity of ecosystem, sustains agricultural productivity and is a native hub of microorganisms. Majority of soil microbiome comprises plant growth-promoting rhizobacteria that establish near the root region of soil and improve crop yield by upgrading soil health. Rhizobacteria establish both associative and symbiotic relationship with host plant and can also survive as free-living organism. Genus *Bacillus*, *Pseudomonas*, *Arthrobacter*, *Azospirillum*, *Burkholderia*, *Azotobacter*, *Klebsiella*, *Alcaligenes*, and *Serratia* belong to PGPR category (Saharan and Nehra 2011). PGPR conserves soil fertility by performing various tasks such as solubilizing nutrients from their source in soil like potassium, phosphorous, iron and zinc solubilizing bacteria that accelerate release of nutrient into the soil when in direct proximity with the nutrient source either by production of enzymes, organic acids, siderophores or by upgrading the root morphology like root length and root hair growth (Saha et al. 2016; Hodge 2017). Symbiotic nitrogen fixation by PGPR like *Rhizobia*, *Bradyrhizobia* in leguminous plants is an important function of PGPR. Toxic pesticides in soil can be degraded by microorganisms by their enzymatic activity through a process called bioremediation. Destruction of xenobiotic compounds in soil is also achieved by such growth-promoting rhizobacteria (Kuppusamy et al. 2016). Such microbes are also used in water treatment plants for waste water treatment. These can also be used as biocontrol agents for protecting soil and plant from pathogen attack. Fungal and bacterial by-products hold soil particles into macroaggregates that upgrade soil structure and stability. It was also studied that arbuscular mycorrhizal hyphae enhanced macroaggregate formation, decelerated their breakdown, and improved turnover of microaggregate formation (Morris et al. 2019).

### 3.2.3 Plant Growth-Promoting Rhizobacteria: A Probiotic for Augmenting Agricultural Productivity

Microorganisms adopt different ways to stimulate plant growth and development. Direct and indirect mechanisms are two predominant modes for improving plant health which include nutrient acquisition, phytohormones secretion, biocontrol activity, and bioremediation as illustrated in Fig. 3.1. PGPR activity relies on host plant species, growth stage of host, defense modes of host plant, soil constituents, endogenous soil microbial community, and environmental factors (Prasad et al. 2019). Table 3.2 depicts diverse PGPR and their growth-promoting attributes in host plants.

## 3.3 Direct Modes of Plant Growth Promotion by PGPR

### 3.3.1 Biological Nitrogen Fixation

Nitrogen is an indispensable and limiting nutrient for productivity of crops and is the constituent of amino acids, lipids, and nucleoproteins (Marschner 1995). This element is not directly accessible to plants and is present in gaseous form (i.e., accounts for 78% of atmosphere) and is available to plants through the process of

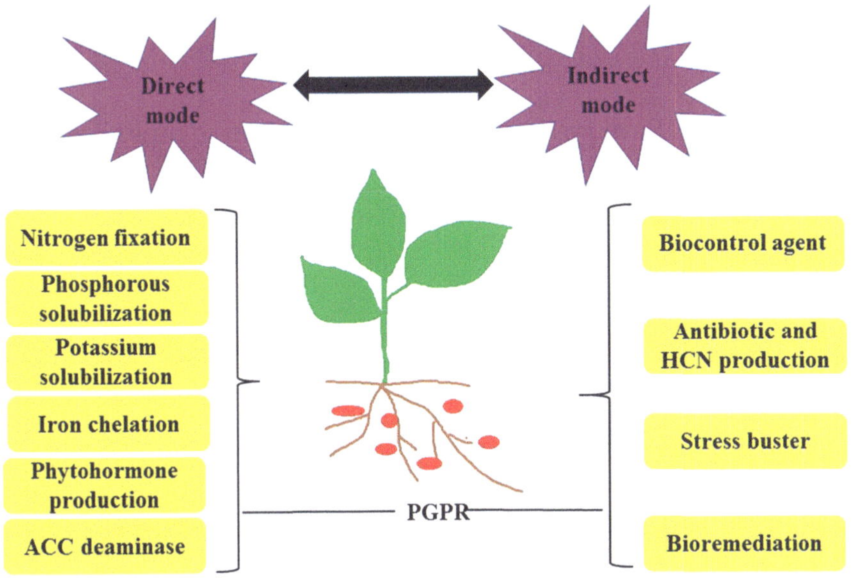


**Fig. 3.1** Diverse mechanism of growth enhancement in plants by plant growth-promoting rhizobacteria

**Table 3.2** Plant growth-promoting traits of diverse PGPR

| Plant growth-promoting rhizobacteria | Host plant | Growth regulation | References |
|---|---|---|---|
| *Pseuomona* sp. P34 | *Triticum aestivum* | Siderophore production, enhancement in root traits and dry matter accumulation of plant | Liu et al. (2019) |
| *Bradyrhizobium diazoefficiens* | *Glycine max* | Symbiotic nitrogen fixation | López et al. (2019) |
| *Bacillus megaterium* CS22 | *Brassica napus* | Phosphorous solubilization and biomass production | Zheng et al. (2019) |
| *Paenibacillus polymyxa* Sx3 | *Oryza sativa* L. | Biocontrol agent. Secondary metabolite fusaricidins and polymyxin P production. Suppression of bacterial leaf blight in host | Abdallah et al. (2019) |
| *Halomonas* | *Oryza sativa* L. | Bioremediation of arsenic under salt stress | Mukherjee et al. (2019) |
| *Bacillus mojavensis* JK07 and *Rhodopseudomonas palustris* | *Zea mays* | Potassium solubilizers, increment in yield and biomass under salinity | Feng et al. (2019) |
| *Bacillus* sp. CP h60 | *Cicer arietinum* | ACC deaminase, IAA, increase in number and dry weight of nodules, root and shoot biomass increment | Ditta et al. (2018) |
| *Bacillus mojavensis* | *Glycine max* | HCN, auxin, siderophore and antioxidant production, phosphate solubilization | Prajakta et al. (2019) |
| *Bacillus methylotrophicus* M4-96 | *Fragaria ananassa* | Auxin, gibberellin production, induce systemic resistance in host | Vicente-Hernández et al. (2019) |
| *Pseudomonas aeruginosa* MML2424 and *Bacillus amyloliquefaciens* MML2522 | *Curcuma longa* | Bacillomycin and fengycin antibiotic production, HCN, siderophores production | Chenniappan et al. (2019) |
| *Serratia nematodiphila* RJ10 | *Vigna mungo, Pisum sativum* | ACC deaminase, siderophores, IAA production, increase in biomass of crop under drought stress condition | Saikia et al. (2018) |

biological nitrogen fixation by nitrogenase-containing microorganisms (Shridhar 2012). Nitrogen fixing organisms are categorized as: (a) symbiotic nitrogen fixers that establish mutualistic association with host plant. Bacteria belonging to genus *Bradyrhizobium*, *Mesorhizobium*, *Rhizobium*, and *Sinorhizobium* associate with leguminous plants. (b) Nonsymbiotic nitrogen fixers which may be free living, endophytes, and associative. Bacteria belonging to genera *Azotobacter*, *Azospirillum*, *Diazotrophicus*, *Azoarcus*, cyanobacteria (*Anabaena*, *Nostoc*), *Gluconacetobacter*, *Acetobacter*, *Burkholderia*, *Enterobacter*, and *Pseudomonas* that encourage the growth of nonleguminous plants (Bhattacharyya and Jha 2012).

Substituting chemicals by nitrogen fixing biofertilizers in fields is a judicious and sustainable practice for eco-friendly agricultural system. Highest input of nitrogen in agricultural land is furnished by nitrogen fixing microbes globally. It was estimated that worldwide up to 30 × $10^9$ Kg N is fixed by *Rhizobia* every year in 250 million ha of land (Kinzig and Socolow 1994). An increment of about 44–72% was discerned in soybean yield through biological nitrogen fixation (Ciampitti and Salvagiotti 2018). In chickpea inoculated with consortia of *Azotobacter* and *Rhizobium* under salt stress improved yield up to 23.9% and 27.9%, respectively. Also, nitrogen, potassium, and phosphorous amount was greater in roots and shoots inoculated with consortia compared to control and plant sodium content was minimal in chickpea co-inoculated with strains alleviating stress impact in host plant (Abdiev et al. 2019).

### 3.3.2 Phosphate Solubilization

Phosphorous is the second most inevitable macronutrient after nitrogen. Although phosphorous pool both organic and inorganic is copious in soil, i.e., 400–1200 mg $kg^{-1}$ soil but the plant available form $\left(HPO_4^{-2}, H_2PO_4^-\right)$ is very scant, i.e., only 0.1% of total phosphorous due to its complexation with cations in soil (Zou et al. 1992). Phosphorous is utmost for metabolic processes like respiration, photosynthesis, energy transfer, signal transduction, and biosynthesis of macromolecules (Anand et al. 2016). To vanquish the inadequacy of phosphorous chemical fertilizers are not reliable as they are expensive and toxic for soil. Phosphate solubilizing bacteria (PSB) are potent substitute of chemicals that solubilize phosphorous in an eco-friendly manner. Figure 3.2 demonstrates various mode of action of PSB for dissolution of insoluble phosphate. PSB secretes organic acids like gluconic acid, carboxylic acid for dissemination of inorganic phosphates like Fe-P, Ca-P by lowering the pH of rhizosphere and chelation of cations bound to phosphate (Kishore et al. 2015).

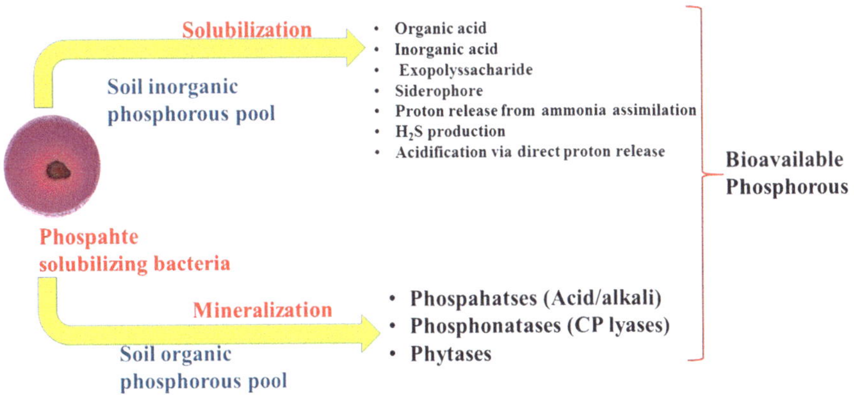


**Fig. 3.2** Modes of action of phosphate solubilizing bacteria for phosphorous dissolution in soil

**Table 3.3** Impact of phosphate solubilizing bacteria on phosphorous uptake and yield in different crops

| Phosphate solubilizing microorgansims | Increase in phosphorous uptake in host plant compared to uninoculated plant (%) | Host plant | Yield increment (%) | References |
|---|---|---|---|---|
| *Bacillus megatherium* var. *phosphaticum* | 7.81 | *Vigna radiata* L. | 21.10 | Kumar and Yadav (2018) |
| *Gluconacetobacter* sp. and *Burkholderia* sp. | 435.71 | *Oryza sativa* L. | 14.75 | Stephen et al. (2015) |
| *Pseudomonas putida* | 30.00 | *Cicer arietinum* L. | 14.00 | Israr et al. (2016) |
| *Rhizobium, Pseudomonas* and *Bacills* | 103.00 | *Pisum sativum* L. | 32.00 | Rani et al. (2016) |
| *Burkholderai, Baciilus, Pseudomonas, Flavobacterium* | 185.00 | *Zea mays* L. | 33.00 | Iqbal Hussain et al. (2013) |
| *Rhizobium meliloti* | 37.50 | Gossypium | 77.00 | Egamberdiyeva et al. (2004) |

Inorganic acid production, siderophores and exopolysaccharides secretions are indirect mechanisms of organic phosphate dissolution. PSBs also secrete enzymes like phosphatases, phytases and phosphonatases for dissolution of organic phosphates (Martínez et al. 2015). Most potent phosphate solubilizing bacteria belong to genera *Bacillus*, *Pseudomonas*, *Burkholderia*, *Arthrobacter*, *Enterobacter*, and *Rhodobacter* (Biswas et al. 2018; Liu et al. 2019; Teng et al. 2019). Table 3.3 represents increment in phosphorous uptake and yield in crops with different phosphate solubilizers. It was evident from the findings that co-inoculation of phosphate solubilizing bacteria *Bacillusaryabhattai* S10 and *Bacillus subtilis* ZM63 improved phosphorous content upto 90% in mungbean and 58% in maize and increased grain number in pods of munbean by 29% compared to control (Ahmad et al. 2019). Likewise, *Pseudomonas species* P34 in wheat enhanced root traits and dry matter accumulation and phosphorous accumulation in inoculated plants was found to be 29.4% higher in contrast to control (Liu et al. 2019). These findings explicit that PSBs are sustainable alternatives to accomplish the agricultural phosphorous demand.

### 3.3.3 Potassium Solubilization

Potassium is the salient macronutrient for plants after nitrogen and phosphorous and is a prerequisite for photosynthesis, protein synthesis and modification, early growth, regulation of water use efficiency and longevity. Total soil potassium content surpasses 20,000 ppm of which only 1–2% is accessible for plants (Sharma et al. 2016). Potassium inadequacy is one of the dominant constrains for agricultural productivity, and as a consequence there is reduction in yield and seed production (Nath et al. 2017). Therefore, it is obligatory to find a remedy for maintaining potassium status in soil and plants. Potential of PGPR to solubilize potassium by secretion of organic acids and acid hydrolysis of potassium from its source can be exploited to increase availability of potassium in soil (Sindhu et al. 2016; Bahadur et al. 2017). *Pseudomonas*, *Bacillus*, *Arthrobacter*, *Thiobacillus*, *Burkholderia*, *Azotobacter*, *Rhizobium*, and *Flavobacterium* species are some popular potassium solubilizers (Meena et al. 2018). Table 3.4 represents the variability in potassium uptake and yield in plants with diverse potassium soublizing biofertilizer. It was evident from the research that three potassium solubilizers isolated from paddy field, i.e., *Rahnellaaquatilis*, *Pantoeaagglomerans*, and *Pseudomonas orientalis* with a blend of half of recommended dose of potassium fertilizer (i.e., 44%) improved grain yield upto 20–52% in rice (Khanghahi et al. 2019). Hence, potassium solubilizers can be used as biofertilizers for enhancing sustainable crop productivity and limiting the use of agrochemicals.

**Table 3.4** Impact of potassium solubilizing bacteria on potassium uptake and yield in different plants

| Potassium solubilizing microorgansims | Increase in potassium uptake in host plant compared to uninoculated plant (%) | Host plant | Yield increment (%) | References |
|---|---|---|---|---|
| *Bacillus megaterium* and *Bacillus mucilaginous* | 40.2 | *Capsicum annuum* | 28.5 | Zhao et al. (2019) |
| *Bacillus mucilaginosus* | 59.6 | *Sorghum vulgare* Pers. | 36.6 | Basak and Biswas (2009) |
| *Arthrobacter* sp. | 24.40 | *Lolium multiflorum* | 27.4 | Xiao et al. (2017) |
| *Enterobacter cloacae Rhizo_33* | 85.00 | *Triticum aestivum* | 19.16 | Ghadam Khani et al. (2019) |
| *Frateuria aurantia* | 39.00 | *Nicotiana* | 16.00 | Subhashini (2015) |

### 3.3.4 Iron Chelation

Iron is the crucial element for all life forms and is involved in fundamental processes like respiration, nitrogen fixation, and photosynthesis. Iron is present in insoluble form in nature, i.e., $Fe^{3+}$ ferric ion which is not readily absorbed by living forms (Ammari and Mengel 2006). For availability of iron to plant, PGPR secrete high affinity low molecular weight iron chelating compounds under iron limiting conditions in the rhizosphere named as siderophores. These are ferric ion complexing agents that bind the $Fe^{3+}$ ion and translocate it inside the bacterial cell where the iron–siderophore complex gets reduced to soluble $Fe^{2+}$ form. Plants acquire iron through production of siderophores by themselves and also from the bacterial siderophores by ligand exchange reaction or direct uptake of iron–siderophore complex. Currently more than 500 siderophores are reported excreted by both plants and microorganisms (Hider and Kong 2010). Microbes utilize the concept of iron chelation for competition against other microbes for nutrition, habitat, and host plant. *Grimontella*, *Enterobacter*, and *Burkholderia* are prominent siderophore producers. Bacteria like *Pseudomonas putida* utilize siderophores produced by other microorganisms to improve iron uptake in their natural niche (Rathore 2014). It was also studied that *Bacillus megaterium*, *Bacillus subtilis*, *Rhizobium radiobacter*, and *Pantoeaallii* secrete siderophores in the range of 80–140 μmol $L^{-1}$ in alkaline condition that aided organisms to survive under stress environment (Ferreira et al. 2019). Table 3.5 presents the potent iron solubilizing bacteria and their impact on

**Table 3.5** Influence of iron and zinc solubilizers on micronutrient uptake and yield parameters

| Iron and zinc solubilizing microorgansims | Increment in zinc content in crop (%) | Increment in iron content in crop (%) | Host plant | Yield increment (%) | References |
|---|---|---|---|---|---|
| *Bacillus aryabhattai* (ZM31) and *Bacillus subtilis* (ZM63) | 25.9 (shoot) 68.0 (grains) | 48.11 (shoot) 78.0 (grains) | *Zea mays* | 41.23 | Mumtaz et al. (2018) |
| *Pseudomonas fluorescens* | 32.8 (shoot) 17.2 (grains) | 49.5 (shoot) 60.88 (grains) | *Oryza sativa* L. | 40.7 | Sharma et al. (2013, 2015) |
| *Enterobacter* B 17 | 21.5 (shoot) 11.21 (grains) | 60.3 (shoot) 52.3 (grains) | *Oryza sativa* L. | 23.3 | Sharma et al. (2013, 2015) |
| *Enterobacter* BN 30 | 32.1 (shoot) 10.5 (grains) | 16.6 (shoot) 32.8 | *Oryza sativa* L. | 30.4 | Sharma et al. (2013, 2015) |
| *Pseudomonas putida* | 10.84 (shoot) 18.08 (grains) | 48.9 (shoot) 53.2 (grains) | *Oryza sativa* L. | 25.0 | Sharma et al. (2013, 2015) |

yield and micronutrient uptake. Siderophores producing PGPR are blessings to conquer iron deficiency in plants. Nevertheless, more research on siderophores and their mode of action for iron uptake is required to exploit the potential of iron chelating microorganisms.

### *3.3.5 Zinc Solubilization*

One of the pivotal micronutrients for plant growth and development is zinc which is required in low concentration (i.e., 5–100 mg $kg^{-1}$ in tissues). Zinc is an important cofactor of enzymes such as carbonic anhydrase, RNA polymerase, alcohol dehydrogenase, and superoxide dismutase (Cakmak 2000). It is essential for maturity, vigor, and yield of plants. Zinc scarcity results in membrane disintegration, reduction in synthesis of chlorophyll, growth regulators, cytochromes, enzymes, nucleotides, and increased sensitivity to high temperature (Singh et al. 2005). Symptoms of zinc deficiency include chlorosis of leaves, stunted growth, and spikelet sterility. Soil environment affects zinc prevalence in crops. Zinc scarcity is also known as "hidden scarcity" as the yield and quality of the product is damaged without the appearance of symptoms (Alloway 2004). High pH, organic matter, high percentage of phosphorous and iron, magnesium-to-calcium ratio, and high bicarbonate content are factors that influence zinc content in soil (Wissuwa et al. 2006; Li et al. 2016). Zinc scarcity in crops is due to low zinc solubility instead of zinc availability in soil (Gontia-Mishra et al. 2016).

Application of inorganic zinc in soil partially fulfills demand of plants as 96–99% of zinc fertilizer is converted to insoluble form within the 7 days of application which is not assimilated by plants and they become deficient in zinc (Saravanan et al. 2004). In this regard, there is a need of an organic solution that improves the micronutrient deficiency in plants as well as humans. Zinc solubilizing PGPR are crucial for zinc availability and bio-fortification of zinc in grains as depicted in Table 3.5. *Pseudomonas*, *Bacillus*, *Burkholderia*, *Gluconacetobacter*, and *Serratia* are common zinc solubilizing PGPR. These microorganisms solubilize zinc through production of organic acids (gluconic acid, 5-keto-gluconic acid), exchange reactions, and chelation by siderophores. It was evident from recent study that application of zinc solubilizing bacterial consortia, i.e., *Pseudomonas striata*, *Bacillus polymyxa*, *Pseudomonas fluroscence*, *Bacillus megaterium*, *Gluconoacetobacter diazotrophicus*, and *Aspergillus awamori* with 100% recommended dose of zinc (20 kg $ha^{-1}$) improved agronomic efficiency and zinc use efficiency in groundnut (Raut et al. 2019).

### 3.3.6 Phytohormone Production

Plant growth regulators are organic compounds that are synthesized in specific parts of the plant and are translocated to various sites of plants or are active in the source tissue. These are the signalling molecules that regulate the overall growth and development of plants at very low concentration (less than 1 mM) (Damam et al. 2016). Phytohormones are produced by plants themself or exogenously produced by plant-associated PGPR. Indole acetic acid, gibberellins, and cytokinins are phytohormones secreted by microorganisms that improve plant growth. Indole acetic acid or auxin is involved in enlargement of cell, cell differentiation and division, apical dominance, flowering, root initiation, stimulation of seed germination, xylem development rate, mediation of tropic response (phototropic, geotropic), and formation of photosynthetic pigments in plants. Auxins from PGPR escalate root surface area and length to enable more nutrient accession for plants. It is crucial for phytostimulation and pathogenesis (Nath et al. 2017).

Auxin secretion by bacteria enables them to detoxify excess tryptophan in bacterial cell that are detrimental for them as tryptophan is the precursor for auxin production in bacteria and is a major amino acid in root exudates (Etesami et al. 2009). Gibberellin stimulates root and stem growth, triggers germination of seed, root elongation and expansion (Yaxley et al. 2001). Cytokinins are responsible for cell division, differentiation, and enlargement. It also regulates axillary bud growth, leaf senescence, and apical dominance. It also stimulates shoot growth (Riefler et al. 2006). Ethylene is a versatile plant hormone. It triggers fruit ripening, leaf abscission, inhibits root elongation, stimulates the production of other phytohormones and causes defoliation (Bhattacharyya and Jha 2012). *Bacillus*, *Azospirillum*, *Microbacterium*, *Pseudomonas*, *Rhizobium*, *Serratia*, and *Paenibacillus* are the dominant phytohormones producing bacteria (Holland 2019; Kashyap et al. 2019; Pal et al. 2019). It was discerned that *Pseudomonas* sp. AZ5 and *Bacillus* sp. AZ17 produced 17.8 $\mu g\ ml^{-1}$ and 19 $\mu g\ ml^{-1}$ indole acetic acid, respectively, and also enhanced grain yield by 17.47 and 17.34%, respectively, in chickpea in contrast to control. Increment in zinc content (26.12%) and phosphorous content (25.69%) was observed in chickpea plant inoculated with *Bacillus* sp. Moreover, nodule dry weight, nodule number, shoot biomass also improved in inoculated plants in comparison to uninoculated plants (Zaheer et al. 2019). *Bacillus tequilensis* is reported to produce auxin, abscisic acid, and gibberellin ($GA_1$, $GA_3$, $GA_5$, $GA_{18}$, and $GA_{19}$) and improve shoot biomass, maintain photosynthetic pigment and leaf ultrastructure under heat stress in soybean. It also reduces the level of stress abscisic acid and improves salicylic acid and jasmonic acid content in the rhizosphere (Kang et al. 2019). Such studies reveal that phytohormones producing PGPR upgrade overall growth attributes and crop productivity.

### *3.3.7 Aminocyclopropane-1-Carboxylate (ACC) Deaminase*

Some PGPR have the potential to produce a stress buster enzyme ACC deaminase. This enzyme improves overall growth of plants by reducing ethylene levels in stress environment. Stress hormone is a synonym for ethylene hormone. Although ethylene promotes plant growth at low concentration (~10 μg $L^{-1}$) in normal surrounding, under stress environment ethylene is produced in higher amounts, and beyond a threshold level (~25 μg $L^{-1}$) it retards overall development of plants (Gamalero and Glick 2015). ACC deaminase converts ethylene precursor ACC to α ketobutyrate and ammonia thereby lowering stress ethylene levels in plants and encouraging the survival and development of plants under stress condition (Olanrewaju et al. 2017) as explained in Fig. 3.3. The capability of *Enterobacter* sp. with ACC deaminase activity was investigated in rice under salinity conditions. It was evident that the strain induces increment in root shoot biomass in rice seedling under salinity. Antioxidants like catalase, superoxide dismutase, polyphenol oxidase also shoot up in the seedlings when inoculated with the strain. Moreover, ethylene production was lowered in the seedlings treated with Enterobacter and 150 mM salt concentration in contrast to control condition without stress and inoculum (Sarkar et al. 2018). Inoculation of *Bacillus amyloliquefaciens* with ACC deaminase activity (313.2 ± 34.3 μmol α-ketobutyrate $mg^{-1}$ protein $h^{-1}$) along with biochar (30 mg $ha^{-1}$) in wheat results in 77% increment in grain yield and 75% increment in straw yield under severe drought condition (Zafar-ul-Hye et al. 2019).

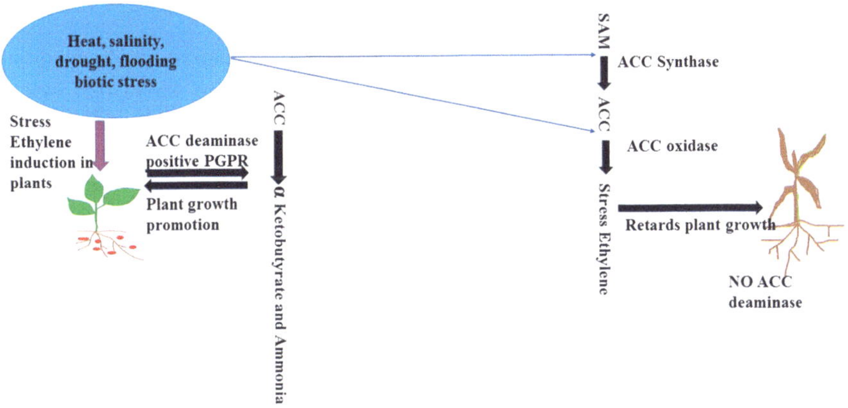


**Fig. 3.3** A schematic model of ACC deaminase positive PGPR action under stress condition (*ACC* 1-amino cyclopropane-1-carboxylic acid, *SAM* *S*-adenosyl methionine)

## 3.4 Indirect Modes of Plant Growth Promotion by PGPR

### *3.4.1 Biocontrol Weapon*

Microorganisms exhibit antibacterial and antifungal attributes, and as a consequence they can be exploited as biocontrol agents (Rani et al. 2017). Such tiny life forms trigger systemic acquired resistance in host plant secretion of hydrolytic enzymes, toxic secondary metabolites, and production of antibiotics (Rani et al. 2017; Shine et al. 2018). These organisms regulate hormone levels like auxin, ethylene in host plant and also lower the damage level in host by pathogens. Major antifungal metabolites secreted by biocontrol organisms that suppress fungal pathogenesis in host are daucans, terpenoids, polyketides, peptaibols, viscosinamide, pyrrolnitrin, phenazines, and gliovirin (Tomer et al. 2016). *Bacillus*, *Pseudomonas*, *Streptomyces*, *Serratia*, and *Trichoderma* are promising biocontrol agents. *Bacillus* sp. B25 exhibits biocontrol effect in maize against fungus *Fusarium verticillioides* that causes ear and stalk rot and aggregation of mycotoxin in maize (Maldonado-Mendoza et al. 2018). *Pseudomonas fluorescens* Pf7 is a biocontrol tool against fungus *Aspergillus flavus* which is a source of aflatoxin contamination in groundnut which causes teratogenic, hepatotoxic, immunosuppressive, and carcinogenic impact in consumer (Teja et al. 2019).

### *3.4.2 Antibiotic and Hydrogen Cyanide (HCN) Production*

To impede the devastating impact of phytopathogen some bacteria produce antibiotics to suppress the growth of pathogen. Antibiotics are low molecular weight molecules that are secreted by PGPR and are effective against other pathogenic microbes (Kundan et al. 2015). Outcome of antibiotic is both biocidal (kills the pathogen) and biostatic (inhibits the metabolism of pathogen). Antibiotics targets crucial metabolic reactions of its target like cell wall synthesis, protein synthesis, and DNA replication (Bhattacharyya et al. 2016).

Phytopathogen also develops resistance mechanisms against antibiotics. Likewise, the antibiotic secreting PGPR also displays varying response in different environmental conditions. As a consequence there should be more emphasis on how the pathogen antagonist strains can be formulated or cultured to improve its antagonistic effect (Glick 2015). Batumin is a polyene antibiotic possessing anticancer activity is produced by *Pseudomonas batumici* (Soldatkina et al. 2018). *Pseudomonas parafulva* produces phenazine-1-carboxylic acid antibiotic (Zhang et al. 2019). It was studied that *Bacillus subtilis* Bj-1 produced antibiotics like subtilin, fengycin, and bacilysin that inhibited the growth of *Magnaporthe oryzae* which is a rice blast fungus. Filtrate of *bacillus* strain restricted the fungal mycelial proliferation (He et al. 2019).

HCN is used to control weeds. These are secondary metabolites that are generated in minute amount by some PGPR. HCN results in cell death by restricting energy transfer and electron transfer reactions in the cell (Kundan et al. 2015). HCN generating PGPR are exploited as weed biocontrol weapon. It can also behave as fungal antagonist and is more toxic due to its ability to bullseye major components of metabolic reactions like cytochrome c oxidase (Nandi et al. 2017). HCN positive *Pseudomonas aeruginosa* H6 showed inhibition of *Pennisetum purpureum* and Spiny *amaranthus* weeds. Restriction in root hair formation, browning of roots, necrotic lesions on leaves was prominent symptoms on weeds. HCN effect was correlated with growth hindrance. Volatile HCN also arrested the germination of weed seeds when the seeds were treated with *Pseudomonas* generated metabolites (Lawrance et al. 2019).

### *3.4.3 Stress Repulsion*

Stress is any factor that has negative impact on growth, survival, and productivity of plants. Factors can be abiotic like high temperature, drought, flood, salinity etc., biotic factors include bacteria, fungi, insects, animals etc. PGPR adopts various methods to cope up with the abiotic stress environment like production of ACC deaminase and antioxidants, regulation of phytohormones in host plant, siderophores secretion and solubilization of nutrients that overall improve crop economic yield as discussed earlier in the manuscript.

For management of biotic stress PGPR-induced antibiotic production acts as biocontrol agent, secretes secondary metabolites, volatile compounds that are toxic to pathogens. In addition to these strategies PGPR also triggers defense mechanism in host plant known as induced systemic resistance.

Induced systemic resistance (ISR) is the capability of PGPR to induce resistance mechanism in their host against the phytopathogens (Yang et al. 2009). ISR is turned on when there is a pathogen attack and is active against a diversity of pathogen (Kundan et al. 2015). ISR regulates ethylene and jasmonate levels that restore the dormant resistant mechanism in host plant. A cascade of signalling network controls ISR response in plants and these signals are regulated by phytohormones (Pieterse et al. 2014). Many PGPR have been documented to induce ISR response in their host plants. For instance, halotolerant *Klebsiella* sp. MBE02 had growth-promoting effect and stimulated ISR in peanut. RNA sequencing revealed that inoculation of strain with host upregulated genes involved in ethylene, jasmonic acid signaling and plant defense action against *Aspergillus* fungus (Sharma et al. 2019). *Bacillus amyloliquefaciens* Ba13 triggers systemic resistance in tomato against tomato yellow leaf curl virus. Disease severity, rate of infection, and virus amount in leaf were lowered at rate of 48–52% under bacterial treatment. Strain also improved shoot and root biomass in tomato. Upregulation of pathogenesis related genes like PR1 and PR2 and increment in the activity of glucanase, phenylammoni a lyase, chitinase and polyphenol oxidase was observed in virus-infected tomato plant. According to culture-dependent analysis, there was reduction in pathogenic fungi in the rhizosphere in comparison to growth-promoting *bacillus* strain (Guo et al. 2019).

### *3.4.4 Bioremediation*

Bioremediation is a practice by which living forms or their by-products are exploited either organically or artificially to detoxify the immobilized pollutants from the nature (Uqab et al. 2016). Many PGPR have the potential to remediate the toxic chemicals like xenobiotic compounds, heavy metals, herbicides, pesticides, and organic solvents from the environment which is poisonous for soil, plants, humans, and animals. *Pseudomonas taiwanensis* was found to be arsenic and cadmium resistant and exhibited 10 ppm resistance to both the heavy metals. High activity of antioxidants like catalase, glutathione *S*-transferase, superoxide dismutase was observed in the bacterial strain due to presence of heavy metals. Thiol, hydroxyl groups and amides were also accumulated in the bacteria under heavy metal stress as adaptive response in contrast to the normal condition (Satapute et al. 2019). *Acinetobacter junii* and *Bacillus flexus* are arsenic-tolerant PGPR and possess arsC gene that encodes metalloregulatory protein which detoxify arsenic (Marwa et al. 2019).

## 3.5 Conclusion

Microorganisms are a blessing for humans. Microbes are used in various sectors like medical, food, electronics, agriculture, textile, and biotechnology because of their metabolic diversity. There are many plant growth-promoting microorganisms like bacteria, fungi, archaebacteria, and mycorrhiza that improve soil health and facilitate crop productivity under unfavorable circumstances. Exploitation of microbial application for agricultural development is an economical step for overall progress of a nation. Rather than focusing on cost-ineffective agrochemicals, eco-friendly biofertilizers should be commercialized among the farmers. In current scenario where genetically engineered products are not acceptable PGPR can be illuminated as potent alternative for constructing disease-resistant and nutrient-rich crop. Although microorganisms are applied in every field more emphasis should be there on the hidden aspects of microbial diversity, their functions, applications, and genetic manipulation for future sustainable agricultural system.

## References


Abdallah Y, Yang M, Zhang M, Masum MM, Ogunyemi SO, Hossain A, An Q, Yan C, Li B (2019) Plant growth promotion and suppression of bacterial leaf blight in rice by Paenibacillus polymyxa Sx3. Lett Appl Microbiol 68(5):423–429

Abdiev A, Khaitov B, Toderich K, Park KW (2019) Growth, nutrient uptake and yield parameters of chickpea (Cicer arietinum L.) enhance by Rhizobium and Azotobacter inoculations in saline soil. J Plant Nutr 3:1–2

Ahmad M, Adil Z, Hussain A, Mumtaz MZ, Nafees M, Ahmad I, Jamil M (2019) Potential of phosphate solubilizing Bacillus strains for improving growth and nutrient uptake in mungbean and maize crops. Pak J Agric Sci 56(2):283–289

Alloway B (2004) Zinc in soils and crop nutrition. In: Areas of the world with zinc deficiency problems. International Zinc Association, Brussels, pp 1–16

Ammari T, Mengel K (2006) Total soluble Fe in soil solutions of chemically different soils. Geoderma 136(3–4):876–885

Anand KU, Kumari BA, Mallick MA (2016) Phosphate solubilizing microbes: an effective and alternative approach as biofertilizers. J Pharm Pharm Sci 8:37–40

Azzi V, Kanso A, Kazpard V, Kobeissi A, Lartiges B, El Samrani A (2017) Lactuca sativa growth in compacted and non-compacted semi-arid alkaline soil under phosphate fertilizer treatment and cadmium contamination. Soil Tillage Res 165:1–10

Bahadur I, Maurya BR, Meena VS, Saha M, Kumar A, Aeron A (2017) Mineral release dynamics of tricalcium phosphate and waste muscovite by mineral-solubilizing rhizobacteria isolated from indo-gangetic plain of India. Geomicrobiol J 34(5):454–466

Basak BB, Biswas DR (2009) Influence of potassium solubilizing microorganism (Bacillus mucilaginosus) and waste mica on potassium uptake dynamics by sudan grass (Sorghum vulgare Pers.) grown under two Alfisols. Plant Soil 317(1–2):235–255

Bhattacharyya PN, Jha DK (2012) Plant growth-promoting rhizobacteria (PGPR): emergence in agriculture. World J Microbiol Biotechnol 28(4):1327–1350

Bhattacharyya PN, Goswami MP, Bhattacharyya LH (2016) Perspective of beneficial microbes in agriculture under changing climatic scenario: a review. J Phytology 14:26–41

Biswas JK, Banerjee A, Rai M, Naidu R, Biswas B, Vithanage M, Dash MC, Sarkar SK, Meers E (2018) Potential application of selected metal resistant phosphate solubilizing bacteria isolated from the gut of earthworm (Metaphire posthuma) in plant growth promotion. Geoderma 330:117–124

Cakmak I (2000) Possible roles of zinc in protecting plant cells from damage by reactive oxygen species. New Phytol 146:185–205

Chenniappan C, Narayanasamy M, Daniel GM, Ramaraj GB, Ponnusamy P, Sekar J, Ramalingam PV (2019) Biocontrol efficiency of native plant growth promoting rhizobacteria against rhizome rot disease of turmeric. Biol Control 129:55–64

Ciampitti IA, Salvagiotti F (2018) New insights into soybean biological nitrogen fixation. Agron J 110(4):1185–1196

Dakora FD, Phillips DA (2002) Root exudates as mediators of mineral acquisition in low-nutrient environments. In: Food security in nutrient-stressed environments: exploiting plants' genetic capabilities. Springer, Dordrecht, pp 201–213

Damam M, Kaloori K, Gaddam B, Kausar R (2016) Plant growth promoting substances (phytohormones) produced by rhizobacterial strains isolated from the rhizosphere of medicinal plants. Int J Pharm Sci Rev Res 37(1):130–136

Ditta A, Imtiaz M, Mehmood S, Rizwan MS, Mubeen F, Aziz O, Qian Z, Ijaz R, Tu S (2018) Rock phosphate-enriched organic fertilizer with phosphate-solubilizing microorganisms improves nodulation, growth, and yield of legumes. Commun Soil Sci Plant Anal 49(21):2715–2725

Doran JW, Safley M (1997) Defining and assessing soil health and sustainable productivity. In: Biological indicators of soil health. CAB International, New York. https://www.ars.usda.gov/research/publications/publication/?seqNo115=78601

Dubey RK, Tripathi V, Dubey PK, Singh HB, Abhilash PC (2016) Exploring rhizospheric interactions for agricultural sustainability: the need of integrative research on multi-trophic interactions. J Clean Prod 115:362–365

Egamberdiyeva D, Juraeva D, Poberejskaya S, Myachina O, Teryuhova P, Seydalieva L, Aliev A (2004) Improvement of wheat and cotton growth and nutrient uptake by phosphate solubilizing bacteria. In: Proceeding of 26th annual conservation tillage conference for sustainable agriculture, Auburn, pp 58–65

Etesami H, Alikhani HA, Akbari AA (2009) Evaluation of plant growth hormones production (IAA) ability by Iranian soils rhizobial strains and effects of superior strains application on wheat growth indexes. World Appl Sci J 6(11):1576–1584

Feng K, Cai Z, Ding T, Yan H, Liu X, Zhang Z (2019) Effects of potassium-solubulizing and photosynthetic bacteria on tolerance to salt stress in maize. J Appl Microbiol 126(5):1530–1540

Ferreira CM, Vilas-Boas Â, Sousa CA, Soares HM, Soares EV (2019) Comparison of five bacterial strains producing siderophores with ability to chelate iron under alkaline conditions. AMB Express 9(1):78

Gamalero E, Glick BR (2015) Bacterial modulation of plant ethylene levels. Plant Physiol 169 (1):13–22

Ghadam Khani A, Enayatizamir N, Norouzi Masir M (2019) Impact of plant growth promoting rhizobacteria on different forms of soil potassium under wheat cultivation. Lett Appl Microbiol 68(6):514–521

Glick BR (2015) Biocontrol mechanisms. In: Lugtenberg B (ed) Beneficial plant-bacterial interactions. Springer, Heidelberg, pp 159–188

Gontia-Mishra I, Sapre S, Sharma A, Tiwari S (2016) Alleviation of mercury toxicity in wheat by the interaction of mercury-tolerant plant growth-promoting rhizobacteria. J Plant Growth Regul 35(4):1000–1012

Guo Q, Li Y, Lou Y, Shi M, Jiang Y, Zhou J, Sun Y, Xue Q, Lai H (2019) Bacillus amyloliquefaciens Ba13 induces plant systemic resistance and improves rhizosphere microecology against tomato yellow leaf curl virus disease. Appl Soil Ecol 137:154–166

He Y, Zhu M, Huang J, Hsiang T, Zheng L (2019) Biocontrol potential of a Bacillus subtilis strain BJ-1 against the rice blast fungus Magnaportheoryzae. Can J Plant Pathol 41(1):47–59

Hider RC, Kong X (2010) Chemistry and biology of siderophores. Nat Prod Rep 27(5):637–657

Hodge A (2017) Accessibility of inorganic and organic nutrients for mycorrhizas. In: Johnson SC, Gehring C, Jansa J (eds) Mycorrhizal mediation of soil. Elsevier, Amsterdam, pp 129–148

Holland A (2019) Evaluation of Paenibacillus PGPR strains for growth promotion and biocontrol of rice sheath blight. http://hdl.handle.net/10415/6657

Iqbal Hussain M, Naeem Asghar H, Javed Akhtar M, Arshad M (2013) Impact of phosphate solubilizing bacteria on growth and yield of maize. Soil Environ 32(1):71–78

Israr D, Mustafa G, Khan KS, Shahzad M, Ahmad N, Masood S (2016) Interactive effects of phosphorus and Pseudomonas putida on chickpea (Cicer arietinum L.) growth, nutrient uptake, antioxidant enzymes and organic acids exudation. Plant Physiol Biochem 108:304–312

Kang SM, Khan AL, Waqas M, Asaf S, Lee KE, Park YG, Kim AY, Khan MA, You YH, Lee IJ (2019) Integrated phytohormone production by the plant growth-promoting rhizobacterium Bacillus tequilensis SSB07 induced thermotolerance in soybean. J Plant Interact 14(1):416–423

Kashyap BK, Solanki MK, Pandey AK, Prabha S, Kumar P, Kumari B (2019) Bacillus as plant growth promoting rhizobacteria (PGPR): a promising green agriculture technology. In: Ansari R, Mahmood I (eds) Plant health under biotic stress. Springer, Singapore, pp 219–236

Keiluweit M, Bougoure JJ, Nico PS, Pett-Ridge J, Weber PK, Kleber M (2015) Mineral protection of soil carbon counteracted by root exudates. Nat Clim Chang 5(6):588

Khadeejath Rajeela TH, Gupta A, Gopal M, Hegde V, Thomas GV (2018) Evaluation of combinatorial capacity of coconut and cocoa plant growth promoting rhizobacteria (PGPR) with biocontrol agent Trichoderma harzianum. Curr Investig Agric Curr Res 3(4):404–409

Khanghahi MY, Pirdashti H, Rahimian H, Nematzadeh GH, Sepanlou MG, Salvatori E, Crecchio C (2019) Leaf photosynthetic characteristics and photosystem II photochemistry of rice (Oryza sativa L.) under potassium-solubilizing bacteria inoculation. Photosynthetica 57(2):500–511

Kinzig A, Socolow RH (1994) Human impacts on the nitrogen cycle. Phys Today 47(11):24–31

Kishore N, Pindi PK, Reddy SR (2015) Phosphate-solubilizing microorganisms: a critical review. In: Bahadur B, Venkat Rajam M, Sahijram L, Krishnamurthy K (eds) Plant biology and biotechnology. Springer, New Delhi, pp 307–333

Kumar S, Yadav SS (2018) Effect of phosphorus fertilization and bio-organics on growth, yield and nutrient content of mungbean (Vigna radiata (L.) Wilczek). Res J Agric Sci 9(6):1252–1257

Kundan R, Pant G, Jadon N, Agrawal PK (2015) Plant growth promoting rhizobacteria: mechanism and current prospective. J Fertil Pestic 6(2):9

Kuppusamy S, Thavamani P, Megharaj M, Venkateswarlu K, Lee YB, Naidu R (2016) Pyrosequencing analysis of bacterial diversity in soils contaminated long-term with PAHs and heavy metals: implications to bioremediation. J Hazard Mater 317:169–179

Lawrance S, Varghese S, Varghese EM, Asok AK (2019) Quinoline derivatives producing Pseudomonas aeruginosa H6 as an efficient bioherbicide for weed management. Biocatal Agric Biotechnol 18:101096

Lazcano C, Gómez-Brandón M, Revilla P, Domínguez J (2013) Short-term effects of organic and inorganic fertilizers on soil microbial community structure and function. Biol Fertil Soils 49 (6):723–733

Li M, Ahammed GJ, Li C, Bao X, Yu J, Huang C, Yin H, Zhou J (2016) Brassinosteroid ameliorates zinc oxide nanoparticles-induced oxidative stress by improving antioxidant potential and redox homeostasis in tomato seedling. Front Plant Sci 7:615

Liu X, Jiang X, He X, Zhao W, Cao Y, Guo T, Li T, Ni H, Tang X (2019) Phosphate-solubilizing Pseudomonas sp. strain P34-L promotes wheat growth by colonizing the wheat rhizosphere and improving the wheat root. J Plant Growth Regul 38(4):1314–1324

López MF, Hegel VA, Torres MJ, García AH, Delgado MJ, López-García SL (2019) The Bradyrhizobium diazoefficiens two-component system NtrYX has a key role in symbiotic nitrogen fixation of soybean plants and cbb 3 oxidase expression in bacteroids. Plant Soil 440 (1–2):167–183

Madhukar SM, Raha P, Singh RK (2018) Identification of amino acids and sugars in root exudate of mungbean (Vignaradiata L.). J Pharmacogn Phytochem 7(2):1676–1680

Maldonado-Mendoza IE, Ibarra-Laclette E, Blom J (2018) Genomic analysis of Bacillus sp. strain B25, a biocontrol agent of maize pathogen Fusarium verticillioides. Curr Microbiol 75 (3):247–255

Marschner H (1995) In: Marschner P (ed) Mineral nutrition of higher plants. Academic, London

Martínez OA, Crowley DE, Mora ML, Jorquera MA (2015) Short-term study shows that phytate-mineralizing rhizobacteria inoculation affects the biomass, phosphorus (P) uptake and rhizosphere properties of cereal plants. J Soil Sci Plant Nutr 15(1):153–166

Marwa N, Singh N, Srivastava S, Saxena G, Pandey V, Singh N (2019) Characterizing the hypertolerance potential of two indigenous bacterial strains (Bacillus flexus and Acinetobacter junii) and their efficacy in arsenic bioremediation. J Appl Microbiol 126(4):1117–1127

Meena VS, Maurya BR, Meena SK, Mishra PK, Bisht JK, Pattanayak A (2018) Potassium solubilization: strategies to mitigate potassium deficiency in agricultural soils. GJBAHS 7:1–3

Morris EK, Morris DJ, Vogt S, Gleber SC, Bigalke M, Wilcke W, Rillig MC (2019) Visualizing the dynamics of soil aggregation as affected by arbuscular mycorrhizal fungi. ISME J 11:1

Mukherjee P, Mitra A, Roy M (2019) Halomonas rhizobacteria of Avicennia marina of Indian Sundarbans promote rice growth under saline and heavy metal stresses through exopolysaccharide production. Front Microbiol 10:1207

Mumtaz MZ, Ahmad M, Jamil M, Asad SA, Hafeez F (2018) Bacillus strains as potential alternate for zinc biofortification of maize grains. Int J Agric Biol 20:1779–1786

Nandi M, Selin C, Brawerman G, Fernando WD, de Kievit T (2017) Hydrogen cyanide, which contributes to Pseudomonas chlororaphis strain PA23 biocontrol, is upregulated in the presence of glycine. Biol Control 108:47–54

Nath D, Maurya BR, Meena VS (2017) Documentation of five potassium-and phosphorus-solubilizing bacteria for their K and P-solubilization ability from various minerals. Biocatal Agric Biotechnol 10:174–181

Olanrewaju OS, Glick BR, Babalola OO (2017) Mechanisms of action of plant growth promoting bacteria. World J Microbiol Biotechnol 33(11):197

Pal AK, Mandal S, Sengupta C (2019) Exploitation of IAA producing PGPR on mustard (Brassica nigra L.) seedling growth under cadmium stress condition in comparison with exogenous IAA application. Plant Sci Today 6(1):22–30

Pieterse CM, Zamioudis C, Berendsen RL, Weller DM, Van Wees SC, Bakker PA (2014) Induced systemic resistance by beneficial microbes. Annu Rev Phytopathol 52:347–375

Prajakta BM, Suvarna PP, Raghvendra SP, Alok RR (2019) Potential biocontrol and superlative plant growth promoting activity of indigenous Bacillus mojavensis PB-35 (R11) of soybean (Glycine max) rhizosphere. SN Appl Sci 1(10):1143

Prasad M, Srinivasan R, Chaudhary M, Choudhary M, Jat LK (2019) Plant growth promoting rhizobacteria (PGPR) for sustainable agriculture: perspectives and challenges. In: PGPR amelioration in sustainable agriculture. Woodhead Publishing, pp 129–157

Rani S, Kumar P, Kumar A, Kumar AN, Sewhag M (2016) Effect of biofertilizers on nodulation, nutrient uptake, yield and energy use efficiency of field pea (Pisum sativum L.). J Agrometereol 18(2):330–332

Rani A, Singh R, Kumar P, Shukla G (2017) Pros and cons of fungicides: an overview. Int J Eng Sci Res Technol 3:112–117

Rathore P (2014) A review on approaches to develop plant growth promoting rhizobacteria. Int J Recent Sci Res 5:403–407

Raut AD, Durgude AG, Kadlag AD (2019) Effect of zinc solubilizing bacteria on zinc use efficiency and yield of summer groundnut grown in Entisol. IJCS 7(1):1710–1713

Riefler M, Novak O, Strnad M, Schmülling T (2006) Arabidopsis cytokinin receptor mutants reveal functions in shoot growth, leaf senescence, seed size, germination, root development, and cytokinin metabolism. Plant Cell 18(1):40–54

Saha M, Maurya BR, Meena VS, Bahadur I, Kumar A (2016) Identification and characterization of potassium solubilizing bacteria (KSB) from Indo-Gangetic Plains of India. Biocatal Agric Biotechnol 7:202–209

Saharan BS, Nehra V (2011) Plant growth promoting rhizobacteria: a critical review. Life Sci Med Res 21(1):30

Sahu M, Adak T, Patil NB, Gowda GB, Yadav MK, Annamalai M, Golive P, Rath PC, Jena M (2019) Dissipation of chlorantraniliprole in contrasting soils and its effect on soil microbes and enzymes. Ecotoxicol Environ Saf 180:288–294

Saikia J, Sarma RK, Dhandia R, Yadav A, Bharali R, Gupta VK, Saikia R (2018) Alleviation of drought stress in pulse crops with ACC deaminase producing rhizobacteria isolated from acidic soil of Northeast India. Sci Rep 8(1):3560

Saravanan VS, Subramoniam SR, Raj SA (2004) Assessing in vitro solubilization potential of different zinc solubilizing bacterial (ZSB) isolates. Braz J Microbiol 35(1–2):121–125

Sarkar A, Ghosh PK, Pramanik K, Mitra S, Soren T, Pandey S, Mondal MH, Maiti TK (2018) A halotolerant Enterobacter sp. displaying ACC deaminase activity promotes rice seedling growth under salt stress. Res Microbiol 169(1):20–32

Satapute P, Paidi MK, Kurjogi M, Jogaiah S (2019) Physiological adaptation and spectral annotation of arsenic and cadmium heavy metal-resistant and susceptible strain Pseudomonas taiwanensis. Environ Pollut 251:555–563

Sharma A, Shankhdhar D, Shankhdhar SC (2013) Enhancing grain iron content of rice by the application of plant growth promoting rhizobacteria. Plant Soil Environ 59(2):89–94

Sharma A, Patni B, Shankhdhar D, Shankhdhar SC (2015) Evaluation of different PGPR strains for yield enhancement and higher Zn content in different genotypes of rice (Oryza sativa L.). J Plant Nutr 38(3):456–472

Sharma A, Shankhdhar D, Shankhdhar SC (2016) Potassium-solubilizing microorganisms: mechanism and their role in potassium solubilization and uptake. In: Meena V, Maurya B, Verma J, Meena R (eds) Potassium solubilizing microorganisms for sustainable agriculture. Springer, New Delhi, pp 203–219

Sharma S, Chen C, Navathe S, Chand R, Pandey SP (2019) A halotolerant growth promoting rhizobacteria triggers induced systemic resistance in plants and defends against fungal infection. Sci Rep 9(1):4054

Shine MB, Xiao X, Kachroo P, Kachroo A (2018) Signaling mechanisms underlying systemic acquired resistance to microbial pathogens. Plant Sci 279:81–86

Shridhar BS (2012) Nitrogen fixing microorganisms. Int J Microbiol Res 3:46–52

Sindhu SS, Parmar P, Phour M, Sehrawat A (2016) Potassium-solubilizing microorganisms (KSMs) and its effect on plant growth improvement. In: Meena V, Maurya B, Verma J, Meena R (eds) Potassium solubilizing microorganisms for sustainable agriculture. Springer, New Delhi, pp 171–185

Singh JS (2015) Microbes: the chief ecological engineers in reinstating equilibrium in degraded ecosystems. Agr Ecosyst Environ 203:80–82

Singh BK, Millard P, Whiteley AS, Murrell JC (2004) Unravelling rhizosphere–microbial interactions: opportunities and limitations. Trends Microbiol 12(8):386–393

Singh B, Natesan SK, Singh BK, Usha K (2005) Improving zinc efficiency of cereals under zinc deficiency. Curr Sci 10:36–44

Soldatkina MA, Klochko VV, Zagorodnya SD, Rademan S, Visagie MH, Lebelo MT, Gwangwa MV, Joubert AM, Lall N, Reva ON (2018) Promising anticancer activity of batumin: a natural polyene antibiotic produced by Pseudomonas batumici. Future Med Chem 10(18):2187–2199

Stephen J, Shabanamol S, Rishad KS, Jisha MS (2015) Growth enhancement of rice (Oryza sativa) by phosphate solubilizing Gluconacetobacter sp. (MTCC 8368) and Burkholderia sp.(MTCC 8369) under greenhouse conditions. 3 Biotech 5(5):831–837

Subhashini DV (2015) Growth promotion and increased potassium uptake of tobacco by potassium-mobilizing bacterium Frateuria aurantia grown at different potassium levels in vertisols. Commun Soil Sci Plant Anayl 46(2):210–220

Takoutsing B, Weber J, Aynekulu E, Martín JA, Shepherd K, Sila A, Tchoundjeu Z, Diby L (2016) Assessment of soil health indicators for sustainable production of maize in smallholder farming systems in the highlands of Cameroon. Geoderma 276:64–73

Teja MR, Kumar KV, Sudini H (2019) Pseudomonas fluorescens Pf7: a potential biocontrol agent against Aspergillus flavus induced aflatoxin contamination in groundnut. Adv Res 24:1–7

Teng Z, Shao W, Zhang K, Huo Y, Li M (2019) Characterization of phosphate solubilizing bacteria isolated from heavy metal contaminated soils and their potential for lead immobilization. J Environ Manage 231:189–197

Tomer S, Suyal DC, Goel R (2016) Biofertilizers: a timely approach for sustainable agriculture. In: Choudhary D, Varma A, Tuteja N (eds) Plant-microbe interaction: an approach to sustainable agriculture. Springer, Singapore, pp 375–395

Uqab B, Mudasir S, Nazir R (2016) Review on bioremediation of pesticides. J Bioremed Biodegr 7 (3):343

Vicente-Hernández A, Salgado-Garciglia R, Valencia-Cantero E, Ramírez-Ordorica A, Hernández-García A, García-Juárez P, Macías-Rodríguez L (2019) Bacillus methylotrophicus M4-96 stimulates the growth of strawberry (Fragaria × ananassa 'Aromas') plants in vitro and slows Botrytis cinerea infection by two different methods of interaction. J Plant Growth Regul 38 (3):765–777

Wang P, Wang TY, Wu SH, Wen MX, Lu LM, Ke FZ, Wu QS (2019) Effect of arbuscular mycorrhizal fungi on rhizosphere organic acid content and microbial activity of trifoliate orange under different low P conditions. Arch Agron Soil Sci 19:1–4

Weidenhamer JD, Montgomery TM, Cipollini DF, Weston PA, Mohney BK (2019) Plant density and rhizosphere chemistry: does marigold root exudate composition respond to intra-and interspecific competition? J Chem Ecol 27:1–9

Wissuwa M, Ismail AM, Yanagihara S (2006) Effects of zinc deficiency on rice growth and genetic factors contributing to tolerance. Plant Physiol 142(2):731–741

Xiao Y, Wang X, Chen W, Huang Q (2017) Isolation and identification of three potassium-solubilizing bacteria from rape rhizospheric soil and their effects on ryegrass. Geomicrobiol J 34(10):873–880

Yang J, Kloepper JW, Ryu CM (2009) Rhizosphere bacteria help plants tolerate abiotic stress. Trends Plant Sci 14(1):1–4

Yaxley JR, Ross JJ, Sherriff LJ, Reid JB (2001) Gibberellin biosynthesis mutations and root development in pea. Plant Physiol 125(2):627–633

Zafar-ul-Hye M, Danish S, Abbas M, Ahmad M, Munir TM (2019) ACC deaminase producing pgpr bacillus amyloliquefaciens and agrobacterium fabrum along with biochar improve wheat productivity under drought stress. Agronomy 9(7):343

Zaheer A, Malik A, Sher A, Qaisrani MM, Mehmood A, Khan SU, Ashraf M, Mirza Z, Karim S, Rasool M (2019) Isolation, characterization, and effect of phosphate-zinc-solubilizing bacterial strains on chickpea (Cicer arietinum L.) growth. Saudi J Biol Sci 26(5):1061–1067

Zhang Y, Chen P, Ye G, Lin H, Ren D, Guo L, Zhu B, Wang Z (2019) Complete genome sequence of Pseudomonas parafulva PRS09–11288, a biocontrol strain produces the antibiotic phenazine-1-carboxylic acid. Curr Microbiol 76(9):1087–1091

Zhao Y, Zhang M, Yang W, Di HJ, Ma L, Liu W, Li B (2019) Effects of microbial inoculants on phosphorus and potassium availability, bacterial community composition, and chili pepper growth in a calcareous soil: a greenhouse study. J Soils Sediments 19(10):3597–3607

Zheng BX, Ding K, Yang XR, Wadaan MA, Hozzein WN, Peñuelas J, Zhu YG (2019) Straw biochar increases the abundance of inorganic phosphate solubilizing bacterial community for better rape (Brassica napus) growth and phosphate uptake. Sci Total Environ 647:1113–1120

Zou X, Binkley D, Doxtader KG (1992) A new method for estimating gross phosphorus mineralization and immobilization rates in soils. Plant Soil 147(2):243–250

# Chapter 4
# Vermicompost and Soil Health

**Ranjit Chatterjee, Ankita Debnath, and Subhalaxmi Mishra**

**Abstract** The greatest challenge in the coming years is to fulfill the demand for safe food, healthy soil, and a pollution-free environment for the growing populations of the world. Indiscriminate use of inorganic fertilizers and pesticides in conventional crop production is considered to be one of the prime factors for deterioration in crop productivity, degradation of soil health, and serious threats to the environment and human health that largely affect the sustainability of the agricultural production system. With increasing awareness of the ill effects of conventional farming/chemical faming, recent years have seen renewed interest in the sustainability of our food production system by revitalizing and restoring soil fertility and reviving microbial activity to make the soil lively and healthy. Consequently, awareness has been generated on recycling of available organic residues for the production of quality organic manures. Vermicompost is a nutrient-rich, microbiologically active organic amendment that is obtained in the form of castings by earthworms of ingested biomass after undergoing physical, chemical, and microbial transformation. Several research findings have established the beneficial effect of vermicompost on soil health through improvement in the physical, chemical, and biological properties and subsequently better crop growth and yield. Microbial enrichment of vermicompost by addition of biofertilizers and bioinoculants further enhances the microbial population of the soil, nutrient mineralization uptake, and availability of nutrients. Vermicomposting is gradually emerging as a potential technology for recycling available organic wastes as a source of quality organic manures, and vermicompost is rising as a promising organic fertilizer for maintaining good soil health and crop growth as well as reducing dependence on nonrenewable resources. The present work highlights the beneficial role of vermicompost and the preparation and utilization of vermicompost with special emphasis on how vermicompost can address the emerging soil health problems for sustainability in soil health and the ecosystem.

R. Chatterjee (✉) · A. Debnath
Department of Vegetable and Spice Crops, Uttar Banga Krishi Viswavidyalaya (UBKV), Pundibari, Cooch Behar, West Bengal, India

S. Mishra
Department of Vegetable Science, OUAT, Bhubaneswar, Odisha, India

 
B. Giri, A. Varma (eds.), *Soil Health*, Soil Biology 59,
https://doi.org/10.1007/978-3-030-44364-1_4

**Keywords** Organic manure · Organic wastes · Soil health · Soil properties and vermicompost

## 4.1 Introduction

In the post-Green Revolution era, agricultural practices are directed toward growing more crops through use of higher levels of chemical fertilizers and pesticides. Again, rapid urbanization, industrialization, and infrastructure development are gradually shrinking cultivated land areas. Indiscriminate use of chemical fertilizers tends to stagnate or reduce crop yields as well as decreasing the quality of the harvested products. Application of nitrite and nitrate forms of nitrogenous fertilizer severely pollute groundwater and food items, resulting in serious environmental threats (Bhattacharya 2004). Excessive use of such synthetic inputs seriously affects soil health by depleting the microbial populations of the soil. Pressure for more food production for the ever-increasing population forces extensive soil cultivation, which threatens soil health and depletes natural resources. In addition, the increasing gap between nutrient removal by crops and supplied nutrients has caused nutrient depletion of the soil, a serious threat to the sustainability of crop production.

## 4.2 Problem of Conventional Crop Production and Need for Healthy Soil

Conventional crop production in a broad sense refers to the agriculture production system that includes the use of synthetic chemical fertilizers, pesticides, herbicides, and other growth chemicals. This farming practice is highly resource- and energy intensive and can involve the use of genetically modified organisms, heavy irrigation, fertilizers, and intensive tillage. Such production systems always promote monoculture or a single type of crop cultivation, which reduces farm diversity and causes reduction in the diversity of flora and fauna in the farming system. To obtain higher return, large amounts of fertilizers and pesticides are used in monoculture. These synthetic inputs have negative effects on soil health and are contrary to sustainability.

Tillage is one of the important practices for conventional crop production. It is used to loosen the soil, kill weeds, dry the soil, and to mix in organic matter, but it reduces the level of soil aggregate stability, which indicates a decline in organic matter content, biological activity, and soil nutrient cycling (Beilen 2016). Tillage operations expose earthworms, the beneficial indicator of soil fertility, so greatly that they are killed and eaten by birds. Thus, the earthworm biomass is drastically reduced in tilled farmland. Also, disturbance caused by tillage in the soil rhizosphere affects many beneficial fungi and bacteria near the root zone of the plants.

The use of synthetic fertilizers is the key to success in conventional farming. Although this practice increases the yield and productivity of crops, it has negative effects on soil health. Excessive fertilizer input affects the populations of soil microorganisms and earthworms. Heavy use of nitrogen fertilizer results in losses of nutrient cations and promotes soil acidification (Vitousek et al. 1997), and the availability of many trace elements is drastically reduced. Use of fertilizers can cause toxicity and disorders in the plants. Pesticides including herbicides, fungicides, and insecticides are used to control pest infestations. However, a fungicide also kills beneficial fungi and affects the soil bacteria. Insecticides such as the organophosphates reduce the bacteria population (Pandey and Singh 2004). Herbicides are highly toxic to earthworms and also affect enzymatic activity in soil microorganisms (Bunemann et al. 2006). After field application, most of these pesticides are either mixed in groundwater or volatilized to the atmosphere, causing environmental pollution. Pesticide residues in the food chain are an emerging concern for consumers.

Soil health denotes the capacity of soil to function as a vital living system, within ecosystem and land use boundaries, to sustain biological productivity, promote environmental quality, and maintain plant and animal health. Soil health and quality determine agricultural sustainability (Acton and Gregorich 1995) as well as environmental quality (Pierzynski et al. 1994). In conventional crop production, the fertility of the soil declines over time from the continuous extraction of nutrients, with an increasing gap between removal of nutrients by the crops and nutrient replenishment in the soil. Further, soil acidification results from the continuous use of acid-producing chemical fertilizers and soil compaction caused by heavy tillage machinery and continuous cropping. Earthworm populations in the soil are decreasing at an alarming rate in the conventional agricultural production system. Agricultural operations such as the use of heavy tillage machinery and higher doses of chemical fertilizers and soil pesticides disturb the existence of the earthworms and greatly contribute to the reduction of earthworm populations in the soil. The practice of continuous cropping makes the conventional system unsuitable for the growth and proliferation of earthworms. Soil microbes have a vital role in regulating organic matter decomposition and nutrient cycling in the soil. A balanced ratio of microbial biomass and activity is vital for release of nutrients to plants and microbial growth. Soils rich in organic matter promote microbe colonization with increased soil physicochemical and biological properties that lead to healthy fertile soil.

## 4.3 Vermicompost and Its Characteristics

Vermicompost is a nutrient-rich, microbiologically active organic amendment that is obtained in the form of castings of ingested biomass by earthworms after undergoing physical, chemical, and microbial transformations. Vermicompost is a combination of the earthworm casts, partially decomposed bedding materials, seeds or cocoons, humic substances, and associated microorganisms that is stabilized to small particles

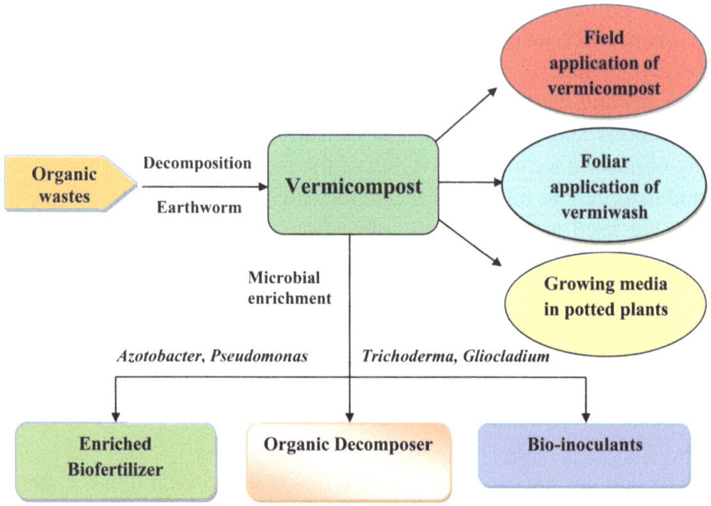


**Fig. 4.1** Schematic diagram of vermicompost production and utilization

of a peat-like material with high porosity, and with more water-holding capacity and low C/N ratio with a very large surface area, which facilitate holding and retaining plant nutrients (Dominguez 2004).

Vermicompost is a good source of nitrogen (N) (Bansal and Kapoor 2000), phosphorus (P) (Pramanik et al. 2007), and potassium (K) as well as vitamins, antibiotics, humic acid (Arancon et al. 2006), N-fixing and P-solubilizing bacteria, enzymes such as protease, amylases, lipases, cellulases, and chitinases, and growth-promoting substances such as auxins and gibberellins (Sinha et al. 2009; Lazcano and Dominguez 2011). The process of breaking down complex organic substrates into a stabilized humus-like substance through the action of earthworms is called vermicomposting. The presence of earthworms increases the natural biodegradation and decomposition of organic substrates from 60% to 80% and upgrades the value of the compost. During the process of decomposition, earthworms have a major function in fragmenting and conditioning the substrates, increasing surface area for the growth of microbes, and altering biological activities (Dominguez and Edwards 2004), whereas the microbes create a humified condition that converts the oxidized instable organic matter into more stable forms (Lemtiri et al. 2014). During conversion of organic residues, earthworms act as a crusher, grinder, aerator, chemical degrader, and a biological stimulator (Sinha et al. 2002). Earthworms and microorganisms work together synergistically to accelerate the decomposition of organic substrates (Fig. 4.1).

Earthworms consume decomposable organic wastes that are egested as casts after physical, chemical, and biological transformation within the earthworm gut. After ingestion of the substrate, the earthworm increases decomposition of the organic substrate and modifies the physical and chemical properties of the material, which gradually creates effects similar to composting in which the unstable organic matter is oxidized and stabilized aerobically with increased decomposition and humification. Most of the epigeic earthworms can consume amounts of organic material greater than their body weight. Among the epigeic earthworms, three major genera are commercially exploited for vermicompost production: *Eisenia foetida* (*fetida*), *Eudrilus eugeniae*, and *Perionyx excavatus*. The earthworm *Eisenia foetida* is considered highly suitable for decomposition of organic residues compared to others because it has a faster growth rate, higher reproductive potential, and can consume various types of organic matter at different stages of decomposition. In addition, it can tolerate a wider temperature range and can live in organic wastes with different moisture conditions (Ismail 1997). The earthworm *Eisenia fetida* is reported to consume organic matter at a rate equal to their body weight every day. Under optimum temperature (20 °C–30 °C) and moisture (60–70%), about 5 kg worms can recycle 1 ton of substrate into vermicompost within 30 days. Upon vermicomposting, the volume of solid wastes is reduced to approximately one third of the initial bulk. Singleton et al. (2003) isolated the bacterial flora associated with the intestine and vermicasts of earthworms and reported such genera as *Pseudomonas*, *Mucor*, *Paenibacillus*, *Azoarcus*, *Burkholderia*, *Spiroplasm*, *Acaligenes*, and *Acidobacterium* have the potential to degrade diverse organic substrates. Karmakar et al. (2009) recorded beneficial microorganisms such as *Actinomycetes*, *Azotobacter*, *Nitrobacter*, *Nitrosomonas*, and *Aspergillus* in the guts of earthworms. Yasir et al. (2009) showed that changes in the bacterial community have a major effect during vermicomposting. Apart from bacteria, fungi, especially cellulolytic fungi, also are important during decomposition. The population of cellulolytic fungi was found to be increased during vermicomposting of different organic residues. The enzyme cellulose produced by these fungi has a major role in decomposition of cellulolytic materials of the organic wastes.

Vasanthi and Kumaraswamy (1999) reported that nutrient composition of the end product was improved after vermicomposting as compared to ordinary compost from the same organic materials. Chatterjee et al. (2005) studied the quality of traditional composting and earthworm-mediated vermicomposting and observed that highest carbon mineralization rate, lowest easily mineralizable carbon concentration, and neutral pH in the final compost makes the vermicompost better compared to traditional composting. Bhatnagar and Palta (1996) found that earthworm-accelerated vermicomposting is very efficient in breaking the complex decomposable organic waste and that it was 2 to 5 times faster than conventional methods of composting. In traditional composting the full conversion period may be as long as 6 months. However, 10 kg earthworms (10,000 worms) would convert 1 ton organic wastes per month within a 5 $m^2$ composting pit. One million earthworms in an enclosure of 22.6 m × 22.6 m can decompose 250 tonnes organic wastes per month. Kale (1998) reported that on average vermicompost contains 9.15–17.98% organic C, 0.5–1.5% total N, 0.1–0.3% available P, 0.15–0.56% available K, 0.06–0.3% available Na, 2–9.5 ppm Cu, 2–9.3 ppm Fe, 5.7–11.5 ppm Zn, and 128–548 ppm available S. Sharma and Madan (1983) found that earthworm castings contain as much as 5 times more nitrate N, 14 times more Ca, 3 times more

Mg, and 11 times more K than that of topsoil at 15 cm depth. Reddy and Reddy (1998) stated that vermicompost contains 1.98% N, 1.23% P, 1.59% K, and total Zn 132.0 mg/kg, Cu 70.5 mg/kg, Fe 144.2 mg/kg, and Mn 317.5 mg/kg. Vasanthi and Kumaraswamy (1999) evaluated the nutrient composition of vermicompost prepared from different plant wastes and reported the nitrogen content of 2.99% (*Ipomoea* weeds), 2.83% (banana wastes), 2.99% (*Parthenium* weeds), 2.67% (sugarcane trash), and 2.61% (neem leaves). They also reported the phosphorus content of 1.37% (*Ipomoea* weeds), 1.18% (banana wastes), 1.20% (*Parthenium* weeds), 1.06% (sugarcane trash), and 1.17% (neem leaves). They found the potassium content of vermicompost was 1.46% (*Ipomoea* weeds), 1.32% (banana wastes), and 1.19% (*Parthenium* weeds).

## 4.4 Preparation and Utilization of Vermicompost

A wide variety of organic substrates have been extensively studied for their suitability for vermicomposting. Organic wastes including kitchen wastes, garden wastes, farm wastes, animal wastes, spent wash wastes, broiler ash, municipal waste, institutional wastes, sewage sludge from water treatment plants, industrial wastes from the paper industry, sericulture industry, sugarcane industry, and guar gum industry were all found suitable for large-scale vermicompost production (Garg et al. 2006; Arancon et al. 2008). However, the physical and chemical characteristics of the wastes are important for suitability as substrate of vermicomposting and modulating the microbiological decomposition that determines the final quality of the vermicompost and (Brink 1995). The chemical characteristics of the organic residues may be considered in terms of nutrient quality and quantity. Zibiliske (1998) reported that the relative quantity of C, N, P, S, and other nutrients, and also the quality of the waste, is important to determine the decomposition rate. Cellulose- and lignin-rich substrates have similar percentages of carbon but lignin undergoes decomposition much more slowly than does cellulose, and the quantity of lignin as a carbon source is much lower than that of cellulose (Chang 1997) (Fig. 4.2).

Gaur (1999) carried out a comparative study on nitrogen content and C:N ratio of green vegetable crop residues and field crop residues and stated that vegetable residues contain a greater amount of nitrogen and have a minimum stable C:N ratio with lesser amounts of resistant materials such as lignin and are therefore highly suitable for microbial decomposition. Biradar and Patil (2001) evaluated the suitability of some weed species for vermicomposting with the earthworm *Eudrilus eugeniae*. They found maximum vermicompost yield and more clitellate and non-clitellate worms with the weed *Cassia sericea*. High lignin content in coconut leaves and petioles makes these among the most recalcitrant organic matter, resisting natural decomposition. By employing a local strain of earthworm, *Eudrilus* sp., degradation of coconut leaves could be hastened to 3 months and the leaves recycled to valuable vermicompost (Prabhu et al. 1998). Ndegwa and Thompson

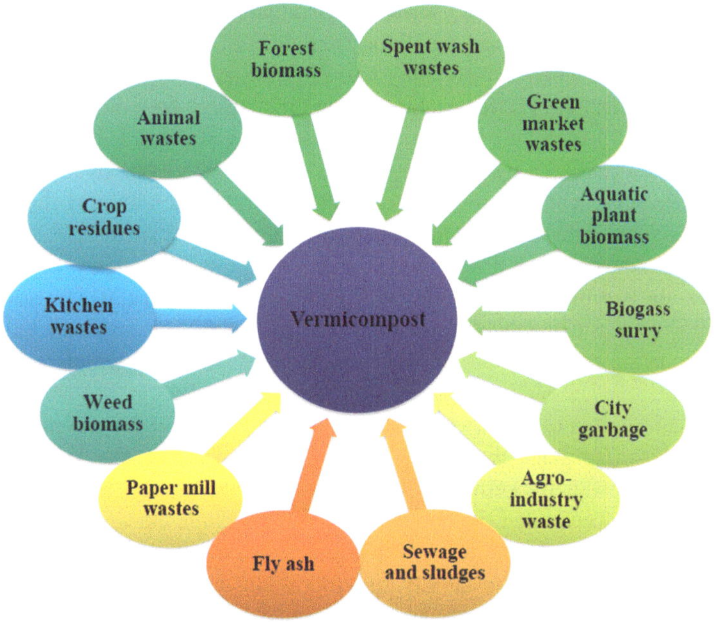


**Fig. 4.2** Diverse organic wastes suitable for vermicomposting

(2001) studied the decomposition of biosolids through vermicomposting and found that integrating pre-composting followed by vermicomposting with the earthworm *Eisenia foetida* shortened stabilization time and resulted in a product that was more stable and homogeneous in quality. Nirmalnath et al. (2001) analyzed the changes in microbial populations in vermicompost during recycling of different crop residues. They recorded the highest vermicompost yield with maximum microbial populations such as bacteria ($73 \times 10^5$), actinomycetes ($100 \times 10^4$), and phosphate-soluble microorganisms ($29 \times 10^4$) with vermicompost composed solely of cow dung. Raghavendra and Bano (2001) found that green leaves of perennial legumes such as *Pongamia pinnata* and *Leucaena leucocephala* are highly effective for vermicomposting when combined with cow dung slurry at 4:1 proportion in the presence of the earthworms *Eudrilus eugeniae* and *Eisenia foetida* resulted in rapid humification with decreased C:N ratio of 10–15% along with increased macro- and micronutrients.

Talukdar et al. (2001) studied suitability of seven different biowastes, namely kitchen wastes, crop residues, cattle shed wastes, water hyacinths, city garbage, sugarcane bagasse, and wastepaper for vermicomposting in the presence of the locally available earthworm *Amynthas diffringens*. They found the highest decomposition and earthworm populations in cattle shed wastes followed by kitchen wastes and crop residues. Shweta and Sharma (2003) studied the recycling of locally available organic wastes by using the indigenous earthworm *Lampito mauritii*. The substrates used were kitchen wastes, leaf litter, cow dung, buffalo dung, oil cakes, and agricultural crop wastes individually or in combination. The vermicompost of leaf litter had the highest numbers of earthworms with minimum earthworm weight, whereas cow dung had the least numbers of earthworms with maximum body weight. However, the optimum increase in both number and weight was found when a mixture of substrate and cow dung was used. Jeevendran et al. (2016) studied the suitability of temple wastes flower composting by the earthworm *Eudrilus eugeniae*. The different flower wastes were mixed with cow dung slurry. After 60 days the highest zoomass productivity was observed when 100% cow dung was used. However, 25% flower waste with 75% cow dung slurry was the best option for the vermicompost production. Gurav and Pathade (2011) also studied temple flower waste management through vermicomposting and concluded that vermicomposting is an outstanding and eco-friendly method of temple solid waste management.

Singh and Sharma (2002) tried to integrate composting and vermicomposting for decomposition of organic residues. The results revealed that the combination of composting followed by vermicomposting reduced the overall time required for composting and accelerated the composting of lignocellulosite wastes besides producing a nutrient-enriched compost product. Pulikeshi et al. (2003) studied the seasonal variation in compostability of crop waste and cow dung manure and production of worm biomass by the earthworm *Eisenia foetida*. They found that the environmental factors prevailing in different seasons directly influenced the life activities of the earthworm and indirectly the compostability of the organic substrate. They concluded that the amount of vermicompost produced by the worm activity mainly depended on environmental factors, followed by the nature of the organic wastes.

## 4.5 How Vermicompost Can Address the Soil Health Problem

Soil health generally refers to a state of dynamic equilibrium between flora and fauna and their surrounding soil environment in which all the metabolic activities of the former proceed optimally without any hindrance, stress, or impedance from the latter (Goswami and Rattan 1992). Soil fertility is the capacity of the soil to supply sufficient quantities and proportions of essential plant nutrients required for optimal growth of specified plants as governed by the chemical, physical, and biological attributes of soil. Soil health and fertility would be the determining factors for the sustainability of the production system in the long run.

### 4.5.1 Beneficial Effect of Vermicompost on Soil Physical Properties

The physical properties of a soil are denoted by such characteristics as structure, texture, bulk density, porosity, soil aeration, soil temperature, and water-holding capacity. The consistent addition of vermicompost improves soil structure, aeration, infiltration, and drainage (Arancon et al. 2008). Vermicompost injection to the soil can influence the degree of soil aggregation, and can reduce bulk density and increase total porosity, moisture-holding capacity, cation-exchange capacity, and oxidation potential of soil caused by increased microbial activity (Manivannan et al. 2009). Vermicompost-rich soil increases the earthworm populations, encouraging extensive burrowing and producing loose and porous soil wherein the micropores improves water absorption, drainage, and aeration to the root rhizosphere. The burrowing action of worms also significantly contributes to the permeability of water in the soil. Castings have shown to hold nine times their weight in water, further enriching the fertility of the soil (Girde et al. 2016). The water-holding capacity of vermicompost-rich soil is increased by the increase in colloidal materials such as earthworm mucus and polysaccharides that act as water-absorbing agents in the vermicompost. Ibrahim et al. (2015) evaluated the effect of vermicompost along with water treatment residuals on changes in selected physical properties of saline sodic soil and wheat yield: addition of vermicompost and water treatment residuals had significant positive effects on the soil physical properties and improved the wheat grain yield. A study by Zucco et al. (2015) showed that soils with high vermicompost rates produced taller plants with greater numbers of leaves and flowers, higher leaf chlorophyll content, greater plant biomass, and more total leaf area compared to soils with low vermicompost rates. Tomatoes grown in sandy soil amended with vermicompost generally had the greatest growth responses compared to clay or silt loam soils, with the silt loam soil generally providing the least response. Azarmi et al. (2008) evaluated the effects of vermicompost on soil physical properties in tomato fields and showed that soils amended with vermicompost had significantly lower bulk density in comparison to control plots. The increased rates of vermicompost further reduced soil bulk density. Compost addition caused a significant decrease of bulk density because greater porosity was added to the soil (Bazzoffi et al. 1998). The greater porosity in soil treated with vermicompost resulted from an increase in the number of rounded pores (Marinari et al. 2000). Pagliai et al. (1980) stated that the increase in porosity has been credited to increased numbers of pores in the 30–50 μm and 50–500 μm size ranges and reduction in number of pores greater than 500 μm. Manivannan et al. (2009) compared the efficacy of vermicompost in comparison to inorganic fertilizers on the physicochemical and biological characteristics of the soils. Results showed that the application of vermicompost (5 t/ha) significantly enhanced pore space, water-holding capacity, and cation-exchange capacity and also reduced bulk density (Fig. 4.3).

Soil rich in vermicompost acts as a warehouse of organic carbon and most of the available plant nutrients and as a source of energy for microorganisms. The

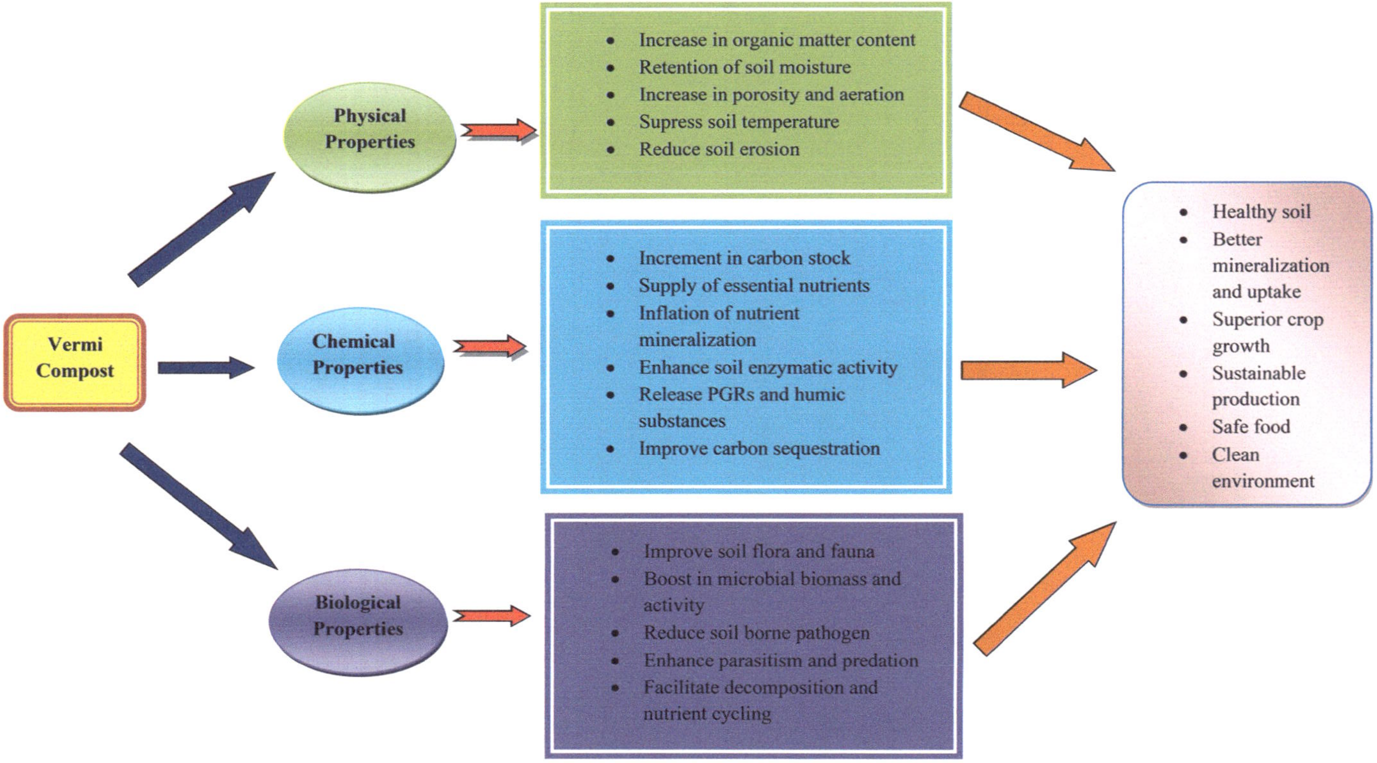


**Fig. 4.3** Beneficial role of vermicompost for soil health and sustainability

vermicast is a very rich source of nitrogen, calcium, phosphorus, and magnesium as well as a source of important micronutrients for plant growth. Ghosh et al. (1999) found that the level of transformation of the nutrient phosphorus from the organic to inorganic state and subsequently into available form was higher in earthworm-inoculated compost. Sudhakar et al. (2002) stated that vermicompost contains microsites rich in available carbon, nitrogen, and water-soluble phosphorus and contains two to three times more available potassium than the surrounding soils, which encourages better crop growth. Moreover, vermicompost promotes nitrogenase activity in the soil rhizosphere that promotes nitrogen fixation in legume plants. In addition, the enzyme phosphatase present in vermicompost increases phosphorus availability to plants (Padmavathiamma et al. 2008). The humic substance present in vermicompost benefits the soil by transferring humified components to the soil. As well as increasing plant growth, humic acid increases root hair proliferation, mineral nutrient release, and is involved in cell respiration, photosynthesis, oxidative phosphorylation, protein synthesis, and various enzymatic reactions (Lim et al. 2015; Masciandaro et al. 2010). The pH of soil increases in acid soil with the application of vermicompost. However, conflicting results are reported on changes in soil pH for different soil types (Valdez-Perez et al. 2011). The exchangeable $Na^+$ and $Ca^+$ ions increase with vermicompost use, which lowers the electrical conductivity of soil (Oo et al. 2013). Vermicompost application reduced nutrient loss and decreases the risk of $NO_3^-$ leaching in soil (Masciandaro et al. 2010). Earthworms in soil increased the delivery of organic carbon and nitrogen to soil aggregates to provide soil organic matter stabilization (Grdisa and Grdisa 2013), and the reduced C/N ratio returned nitrogen in the available form to plants (Basheer and Agrawal 2015).

Manyuchi et al. (2013) analyzed the effect of the vermicompost and vermiwash and their combined effects on soil pH, electrical conductivity, and nitrogen, phosphorus, and potassium content. Increasing the vermicompost level decreased soil pH, electrical conductivity, and nitrogen content with a slight increase in phosphorus content. Also, increasing the vermiwash quantity increased soil pH, electrical conductivity, and potassium content but resulted in reduced nitrogen and phosphorus content. Increasing the vermicompost and vermiwash application duration and their combined effects resulted in decreased soil nitrogen content, possibly caused by decomposition of the organic compounds by microorganisms present in the organic manure. The biological activity of microorganisms helps convert the phosphorus into soluble phosphates, and so the total phosphate content increases. However, increasing the vermiwash quantity resulted in decreased phosphorus content, possibly because phosphorus was in soluble form and was readily absorbed by the soil. Increasing the vermicompost and vermiwash application duration and their combined effects resulted in increased soil potassium content, possibly because the loam-clay soil has good nutrient-absorbing properties. Furthermore, the microorganisms present in the vermiwash and vermicompost resupplied the soil with more potassium ions, so that the total content increased.

Azarmi et al. (2008) evaluated the effects of vermicompost on soil chemical properties in a tomato field, showing that application of vermicompost (15 t/ha) significantly increased soil content of total organic carbon and of total N, P, K, Ca, Zn, and Mn compared with a control plot, probably from the slow release of nutrients by the vermicompost in the presence of soil microorganisms. The soils treated with vermicompost had significantly more electrical conductivity in comparison to unamended plots. The addition of vermicompost in soil resulted in decrease of soil pH. The marked decrease in total nitrogen in soils without vermicompost application in comparison with vermicompost-treated soils may have occurred because the larger amounts of total C and N in sheep manure vermicompost could provide a larger source of N for mineralization (Arancon et al. 2006). There have been other reports of increased nitrogen in soil after application of vermicompost (Nethra et al. 1999). Marinari et al. (2000) showed similar increases in soil phosphorus after application of organic amendments. The enhancement of phosphatase activity and physical breakdown of material resulted in greater mineralization (Sharpley and Syres 1977). The soil-available potash increased significantly with the increasing rate of vermicompost application. Use of vermicompost at rates of 15, 10, and 5 t/ha increased available potash in these treatments to 58%, 46%, and 34%, respectively, as compared to control plots. The feeding habits of earthworms on organically rich substances which break down during passage through the gut, biological grinding, together with enzymatic influence on finer soil particles, were likely responsible for increasing the different forms of potash (Rao et al. 1996). The increase of soil organic matter resulted in decreased K fixation and subsequent increased K availability (Olk and Cassman 1993).

### 4.5.2 Beneficial Effect of Vermicompost on Soil Biological Properties

Soil health quality is significantly influenced by the flora and fauna and microbial diversity present in the soil. Soil microorganisms are the living part of soil organic matter present in the soil. The soil organisms respond sensitively to beneficial soil and ecosystem functions including decomposition and nutrient cycling and suppression of noxious and pathogenic organisms (Doran and Zeiss 2000). High levels of organic matter content in the soil ensure greater microbial activity and greater soil nitrogen-supplying power than in inorganically managed soil. Application of vermicompost promotes greater microbial biomass and diversity in the soil. Earthworm burrows lined with earthworm casts act as an excellent medium for harboring nitrogen-fixing bacteria in the soil. Bacterial growth, particularly of gram-negative bacteria, was increased significantly after application of high doses of vermicompost (Lazcano and Dominguez 2011). Liberal application of vermicompost increased the population of soil microbes such as *Azotobacter chroococcum*, *Azotobacter vinelandii*, *Bacillus stearothermophilus*, *Bacillus megaterium*, *Pseudomonas putida*, and *Bacillus subtilis*. Ansari and Ismail (2012) reported that vermicompost-stimulated microbial activities transformed the nitrogen into mucoprotein that prevents leaching of nitrogen in the soil and also lowers the C/N ratio. Vermicompost-rich soil enhances the breakdown of organic matter in soil and release of plant nutrients. It also secretes plant growth hormones and beneficial soil microbes that increase biological resistance in crop plants and contribute toward crop productivity. Gopalakrishnan et al. (2011) reported that coconut leaf vermicompost reserved a diverse pool of plant-beneficial microorganisms that contained significantly high populations of fungi, free-living nitrogen fixers, phosphate solubilizers, fluorescent pseudomonads, and silicate solubilizers that ensure effective nutrient mineralization, improving the water-holding capacity of the soil, enhancing the soil cation-exchange capacity, and inducing systemic and acquired resistance in plants toward pests and diseases, thus increasing the fitness of the soils for sustainable production.

Manivannan et al. (2009) studied the efficacy of vermicompost in comparison to inorganic fertilizers on the physicochemical and biological properties of the soils. Microbial activity and total population were significantly enhanced in both soils treated with vermicompost and those with vermicompost supplemented with NPK fertilizer. Their findings showed that NPK application significantly reduced the total microbial population and activity and to values even less than the values observed in the absolute control field. Organic residues were found to increase the size, biodiversity, and activity of the soil microbial population in soil (Albiach et al. 2000). Zink and Allen (1998) observed that regular use of the compost increased the microbial population as well as its activity in the soil. Goyal et al. (1999) also observed that soil organic matter level, soil microbial biomass, and activities were increased with the use of organic fertilizer compared to inorganic fertilizers. The greater pore volume in vermicompost-amended soils increased the availability of both water and nutrients to microorganisms in the soils (Scott et al. 1996). Use of inorganic fertilizer had resulted in reduction of microbial populations that might be caused by reduced organic carbon content in the soil, compaction, decreased porosity, reduced water-holding capacity, reduced micronutrients, and increased acidity in soil.

### 4.5.3 Beneficial Effect of Vermicompost on Suppression of Plant Diseases and Pests

Vermicompost can suppress a wide range of soil-borne microbial diseases, insect pests, and plant parasitic nematodes, especially plant pathogens such as *Pythium*, *Rhizoctonia*, *Verticillium*, and *Plectosporium*. Edwards et al. (2006) stated that disease suppression might be caused by biological suppressive agents in the vermicompost. Arancon et al. (2007) observed significant reduction in the populations of spider mites (*Tetranychus urticae*), mealy bugs (*Pseudococcus* sp.), and aphids (*Myzus persicae*) after the addition of food waste vermicompost to several vegetable crops (tomato, cucumber, cabbage, bush beans, eggplant). Use of sufficient vermicompost in soil enhances the microbial biomass and changes the diversity and abundance of soil-borne fauna (Gunadi et al. 2002; Arancon et al. 2006), and thus a diverse range of organisms may act as biocontrol agents. Vermicompost also shows disease suppression through inhibition of the growth of pathogens such as *Rhizoctonia*, *Phythium*, and *Verticillium* (Atiyeh et al. 2002; Singh et al. 2008).

### 4.5.4 Beneficial Effect of Vermicompost on Suppression of Heavy Metal Toxicity

Vermicompost has the ability to bioaccumulate toxic compounds to decrease soil pollution (Kumar et al. 2015; Romero-Freire et al. 2015). Earthworms can bioaccumulate heavy metals such as cadmium (Cd), mercury (Hg), lead (Pb), copper (Cu), manganese (Mn), calcium (Ca), iron (Fe), and zinc (Zn) in tissues without affecting their physiology. In addition, *Lumbricus terrestris*, *Lumbricus rubellus*, and *Dendrobaena rubida* bioaccumulate lead (Pb) and cadmium (Cd) in their tissues (Sinha et al. 2010). *Azotobacter vinelandii* can fix nitrogen and also mobilize cadmium, mercury, and lead in the soil (Mary et al. 2015).

## 4.6 Limitation of Large-Scale Use of Vermicompost

Vermicompost is produced during the biological degradation of different organic wastes by earthworms and microorganisms. The physical, chemical, and biological characteristics of vermicompost largely depend on the quality of substrate composition, earthworm species used, production process, and the age of the vermicompost (Roberts et al. 2007; Warman and Anglopez 2010). Obtaining a regular supply of easily decomposable quality organic substrates is a serious challenge for growers. Sometimes it becomes a difficult process because of the nonavailability of continuous organic waste and water (Singh and Singh 2017). Again, earthworm mortality during vermicomposting poses threats to the production system. Maintaining appropriate temperature, pH, and moisture conditions can make the vermicompost production process more exhaustive and complicated.

Vermicomposting of biodegradable municipal solid wastes poses threats of harmful pathological load and heavy metal toxicity. Use of such vermicompost is restricted to gardens, parks, and roadside vegetation maintenance only. Huge amounts of biodegradable medical wastes are generated every year but suitable literature and research findings are still lacking for their wider utilization as vermicompost substrates.

Farmers mostly prefer to use the organic inputs of agricultural fields but their supplies are limited. Human excreta can be utilized as valuable organic input as a substitute for crop-based organic wastes for large-scale vermicomposting. However, only scanty research reports are available regarding the safe recycling of human excreta into vermicompost.

Earthworms are highly sensitive to temperature variation at the soil surface. They are found abundantly in both temperate and tropical soils. Sensitivity comparisons of tropical and temperate species have shown contradictory results with changes in soil and climate conditions (de Silva et al. 2009).

Easy availably of chemical fertilizers at subsidized rates is a matter of concern for the large-scale use of organic manure, particularly vermicompost, in crop fields because the mineralization rate and nutrient release patterns of vermicompost are much slower than those of chemical fertilizers. The visible effect of fertilizers becomes very prominent and is long lasting. Much effort is needed to convince the farming community about the benefits of vermicompost in soil health.

Several reports have suggested the positive impact of vermicompost on plant growth. However, a few reports have pointed out the negative impacts of vermicompost on crop production. Application of vermicompost in the nursery can reduce the germination of plants perhaps because high concentrations reduce aeration and porosity in the growing media. Similarly, increased salt concentrations as well as elevated concentrations of heavy metals and phytotoxic substances in vermicompost have detrimental effects on plants (Atiyeh et al. 2001). The use of immature vermicompost also inhibits seed germination and plant growth and causes root destruction (Abbasi et al. 2009). Sometimes the presence of heavy metals such as lead and cadmium in vermicompost may damage the soil system and contaminate parts of the plant and food chains (Godoi et al. 2014). Only very preliminary research results are available on these aspects. More heavy metal-specific and crop response research findings are required for future research directions.

## 4.7 Conclusion

With alarm about deteriorating soil health and chemicalization of the modern crop production system, awareness has been generated for liberal application of organic manures for restoration of soil fertility and reviving microbial activity to make the soil more healthy and life supporting for the sake of sustainable crop production.

Vermicompost has emerged as a promising organic fertilizer and alternative to inorganic fertilizers for successful crop growth and yield. The water-soluble components of vermicompost such as humic acid, growth regulators, vitamins, micronutrients, and beneficial microorganisms increase the availability of plant nutrients, resulting in higher yield and produce of better quality. Regular application of vermicompost improves the physical, chemical, and biological properties of the soil. Besides the supply of plant growth-promoting substances, the vermicompost carries beneficial soil microorganisms and thus improves soil biological function in promoting plant growth and soil health. In the current context of decreasing livestock populations and the scarcity of domestic organic manure, vermicomposting can be a potential technology for recycling available organic wastes as a source of quality organic manures to maintain good soil heath and to reduce dependence on nonrenewable resources. In conventional agriculture, extensive research has been carried out on plant response to applied fertilizers, but only scant information is available regarding the in-depth interaction of vermicompost application and plant response as well as soil health. More research is needed for microbial enrichment of vermicompost to increase the nutrient content, solubility, and mobility for availability throughout the growth period for higher yield, better quality, and sustainability in soil health and the ecosystem.

## References

Abbasi T, Gajalakshmi S, Abbasi SA (2009) Towards modeling and design of vermicomposting systems: mechanisms of composting/vermicomposting and their implications. Indian J Biotechnol 8:177–182

Acton DF, Gregorich LJ (1995) The health of our soils: toward sustainable agriculture in Canada. Agriculture Agri-Food Canada, CDR Unit, Ottawa

Albiach R, Canet R, Pomares R, Ingelmo F (2000) Microbial biomass content and enzymatic activities after the application of organic amendments to a horticultural soil. Bioresour Technol 75:43–48

Ansari AA, Ismail SA (2012) Role of earthworms in vermitechnology. J Agric Technol 8(2):403–415

Arancon NQ, Edwards CA, Bierman P (2006) Influences of vermicompost on field strawberries: effects on soil microbiological and chemical properties. Bioresour Technol 97:831–840

Arancon NQ, Edwards CA, Yardim EN, Oliver TJ, Byrne RJ, Keeney G (2007) Suppression of two-spotted spider mite (*Tetranychus urticae*), mealy bug (*Pseudococcus* sp.), and aphid (*Myzus persicae*) populations and damage by vermicompost. Crop Prot 26:29–39

Arancon NQ, Edwards CA, Babenko A, Cannon J, Galvis P, Metzger JD (2008) Influences of vermicomposts, produced by earthworms and microorganisms from cattle manure, food waste and paper waste, on the germination, growth and flowering of petunias in the greenhouse. Appl Soil Ecol 39:91–99

Atiyeh RM, Edwards CA, Subler S, Metzger JD (2001) Pig manure vermicompost as a component of a horticultural bedding plant medium: effects on physiochemical properties and plant growth. Bioresour Technol 78:11–20

Atiyeh RM, Arancon N, Edwards CA, Metzger JD (2002) The influence of earthworm-processed pig manure on the growth and productivity of marigolds. Bioresour Technol 81:103–108

Azarmi R, Giglou MT, Taleshmikail RD (2008) Influence of vermicompost on soil chemical and physical properties in tomato (Lycopersicum esculentum) field. Afr J Biotechnol 7(14):2397–2401

Bansal S, Kapoor KK (2000) Vermicomposting of crop residues and cattle dung with *Eisenia fetida*. Bioresour Technol 73:95–98

Basheer M, Agrawal OP (2015) Effect of some additives on vermicomposting of garden waste using *Eudrilus eugeniae*, an epigeic earthworm. World J Zool 10:153–160

Bazzoffi P, Pellegrini S, Rocchini A, Morandi M, Grasselli O (1998) The effects of urban refuse compost and different tractors tyres on soil physical properties, soil erosion and maize yield. Soil Tillage Res 48:275–286

Beilen VN (2016) Effects of conventional and organic agricultural techniques on soil ecology. Cent Dev Strategy 1:1–2
Bhatnagar RK, Palta RK (1996) Earthworm: vermiculture and vermicomposting. Kalyani Publishers, Ludhiana
Bhattacharya P (2004) Organic food production in India. Agrobios, Jodhpur
Biradar AP, Patil MB (2001) Studies on utilization of prominent weeds for vermiculturing. Indian J Weed Sci 33:229–230
Brink N (1995) Composting of food wastes with straw and other carbon sources for nitrogen catching. Acta Agric Scand B Soil Plant Sci 45:118–123
Bunemann EK, Schwenke GD, Van Zwieten L (2006) Impact of agricultural inputs on soil organisms: a review. Aust J Soil Res 44:379–406
Chang ST (1997) Microbial biotechnology—integrated studies on utilization of solid organic wastes. Resour Conserv Recycling 13:75–82
Chatterjee P, Metiya G, Saha N, Halder M, Muhkerjee D (2005) Effect of substrates and composting methods on the quality of compost. J Inter Academicia 9:66–75
de Silva PMCS, Pathiratne A, Van Gestel CAM (2009) Influence of temperature and soil type on the toxicity of three pesticides to Eisenia andrei. Chemosphere 76:1410–1415
Dominguez J (2004) State of the art and new perspectives on vermicomposting research. In: Edwards CA (ed) Earthworm ecology. CRC Press, Boca Raton, pp 401–424
Dominguez J, Edwards CA (2004) Vermicomposting organic wastes: a review. In: Shakir Hanna SH, WZA M (eds) Soil Zoology for sustainable development in the 21st century. IUCN, Cairo, pp 369–395
Doran JW, Zeiss MR (2000) Soil health and sustainability: managing the biotic component of soil quality. Appl Soil Ecol 15:3–11
Edwards CA, Arancon NQ, Greytak S (2006) Effects of vermicompost teas on plant growth and disease. BioCycle 47:28–31
Garg P, Gupta A, Satya S (2006) Vermicomposting of different types of waste using *Eisenia foetida*: a comparative study. Bioresour Technol 97:391–395
Gaur AC (1999) Microbial technology for composting of agricultural residues by improved methods. ICAR, New Delhi, p 78
Ghosh M, Chattopadhyay GN, Baral K (1999) Transformation of phosphorus during vermicomposting. Bioresour Technol 69:149–154
Girde U, Bhagat P, Lonkar V, Anmalwar A, Rangari DR (2016) Study of vermicasts from earthworms: a review study. Int J Res 3:384–387
Godoi PM, Souza Neta LC, Ferreira Fontes MP, Souza AN, Matos TC, Lima Sachdev R, Santos AV, Guarda Souza MO, Andrade MVAS, Paulo GMM, Ribeiro JN, Ribeiro AVFN (2014) An overview of the environmental applicability of vermicompost: from wastewater treatment to the development of sensitive analytical methods. Sci World J 2014:1–14
Gopalakrishnan S, Pande S, Sharma M, Humayun P, Kiran BK, Sandeep D (2011) Evaluation of actinomycete isolates obtained from herbal vermicompost for the biological control of *Fusarium* wilt of chickpea. Crop Prot 30:1070–1078
Goswami NN, Rattan RK (1992) Soil health-key to sustained agricultural productivity. Fertiliser News 37:53–60
Goyal S, Inubushi K, Kato S, Xu HL, Umemura H (1999) Effect of anaerobically fermented manure on the soil organic matter, microbial properties and growth of spinach under greenhouse conditions. Ind J Microbiol 39:211–216
Grdisa M, Grdisa K (2013) Earthworms: role in soil fertility to the use in medicine and as a food. Indones Scholar J 10:38–45
Gunadi B, Edwards CA, Arancon NA (2002) Changes in trophic structure of soil arthropods after the application of vermicomposts. Eur J Soil Biol 38:161–165
Gurav MV, Pathade GR (2011) Production of vermicompost from temple waste (Nirmalya): a case study. Univ J Environ Res Technol 2:182–192

Ibrahim MM, Mahmoud EK, Ibrahim DA (2015) Effects of vermicompost and water treatment residuals on soil physical properties and wheat yield. Int Agrophys 29:157–164
Ismail SA (1997) Vermicology: the biology of earthworms. Orient Longman, Chennai, p 92
Jeevendran S, Muthu MN, Kamalraj R, Poyyamoli G (2016) Estimation of zoomass productivity of epigeic earthworm species *Eudriluse ugeniae* fed with temple flower waste and cow dung mixtures. Int J Adv Res 4:403–409
Kale RD (1998) Earthworm: nature's gift for utilization of organic waste. In: Edward CA (ed) Earthworm ecology. CRC Press, Boca Raton, pp 355–376
Karmakar S, Gangopadhyay A, Brahmachari K, Bandyopadhyay PK (2009) Soil health management by applying vermicompost prepared from organic wastes. J Int Acad 13:412–417
Kumar T, Bhargava R, Prasad KSH, Pruthi V (2015) Evaluation of vermifiltration process using natural ingredients for effective wastewater treatment. Ecol Eng 75:370–377
Lazcano C, Dominguez J (2011) The use of vermicompost in sustainable agriculture: impact on plant growth and soil fertility. In: Miransari M (ed) Soil nutrients. Nova Science, New York, pp 1–23
Lemtiri A, Colinet G, Alabi T, Cluzeau D, Zirbes L, Haubruge E, Francis A (2014) Impacts of earthworms on soil components and dynamics: a review. Biotechnol Agron Soc Environ 18 (1):121–133
Lim SL, Wu TY, Lim PN, Shak KPY (2015) The use of vermicompost in organic farming: overview, effects on soil and economics. J Sci Food Agric 95:1143–1156
Manivannan S, Balamurugan M, Parthasarathi K, Gunasekaran G, Ranganathan LS (2009) Effect of vermicompost on soil fertility and crop productivity of beans (*Phaseolus vulgaris*). J Environ Biol 3:275–281
Manyuchi MM, Chitambwe T, Phiri A, Muredzi P, Kanhukamwe Q (2013) Effect of vermicompost, vermiwash and application time on soil physicochemical properties. Int J Chem Environ Eng 4:216–220
Marinari S, Masciandaro G, Ceccanti B, Grego S (2000) Influence of organic and mineral fertilizers on soil biological and physical properties. Bioresour Technol 72:9–17
Mary LCL, Sujatha R, Chozhaa AJ, Navas PMA (2015) Influence of organic manures (biofertilizers) on soil microbial population in the rhizosphere of mulberry (Morus indica). Int J Appl Sci Biotechnol 3(1):61–66
Masciandaro G, Bianchi V, Macci C, Doni S, Ceccanti B, Iannelli R (2010) Potential of on-site vermicomposting of sewage sludge in soil quality improvement. Desalin Water Treat 23(1–3):123–128
Ndegwa PM, Thompson SA (2001) Integrating composting and vermicomposting in the treatment and bioconversion of biosolids. Bioresour Technol 76:107–112
Nethra NN, Jayaprasad KV, Kale RD (1999) China aster (*Callistephus chinensis* L.) cultivation using vermicompost as organic amendment. Crop Res 17:209–215
Nirmalnath PJ, Biradar AP, Patil AB, Patil MB (2001) Vermicompost microflora as influenced by different crop residues. Karnataka J Agric Sci 14:68–69
Olk DC, Cassman KG (1993) Reduction of potassium fixation by organic matter in vermiculitic soils. In: Mulongoy K, Merckx R (eds) Soil organic matter dynamics and sustainability of tropical agriculture. Wiley, Chichester, pp 307–315
Oo AN, Iwai CB, Saenjan P (2013) Soil properties and maize growth in saline and nonsaline soils using cassava–industrial waste compost and vermicompost with or without earthworms. Land Degrad Dev 26:300–310
Padmavathiamma PK, Li LY, Kumari UR (2008) An experimental study of vermin biowaste composting for agricultural soil improvement. Bioresour Technol 99:1672–1681
Pagliai M, Guidi G, La Marca M (1980) Macro and micro morphometric investigation on soil–dextran interactions. J Soil Sci 31:493–504
Pandey S, Singh D (2004) Total bacterial and fungal population after chlorpyrifos and quinalphos treatments in groundnut (*Arachis hypogaea* L.) soil. Chemosphere 55:197–205

Pierzynski GM, Sims JT, Vance G (1994) Soils and environmental quality. Lewis Publishers, CRC Press, Boca Raton

Prabhu SR, Subramanian P, Biddappa CC, Bopaih BM (1998) Prospects of improving coconut productivity through vermiculture technology. Indian Cocon J 29:79–84

Pramanik P, Ghosh GK, Ghosal PK, Banik P (2007) Changes in organic-C, N, P and K and enzyme activities in vermicompost of biodegradable organic wastes under liming and microbial inoculants. J Bioresour Technol 98:2485–2494

Pulikeshi MB, Amoji SD, Shagoti UM, Biradar VA (2003) Seasonal variation in compostability and production of vermiprotein by *Eisenia foetida*. J Environ Biol 24:165–171

Raghavendra MA, Bano K (2001) Studies on the manorial value of vermicompost from different green leaves. Curr Res (UAS, Bangalore) 30:35–37

Rao S, Subba Rao A, Takkar PN (1996) Changes in different forms of K under earthworm activity. National Seminar on Organic Farming and Sustainable Agriculture, India, pp 9–11

Reddy BG, Reddy MS (1998) Effect of organic manures and nitrogen levels on soil available nutrient status in maize–soybean cropping system. J Indian Soc Soil Sci 46:474–476

Roberts P, Jones GE, Jones DL (2007) Yield Responses of wheat (Triticum aestivum) to vermicompost. Compost Sci Util 15:6–15

Romero-Freire A, Peinado FJ, Ortiz MD, Van Gestel CA (2015) Influence of soil properties on the bioaccumulation and effects of arsenic in the earthworm Eisenia Andrei. Environ Sci Pollut Res 22(19):15016–15028

Scott NA, Cole ET, Huffman SA (1996) Soil textural control on decomposition and soil organic matter dynamics. Soil Sci Soc Am J 60:1102–1109

Sharma N, Madan M (1983) Earthworm for soil health and population control. J Sci Ind Res 42:575–583

Sharpley AN, Syres JK (1977) Seasonal variations in casting activity and in the amounts and release to solution of phosphorous forms in earthworm casts. Soil Biol Biochem 9:227–231

Shweta, Sharma M (2003) Biomass and vermicompost production by the earthworm, *Lampito mauritii* in different organic wastes. J Appl Zool Res 14:98–100

Singh A, Sharma S (2002) Composting of a crop residues through treatment with microorganisms and subsequent vermicomposting. Bioresour Technol 85:107–111

Singh A, Singh GS (2017) Vermicomposting: a sustainable tool for environmental equilibria. Environ Qual Manag 27:23–40

Singh R, Sharma RR, Kumar S, Gupta RK, Patil RT (2008) Vermicompost substitution influences growth, physiological disorders, fruit yield and quality of strawberry (*Fragaria* × *ananassa* Duch.). Bioresour Technol 99:8507–8511

Singleton DR, Hendrix BF, Coleman DC, Whitemann WB (2003) Identification of uncultured bacteria tightly associated with the intestine of the earthworms *Lumbricus rubellus*. Soil Biol Biochem 35:1547–1555

Sinha RK, Herat S, Agarwal S, Asadi R, Carretero E (2002) Vermiculture technology for environmental management: study of action of earthworms *Eisenia fetida*, *Eudrilus eugeniae* and *Perionyx excavatus* on biodegradation of some community wastes in India and Australia. Environmentalist 22:261–268

Sinha RK, Herat S, Valani DB (2009) Earthworms: the 'unheralded soldiers of mankind' and 'farmer's friend' working day and night under the soil: reviving the dreams of Sir Charles Darwin for promoting sustainable agriculture. Am-Euras J Agric Environ Sci 5(S):1–55

Sinha RK, Agarwal S, Chauhan K, Valani D (2010) The wonders of earthworms and its vermicompost in farm production: Charles Darwin's 'friends of farmers', with potential to replace destructive chemical fertilizers from agriculture. Agric Sci 1(2):76–94

Sudhakar G, Christopher LA, Rangasamy A, Subbian P, Velayuthan A (2002) Effect of vermicompost application on the soil properties, nutrient availability, uptake and yield of rice: a review. Agric Rev 23:127–133

Talukdar MC, Das J, Goswami B (2001) Biowastes as source of vermicompost. J Agric Sci Soc NE India 14:171–175

Valdez-Perez MA, Fernández-Luqueno F, Franco-Hernandez O, Cotera LBF, Dendooven L (2011) Cultivation of beans *(Phaseolus vulgaris* L.) in limed or unlimed wastewater sludge, vermicompost of inorganic amended soil. Sci Hortic 28:380–387

Vasanthi D, Kumaraswamy K (1999) Efficacy of vermicompost to improve soil fertility and rice yield. J Indian Soc Soil Sci 47:268–272

Vitousek P, Aber J, Howarth R, Likens G, Matson P, Schindler D (1997) Human alteration of the global nitrogen cycle: sources and consequences. Ecol Appl 7:737–750

Warman PR, AngLopez MJ (2010) Vermicompost derived from different feedstock as a plant growth medium. Bioresour Technol 101:4479–4483

Yasir M, Aslam Z, Kim S, Lee S, Jeon CO (2009) Bacterial community composition and chitinase gene diversity of vermicompost with antifungal activity. Bioresour Technol 100:4396–4403

Zibiliske LM (1998) Composting of organic wastes. Lewis, Boca Raton

Zink TA, Allen MF (1998) The effects of organic amendments on the restoration of a disturbed coastal sage scrub habitat. Restor Ecol 6:52–58

Zucco MA, Walters SA, Chong SK, Klubek BP, Masabni JG (2015) Effect of soil type and vermicompost applications on tomato growth. Int J Recycl Org Waste Agric 4:135–141

# Chapter 5
# Impact of Agricultural Practices on Soil Health

**Elizabeth Temitope Alori, Aruna Olasekan Adekiya, and Kehinde Abodunde Adegbite**

**Abstract** Conventional agricultural systems such as the use of chemical fertilizers, mechanical tillage, and irrigation, although increasing crop production, cause soil erosion, loss of biodiversity, and decline in soil structure and soil organic matter. However, there is a need to increase crop production to meet the increasing demands for food by the growing human population without jeopardizing soil health. The present review discusses the significance of a healthy soil and the effects of some agricultural inputs and practices on soil health. Strategies to increase agricultural food production without jeopardizing soil health are also itemized.

**Keywords** Fertilization · Soil fertility · Agricultural practices · Soil management · Food production

## 5.1 Introduction

To improve or increase food and fibre production to meet the demands of the increasing human population, a variety of agricultural management processes are imposed on the soil ecosystem, including artificial inputs such as pesticides, fertilizers, herbicides, and tillage. These practices and inputs supplement or even 'substitute' for biological functions that are seen as inadequate or inefficient for achieving required levels of production. This approach distorts the natural balance of the soil ecosystem and may compromise the output of other environmental services (Kibblewhite et al. 2008), which to a large extent, results in loss of other ecosystem functions and thus renders the soil unhealthy.

A healthy soil is that which is physically, nutritionally, and biologically balanced, productive, and stable, and can withstand environmental impacts without loss of fertility, structure, and biological activity (Kibblewhite et al. 2008). A healthy soil is one that has continued capacity to function as a vital living ecosystem for sustaining

E. T. Alori (✉) · A. O. Adekiya · K. A. Adegbite
Crop and Soil Sciences Department, Landmark University, Omu-aran, Kwara State, Nigeria



B. Giri, A. Varma (eds.), *Soil Health*, Soil Biology 59,
https://doi.org/10.1007/978-3-030-44364-1_5

plants, animals, and humans (Natural Resource and Conservation Service 2009). However, these soil-based biological processes may become disturbed or altered by factors such as addition of agricultural inputs, improper land cultivation, and irrigation. More specifically, underuse, overuse, and adequate use of crop production inputs influence the health of any soil. Several authors have reported the effects of some agricultural inputs and practices on soil nutrient and plant nutrient composition and on the growth and yield of some crop plants and nontarget soil organisms (Adekiya et al. 2016; Hagner et al. 2019). This chapter therefore discusses the significance of a healthy soil and the effects of some agricultural inputs and practices on soil health. Strategies to increase agricultural food production without jeopardizing soil health are also presented.

## 5.2 Importance of a Healthy Soil

Soil is an ecosystem containing living organisms that require the basic necessities of life (food, shelter, water) to produce food, shelter, and fibre for humans. Healthy soil provides nutrients for plant growth, absorbs and holds rainwater for use during drier periods, filters and buffers potential pollutants from leaving our fields, and serves as a firm foundation for agricultural activities. Healthy soil gives us clean air and water, bountiful crops and forests, productive grazing lands, diverse wildlife, and beautiful landscapes (Burns et al. 2006).

For a soil to provide the foregoing services, it must be able to effectively and efficiently accommodate active and diverse populations of beneficial organisms, with minimum populations of plant pests and pathogens. It must contain high levels of relatively fresh residues that provide beneficial organisms with food. The decomposed organic matter content of a healthy soil must be very high as this will help to retain both water and readily leachable nutrients. The soil should contain low levels of toxic compounds such as soluble aluminum and only low to moderate concentrations of salt. A healthy soil can only support adequate levels of nutrients because excessive nutrients can make the crop more attractive to insect pests or can increase the threat of surface or subsurface water pollution. It has a sufficiently porous surface, with many pores connected to subsoil to permit easy entry by rainfall or irrigation water. Similarly, it has good tilth to allow plant roots to easily penetrate large volumes of soil. A healthy soil will provide physical stability and support for plants and regulate water; the soil helps control where rain, snowmelt, and irrigation water go. Water and dissolved solutes flow over the land or into and through the soil. A good soil sustains plant and animal life and filters and buffers potential pollutants.

## 5.3 Agriculture Practices and Their Impact on Soil Health

The form and extent of substitutions in agricultural practice are potential hazards to soil health. Some of the most frequent practices include the use of chemical pesticides (substituting for biological pest control), mechanical tillage (substituting for biological regulation of soil structure), and inorganic fertilizers (substituting for organically and biologically driven nutrient cycles). The bypass or modification of any particular biological function has been found to have significant consequences on other functions that were not been targeted (Sivaramanan and Kotagama 2019). However, the intensity of agricultural intervention varies enormously across different farming systems and thus may be expected to have both quantitatively and qualitatively different impacts on the soil health system. Different soils in different climatic and topographic situations may be more or less resilient to the introduction of intensive agriculture. Flat alluvial soils in areas without extremes of climate are less likely to degrade quickly compared with shallow soils on steep slopes where rainfall may be intense (Kibblewhite et al. 2008). Agricultural interventions such as the use of pesticides, powered tillage, and the use of inorganic sources of nutrients impact the biological communities of soils, damage habitats, and disrupt their functions to varying extents (Kibblewhite et al. 2008). The following are some agricultural practices with negative impacts on soil health.

### *5.3.1 Clearing of Native Vegetation Changes the Soil Water Balance*

Clearing of such vegetation also, to a great degree, changes negatively all the major soil properties whereby we describe its health. After a period of continuous cultivation, the soil reaches a new, dynamic, equilibrium. The consequences of this transition have been documented to include decline in soil organic matter content (Leigh and Johnston 1994); loss of ion-exchange capacity, which is concomitant with a decline in soil organic matter that reduces the capacity of the soil system to retain nutrients which would otherwise be leached to groundwater (Kibblewhite et al. 2008); soil erosion, as lack of surface vegetation leads to moderate to severe gully erosion; and the soil food web may also be substantially changed (Kibblewhite et al. 2008). In the Brazilian Amazon, large areas of forest have been converted to cattle pasture. Soil fauna studies showed that many of the main species of macrofauna present in forest soils are not found in the pastures. In particular, the earthworm community changed from one commonly characterized by about six endemic species to one dominated by the opportunistic exotic species *Pontoscolex corethrurus*. This is a species which, in contrast to many of the native worms, produces highly compact casts that have the effect of decreasing soil macroporosity, resulting in a surface layer which quickly becomes saturated and develops anaerobic conditions in the rainy season, which in turn stimulates methane emission and denitrification (Barros et al. 2004).

### 5.3.2 Mechanical Tillage

Mechanical tillage disrupts the spatial integrity of the soil fabric, particularly at meso- and macrofaunal scales. To some extent, tillage is intended to substitute for biological ploughing, and it is well known that earthworms are killed during this process (Landers et al. 2001). Soil levels of O, M, N, P, K, Ca, and Mg are reduced with increased tillage intensity (Adekiya et al. 2016).

### 5.3.3 Irrigation

Irrigation water contains dissolved mineral salts, with the concentration and composition varying depending on the water resource being used. Too much salt can reduce water infiltration in soils, thereby reducing crop production, and too little salt can also result in a chemically compacted soil (McKenzie 2010). Irrigated soils have reduced water entry and infiltration rates and increased runoff and soil losses (Mon et al. 2007). Similarly, a prolonged period of supplementary irrigation (10 years) caused soil sodication and alkalinisation (Mon et al. 2007).

### 5.3.4 Inorganic Fertilization

Inorganic fertilizers are synthetically made soil enhancers used to raise the level of nutrients found in the soil. Although inorganic or chemical fertilizers improve the growth and yield of crops in a relatively short period of time, there are certain disadvantages of using chemical fertilizers. Masto et al. (2007) reported increased concentrations of nitrogen, potassium, and phosphate in farmlands as a result of extensive use of inorganic fertilizers. Inorganic fertilizer made the soil more acidic and decreased the soil aggregates, so that the soil was more prone to erosion (Ozlu and Kumar 2018). The continuous use of chemical fertilizers results in groundwater pollution because chemical fertilizers are highly soluble and are therefore absorbed by the ground more rapidly than they are absorbed by the intended plants. Consequently, these chemicals react with clay to create hard layers of soil known as hardpan that hinder the penetration of plant roots into the soil (Sarfaraz 2019). In addition, chemical fertilizers destroy soil crumbs; the result is a highly compacted soil with reduced drainage and air circulation (Melkamu and Alemayehu 2017). The use of inorganic fertilizers also jeopardizes the health of soil beneficial microorganisms such as bacteria that fix nitrogen balance in the soil (Sarfaraz 2019). Inorganic P and N fertilizers decrease arbuscular mycorrhizal fungi (AMF) root colonization of pastures (Ryan et al. 2000). As reported by Melkamu and Alemayehu (2017), application of N or S fertilizers caused a decrease in microbial C, and this was consequently followed by a decrease in soil pH.

### 5.3.5 *Application of Chemical Pesticides*

Soil microbial diversity may be changed following pesticide use, and such changes may affect soil fertility (Lo 2010). Pesticides can contaminate soil so that it is toxic to a host of other organisms including beneficial soil organisms (Aktar et al. 2009). Soil beneficial microorganisms such as AMF improve water access and soil minerals for plants, improve drought tolerance, and help with resistance against pathogens; however, application of a chemical herbicide (glyphosate, or its metabolite AMPA) reduces the spore viability and root colonisation of AMF (Druille et al. 2013). Herbicides impair agricultural soil ecosystems (Nicolas et al. 2016). The persistence of herbicides in the soil pose a major threat to organisms living in the soil that are beneficial to crop production and support an important number of ecosystem services (Thiour-Mauprivez et al. 2019). Herbicides used by herbigation increase the possibility of water and soil contamination by these toxins (Noshadi and Homaee 2018). Also, Silambarasan et al. (2017) recorded lower microbial populations 5 days after application of herbicides. Table 5.1 outlines some common agricultural chemical inputs and their effects on soil life.

## 5.4 Strategies for Improved Soil Health

Soil health can be maintained or improved by engaging in agricultural practices that are based on the principles of minimizing soil disturbances, keeping the soil covered, maximizing plant diversity, and maximizing the period of living root growth, keeping in mind the soil natural characteristics such as texture, natural drainage class, and slope (Woodyard and Kladivko 2017). Managing soil for improved health demands a long-term commitment to using combinations of soil-enhancing practices (SARE 2012). The following agricultural practices can enhance improved soil health.

### 5.4.1 *Cover Cropping*

The practice of cover cropping could include the use of living vegetation or crop residue. Cover crop roots improve soil aggregation and reduce erosion. Cover crop residue also reduces the impact of raindrops on the soil surface and serves as a

**Table 5.1** Some chemical inputs and their effects on soil health

| S/N | Chemical input | Name of chemical input | Effects | References |
|---|---|---|---|---|
| 1 | Herbicide | Glyphosate | Toxic to the soil fungus *Aspergillus nidulans* | Nicolas et al. (2016) |
| 2 | Herbicide | Glyphosate | Reduces the spore viability of AMF | Druille et al. (2013) |
| 3 | Herbicide | Glyphosate | Increased frequency of the soil-borne fungus *Fusarium solani* | Sanogo et al. (2000) |
| | Fungicide | Mancozeb | Total fungi, actinomycetes, and *Pseudomonas* bacteria were significantly reduced | Magarey and Bull (2003) |
| 4 | Herbicide | Atrazine, pendimethalin | Lower microbial population | Silambarasan et al. (2017) |
| 5 | Herbicide | Pendimethalin | Significant reduction of soil microbe population | Nalini et al. (2013) |
| 6 | Herbicide | Triclopyr | Inhibits soil bacteria that transform ammonia into nitrite | Pell et al. (1998) |
| 7 | Insecticide | Methamidophos | Significantly decrease microbial biomass by 41–83% | Wang et al. (2006) |
| | Herbicides | Glyphosate | Reduces the growth and activity of free-living nitrogen-fixing bacteria in soil | Santos and Flores (1995) |
| 8 | Herbicide | 2,4-D | Inhibits the transformation of ammonia into nitrates by soil bacteria | Martens and Bremner (1993) |
| 9 | Fungicide | Butachlor | Reduced the population of *Azospirillum* and aerobic nitrogen fixers in non-flooded soil | Lo (2010) |
| 10 | Insecticide | Fenamiphos | Was detrimental to nitrification bacteria | Lo (2010) |

habitat and food source for soil microbes. As organisms decompose the residue, nutrients are released back into the soil (Woodyard and Kladivko 2017). The residue protects soils from moisture and temperature extremes and allows earthworms to adjust gradually to decreasing temperatures, reducing their mortality (SARE 2012). Intensive use of cover crops supplies nitrogen to the succeeding crops, soaks up leftover soil nitrates, increases soil organisms, and improves crop health as it reduces runoff, erosion, and soil compaction (SARE 2012).

### 5.4.2 *Crop Rotation*

The practice of crop rotation can help to manage soil and soil fertility, reduce soil loss, increase nutrients available for crop use, improve the workability of the soil, reduce soil crusting, increase water available for plants, reduce erosion and sedimentation, and recycle nutrients in the soil, hence improving soil health (NRSC 2009). However, a strong strategy for long-term resiliency is to increase plant diversity in the rotation system (Woodyard and Kladivko 2017).

### 5.4.3 *Conservation Tillage*

Reducing tillage to either no-till or strip-till minimizes disruptions to soil aggregates by not constantly breaking them up. Minimal tillage maintains natural aggregates and helps prevent loose soil particles from washing or blowing away easily. Residue decomposes more slowly under a reduced tillage system because fewer aggregates are broken up with less intensive tillage and less organic matter is therefore exposed to decomposition (Adekiya et al. 2016; Woodyard and Kladivko 2017). More reduced tillage also can make soil temperatures slightly cooler, and the lower temperatures help organic matter accumulate because the residue is not broken down as quickly. Reducing tillage can increase soil organism diversity and activity. Reduced tillage does not disrupt earthworm burrowing and helps protect the network created by mycorrhizal fungi that connects them to their host plant (Woodyard and Kladivko 2017). Leaving residue on the soil surface also acts as a barrier against raindrops and wind that could cause erosion (Woodyard and Kladivko 2017). Excessive tillage destroys the food sources and microniches on which beneficial soil organisms depend (SARE 2012).

### 5.4.4 *Use of Microbial Inoculants*

Microbial inoculants are involved in many natural biological and chemical processes in the ecosystem such as nutrient cycling and biological control of pathogens, hence improving nutrient availability. Microbial inoculant application creates suitable conditions for the development of beneficial microorganism, thereby increasing biodiversity.

Inoculants also improve the physical properties of the soil such as better structure and aggregation of soil particles, reduced soil compaction, and increased spore spaces and water infiltration. The antioxidant properties of microbial inoculants promote decomposition of organic matter and increase humus content in the soil matrix and could, therefore, replace the use of chemicals in agriculture (Alori et al. 2017).

### 5.4.5 *Application of Organic Manure*

Application of manure increases soil organic carbon (SOC) concentration, which is effective in maintaining or restoring soil organic matter (SOM) (Ozlu et al. 2019). Manure significantly increases total nitrogen compared to fertilizer treatments, and nitrogen is the key to plant growth. Manure also helps keep soil pH in a healthy range for crops by increased soil organic carbon: more carbon means better soil structure (Ozlu and Kumar 2018). Plentiful amounts of organic materials are added from

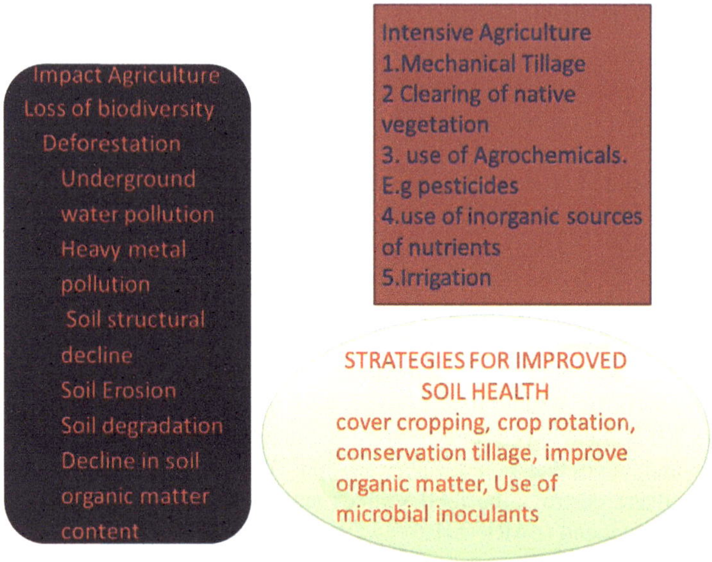


**Fig. 5.1** Some characteristics of intensive agriculture, the impact on soil health, and some strategies for improved soil health

cover crops and other crop residues as well as from off-field sources such as animal manures and composts (SARE 2012). As different organic materials have different effects on soil biological, physical, and chemical properties, it is therefore advisable to use a variety of sources. For instance, well-decomposed compost does not enhance soil aggregation in the short run, but dairy cow manure rapidly stimulates soil aggregation (SARE 2012). In Fig. 5.1, some characteristics of intensive agriculture with negative impacts on soil health and some strategies for improved soil health are itemized.

## 5.5 Benefits of Improved Soil Health Practices

The main perceived benefits driving the adoption of reduced or even zero tillage regimes are improved water and soil conservation, consequent on improved soil protection from the retained crop residues as well as reduced costs in terms of fuel (Kibblewhite et al. 2008). In no-till, however, the enhanced activity of the macrofaunal engineers in soil structure modification 're-substitutes' for the withdrawal of intensive tillage. The origin of changes to the water regime under no-tillage, such as reduced runoff, increased infiltration, and storage, are significantly physical in origin, but results of food web studies show that enhanced activity of the macrofaunal ecosystem engineers also plays a substantial part.

## 5.6 Conclusion

Soil health is related to functional capacity rather than actual service outputs. An integrative approach is also essential for the assessment of soil health. It is not feasible to assess soil health directly on the basis of its delivery of different ecosystem services. Most conventional agricultural practices pose significant risks to the environment. It is therefore important to balance the ecosystem functions in such a way as to secure the target of agricultural production without compromising other ecosystem functions with respect to both present and future needs.

## References

Adekiya AO, Agbede TM, Ojeniyi ST (2016) The effect of three years of tillage and poultry manure application on soil and plant nutrient composition, growth and yield of cocoyam. Exp Agric 52 (3):466–476

Aktar MW, Sengupta D, Chowdhury A (2009) Impact of pesticides use in agriculture: their benefits and hazards. Interdiscipl Toxicol 2(1):1–12

Alori ET, Dare MO, Babalola OO (2017) Microbial inoculants for soil quality and plant fitness. In: Lichtfouse E (ed) Sustainable agriculture review. Springer, Cham, pp 281–308

Barros ME, Grimaldi M, Sarrazin M, Chauvel A, Mitja D, Desjardins D, Lavelle P (2004) Soil physical degradation and changes in macrofaunal communities in Central Amazonia. Appl Soil Ecol 26:157–168

Burns RG, Nannipieri P, Benedetti A, Hopkins DW (2006) Defining soil quality. In: Bloem J, Hopkins DW, Benedetti A (eds) Microbiological methods for assessing soil quality. CAB International, Wallingford, pp 15–22

Druille M, Cabello MN, Omacini M, Golluscio RA (2013) Glyphosate reduces spore viability and root colonization of arbuscular mycorrhizal fungi. Appl Soil Ecol 64:99–103

Hagner M, Mikola J, Saloniemi I, Saikkonen K, Helander M (2019) Effects of a glyphosate-based herbicide on soil animal trophic groups and associated ecosystem functioning in a northern agricultural field. Sci Rep 9(1):8540

Kibblewhite MG, Ritz K, Swift MJ (2008) Soil health in agricultural systems. Philos Trans R Soc Lond B Biol Sci 363(1492):685–701

Landers JN, De C Barros GS-A, Manfrinato WA, Weiss JS, Rocha MT (2001) Environmental benefits of zero-tillage in Brazilian agriculture—a first approximation. In: Garcia TL, Benites L, Martinez VA (eds) Conservation agriculture: a worldwide challenge. XUL, Cordoba, pp 317–326

Leigh RA, Johnston AE (1994) Long-term experiments in agricultural and ecological sciences. In: Proceedings of a conference to celebrate the 150th anniversary of Rothamsted Experimental Station. CAB International, Wallingford

Lo C (2010) Effect of pesticides on soil microbial community. J Environ Sci Health Part B 45:348–359

Magarey RC, Bull JI (2003) Effect of the dithiocarbamate fungicide mancozeb on sugarcane growth and soil biology in yield decline affected soils. Proc Aust Soc Sugar Cane Technol 25:1–15

Martens DA, Bremner JM (1993) Influence of herbicides on transformations of urea nitrogen in soil. J Environ Sci Health Part B 28(4):377–395

Masto RE, Chhonkar PK, Singh D, Patra AK (2007) Soil quality response to long-term nutrient and crop management on a semi-arid Inceptisol. Agric Ecosys Environ 118(1):130–142

McKenzie RH (2010) Agricultural soil compaction: causes and management. https://www.agriculture.alberta.ca
Melkamu YB, Alemayehu M (2017) Impact of crop production inputs on soil health: a review. Asian J Plant Sci 16(3):109–131
Mon R, Irurtia C, Fernando BG, Pozzolo O, Bellora MF, Rivero D, Bomben M (2007) Effects of supplementary irrigation on chemical and physical soil properties in the rolling pampa region of Argentina. Cienc Invest Agrar 34:187–194
Nalini K, Muthukrishan P, Chinnusamy C, Janaki P (2013) Response of soil microflora in herbicide residue of winter irrigated cotton. Crop Res 45:268–271
Nicolas V, Oestreicher N, Vélot C (2016) Multiple effects of a commercial Roundup® formulation on the soil filamentous fungus *Aspergillus nidulans* at low doses: evidence of an unexpected impact on energetic metabolism. Environ Sci Pollut Res 23(14):14393–14404
Noshadi E, Homaee M (2018) Herbicides degradation kinetics in soil under different herbigation systems at field scale. Soil Tillage Res 184:37–44
NRSC (Natural Resources Conservation Service) (2009) Rotations for soil fertility: Small scale solutions for your farm. https://www.nrcs.usda.gov/Internet/FSE_DOCUMENTS/stelprdb1167375
Ozlu E, Kumar S (2018) Response of soil organic carbon, pH, electrical conductivity, and water stable aggregates to long-term annual manure and inorganic fertilizer. Soil Sci Soc Am J 82 (5):1243–1251
Ozlu E, Sandhu SS, Kumar S, Arriaga FJ (2019) Soil health indicators impacted by long-term cattle manure and inorganic fertilizer application in a corn-soybean rotation of South Dakota. Sci Rep 9(1):11776
Pell M, Stenberg O, Torstensson L (1998) Potential denitrification and nitrification tests for evaluation of pesticide effects in soil. Ambio 27(1):24–28
Ryan MH, Small DR, Ash JE (2000) Phosphorus controls the level of colonisation by arbuscular mycorrhizal fungi in conventional and biodynamic irrigated dairy pastures. Anim Prod Sci 40:663–670
Sanogo S, Yang XB, Scherm H (2000) Effects of herbicides on *Fusarium solani* f. sp. glycines and development of sudden death syndrome in glyphosate-tolerant soybean. Dis Control Pest Manag 90(1):57–66
Santos A, Flores M (1995) Effects of glyphosate on nitrogen fixation of free-living heterotrophic bacteria. Lett Appl Microbiol 20(6):349–352
SARE (Sustainable Agriculture Research & Education) (2012) Enhancing biota and improving soil health. https://www.sare.org/Learning-Center/Books/Manage-Insects-on-Your-Farm/Text-Version/Putting-it-All-Together/Enhancing-Biota-and-Improving-Soil-Health
Sarfaraz I (2019) The effects of chemical fertilizers on soil. Hunker. https://www.hunker.com/13427782/the-effects-of-chemical-fertilizers-on-soil
Silambarasan M, Sangli K, Kumar V, Sathyamoorthy NK, Dhananivetha M, Kathiravan V (2017) Effect of drip herbigation on native microbial population in maize. J Pharmacognosy Phytochem 6(5):2696–2698
Sivaramanan S, Kotagama SW (2019) Study on interconnected nature of man-made environmental problems and discovery of keystone environmental problems. In 5th World Congress on Environmental Science, Toronto, Canada
Thiour-Mauprivez C, Martin-Laurent F, Calvayrac C, Barthelmebs L (2019) Effects of herbicide on non-target microorganisms: towards a new class of biomarkers? Sci Total Environ 684:314–325
Wang MC, Gong M, Zang HB, Hua XM, Yao J, Pang YJ, Yang YH (2006) Effect of methamidophos and urea application on microbial communities in soils as determined by microbial biomass and community level physiological profiles. J Environ Sci Health Part B 41:399–341
Woodyard J, Kladivko E (2017) Four strategies to improve your field's soil health. Purdue Extension, Purdue Agronomy, West Lafayette

# Chapter 6
# Contribution of Biochar in Improving Soil Health

**A. O. Adekiya, T. M. A. Olayanju, S. W. Ejue, E. T. Alori, and K. A. Adegbite**

**Abstract** Soils are the premise for agriculture and the medium in which almost all food-generating plants grow and as such should be kept healthy. Healthy soils produce healthy crops that in turn nourish humans and animals. Good management practices are very essential in order to maintain soil health, and one of these practices is application of biochar. Biochar provides a unique opportunity to improve soil fertility and nutrient-use efficiency using locally available and renewable materials in a sustainable way. Application of biochar to the soil leads to several interactions mainly with the soil physical, chemical and biological properties to produce a healthy soil. Due to the unique properties of biochar, which include high concentrations of organic carbon, high porosity, large surface area and presence of micropores, improvement in soil physical and hydraulic properties including soil structure, aggregation, bulk density and water holding capacity would be expected following incorporation into soils. Biochar also improves chemical soil properties by increasing soil pH, cation exchange capacity, base saturation, exchangeable bases, and organic carbon content as well as decreases Al saturation in acid soils and reduces nitrogen leaching, thereby reducing fertilizer and lime requirements and maintaining a healthy soil. Changes to both soil physical and chemical properties as a result of biochar ultimately affect the biological properties of the soil by providing microbes with a more favourable habitat. Also, because of its sustainability and affordability, biochar can be used in soil remediation.

**Keywords** Biochar · Soil health · Soil physical properties · Soil chemical properties · Soil biological properties · Soil remediation

---

A. O. Adekiya (✉) · T. M. A. Olayanju · S. W. Ejue · E. T. Alori · K. A. Adegbite
Landmark University, Omu-Aran, Kwara State, Nigeria



B. Giri, A. Varma (eds.), *Soil Health*, Soil Biology 59,
https://doi.org/10.1007/978-3-030-44364-1_6

## 6.1 Introduction

According to Blanco-Canqui and Lal (2008), about one-third of the world's soils are experiencing a decrease or total loss in productivity since 1970. Therefore, the urgent need to safeguard the soil (which is one of the world's most valuable resources) in order to enhance food security cannot be overemphasized, this is as a result of the ever-increasing demand for food for the fast growing world population which is an ongoing challenge in the world (Godfray et al. 2010). Soil is a critical resource and the way in which it is managed and maintained can improve or degrade its health and quality. According to FAO, it is estimated that 95% of our food is directly or indirectly produced on soils; consequently we can say that soils are the premise for agriculture and the medium in which almost all food-generating plants grow and as such should be kept healthy. Healthy soils produce healthy crops that in turn nourish humans and animals; they are the foundation for profitable, productive and environmentally sound agricultural systems.

Soil health was defined by Doran and Zeiss (2000) as the capacity of soil to function as a vital living system, within ecosystem and land-use boundaries, to sustain plant and animal productivity, maintain or enhance water and air quality, and promote plant and animal health. Healthy soils improve infiltration and water use efficiency, prevent compaction and erosion, recycle nutrients and favour natural biological processes. The productivity of arable systems depends on soil health that is reflected by biotic and abiotic indicators such as the soil organic matter, nutrient status, moisture, and pH that are largely influenced by management practices (Atkinson et al. 2005; Karlen et al. 2003). It can indeed be directly linked to food quality and quantity. A healthy soil provides many functions that support plant growth, including nutrient cycling, biological control of plant pests, and regulation of water and air supply.

These functions are influenced by the interrelated physical, chemical and biological properties of soil, many of which are sensitive to soil management and maintenance practices. Good management practices are very essential in order to maintain soil health and many of these practices are being practiced as well as new ones are being adopted. One of these practices is the incorporation of biochar into the soil.

Biochar is a carbon-rich product obtained from the thermal conversion of biomass (crop residues, wood material, manures and other agricultural wastes) in an oxygen-limited environment (pyrolysis) (Lehmann and Joseph 2009). During pyrolysis, biomass is converted to char (Fig. 6.1), combined gas (mixture of $H_2$, $CO$, $CH_4$ and $CO_2$) and bio-oil with heat energy in the absence of $O_2$. Biochar contains high concentrations of carbon that can be rather recalcitrant to decomposition, so it may stably sequester carbon (Glaser et al. 2002). The type of feedstock used in biochar production influences the efficiency of carbon conversion of biomass to biochar but is not significantly affected by the pyrolysis temperature (within 350–500 °C common for pyrolysis). Addition of biochar to the soil affects soil health (Paz-Ferreiro and Fu 2013; Chintala et al. 2014). In this chapter, we discuss the contribution of biochar in maintaining soil health (Fig. 6.1).

**Fig. 6.1** Biochar from wood material after pyrolysis

## 6.2 Effect of Biochar on Soil Health

Soil health is essential for sustainable agricultural fertility and productivity and it encompasses, chemical, physical and biological properties maintaining the functions of both natural and managed ecosystems (Enriqueta-Arias et al. 2005; Kumar et al. 2014). Biochar provides a unique opportunity to improve soil fertility and nutrient-use efficiency using locally available and renewable materials in a sustainable way. Adoption of biochar management does not require new resources, but makes more efficient and more environmentally conscious use of existing resources. Application of biochar to the soil leads to several interactions mainly with the soil physical, chemical and biological properties (Fig. 6.2) as well as soil matrix, soil microbes, and plant roots (Lehmann and Joseph 2009). These interactions depend on several factors like type and chemical composition of biomass, methods of biochar preparation, physical aspect of biochar and soil environmental condition mainly soil types, soil temperature and moisture (Fig. 6.2).

### *6.2.1 Biochar and Soil Physical Properties*

Soil physical properties directly affect the soil productivity for crop production by determining water holding capacity, aeration and soil strength limitations for root activity (Benjamin et al. 2003). Soil having good structure, porosity, hydraulic conductivity, bulk density and strength provide good medium for growth to beneficial microorganisms, better nutrient and water movement into the soil profile, higher nutrient and water retention and more root growth ultimately providing higher

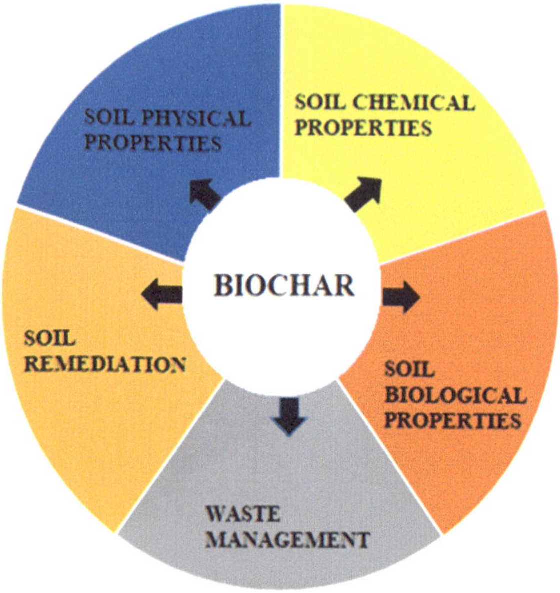


**Fig. 6.2** A schematic diagram showing the contribution of biochar to soil health

yield as compared to degraded soil having poor physical properties (Malkawi et al. 1999). Therefore, soil physical properties need to be maintained as well as improved. Due to the unique physical properties of biochar which include high concentrations of organic carbon, high porosity, large surface area and presence of micropores, improvement in soil physical and hydraulic properties including soil structure, aggregation and water holding capacity would be expected following incorporation into soils (Mukherjee et al. 2011; Chintala et al. 2014). These properties of biochar will potentially alter surface area, pore size distribution, bulk density, water retention, hydraulic conductivity, porosity and penetration resistance of the soil. Liang et al. (2006) confirmed this by stating that the incorporation of biochar can enhance specific surface area up to 4.8 times that of the adjacent soils.

The effect of biochar on soil physical properties according to Burrell et al. (2016) can be divided into direct and indirect effects. The porous nature of biochar has a direct effect on soil physical properties.

Bulk density is an indicator of soil compaction and soil health. It affects rooting depth and its restriction, soil aeration, infiltration, available water, plant nutrient availability, and activity of soil microorganism, which influence key soil processes and productivity (Alghamdi 2018). Therefore, biochar, being a porous material when added to the soil, increases its porosity and therefore reduces bulk density (Adekiya et al. 2019). Hseu et al. (2014) reported that the change in porosity with biochar-treated soils was as a result of formation of macropores and rearrangement of soil particle. An experiment conducted by Adekiya et al. (2019) showed that soil bulk density decreased from 1.49 to 1.20 g $cm^{-3}$ by applying 25 t $ha^{-1}$ biochar and 1.49 to 1.11 g $cm^{-3}$ with 50 t $ha^{-1}$ wood biochar. Biochar also has significant effect on soil porosity, which is usually the reverse of bulk density. The increase in porosity due to biochar may be as a result of biochar still retaining the cell wall structure of the biomass feedstock (Yadav et al. 2018). Compared with the control, Adekiya et al. (2019) found an increase of 32.6 and 24.9% with the application of 50 t $ha^{-1}$ and 25 t $ha^{-1}$ wood biochar, respectively. Mean weight diameter (MWD) indicates prevalence of larger and more stable aggregates and therefore is an index of soil aggregate stability and soil quality and health (Amezketa 1999; Arshad and Coen 1992). Biochar had been reported to increase this soil property. This quality of biochar is due to increase in binding organic substances from the biochar thereby improving the inter-particular aggregate cohesion among the soil particles (Aggelides and Londra 2000; Dexter et al. 2008). The increase in soil aggregate stability with biochar application could be due to high carbon (C) associated with biochar (Alghamdi 2018). The C molecules form bonds with the oxides and the organic matter (OM) serves as food for soil microorganism making the environment favourable for them. The microorganisms can also secrete polysaccharides which increase soil aggregation (Angers et al. 1993).

Biochar applied soils also have its stake in infiltration rate of the soil. Biochar applied soils increased infiltration rate compared with the control. This could be as a result of more pores created in the soil matrix as a result of biochar application because biochar is very porous. Studies by Novak et al. (2016) and Prober et al. (2014) showed that infiltration rate increased following application of biochar.

Tensile strength is a parameter of the soil strength and refers to the inherent ability of soil to resist the disruptive forces that cause fracture or rupture of the soil. Changes in tensile strength, similar to penetration resistance, can influence soil tillability, seedling emergence, root growth and other soil processes (Canqui 2017). Canqui (2017) reported that biochar application can, in general, reduce tensile strength by 42–242% regardless of soil textural class. However, this can only be effective when application is made at a higher rate of >50 Mg $ha^{-1}$ (Chan et al. 2007).

Biochar also affects the soil hydrological properties such as moisture content, water retention and hydraulic conductivity, which invariably according to Mukherjee and Lal (2013) are related to porosity, bulk density and aggregate stability. Biochar increased the moisture content/retention capacity of the soil because of the low bulk density and high porosity of biochar-applied soils thereby increasing the spaces where water could be retained (Adekiya et al. 2019). A long-term column study indicated that biochar-amended Clarion soil retained up to 15% more water, and 13 and 10% more water retention at −100 kPa and −500 kPa soil matric potential, respectively, compared to unamended controls (Laird et al. 2010). Piccolo et al. (1996) demonstrated that coal-derived humic acid substances can increase water retention, available water capacity and aggregate stability of inherently degraded soils.

Comparing with fertilizer application, Peng et al. (2011) reported that biochar amendment to a typical soil Ultisol resulted in better crop growth. Though the application of biochar improves soil physical properties, this is dependent on the pyrolysis conditions and type of biomass used, soil type where biochar is applied and the rate of biochar application.

Addition of biochar to soils may also improve soil physical properties such as soil structure indirectly by providing improved habitat for soil microorganisms (Pietikainen et al. 2000; Van Zwieten et al. 2009a, b), through favourable association with soil organic matter and improved aggregation (Lehmann et al. 2011; Fletcher et al. 2014), or by improving plant growth thereby enhancing rhizosphere effects (Joseph et al. 2010).

### *6.2.2 Biochar and Soil Chemical Properties*

It was observed that application of biochar improves chemical soil properties by increasing soil pH, cation exchange capacity, base saturation, exchangeable bases, organic carbon content as well as decreases Al saturation in acid soils and reducing nitrogen leaching, thereby reducing fertilizer and lime requirements (Glaser et al. 2002; Van Zwieten et al. 2010). Black C has been found (Liang et al. 2006) to affect nutrient retention and plays a key role in a wide range of biogeochemical processes in soils, especially for nutrient cycling. Anthrosols rich in biochar were found to maintain high cation availability (Lima et al. 2002) compared with adjacent forest soils with similar mineralogy despite high leaching conditions in humid tropical Amazonia. Such greater cation could be according to Liang et al. (2006) as a result of a higher charge density per unit surface area which means a higher degree of oxidation of SOM; or as a result of a higher surface area for cation adsorption sites, or a combined effect of both. Glaser et al. (2003) suggested the oxidation of the aromatic C and formation of carboxyl groups to be the main reason for the observed high cation exchange capacity (CEC). Several studies have showed that due to biochar amendments, soil pH increased especially in acidic soil, with greater increase observed in sandy and loamy soils than in clayey soils (Tyron 1948; Yamato et al. 2006; Major et al. 2010a; Yuan and Xu 2010). This was attributed to the high alkaline nature and high base cation concentration of biochar which upon integration in the soil releases protons into the soil solution leading to reduction in soil acidity through proton consumption reaction and higher availability of $CaCO_3$. However, the ameliorating ability of biochar depends on pyrolytic parameter (higher pyrolytic temperature helps to produce alkaline pH), feedstock (Lin et al. 2012) and soil properties (Wang et al. 2014). Nelissen et al. (2012) also reported that incorporation of biochar to soil improves $NH_4^+$ immobilization and subsequently decreases nitrification which in turns conquers the discharge of $H^+$ concentration to the soil and relieves soil acidification. Also, based on the highly porous nature of biochar, large specific surface area, possession of organic materials of variable charge, when added to the soil, has the potential to increase soil CEC, base saturation and other nutrients that have correlation with cation exchange capacity (Glaser et al. 2002).

Studies have also showed that incorporation of biochar increased organic carbon and decreased nitrogenous fertilizer requirement; this is because biochar contains high concentrations of carbon that can be rather recalcitrant to decomposition, so it may stably sequester carbon (Glaser et al. 2002; Widowati et al. 2012). Biochar application elevates total C, organic C, total N, available P, and exchangeable cations like Ca, Mg, Na, and K increase, and Al decreases in soil (Chan et al. 2007; Chen et al. 2008; Major et al. 2010b; Van Zwieten et al. 2010). Major et al. (2010a) reported that nutrient uptake by plants was increased in biochar amended soil, with increase in plant yield and greater availability of Ca and Mg in soil. Also, as a result of the high surface area and high surface charge density of biochar most of the nutrients that are removed due to the harvest are recycled (Liang et al. 2006), and biochar increases the ability of soils to retain nutrients and plant available water and reduces leaching of nutrients and agricultural chemicals (Laird et al. 2010; Lehmann et al. 2003; Glaser et al. 2002). In literatures, biochar had been reported to be effective in adsorbing dissolved soluble nutrients such as ammonium (Lehmann et al. 2002), nitrate (Mizuta et al. 2004), phosphate (Beaton et al. 1960), and other ionic solutes (Radovic et al. 2001). The carboxylate groups found in black carbon provide cation exchange capacity (CEC), increase the O/C ratio, and are the primary source of biochar's high nutrient retention ability (Glaser et al. 2001). Also, biochar has been reported (Oya and Iu 2002; Iyobe et al. 2004) to act as a buffer for ammonia in soils and therefore have the ability to reduce ammonia volatilization from soils.

Several researchers have also revealed that biochar affects nitrogen (N) cycling in soil, which offers potential options for tightening the N cycle in agricultural ecosystems. Lehmann et al. (2003) and Steiner et al. (2008) reported that the use of biochar can improve the efficiency of nitrogen fertilizer, as biochar can reduce the loss of nitrogen and potassium that occurs through leaching (Widowati et al. 2011) due to many proposed mechanisms such as enhanced nutrient retention due to cation and anion exchange reactions, immobilization of N due to labile C fraction of biochar, adsorption of organic N on biochar etc. Also the increase in soil pH resulting from biochar addition enhances $N_2O$ reductase activity and therefore favours completion of $NO_3^-$ reduction to $N_2$ (from $N_2O$) (DeLuca et al. 2009; Van Zwieten et al. 2010). It also helps in the adsorption of $NH_4^+$ that prevents nitrification and denitrification (Chintala et al. 2014). Reduction of leaching loss and consequently higher fertilizer use efficiency should lead to a lower fertilizer requirement per unit yield and usually lower $N_2O$ emission. The immediate beneficial effects of biochar additions for nutrient availability are largely due to higher potassium, phosphorus, and zinc availability, and to a lesser extent, calcium and copper (Lehmann et al. 2003). Biochar was also found to enhance biological N fixation (BNF) biochar amended soils (Krishnakumar et al. 2014). The reason for the improved BNF is most likely a result of nutrient availability in soil (Lehmann et al. 2003) and stimulation of plant–microbe interactions (Nishio and Okano 1991), along with increased nitrogen levels in soil resulting in increased colonization of the host plant roots by arbuscular mycorrhizal fungi (Ishii and Kadoya 1994).

### 6.2.3 Biochar and Soil Biological Properties

The addition of biochar brings about changes to both soil physical and chemical properties (Joseph et al. 2009; Chintala et al. 2013; Major et al. 2010a) which will ultimately affect the biological properties of the soil. These changes which are directly influenced by the inherent properties of biochar (which are dependent on the nature of feedstock and pyrolysis conditions) may ultimately affect soil–plant–microbe interactions (Quilliam et al. 2013). The soil biota is vital for the functioning of soils providing many essential ecosystem services. Different effects on the soil biota are likely to occur from the addition of biochar to the soil.

There is a growing body of knowledge showing that the addition of biochar to the soil results to increase in microbial biomass, with significant changes in microbial community structure and enzyme activities (Steiner et al. 2008; Hammes and Schmidt 2009; Liang et al. 2010; Jin 2010; Chintala et al. 2014). Biochar has a role in changing soil microorganism abundance (Liang et al. 2010; Grossman et al. 2010); this was confirmed in a study by Domene et al. (2014) indicating that microbial abundance increased from 366.1 (control) to 730.5 μg C $g^{-1}$ after application of biochar at the rate of 30 t $ha^{-1}$. It was also reported by Graber (2009) that with increasing rate of biochar application maximum number of culturable colonies of general bacteria, *Bacillus* spp., yeasts and Trichoderma spp. were found.

Biochar improves soil physical and chemical environment thereby providing microbes with a more favourable habitat (Krull et al. 2010; Jaafar et al. 2014). Biochar-amended soil has more suitable pH for the growth of microbes, especially for fungal hyphae (Wuddivira et al. 2009). Owing to its highly porous nature, its high internal surface area and its ability to adsorb soluble organic matter, gases and inorganic nutrients, biochar may provide a highly suitable habitat for microbes to colonize, grow and reproduce particularly for bacteria, actinomycetes and arbuscular mycorrhizal fungi (Thies and Rillig 2009). Studies (Saito and Marumoto 2002; Warnock et al. 2007) showed that the pores of biochar may be a habitat for microbes where they are protected from being grazed upon by their natural predators. These microorganisms secrete polysaccharides which increase soil aggregation (Angers et al. 1993). Certain toxic compounds such as catechol that would otherwise inhibit microbial growth may get adsorbed on biochar surface causing increases in microbial abundance (Chen et al. 2009).

Joseph et al. (2010) indicated that most biochar has a high concentration of macropores that extends from the surface to the interior, and minerals and small organic particles might accumulate in these pores. Other positive effects that have been reported include: enhanced biological nitrogen fixation (Rondon et al. 2007), improved colonization of mycorrhizal fungi, earthworms showing preference to biochar amended soils (Van Zwieten et al. 2010), increased methane uptake (Karhu et al. 2011), potential catalyst in reducing nitrous oxide to nitrogen (Van Zwieten et al. 2009a, b). The relationship between biochar and the soil biota, and their implications on different soil processes have yet not been adequately described. At the moment, there is a wide gap in our knowledge of interactions between the soil biota and biochar. Biochar could affect soil fauna directly or indirectly (Thies and Rillig 2009). Indirectly, soil fauna could be affected by altered biotic resources. Directly, soil fauna could be influenced by ingesting biochar particles. This is the case for geophagous fauna, such as earthworms (Topoliantz and Ponge 2003, 2005).

### *6.2.4 Biochar and Waste Management*

Managing animal and crop wastes from agriculture poses a significant environmental burden that leads to pollution of ground and surface waters (Carpenter et al. 1998; Matteson and Jenkins 2007) affecting soil health. For more effective management and disposal of the crop residues and animal waste, conversion of organic waste to produce biochar is one viable option that can enhance natural rates of carbon sequestration in the soil, reduce farm waste and improve the soil quality (Srinivasarao et al. 2012). The beauty of using biochar for waste management is that not only can energy be obtained in the process of charring, but the volume and especially weight of the waste material is significantly reduced which is an important aspect, for example, in managing farm wastes (Cantrell et al. 2007).

A number of studies have shown that biochar has the potential to bind both organic and inorganic pollutants including pesticides from environment via chelate formation. Moreover, if biochar is produced from waste biomass, environmental pollution can be reduced considerably. As compared to biomass burning, the release of smoke and gases is low in pyrolysis process. So, it could be a good technology for us if we could utilize our municipal waste for biochar production.

### *6.2.5 Biochar and Soil Remediation*

Soil contamination by organic and inorganic substances is now a worldwide problem (Mench et al. 2010) and remediation method that is environmentally safe, economically viable and sustainable is imperative (Lone et al. 2015). In this regard, biochar is the answer because of its sustainability, affordability, adoptability and being environmentally friendly. It was reported (Kasozi et al. 2010) that biochar made from a variety of feedstocks has a strong sorption ability to different types of pesticides and other organic contaminants. Biochar has been demonstrated by several worker to have high surface area and porosity (Adekiya et al. 2019). As a result of its large surface area and porosity, biochar can adsorb organic or inorganic pollutants and thus reduce their bioavailability and toxicity in the sediments and soil (Bielska et al. 2018; Jin et al. 2011). Biochar has been found very useful for restoration and revegetation of mine tailings (Gwenzi et al. 2015). Biochar strongly sorbs salts and ameliorates salt stress effects on plants in agricultural, urban and contaminated soils (Thomas et al. 2013). Coconut charcoal was specifically reported (Cho et al. 1997) to be efficient in promoting oil polluted soil biodegradation. The biochar addition to contaminated soil and waste rock piles may increase soil pH, water holding capacity, and soil fertility, reduce the mobility of plant-available pollutant, and promote revegetation (Fellet et al. 2014; Kelly et al. 2014). Application of biochar alone or when combined with compost into acidic mining waste increased soil pH from 3.33 to 3.63 and 4.07 to 4.77, respectively, as well as organic matter content, base cations, and nitrate availability, but decreased bulk density, Al, Cd, Cu, Pb, Ni and Zn (Beesley et al. 2010; Kelly et al. 2014; Reverchon et al. 2015; Rodríguez-Vila et al. 2014).

## 6.3 Future Outlook

Going by the current state of knowledge, biochar appears to be indispensable soil amendments in improving the physical, chemical and biological properties of the soil and therefore soil's health. However, there should be more research towards the standardization of the amount of biochar to be applied with regards to the type of charred materials (feedstocks), soil types, time and temperature of pyrolysis. Research on biochar use in Africa is still in its infancy (Torres-Rojas et al. 2011); therefore, there should be more research on the use and adoptability of biochar by farmers of sub-Saharan Africa who are hitherto complaisant on the use of biochar for sustaining soil and increasing crop productivity.

## References


Adekiya AO, Agbede TM, Aboyeji CM, Dunsin O, Simeon VT (2019) Effects of biochar and poultry manure on soil characteristics and the yield of radish. Sci Hortic 243:457–446

Aggelides S, Londra P (2000) Effects of compost produced from town wastes and sewage sludge on the physical properties of a loamy and a clay soil. Bioresour Technol 71:253–259

Alghamdi AG (2018) Biochar as a potential soil additive for improving soil physical properties – a review. Arab J Geosci 11:766. https://doi.org/10.1007/s12517-018-4056-7

Amezketa E (1999) Soil aggregate stability: a review. J Sustain Agr 14:83–151

Angers DA, N'dayegamiye A, Ĉoté D (1993) Tillage-induced differences inorganic matter of particle-size fractions and microbial biomass. Soil Sci Soc Am J 57:512–516

Arshad MA, Coen GM (1992) Characterization of soil quality: physical and chemical criteria. Am J Altern Agric 7:25–31. https://doi.org/10.1017/S0889189300004410

Atkinson A, Black K, Dawson L (2005) Prospects, advantages and limitations of future crop production systems dependent upon the management processes. Ann Appl Biol 146:203–215

Beaton JD, Peterson HB, Bauer N (1960) Some aspects of phosphate adsorption by charcoal. Soil Sci Soc Am J 24(5):340–346

Beesley L, Moreno-Jiménez E, Gomez-Eyles JL (2010) Effects of biochar and green waste compost amendments on mobility, bioavailability and toxicity of inorganic and organic contaminants in a multi-element polluted soil. Environ Pollut 158:2282–2287

Benjamin JG, Nielsen DC, Vigil MF (2003) Quantifying effects of soil conditions on plant growth and crop production. Geoderma 116:137–148

Bielska E, Sisquella MA, Aldeieg M, Birch C, O'Donoghue EJ, May RC (2018) Pathogen-derived extracellular vesicles mediate virulence in the fatal human pathogen Cryptococcus gattii. Nat Commun 9(1):1556. https://doi.org/10.1038/s41467-018-03991-6

Blanco-Canqui H, Lal R (2008) Soil Erosion and food security. In: Principles of soil conservation and management. Springer, Dordrecht

Burrell LD, Zehetner F, Rampazzo N, Wimmer B, Soja G (2016) Long-term effects of biochar on soil physical properties. Geoderma 282:96–102

Canqui HB (2017) Review and analysis – soil physics & hydrology: biochar and soil physical properties. Soil Sci Soc Am J 81:687–711. https://doi.org/10.2136/sssaj2017.01.0017

Cantrell K, Ro K, Mahajan D, Anjom M, Hunt PG (2007) Role of thermochemical conversion in livestock waste-to-energy treatments: obstacles and opportunities. Ind Eng Chem Res 46:8918–8927

Carpenter SR, Caraco NF, Correll DL, Howarth RW, Sharpley AN, Smith VH (1998) Nonpoint pollution of surface waters with phosphorus and nitrogen. Ecol Appl 8:559–568

Chan KY, Van Zwieten L, Meszaros I, Downie A, Joseph S (2007) Agronomic values of greenwaste biochar as a soil amendment. Soil Res 45:629–634

Chen BL, Zhou DD, Zhu LZ (2008) Transitional adsorption and partition of nonpolar and polar aromatic contaminants by biochars of pine needles with different pyrolytic temperatures. Environ Sci Technol 42:5137–5143

Chen HL, Yao J, Wang F, Choi MMF, Bramanti E, Zaray G (2009) Study on the toxic effects of diphenol compounds on soil microbial activity by a combination of methods. J Hazard Mater 167:846–851

Chintala R, Schumacher TE, McDonald LM, Clay DE, Malo DD, Clay SA, Papiernik SK, Julson JL (2013) Phosphorus sorption and availability from biochars and soil biochar mixtures. Clean Soil Air Water 41:1–9

Chintala R, Owen R, Kumar S, Schumacher TE, Malo D (2014) Biochar impacts on denitrification under different soil water contents. World Cong Soil Sci 6:157–157

Cho BH, Chino H, Tsuji H, Kunito T, Nagaoka K, Otsuka S, Yamashita K, Matsumoto S, Oyaizu H (1997) Laboratory-scale bioremediation of oil-contaminated soil of Kuwait with soil amendment materials. Chemosphere 35:1599–1611

DeLuca TH, MacKenzie MD, Gundale MJ (2009) Biochar effects on soil nutrient transformations. In: Lehmann J, Joseph S (eds) Biochar for environmental management: science and technology. Earthscan, London, pp 251–270

Dexter A, Richard G, Arrouays D, Czyż E, Jolivet C, Duval O (2008) Complexed organic matter controls soil physical properties. Geoderma 144:620–627

Domene X, Mattana S, Hanley K, Enders A, Lehmann J (2014) Medium-term effects of corn biochar addition on soil biota activities and functions in a temperate soil cropped to corn. Soil Biol Biochem 72:152–162

Doran JW, Zeiss MR (2000) Soil health and sustainability: managing the biotic component of soil quality. Appl Soil Ecol 15:3–11

Enriqueta-Arias M, Gonzalez-Perez JA, Gonzalez-Vila FJ, Ball AS (2005) Soil health—a new challenge for microbiologists and chemists. Int Microbiol 8:13–21

Fellet G, Marmiroli M, Marchiol L (2014) Elements uptake by metal accumulator species grown on mine tailings amended with three types of biochar. Sci Total Environ 468–469:598–608

Fletcher AJ, Smith MA, Heinemeyer A, Lord R, Ennis CJ, Hodgson EM, Farrar K (2014) Production factors controlling the physical characteristics of biochar derived from phytoremediation willow for agricultural applications. Bioenergy Res 7:371–380

Glaser B, Haumaier L, Guggenberger G, Zech W (2001) The 'Terra Preta' phenomenon: a model for sustainable agriculture in the humid tropics. Naturwissenschaften 88(1):37–41

Glaser B, Lehmann J, Zech W (2002) Ameliorating physical and chemical properties of highly weathered soils in the tropics with charcoal - a review. Biol Fertil Soils 35(4):219–230

Glaser B, Guggenberger G, Zech W, Ruivo ML (2003) Soil organic matter stability in Amazonian dark earths. In: Lehmann J et al (eds) Amazonian dark earths: origin, properties, management. Kluwer Academic Publishers, Dordrecht, pp 141–158

Godfray HCJ, Beddington JR, Crute IR, Haddad L, Lawrence D, Muir JF, Pretty J, Robinson S, Thomas SM, Toulmin C (2010) Food security: the challenge of feeding 9 billion people. Science 327:812–818

Graber ER (2009) Biochar for 21st century challenges: carbon sink, energy source and soil conditioner. In: Conference Proceedings, Dahlia Gredinger International Symposium, Haifa, May 2009

Grossman JM, O'Neill BE, Tsai SM, Liang B, Neves E, Lehmann J, Thies JE (2010) Amazonian anthrosols support similar microbial communities that differ distinctly from those extant in adjacent, unmodified soils of the same mineralogy. Microb Ecol 60:192–205

Gwenzi W, Chaukura N, Mukome FND, Machado S, Nyamasoka B (2015) Biochar production and applications in sub-Saharan Africa: opportunities, constraints, risks and uncertainties. J Environ Manag 150:250–261

Hammes K, Schmidt WI (2009) Changes of biochar in soil. In: Lehmann J, Joseph S (eds) Biochar for environmental management: science and technology. Earthscan, London, pp 169–182

Hseu Z-Y, Jien S-H, Chien W-H, Liou R-C (2014) Impacts of biochar on physical properties and erosion potential of a mudstone slope land soil. Sci World J 2014:10. https://doi.org/10.1155/2014/602197. Article ID 602197

Ishii T, Kadoya K (1994) Effects of charcoal as a soil conditioner on citrus growth and vesicular-arbuscular mycorrhizal development. J Jap Soc Hortic Sci 63:529–535

Iyobe T, Asada T, Kawata K, Oikawa K (2004) Comparison of removal efficiencies for ammonia and amine gases between woody charcoal and activated carbon. J Health Sci 50(2):148–153

Jaafar NM, Clode PL, Abbott LK (2014) Microscopy observations of habitable space in biochar for colonisation by fungal hyphae from soil. J Integr Agr 13:483–490

Jin HY (2010) Characterization of microbial life colonizing biochar and biochar-amended soils. Ph. D. Dissertation, Cornell University

Jin HP, Choppala GK, Bolan NS, Chung JW, Chuasavathi T (2011) Biochar reduces the bioavailability and phytotoxicity of heavy metals. Plant Soil 348:439–451

Joseph S, Peacocke C, Lehmann J, Munroe P (2009) Developing biochar classification and test methods. In: Lehmann J, Joseph S (eds) Biochar for environmental management: science and technology. Earthscan, London, pp 107–126

Joseph SD, Camps-Arbestain M, Lin Y, Munroe P, Chia CH, Hook J, van Zwieten L, Kimber S, Cowie A, Singh BP, Lehmann J, Foidl N, Smernik RJ, Amonette JE (2010) An investigation into the reactions of biochar in soil. Aust J Soil Res 48:501–515

Karhu K, Mattila T, Bergstrom I, Regina K (2011) Biochar addition to agricultural soil increased CH4 uptake and water holding capacity – results from a short-term pilot field study. Agric Ecosyst Environ 140:309–313

Karlen D, Ditzler C, Andrews S (2003) Soil quality: why and how? Geoderma 114:145–156

Kasozi GN, Zimmerman AR, Nkedi-Kizza P, Gao B (2010) Catechol and humic acid sorption onto a range of laboratoryproduced black carbons (biochars). Environ Sci Technol 44:6189–6195

Kelly CN, Peltz CD, Stanton M, Rutherford DW, Rostad CE (2014) Biochar application to hardrock mine tailings: soil quality, microbial activity, and toxic element sorption. Appl Geochem 43:35–48

Krishnakumar S, Rajalakshmi AG, Balaganesh B, Manikandan P, Vinoth C, Rajendran V (2014) Review article: impact of biochar on soil health. Int J Adv Res 2(4):933–950

Krull E, Singh B, Joseph S (2010) Preface to special issue: proceedings from the 1st Asia-pacific biochar conference, 2009, gold coast, Australia. Aust J Soil Res 48:i–iv

Kumar S, Nakajima T, Mbonimpa EG, Gautam S, Somireddy UR, Kadono A, Lal R, Chintala R, Rafique R, Fausey N (2014) Long-term tillage and drainage influences on soil organic carbon dynamics, aggregate stability, and carbon yield. Soil Sci Plant Nutr 60:108–118

Laird DA, Fleming PD, Davis DD, Horton R, Wang B, Karlen DL (2010) Impact of biochar amendments on the quality of a typical Midwestern agricultural soil. Geoderma 158:443–449
Lehmann J, Joseph S (2009) Biochar for environmental management: an introduction. In: Lehmann J, Joseph S (eds) Biochar for environmental management: science and technology. Earthscan, London, pp 1–12
Lehmann J, Da Silva JP Jr, Rondon M, Da Silva CM, Greenwood J, Nehls T, Steiner C, Glaser B (2002) Slash-andchar: a feasible alternative for soil fertility management in the Central Amazon? In: Proceedings of the 17th World Congress of Soil Science, symposium no. 13, Paper no. 449. Bangkok (Thailand), pp 1–12
Lehmann J, da Silva JP Jr, Steiner C, Nehls T, Zech W, Glaser B (2003) Nutrient availability and leaching in an archaeological Anthrosol and a Ferralsol of the Central Amazon basin: fertilizer, manure and charcoal amendments. Plant Soil 249:343–357
Lehmann L, Rillig MC, Thies J, Masiello CA, Hockaday WC, Crowley D (2011) Biochar effects on soil biota - a review. Soil Biol Biochem 43:1812–1836
Liang B, Lehmann J, Solomon D, Kinyangi J, Grossman J, O'Neill B, Skjemstad JO, Thies J, Luizão FJ, Petersen J, Neves EG (2006) Black carbon increases cation exchange capacity in soils. Soil Sci Soc Am J 70:1719–1730
Liang B, Lehmann J, Sohi SP, Thies JE, O'Neill B, Trujillo L, Gaunt J, Solomon D, Grossman J, Neves EG, Luizão FJ (2010) Black carbon affects the cycling of non-black carbon in soil. Org Geochem 41:206–213
Lima HN, Schaefer CER, Mello JWV, Gilkes RJ, Ker JC (2002) Pedogenesis and pre-Colombian land use of "Terra Preta Anthrosols" ("Indian black earth") of Western Amazonia. Geoderma 110:1–17
Lin Y, Munroe P, Joseph S, Henderson R, Ziolkowski A (2012) Water extractable organic carbon in untreated and chemical treated biochars. Chemosphere 87:151–157
Lone AH, Najar GR, Ganie MA, Sofi JA, Ali T (2015) Biochar for sustainable soil health: a review of prospects and concerns. Pedosphere 25(5):639–653
Major J, Lehmann J, Rondon M, Goodale C (2010a) Fate of soilapplied black carbon: downward migration, leaching and soil respiration. Glob Change Biol 16:1366–1379
Major J, Rondon M, Molina D, Riha S, Lehmann J (2010b) Maize yield and nutrition during 4 years after biochar application to a Colombian savanna oxisol. Plant Soil 333:117–128
Malkawi A, Alawneh A, Abu-Safaqah O (1999) Effects of organic matter on the physical and the physicochemical properties of an illitic soil. Appl Clay Sci 14(5):257–278
Matteson GC, Jenkins BM (2007) Food and processing residues in California: resource assessment and potential for power generation. Bioresour Technol 98:3098–3105
Mench M, Lepp N, Bert V, Schwitzguébel J-P, Gawronski SW, Schröder P, Vangronsveld J (2010) Successes and limitations of phytotechnologies at field scale: outcomes, assessment and outlook from COST Action 859. J Soils Sediments 10:1039–1070. https://doi.org/10.1007/s11368-010-0190-x
Mizuta K, Matsumoto T, Hatate Y, Nishihara K, Nakanishi T (2004) Removal of nitrate-nitrogen from drinking water using bamboo powder charcoal. Bioresour Technol 95:255–257
Mukherjee A, Lal R (2013) Biochar impacts on soil physical properties and greenhouse gas emissions. Agronomy 3:313–339
Mukherjee A, Zimmerman AR, Harris W (2011) Surface chemistry variations among a series of laboratory-produced biochars. Geoderma 163:247–255
Nelissen V, Rütting T, Huygens D, Staelens J, Ruysschaert G, Boeckx P (2012) Maize biochars accelerate short-term soil nitrogen dynamics in a loamy sand soil. Soil Biol Biochem 55:20–27
Nishio M, Okano S (1991) Stimulation of the growth of alfalfa and infection of mycorrhizal fungi by the application of charcoal. Bull Natl Grassl Res Inst 45:61–71
Novak J, Sigua G, Watts D, Cantrell K, Shumaker P, Szogi A, Johnson MG, Spokas K (2016) Biochars impact on water infiltration and water quality through a compacted subsoil layer. Chemosphere 142:160–167

Oya A, Iu WG (2002) Deodorization performance of charcoal particles loaded with orthophosphoric acid against ammonia and trimethylamine. Carbon N Y 40:1391–1399
Paz-Ferreiro J, Fu SL (2013) Biological indices for soil quality evaluation: perspectives and limitations. Land Degrad Develop 27(1):14–25. https://doi.org/10.1002/ldr.2262
Peng HT, Su HT, Zhang XP, Wang J (2011) An experimental comparison of compressive strengths of soils stabilized with enzyme and ground quicklime. Adv Mater Res 280:9–12
Piccolo A, Pietramellara G, Mbagwu JSC (1996) Effects of coal derived humic substances on water retention and structural stability of mediterranean soils. Soil Use Manag 12:209–213
Pietikainen J, Kiikkila O, Fritze H (2000) Charcoal as a habitat for microbes and its effect on the microbial community of the underlying humus. Oikos 89:231–242
Prober SM, Stol J, Piper M, Gupta V, Cunningham SA (2014) Enhancing soil biophysical condition for climate-resilient restoration in Mesic woodlands. Ecol Eng 71:246–255
Quilliam RS, Glanville HC, Wade SC, Jones DL (2013) Life in the 'charosphere'—does biochar in agricultural soil provide a significant habitat for microorganisms? Soil Biol Biochem 65:287–293
Radovic LR, Moreno-Castilla C, Rivera-Utrilla J (2001) Carbon materials as adsorbents in aqueous solutions. In: Radovic LR (ed) Chemistry and physics of carbon. Marcel Dekker, New York, pp 227–405
Reverchon F, Yang H, Ho TY, Yan G, Wang J, Xu Z, Chen C, Zhang D (2015) A preliminary assessment of the potential of using an acacia-biochar system for spent mine site rehabilitation. Environ Sci Pollut R 22:2138–2144
Rodríguez-Vila A, Covelo EF, Forján R, Asensio V (2014) Phytoremediating a copper mine soil with *Brassica juncea* L., compost and biochar. Environ Sci Pollut 21:11293–11304
Rondon MA, Lehmann J, Ramírez J, Hurtado M (2007) Biological nitrogen fixation by common beans (*Phaseolus vulgaris* L.) increases with bio-char additions. Biol Fert Soils 43:699–708
Saito M, Marumoto T (2002) Inoculation with arbuscular mycorrhizal fungi: the status quo in Japan and the future prospects. Plant Soil 244:273–279
Srinivasarao C, Venkateswarlu B, Lal R, Singh AK, Kundu S, Vittal KPR, Sharma SK, Sharma RA, Jain MP, Chary GR (2012) Soil carbon sequestration and agronomic productivity of an Alfisol for a groundnut-based system in a semiarid environment in southern India. Can J Soil Sci 92:771–785. https://doi.org/10.4141/CJSS2011-098
Steiner C, Glaser B, Teixeira WG, Lehmann J, Blum WEH, Zech W (2008) Nitrogen retention and plant uptake on a highly weathered central Amazonian Ferralsol amended with compost and charcoal. J Plant Nutr Soil Sci 171:893–899
Thies EJ, Rillig MC (2009) Characteristics of biochar: biological properties. In: Lehmann J, Joseph S (eds) Biochar for environmental management: science and technology. Earthscan, London, pp 85–105
Thomas MK, Murray R, Flockhart L, Pintar K, Pollari F, Fazil A, Nesbitt A, Marshall B (2013) Estimates of the burden of foodborne illness in Canada for 30 specified pathogens and unspecified agents, circa 2006. Foodborne Pathog Dis 10:639–648
Topoliantz S, Ponge J-F (2003) Burrowing activity of the geophagous earthworm*Pontoscolex corethrurus* (Oligochaeta: Glossoscolecidae) in the presence of charcoal. Appl Soil Ecol 23:267–271
Topoliantz S, Ponge J-F (2005) Charcoal consumption and casting activity by *Pontoscolex corethrurus* (Glossoscolecidae). Appl Soil Ecol 28:217–224
Torres-Rojas D, Lehmann J, Hobbs P, Joseph S, Neufeldt H (2011) Biomass availability, energy consumption and biochar production in rural households of Western Kenya. Biomass Bioenegy 35:3537–3546
Tyron EH (1948) Effect of charcoal on certain physical, chemical and biological properties of forest soils. Ecol Monogr 18:81–115
Van Zwieten L, KimberS MS, Chan KY, Downie A, Rust J, Joseph S, Cowie A (2009a) Effects of biochar from slow pyrolysis of papermill waste on agronomic performance and soil fertility. Plant Soil 327:235–246

Van Zwieten L, Sigh B, Joseph S, Kimber S, Cowie A, Chan KY (2009b) Biochar and emissions of non-$CO_2$ greenhouse gases from soil. In: Lehmann J, Joseph S (eds) Biochar for environmental management – science and technology. Earthscan, London, pp 227–247

Wang L, Butterly CR, Wang Y, Herath HMSK, Xi YG, Xiao XJ (2014) Effect of crop residue biochar on soil acidity amelioration in strongly acidic tea garden soils. Soil Use Manag 30:119–128

Warnock DD, Lehmann J, Kuyper TW, Rillig MC (2007) Mycorrhizal responses to biochar in soil – concepts and mechanisms. Plant Soil 300:9–20

Widowati UWH, Soehono LA, Guritno B (2011) Effect of biochar on the release and loss of nitrogen from urea fertilization. J Agric Food Technol 1:127–132

Widowati, Asnah, Sutoyo (2012) The effects of biochar and potassium fertilizer on the absorption and potassium leaching. Buana Sains 12:83–90

Wuddivira MN, Stone RJ, Ekwue EI (2009) Structure stability of humid tropical soils as influenced by manure incorporation and incubation duration. Soil Sci Soc Am J 73:1353–1360

Yadav NK, Kumar V, Sharma KR, Choudhary RS, Butter TS, Singh G, Kumar M, Kumar R (2018) Biochar and their impacts on soil properties and crop productivity: a review. J Pharmacogn Phytochemistry 7(4):49–54

Yamato M, Okimori Y, Wibowo IF, Anshori S, Ogawa M (2006) Effects of the application of charred bark of *Acacia mangiumon* the yield of maize, cowpea and peanut, and soil chemical properties in South Sumatra, Indonesia. Soil Sci Plant Nutr 52:489–495

Yuan JH, Xu RK (2010) The amelioration effects of low temperature biochar generated from nine crop residues on an acidic Ultisol. Soil Use Manage 27:110–115

Van Zwieten L, Kimber S, Downie A, Morris S, Petty S, Rust J, Chan KY (2010) A glasshouse study on the interaction of low mineral ash biochar with nitrogen in a sandy soil. Aust J Soil Res 48:569–576

# Chapter 7
# Soil Health and Foliar Fertilisers

**Apostolos Papadopoulos**

**Abstract** Soil health in terms of its ability to sustain crop productivity is linked closely with foliar fertilisers. There is little documented work to provide knowledge on the practical aspect of farming and how the soil health may be affected where foliar fertiliser applications are associated with increase soil health. This chapter explains standard foliar fertiliser practices that provide nutrition and stimulate crop production including root growth. The effects of these inputs have a direct effect to soil health through affecting the rooting distribution, rooting exudates, shrinking and swelling of soil and return of organic matter to the soil. This chapter described how foliarly applied nutrition contributes to soil health and it should be considered as part of the solution for sustainable farming. Furthermore, this chapter will support soil scientist understand the practice of spraying crops with nutrients to correct deficiencies and stimulate plants to increase soil nutrient uptake by stimulating growth. In return, the biomass will enhance soil health and contribute to sustainable farming practices.

**Keywords** Foliar fertilisers · Soil health · Plant symbiosis · Bio-stimulants

## 7.1 Introduction

The definition of "soil health" has been attempted by several authors to produce a universal one with some success. It has been generalised and often it is also referred to as "soil quality" to cover the built environment, living organisms and service to the ecosystem amongst other descriptions (Rinot et al. 2019). This chapter is concerned with soil health referring to agricultural land. Even agricultural land subdivides to several meanings. These include land to produce annual crops for human consumption, fruit production, animal grazing, subsoil composition and

A. Papadopoulos (✉)
Crop Intellect, Lincoln University, Riseholme College, Lincoln, UK
e-mail: apostolos@cropintellect.co.uk

 
B. Giri, A. Varma (eds.), *Soil Health*, Soil Biology 59,
https://doi.org/10.1007/978-3-030-44364-1_7

depth, mineral content, ability to support biological synergists for crop production and carrier of diseases etc. Furthermore, achieving a healthy soil may not be desirable for agriculture. For instance, increasing organic matter, avoidance of agrochemical use, increasing biodiversity and no soil cultivations contribute towards a healthier soil; however to achieve this, a reduction in yield is most likely the price to pay. Finding the right balance for each case is essential to ensure a sustainable system providing a healthy soil whilst satisfying socio-economic and environmental services. Foliar fertilisers may appear as a small part to all that, but they can offer a considerable benefit to soils. Spraying crops has become a common practice to administer agrochemicals and correct micronutrient deficiencies (Aziz et al. 2019). This chapter aims to link soil health with foliar nutrition and explain how these are associated.

## 7.2 Foliar Fertilisers

Soils that are poor in providing the full nutritional package required for successful crop growth have been supported by fertiliser applications. This has allowed to convert poor agricultural soils into productive ones. Typically, the fertiliser applications are solids applied on the soil surface and often incorporated into the top 20–30 cm by soil cultivations. The major elements they supply are nitrogen, phosphorus and potassium including sulphur, magnesium and manganese in some cases. These are applied at various ratios to suit the crop growing and the existing levels in the soil. Nutrient availability which can be obtained by soil analysis offered by laboratories is key to decision making. Fertiliser companies offer specially formulated blends to suit customer needs. The application of the bulk solid fertiliser is typically performed at the start of the growing season which provides adequate time to break down and provide the nutrition required by the crop.

The application of bulk fertiliser to the soil can cause negative effects. These are overdosing, locking other nutrients, toxicity to living organisms, percolation to lower soil profiles, changes in soil pH etc. However, they are responsible for increasing yield and quality in crop production systems and are widely used in most cultivations globally. The application of nitrogen which has been responsible for the most part of producing a larger biomass has also caused issues to crops. The lavish growth develops weaker cells that are more vulnerable to disease and their requirements for micronutrients increases. This results in creating micronutrient deficiencies easier compared to a normal crop growth due to a faster cell division and expansion. Furthermore, elements such as manganese, magnesium and calcium are negatively affected by changes in soil pH and the presence of other nutrients. The imbalance produced by bulk fertilisers can be corrected by foliar applications of crop nutrition.

There are still sceptics who are not convinced that foliar fertilisers provide nutrition to crops. This is not unexpected as the farming world has been an arena that many companies would like to entertain its stability and value. Several companies have appeared offering solutions to farmers for better crop production. In some cases, these have resulted in inconsistent and unproven efficacy imposing a cost to the producer without the promised return. Such companies come and go making the legitimate efforts tougher. Therefore, adaptation of innovation takes a longer time compared to other industries. Furthermore, the commodity prices restrict the value of innovation that can be charged to producers as it must be cost effective (Zhang et al. 2018).

Straight foliar crop nutrition has been widely accepted and used by producers. These are single nutrient products that have been typically used in land known to suffer certain deficiencies due to its soil composition. These products are sometimes offered as multinutrients specific for certain crops. Making a choice of which one to use usually depends on the farmer's relationships with suppliers. There are hundreds of companies offering such foliar fertilisers and this has resulted in their value to be very low with small margins. The UK market is worth somewhere in the region of 50 million euros and the European market around 0.9 billion euros. There is a further section in the micronutrients called specialty fertilisers which are more customised and offer further value than that of straights. The specialty fertilisers include higher concentrations, a variety of nutrients in a balanced level for specific crops, penetrants, stickers, safeners and other components such as bio-stimulants to claim the added value.

## 7.3 Forms of Foliar Fertilisers

The form of the elements supplied is of vital importance to its efficacy. Elements that are highly water soluble are desirable due to their ability to dissolve in water and become available to the plants (Fageria and Baligar 2005). However, as the elements dissolve may also react so formulating these products can become tricky and limit the aim of such products. A good example of such formulations are the nutrient solutions for use in hydroponics. Products are available in water-soluble powder form, suspension concentrates (suspended insoluble and soluble material in solution), fully water soluble etc. Further, there is a practice which is also under the foliar applications and that is concentrated nitrogen in liquid form which is sprayed or dripped on cereals as a top-up in spring time and/or summer in the northern countries to support fast growth. This application has severe effects to the crop initially due to scorching causing a deleterious effect to the normal growth; however, the eventual benefit of the nitrogen counteracts the negative with a significant positive effect.

Foliar fertiliser concentrates are added into water for spraying. The typical volumes range from 1 to 5 lt/ha used in 100–500 lt of water depending on the surface area of the crop and the equipment used to perform the spraying. There is a wide range of equipment for spraying to provide a good coverage and to avoid drift. Another range of products fall under the formulation aids, spraying aids or adjuvants. These are a type of products that have been well studied and proven to support uptake and enhance the use of fertilisers and crop protection products. In most cases, farmers will not use one product to spray but prepare a "hot mix" which is a combination of various fertilisers and crop protection products. Typically, companies supply compatibility information which inform the producer as to what can be mixed within the spraying tank. The higher the solubility of the elements in water the better the efficacy in general. Forms such as nitrates are easily soluble and absorbed by the plants. Other forms such as chlorides although water soluble pose a weak toxicity to crops. Sometime however, this may be desirable for the cultivation as it will retard growth and counteract the effect of applied nitrogen.

## 7.4 Bio-stimulants in Fertilisers

Bio-stimulants have been re-appearing in the market as a new way to add value to fertilisers and enhance crop yields. They have been previously used when looking at historical agriculture, using plant mulches from brassica, leaves, nettles, seaweed or a range of other material which demonstrated an improvement when applied to plants (Zulfiqar et al. 2019). It was realised that it is not just the nutritional value of the mulch that enhances growth but other chemical components including hormones, terpenes, phenols etc. The most common bio-stimulant used today is seaweed. However, its use is not easy and straight forward and it is a good way to demonstrate the requirements of expertise and understanding of how to use them within this chapter. What makes seaweed work as a bio-stimulant is the chemistry that it contains. This chemistry varies widely depending on the region of where the seaweed has grown, the season that the seaweed is collected, the species of the seaweed and the weather pattern in that year. These variations can alter the chemical composition of the seaweed which in return will alter the effects they will impose on the sprayed crop. Furthermore, the ideal application dose will vary per crop and as it is typical with bio-stimulatory effects there is an optimum application dose and anything below or above will not have the desirable effects. The way seaweed is applied it is on a "spray and pray" basis resulting in inconsistent efficacy. This has resulted in loss of trust in seaweed and use of bio-stimulants in general.

Phosphites are another example as they have been considered bio-stimulants. Historically, they have been applied as fungi-stats or even as fungicides. The bio-stimulant effect relates mostly to root development. Applied as a starter fertiliser at planting will increase rooting and therefore establishment to support overwintering. A further application is made in spring to stimulate root growth again. The increase in rooting will result in the plants being able to access more soil nutrients and impose more changes to the soil. These changes relate to enhancing bacteria to colonise roots, swelling and shrinking cycles that alter soil structure and physically creating spaces in the soil. Furthermore, as the crops balance the below growth with the above, more biomass will return more organic matter to the soil. Greater organic matter will enable more soil aggregation and result in a stronger, more pronounced soil structure offering water percolation, crop anchorage, better recovery from soil traffic and reduced soil erosion.

## 7.5 Foliar Fertilisers and Soil Health

Any appropriate foliar nutrition with or without bio-stimulants that increase crop biomass is likely to improve soil health as a result. Promoting a healthier crop will reduce the incidence of diseases resulting in less infested crop residue in the soil. More rooting and biomass will enhance the soil biota and soil structure improving soil quality. However, there is a branch of bio-stimulants that are bacterial based, and they are marketed under soil enhancers or soil improvers. These claim to promote the multiplication of soil bacteria or apply concentrated bacteria to the soil. The focus of such applications is to enhance crop growth. These products are efficacious in many instances; however, it has to be realised that the enhanced bacteria populations in most cases are breaking down the organic matter at a faster rate than normal. This releases more nutrients to crops resulting in higher yield and healthier plants. However, breaking down of the organic matter, particularly the old organic matter which is responsible for the strength of soil structure, in the long term will deteriorate soil resulting in a poorer "soil health" (Papadopoulos et al. 2009). Endophytic bacteria are different to the ones applied in the soil and therefore, the negative effects are not observed.

The ideal use of foliar fertilisers to reflect improvements in soil health require expert knowledge on how to use them for each case. The soil properties and composition should be considered and the fertiliser and bio-stimulant product well understood following closely the instruction of the manufacturer for the specific crop. The aim should be to complement the soil nutrient content and the bulk fertiliser applied to the requirements of the crop considering potential lock up of micronutrients or inadequacy of the soil to fulfil the crop's needs. Foliar fertilisers containing micronutrients are more efficient than those of soil applied as when bulk fertilisers are applied to the soil for root uptake, complexes are formed changing the solubility of the nutrients and reduce their availability to the plants. Enhancing healthy crop growth in a well-understood system can enhance soil health by adding more organic matter, improving soil structure, enriching soil composition, increasing soil biological activity and reducing the potential of diseases in soil residues.

## 7.6 Conclusion/Recommendations

Professional advice is recommended to enable a good understanding of the circumstances for the correct choice of products, timing and dosage for an optimum outcome. It is essential to build a system where intensive crop production is performed in a sustainable manner incorporating practices that return organic matter to the soil and feeds the plants rather than only the soil. Reducing the environmental impact of bulk fertilisers and in particular nitrogen is essential for farming sustainability. Nitrous oxide production from farmed soil, leaching of nitrogen to underground water courses and the environmental impact of producing nitrogen are considered deleterious, contribute to climate change and reduce soil health.

## References

Aziz MZ, Yaseen M, Abbas T, Naveed M, Mustafa A et al (2019) Foliar application of micronutrients enhances crop stand, yield and the biofortification essential for human health of different wheat cultivars. J Integr Agric 18(6):1369–1378

Fageria NK, Baligar VC (2005) Nutrient availability. In: Hille D et al (eds) Encyclopaedia of soils in the environment, vol 3, 1st edn. Academic Press, San Diego, pp 63–70. isbn-10:0123485304

Papadopoulos A, Bird NRA, Whitmore AP, Mooney SJ (2009) Investigating the effects of organic and conventional management on soil aggregate stability using X-ray computed tomography. Eur J Soil Sci 60(3):360–368

Rinot O, Levy GJ, Steinberger Y, Svoray T, Eshel G (2019) Soil health assessment: a critical review of current methodologies and a proposed new approach. Sci Total Environ 648:1484–1491

Zhang Y, Wang S, Wang H, Wang R, Wang X, Li J (2018) Crop yield and soil properties of dryland winter wheat-spring maize rotation in response to 10-year fertilization and conservation tillage practices on the loess plateau. Field Crop Res 225:170–179

Zulfiqar F, Casadesús A, Brockman H, Munné-Bosch S (2019) An overview of plant-based natural biostimulants for sustainable horticulture with a particular focus on moringa leaf extracts. Plant Sci (in press):110194. https://doi.org/10.1016/j.plantsci.2019.110194

# Chapter 8
# Wild Plants from Coastal Habitats as a Potential Resource for Soil Remediation

**Gederts Ievinsh, Una Andersone-Ozola, Zaiga Landorfa-Svalbe, Andis Karlsons, and Anita Osvalde**

**Abstract** The aim of the present review is to stimulate an interest in experimental exploration of native coastal plants as a valuable resource for development of and use in environmental remediation technologies. An attempt is made to analyze wild plant species from different coastal habitats in respect to the functional properties useful for soil remediation purposes. Several plant species from a number of coastal plant genera are described as potential models for further studies aimed at practical environmental phytoremediation, including *Armeria maritima*, *Rumex hydrolapathum*, *Ranunculus sceleratus*, *Anthyllis maritima*, *Alyssum montanum* subsp. *gmelinii*, and *Sedum maximum*. It is concluded that a large number of species of vascular plants native to coastal habitats of the temperate zone have potential for use in different phytoremediation systems, but this potential needs to be systemically and comparably explored experimentally. Important aspects to consider are plant responses to polymetalic substrates or wastewaters, effect of substrate moisture, and effect of edaphic conditions on phytoextraction capacity with emphasis on nitrogen fertilization.

**Keywords** *Alyssum* · *Anthyllis* · *Armeria* · Coastal plants · Metal hyperaccumulators · Phytoremediation · *Ranunculus* · *Rumex* · Salinity · *Sedum*

---

G. Ievinsh (✉) · U. Andersone-Ozola · Z. Landorfa-Svalbe
Department of Plant Physiology, Faculty of Biology, University of Latvia, Riga, Latvia
e-mail: gederts.ievins@lu.lv

A. Karlsons · A. Osvalde
Laboratory of Plant Mineral Nutrition, Institute of Biology, University of Latvia, Salaspils, Latvia



B. Giri, A. Varma (eds.), *Soil Health*, Soil Biology 59,
https://doi.org/10.1007/978-3-030-44364-1_8

## 8.1 Introduction

As early as in the 1970s it was recognized that submersed macrophytes are important targets for absorption and accumulation of chemical contaminants, both heavy metals and toxic organic substances (Guilizzoni 1991). This trend has developed up to the present day, as several recent experimental studies have focused on tolerance and accumulation potential of wetland plant species toward trace elements (Alfadul and Al-Fredan 2013; Bonanno et al. 2018; Klink 2017; Polechońska et al. 2017; Yang et al. 2017) and organic pollutants (Chandra et al. 2017), but several reviews have analyzed metal tolerance and accumulation in wetland plants (Teuchies et al. 2013; Yang and Ye 2009).

Another branch of studies since the 1970s has centered on metallophyte plants (for recent reviews, see Bothe 2011; Bothe and Słomka 2017) and phenomenon of hyperaccumulation of heavy metals (for recent reviews, see Reeves et al. 2017, 2018). While metallophytes are simply plants found natively growing on heavy metal-enriched soils, so-called hyperaccumulation criteria need to be taken into account when defining plants—hyperaccumulators: a certain threshold concentration of a particular element needs to be exceeded in aerial parts of a plant growing in a natural habitat (Reeves et al. 2018). A recent scientific discussion on these criteria (Goolsby and Mason 2015, 2016; van der Ent et al. 2015) accentuated several problems associated with this concept, especially when needs of practical biotechnological application are taken into account.

Both wetland plants as well as heavy metal hyperaccumulators have been considered for use in different plant-assisted soil remediation techniques. While wetland plants have been used mostly for purposes of construction of artificial wetland systems (Marchand et al. 2010), plant resources for possible soil phytoremediation activities have been looked for mainly among native species growing on contaminated soils, as many endemic metallophyte species or ecotypes are heavy metal hyperaccumulators (Favas et al. 2014). In addition, facultative metallophytes, growing on metal-contaminated soils, also might possess general tolerance mechanisms and high accumulation capacity, together with relatively high growth rates (Nadgórska-Socha et al. 2015). As an example of this type of approach, classification and identification study of plant species with potential value for phytoextraction needs to be mentioned, leading to identification of accumulators and hyperaccumulators of Cd, Zn, and Pb within 21 plant species native to ancient silver-mining site (Yang et al. 2014).

One more approach to phytoremediation has emerged recently due to enormous practical interest in studies of plant salt tolerance. Extensive accumulation of information on functional characteristics of salt-tolerant plant species (halophytes) has led to a notion that there are common physiological properties linking NaCl tolerance with heavy metal tolerance. Several review papers recently have analyzed a potential use of halophytic plant species in remediation of heavy metal-affected soils (Anjum et al. 2016; Liang et al. 2017; Lutts and Lefèvre 2015; Moray et al. 2016; Nikalje and Suprasanna 2018; Sruthi et al. 2017; Van Oosten and Maggio 2015) or organic pollutant-affected soils (Shiri et al. 2015), but mostly saline soils contaminated with heavy metals are considered. However, concept of halophytes is somehow dubious (for a recent review, see Grigore 2019 and references within) and the term usually is simply used to indicate salt-tolerant plants without even specifying the level of tolerance.

In addition, another ecophysiological group of wild plants with specific mineral nutrition needs is interesting in the context of environmental remediation, i.e., nitrophilous species. Nitrophilous plant species need relatively higher soil nitrate concentration for optimum growth in comparison to the majority of common crop species. On the other hand, nitrophilous species have higher tolerance to increased soil nitrate, showing no signs of toxicity at even extremely high substrate N concentrations. Surplus nitrate can be used for increased plant growth or stored in stems and leaves (Lee et al. 1986). It is important to note that high tolerance to nitrate do not automatically make the plant nitrophilous, as many plant species growing natively on soils with high nitrate content does not show increased growth by N addition (Moreau et al. 2013). Many native wetland and coastal species are nitrophilous (Lee and Ignaciuk 1985) as they are well adapted to thrive on N-rich organic deposits and have high rates of biomass accumulation as a result of that.

The main aim of the present review is to stimulate an additional interest in experimental exploration of native coastal plants as a valuable resource for development of and use in environmental remediation technologies. Combining the above, instead of focusing either on "halophytes" or exclusively on wetland plants, in the present review we will attempt to analyze wild plant species from different coastal habitats (even if they are not coastal-specific by their general distribution) in respect to the functional properties useful for soil remediation purposes. Several individual coastal plant species of temperate zone will be described in detail, pointing to their suitability for different remediation systems of degraded soils. Due to space limitations, readers are directed to recent important reviews on particular specific topics.

## 8.2 Coastal Habitats as Highly Heterogeneous Environments: Consequences for Tolerance-Related Plant Adaptations

Environmental heterogeneity is one of the fundamental aspects in coastal habitats, which has crucial importance for physiological adaptation of coastal plants, leading to high degree of phenotypic plasticity both at morphological and biochemical levels (Ievinsh 2006). Most important in the context of the present review is the ability of coastal plants to persist on soils with extremely variable and different concentration range for plant-available essential mineral nutrients. Sea water-affected wetlands are habitats of extreme manifestation of this type of heterogeneity. Extremely high seasonal and spatial variability in plant-available mineral elements has been documented in several coastal wetlands or among sites with particular coastal species (Andersone-Ozola et al. 2017; Karlsons et al. 2011, 2017). Especially high concentration variability was noticed for N, Ca, Cu (variation coefficient 55–63%), Fe, Cl, Mn (90–105%), and S (121%, Karlsons et al. 2017).

In wetland conditions, metals concentrate in rhizosphere near plant roots, which is regarded as mostly plant-induced characteristic (Kissoon et al. 2010). Typical coastal plant species have been adapted to take up surplus mineral nutrients in roots and either translocate them to shoots or store in roots without any negative consequences for plant metabolism and growth (Andersone-Ozola et al. 2017; Karlsons et al. 2011). Active mycorrhizal symbiosis in plants from coastal habitats can have additional functions besides mineral nutrient acquisition, more likely related to accumulation of excess metal ions (Bothe 2012; Hildebrandt et al. 2007).

The same mechanisms of active membrane transport, translocation between organs, ligand systems, and intercellular sequestration are used both for essential surplus elements and heavy metals, especially given the fact that many soil-contaminating heavy metals are essential plant elements, like Mn and Zn. From a functional side, cellular compartmentation of chemicals in vacuole is an adaptive trait for protecting cell metabolism shared both by salt accumulating and heavy metal accumulating species. Maintenance of cellular osmotic balance through synthesis of osmolites (compatible solutes) is vital for establishing cellular homeostasis in these cases (Slama et al. 2015). In addition to osmolytes, a wide range of different protective substances are important for securing cellular integrity. Further, removal capacity of ballast elements from photosynthetic tissues through accumulation in older leaves with further initiation of senescence sequence or in specialized glands can be useful characteristic in both cases.

Due to extreme spatial and temporal variability of different environmental factors in coastal habitats, coastal plants have high levels of general stress protection, one of the manifestations of metabolic phenotypic plasticity, which is evident as an ability of individual plants to promptly mount adequate protective cellular biochemical responses in suboptimal conditions (Ievinsh 2006). These characteristics have special importance in maintenance of cellular homeostasis also in conditions of heavy metal stress. From another point of view, as a result of taxonomic and phylogenetic analysis, it was concluded that halophytes and metal hyperaccumulators are more closely related in Asteraceae, Amaranthacea, Fabacea, and Poaceae, than if these two functions have evolved independently (Moray et al. 2016).

It has not been much experimentally assessed, in particular, on comparative basis, but there is a reason to propose a certain adaptive biochemical flexibility of coastal species at the level of enzymatic antioxidative protection (the idea has been suggested, for example, in Spanò et al. 2013). Antioxidative defense responses are important cellular mechanisms against consequences of environmental-induced endogenous oxidative stress. Among other factors, both salinity (Miller et al. 2010) and heavy metals (Mithofer et al. 2004) increase production of reactive oxygen species. In several halophyte species, a causal relationship between plant salt tolerance and status of enzymatic antioxidative system has been established (Alhdad et al. 2013; Canalejo et al. 2014; Radić et al. 2013). Overexpression of enzymatic components of antioxidative system can lead to salt tolerance even in glycophytic species (Kavitha et al. 2010; Singh et al. 2014). Similar relationship with enzymatic antioxidative system has been found also for heavy metal tolerance (Freeman et al. 2004; Sharma et al. 2010). In addition, indirect evidence for physiological relationship between salt and heavy metal responses can be found from studies showing a certain level of induced cross tolerance (Ellouzi et al. 2013).

It is suggested that coastal plant species represent one of three different groups: (i) species requiring saline soils or halophytes; (ii) other species with preference for coastal conditions; and (iii) widespread species that are able to grow in coastal habitats (Ratcliffe 1977). Further analysis of classification of coastal plant species is out of the scope of the present review, but it needs to be emphasized that in general, accessions from coastal habitats of plant species represent either ecotypes or physiological types of individuals well-adapted to chemical soil heterogeneity and variability in other abiotic factors. As a result, coastal accessions of widespread species could have greater potential to be used in environmental phytoremediation systems in comparison to inland accessions.

## 8.3 Principles of Soil Remediation Using Plants: Important Functional Characteristics

Different plant functional properties associated with chemical uptake, translocation, metabolization, and accumulation can be used for practical purposes in various systems of plant-assisted remediation (Fig. 8.1). By definition, phytoremediation represents a set of environmental biotechnology approaches that use plants for treatment of contaminated lands and waters (Ali et al. 2013; Padmavathiamma and Li 2007; Yadav et al. 2018). The fact that plants are able to take up soluble inorganic substances (containing even nonessential mineral elements) from soil solution is a critical basic mechanism for any application of phytoremediation. Even before that, active release of organic substances from plant roots, root exudates, acts as a starting point for chemical and physical modification of metals as well as organic compounds in rhizosphere leading to changes in their bioavailability as well as stimulating microbiological activity (Jones et al. 2004). Further, contaminants can be chemically modified by plant exudates or taken up and immobilized or degraded by soil microorganisms (phytostabilization, phytoimmobilization, rhizodegradation, Fig. 8.1). Several principal types of phytoremediation use ability of plants to translocate metals from roots and accumulate them in aboveground parts, e.g., phytovolatilization, phytodesalination, and phytoextraction. From the point of particular technological applications, plants can be used for mineral element uptake from contaminated natural or seminatural substrate with various moisture level, from dry to moderately wet or wastewater-impregnated (artificial wetland), or used in a floating system of rafts for purification of waterbodies.

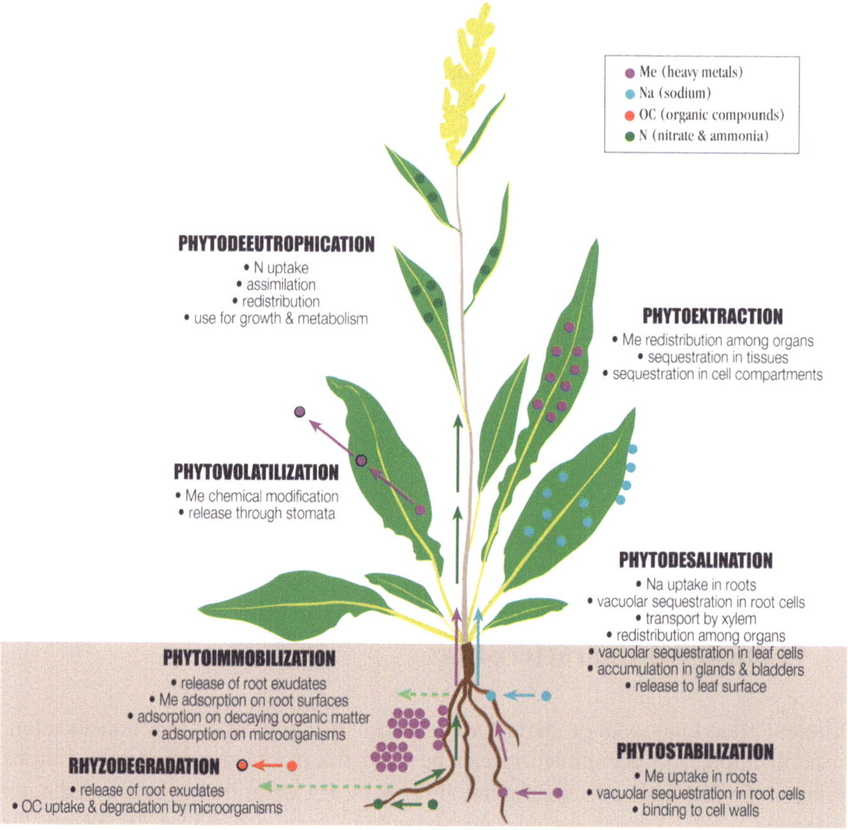


**Fig. 8.1** Schematic overview of different types of phytoremediation with emphasis on important functional plant traits involved

Besides soil metal availability for plants, as based on the mechanisms similar to those participating in mineral nutrition (Clemens 2006; Nagajyoti et al. 2010; Ovečka and Takáč 2014), physiological mechanisms of uptake, translocation and accumulation at subcellular, tissue and organ level are important characteristics of plants used in phytoremediation systems. Both apoplastic and symplastic routes of root transport are relevant for uptake of metals in accumulating plants (Maestri et al. 2010). Metal ligands are important participants for both efficient translocation and intracellular transport as well as cellular sequestration (Bothe 2011; Callahan et al. 2006; Haydon and Cobbett 2007). Both presence and efficient control of cellular membrane transporter systems allows entry of metal ions into cell cytoplasm through endoplasmatic membrane, through tonoplast into vacuole or into specialized long-distance transport systems. It is interesting to note that lack of specificity of cellular transporters is regarded as one of basic reasons for ability of nonessential element uptake (Gallego et al. 2012).

One problem of efficient phytoremediation is based on the fact that a particular plant species needs to have both high tolerance as well as accumulation potential in respect to the element of interest, but these traits seem to be independently genetically encoded and regulated (Goolsby and Mason 2015). Therefore, from a physiological perspective, four general categories of plants can be recognized: non-tolerant non-accumulators (the majority of plant species belong to this group), tolerant non-accumulators, non-tolerant accumulators (only some species are known so far), and tolerant accumulators (Goolsby and Mason 2015). Another problem is related to the fact that as opposed to intensity of accumulation, which is relatively easy to demonstrate over a gradient of substrate concentration of the metal of interest, characterization of metal tolerance is a complex task, and ideally needs to include also aspects of reproduction.

An intriguing question with an immense practical importance for choosing the right species for practical environmental remediation needs is related to whether only native hyperaccumulating metallophyte species from high-metal contaminated or natural soils are useful as efficient chemical removers. As opposed to plants from mining areas, where metal-tolerant and accumulating ecotypes have evolved (as established by Antonovics et al. 1971), wetland plants are reported to have mostly general, species-wide type of tolerance (innate tolerance), as shown by example of *Typha latifolia* as early as in the 1970s (McNaughton et al. 1974) and supported by numerous further studies (reviewed in Yang and Ye 2009). This has led to assumption that all wetland plant species may be relatively tolerant to metals because of biogeochemical peculiarities in wetland rhizosphere (Yang and Ye 2009). Evolution of metal tolerance in plant populations affected by increasing emissions of metals can be surprisingly fast. Some *Agrostis* species can even need only 5 years to show metal tolerance symptoms through local genetic adaptation (Ernst 2006). It is still an open question if there are particular sea coast-specific ecotypes of common, facultative coastal species, having higher tolerance to $Na^+$ or heavy metals as their non-coastal populations. Preliminary analysis of large number of coastal plant species has shown that there is a large variability of tissue $Na^+$ and $K^+$ concentration between groups of plants of the same species growing at different coastal sites, irrespective of spatial or temporal substrate $Na^+$ variation (Samsone and Ievinsh 2018; Ievinsh et al. 2019). It is argued that plant species with large ecological amplitude in principle have a significant potential to evolve metal resistance (Ernst et al. 2004). Therefore, it is highly possible that ecotypes highly tolerant to chemical contamination have evolved in coastal areas.

In addition, wetland plants usually have high demand for nitrogen, making them attractive as potential denitrificators of nitrogen-polluted waters. Moreover, wetland plant species from coastal habitats (flooded beaches, salt marshes, and wet meadows) directly affected by seawater salinity, have remarkable potential to tolerate high electrolyte concentration in leaf mesophyl tissues as a basic mechanism for nutrient (chemical stress) resistance (Ievinsh et al. 2019).

When describing potential plants-metallophytes, it was noted that particular species or accessions can be classified in respect to the character of a metal accumulation in aerial organs with increasing concentration of the metal in substrate.

Based on the type of the response, plants can be classified as accumulators (relatively high constant shoot metal concentration over a wide range of substrate concentration), indicators (correlation between soil metal concentration and that in plant shoots), or accumulators (relatively constant low shoot metal concentration over a wide range of substrate concentration) (Baker 1981). For practical purposes, accumulation of metals in different plant parts in respect to initial soil metal concentration are evaluated by means of different indexes (reviewed in detail in Buscaroli 2017). In general, only species with relatively high rate of transport to aerial parts are considered useful for phytoextraction techniques, but these accumulating metals in roots can be used for phytostabilization purposes or simply as metal-tolerant plants for revegetation of degraded or contaminated lands. An extreme example of metal accumulators are hyperaccumulators, plants growing natively on heavy metal-rich soils and accumulating heavy metals in shoots and are defined based on their ability to have 50–100 times higher concentration of the element of interest as compared to neighboring plant species (Maestri et al. 2010).

In addition to plant-related factors, both abiotic and biotic processes in rhizosphere is of great importance for environmental remediation (Borymski and Piotrowska-Seget 2014). Role of plant root-associated microorganisms is especially important for degradation of organic contaminants due to extremely limited uptake ability of plants for complex organic molecules. In these cases (as for rhizodegradation, Fig. 8.1) plant acts as a source of reduced carbon-rich root exudates supporting respiration needs of contaminant-degrading microorganisms.

## 8.4 Selected Coastal Species from Different Habitats with a High Potential for Phytoextraction

### *8.4.1* **Armeria maritima** *and Other* **Armeria** *Species*

When *Armeria maritima* populations from salt marshes and inland sites were compared in respect to salinity resistance, it appeared that the two populations represented different ecotypes with differences in responses to NaCl and salt tolerance (Köhl 1997). Inland population showed relatively high salinity tolerance, with ability of plants to survive at 200 mM NaCl for several months, but root and leaf growth was significantly reduced. In contrast, plant growth was not affected for salt marsh ecotype of *A. maritima*. Most importantly, allocation of Na to aerial parts in inland ecotype started at higher salinity than that in salt marsh ecotype. The presence of salt glands on leaves of *Armeria maritima* can serve for excretion of NaCl (Rozema et al. 1981) and it is assumed that they can serve similar purpose in the case of heavy metal-resistant ecotypes (Bothe 2011).

Taxonomically and biologically highly related taxon, *Armeria maritima* subsp. *halleri* (syn. *Armeria halleri*), is a well-known key component of metal-resistant vegetation on inland metallipherous soils, *Armerietum halleri*, like in mountains of

Harz region in Northern Germany (Ernst et al. 2004). The subspecies is highly tolerant to Cd, Cu, Pb, and Zn and grows only on heavy metal heaps (Bothe 2011). In natural conditions growing on polymetallic soil, accumulation potential in leaves of *A. maritima* subsp. *halleri* was relatively moderate in comparison to hyperaccumulating species *Arabidopsis halleri* (Ernst et al. 2004). In another study, *A. maritima* subsp. *halleri* growing naturally in soil with plant-available concentration of Zn and Pb of 1388 and 5966 mg $kg^{-1}$, respectively, accumulated 7903 mg $kg^{-1}$ Zn and 9659 mg $kg^{-1}$ Pb in roots, and only 1360 mg $kg^{-1}$ Zn and 532 mg $kg^{-1}$ Pb in leaves (Heumann 2002). Both Zn and Pb were efficiently excluded from flower stalks and flowers. However, when the same plant material was cultivated in hydroponics with high Zn concentration, roots and leaves accumulated 11,354 and 5179 mg $kg^{-1}$ of Zn.

*A. maritima* is also known as a Cu-tolerant plant, growing in a copper-rich bog in soils with 6486 mg $kg^{-1}$ Cu concentration (Brevin et al. 2003). These plants accumulated most Cu in decaying leaves (9000 mg $kg^{-1}$) as compared to 1818 and 171 mg $kg^{-1}$ in roots and living leaves, respectively. Other species of the genus, *Armeria canescens*, *Armeria rumelica*, and *Armeria alpina*, did not show hyperaccumulation potential for any heavy metals in natural conditions growing on soils with variable but high heavy metal (Mn, Zn, Cu, Ni, Cr, Cd, Pb) concentration (Tomović et al. 2018).

We used seeds from coastal population of *A. maritima*, growing in dry calcareous coastal meadow in the Southern Sweden, for establishing heavy metal tolerance and accumulation potential in controlled conditions (Andersone-Ozola et al. 2020b). Plant growth was not affected by increased substrate concentration of Mn (up to 1.0 g $L^{-1}$), Zn (up to 1.0 g $L^{-1}$), Cd (up to 0.1 g $L^{-1}$), and Pb (up to 0.5 g $L^{-1}$). At the highest concentration, Mn was preferentially accumulated in leaves (12,250 mg $kg^{-1}$ DM) and efficiently excluded from roots (2881 mg $kg^{-1}$ DM), flower stalks (1723 mg $kg^{-1}$), and flowers (1708 mg $kg^{-1}$). For Zn, metal concentration in leaves reached 14,750 mg $kg^{-1}$, followed by roots (7191 mg $kg^{-1}$), but in flower stalks and flowers it was only 1390 and 477 mg $kg^{-1}$, respectively. Cd concentration was similar for leaves and roots, and it was efficiently excluded from flower stalks and flowers. Consequently, coastal accessions of *A. maritima* possess high tolerance against several individual heavy metals and showed excellent metal accumulation potential in comparison to accessions of *A. maritima* subsp. *halleri* from metallipherous soils. This idea of general heavy metal tolerances of this species has been supported by other studies comparing heavy metal tolerance of *A. maritima* from metallicolous and nonmetallicolous populations (Parys et al. 2014).

In further studies with subspecies or ecotypes of *A. maritima*, it is crucial to consider optimum moisture level for particular genotypes and its effect on heavy metal tolerance and accumulation potential. It is not completely clear whether significant differences in drought and flood tolerance exist between accessions native to dry meadow vs. wetland conditions. It has been suggested though that plants from coastal populations of *A. maritima* are relatively drought tolerant due to high capacity of antioxidative defense system (Buckland et al. 1991).

### *8.4.2* **Rumex hydrolapathum** *and Other Rumex Species*

Many *Rumex* species are typical elements of dryland vegetation (i.e., *Rumex acetosella*, *Rumex thyrsiflorus*), but several are typical wetland species, some of them with high flooding tolerance (i.e., *Rumex maritima*, *Rumex palustris*; van der Sman et al. 1993). As an extreme example, *Rumex hydrolapathum* is suggested to represent an emergent macrophyte, as the species grows only in permanently wet or flooded conditions (Sager and Clerc 2006).

One may ask whether specialized metal-tolerant populations exist among *Rumex* species. Two populations of *Rumex dentatus*, a pseudometallophytic species, from metalliferous (Cu-contaminated) and non-metalliferous sites were compared in controlled conditions for their response to Cu treatment (Huang et al. 2011). Significant differences between the two populations were found in respect to phenology of reproduction as well as resource allocation to reproductive structures. In particular, plants from metalliferous population allocated more resources to reproduction in conditions of high substrate Cu concentration (500 mg $kg^{-1}$). After 2 months of culture in controlled conditions at 300 mg $kg^{-1}$ Cu, plants from both populations accumulated similar concentration of Cu in roots (>600 mg $kg^{-1}$ DM), but shoot Cu concentration was higher in plants from non-metalliferous site (486 mg $kg^{-1}$) than in plants from metalliferous site (180 mg $kg^{-1}$) (Liu et al. 2004). *Rumex hastatus* represents another copper-accumulating species of the genus, with leaf copper concentration for plants growing in natural contaminated soil ranging from 20 to 120 mg $kg^{-1}$ DM (Tang and Fang 2001).

In a similar experiment, phytoextraction potential of two *Rumex acetosa* accessions from metalliferous (Pb- and Zn-contaminated) and non-metalliferous sites were compared (Barrutia et al. 2009). Plants from contaminated site accumulated significantly higher concentration of Zn (3079 mg $kg^{-1}$) in shoots in comparison to plants from non-contaminated site (573 mg $kg^{-1}$) when cultivated on mine soil, but Pb was preferentially excluded in roots. *R. acetosa* has been relatively thoroughly explored as a candidate species for assisted phytoextraction due to relatively high biomass, fast growth as well as adaptation to different soils (Barrutia et al. 2010). However, metal tolerance and accumulation potential of other *Rumex* species was shown to be relatively low. Thus, *Rumex obtusifolius* can be characterized as heavy metal excluder, relatively sensitive to high substrate heavy metal concentration (Vondráčková et al. 2014). When cultivated in controlled conditions in natural soil with 2688 mg $kg^{-1}$ Mn, 3305 mg $kg^{-1}$ Zn, and 3305 mg $kg^{-1}$ Pb, *R. obtusifolius* plants accumulated 228 and 183 mg $kg^{-1}$ Mn, 1479 and 1260 mg $kg^{-1}$ Zn, and 235 and 142 mg $kg^{-1}$ Pb, in roots and leaves, respectively. Moreover, total leaf biomass of plants growing in contaminated soil was only 0.1 g per plant as compared to 5.8 g in non-contaminated soil.

*R. hydrolapathum* accession from a sea-affected wetland (Samsone and Ievinsh 2018) was used in a study in controlled conditions to evaluate effect of biogenous heavy metals Zn and Mn (Ievinsh et al. 2020a). In *R. hydrolapathum* plants, Zn was preferentially accumulated in older leaves, reaching 1800 mg $kg^{-1}$ DM, with Zn

**Fig. 8.2** Growth and development of *Rumex hydrolapathum* plants as affected by different concentrations of heavy metals (Mn, Zn) in substrate (g $L^{-1}$). Three-week-old (above) and 5-week-old (below) average plants are shown. Courtesy G. Ievinsh

concentration 700 mg $kg^{-1}$ DM in actively photosynthesizing mature leaves, but it was effectively excluded from young leaves and roots. In contrast, Mn concentration reached 6000 mg $kg^{-1}$ DM and was similar in all leaf types except young leaves where it reached only 300 mg $kg^{-1}$ DM, with significantly lower level (below 500 mg $kg^{-1}$ DM) in roots. High tolerance of *R. hydrolapathum* against biogenous heavy metals Zn and Mn was associated with ability to initiate programmed death of older leaves, accumulating high concentration of heavy metals, together with stimulation of development and growth of new leaves (Fig. 8.2).

Plants from coastal accession of *R. hydrolapathum* showed partial tolerance against Ni and full tolerance against Cd in soil culture in controlled conditions. In the case of Ni, both shoot and root growth were depressed by 50% at 1 g $L^{-1}$ Ni concentration, with Ni concentration in leaves and roots reaching 140 and 60 mg $kg^{-1}$ DM, respectively. *R. hydrolapathum* plants exhibited no signs of growth inhibition even when substrate Cd concentration reached 50 mg $L^{-1}$. In these conditions, older leaves accumulated more than 60 mg $kg^{-1}$ Cd, with Cd concentration in actively photosynthesizing leaves being 20 mg $kg^{-1}$, and both in roots and young leaves only 10 mg $kg^{-1}$. Costal accessions of both *R. hdrolapathum* and *R. crispus* plants were highly tolerant to increased substrate concentration of Pb, with no signs of growth

**Fig. 8.3** Growth and development of *Rumex hydrolapathum* plants as affected by different levels of universal fertilizer (UF) and additional treatment with $NO_3$-N and $NH_4$-N. Two-month-old average plants are shown. Courtesy G. Ievinsh

inhibition even at substrate Pb concentration of 1.0 g $L^{-1}$, when $Pb(CH_3COO)_2$ was used (Ievinsh et al. 2020b). Moreover, when Pb was used in a form of $Pb(NO_3)_2$, plant growth was significantly stimulated even at the highest Pb concentration.

Wetland-adapted species usually show more vigorous growth under wetland conditions in comparison to dryland, and their element uptake capability is relatively larger. In a study with *Rumex crispus*, flooding induced 2.5-fold increase in accumulation of chemical elements, with particularly large increase for concentration in Mn and K both in shoots and roots, and Ca in shoots (Kissoon et al. 2010).

As many *Rumex* species are described as nitrogen-demanding (Hejcman et al. 2012 and references within), it seems to be reasonable to assume that wetland-adapted *Rumex* species are obligate nitrophilous and therefore can be used in different denitrification systems. Indeed, when effect of $NO_3$- and $NH_4$-N fertilizers on different background levels of mineral nutrient availability on development and biomass accumulation of *R. hydrolapathum* was compared, it was found that even at optimum supply of a universal fertilizer, N treatment resulted in a further stimulation of growth and development, indicating increased requirement for N to achieve maximum physiological performance (Fig. 8.3; Ievinsh et al. 2020b).

As a *Rumex* species with largest individuals, well-adapted to emergent wetland conditions, and with high regrowth capacity after cutting, *R. hydrolapathum* seems to be an excellent choice for practical use in artificial wetlands and other systems of water purification, using principle of phytoextraction, phytodesalination, and phytodeeutrophication. Other wetland-adapted *Rumex* species (*R. maritima*, *Rumex longifolius*, *R. crispus*) from sea-affected habitats are good candidates for further practically oriented phytoremediation studies. In addition, high biomass species *Rumex confertus* seems to be a good choice for development of phytoremediation systems for dry or moderately wet conditions.

### *8.4.3* **Ranunculus sceleratus** *and Other Ranunculus Species*

*Ranunculus sceleratus* is a semiaquatic circumpolar species of Northern Hemisphere native to wetland habitats. The species is extremely resistant to soil waterlogging and moderate flooding (He et al. 1999). Its survival strategy is based on leaf petiole elongation as a result of flooding in order to regain leaf blade contact with aerial environment to support photosynthesis (Smulders and Horton 1991). Another feature that support flooding tolerance is constitutive presence of aerenchyma in roots of *R. sceleratus* (He et al. 1999).

*R. sceleratus* has been used in constructed wetland systems because of flooding tolerance as well as ability to accumulate chemical pollutants, including heavy metals as well as excess nutrients. When used in a sewage purification system with 15 L volume, 10 *R. sceleratus* plants can remove up to 93% of total nitrogen and 89% of total phosphorus within 7 months (Zuo et al. 2014).

Naturally growing *R. sceleratus* individuals on a coast of Lake Mariut (Egypt) were analyzed in respect to accumulation of Zn, Mn, Cu, Ni, Cd, and Pb in both roots and shoots (Farahat and Galal 2018). These plants preferentially accumulated metals in roots, with shoot concentration being only negligible for Ni and Cd, or 6, 10, 12, and 31% from that in roots for Mn, Pb, Cu, and Zn, respectively. The highest accumulation potential was evident for Mn, with average concentration in roots reaching 4000 mg $kg^{-1}$.

*R. sceleratus* plants, grown from seeds collected in a sea-affected wetland, showed extreme NaCl tolerance in controlled conditions and an ability to successfully reproduce, with only 29 and 39% decrease in total shoot biomass at substrate Na concentration 5 and 10 g $L^{-1}$, respectively (Landorfa-Svalbe et al. 2019). Plants in saline soil preferentially accumulated Na in stems, followed by rosette leaf petioles, stem leaves, and roots. The lowest Na concentration was in leaf blades and flowers. Maximum Na concentration was found in stems of *R. sceleratus* plants grown in the presence of 10 g $L^{-1}$ Na, reaching 90.8 g $kg^{-1}$ on a dry mass basis. However, highest Na concentration on a basis of $H_2O$ content was evident in leaf blades of 10 g $L^{-1}$ Na-treated plants, reaching 6.32 mol $L^{-1}$.

Growth of *R. sceleratus* plants was significantly stimulated by nitrates, even in a form of Na and Pb salts, when substrate Na and Pb concentration was 4 and 1 g $L^{-1}$, respectively (Andersone-Ozola et al. 2020a). In general, plants were highly tolerant to Mn, Zn, Cd, and Pb in substrate and accumulated 7000 mg Mn in leaves, 7000–8000 mg Zn in leaves and roots, 700 mg Cd in roots, and 700 mg Pb in leaves and 1500 mg Pb in roots, per kg dry mass (Fig. 8.4). Consequently, in contrast to the abovementioned study (Farahat and Galal 2018), *R. sceleratus* plants from a salt-adapted coastal accession showed relatively high degree of shoot translocation for Mn, Zn, and Pb, making them good candidates for phytoextraction, phytodesalination, and phytodeeutrophication systems in wet or flooded soil conditions.

**Fig. 8.4** Growth and development of *Ranunculus sceleratus* plants as affected by different concentrations of heavy metals (Mn, Zn, Cd, Pb) in substrate (mg $L^{-1}$). Five-week-old average plants 20 days after the start of the treatment are shown. Courtesy G. Ievinsh

## *8.4.4* Anthyllis maritima *and Other Anthyllis Species*

*Anthyllis maritima* is an endemic coastal species of the Baltic Sea region, taxonomically closely related to the *Anthyllis vulneraria* complex (Puidet et al. 2005). Growing on both semifixed white and fixed gray dunes, it can withstand both moderate sand burial as well as occasional salt spray. So far, heavy metal tolerance of *A. maritima* has not been assessed, but analysis of closely related *Anthyllis* species provides evidence for possible metal tolerance. Moreover, importance of symbiotic nitrogen-fixing bacteria in roots of *Anthyllis* species for efficient environmental adaptation has been pointed out (Ampomah and Huss-Danell 2011).

*Anthyllis vulneraria* from mine sites in France has been identified as Zn accumulator (Escarré et al. 2010). When growing in soil with up to 160,000 mg $kg^{-1}$ Zn and 92,700 mg $kg^{-1}$ Pb, the metal concentration in leaves of some individuals of *A. vulneraria* even reached 10,000 and 1000 mg $kg^{-1}$ for Zn and Pb, respectively. When these plants were cultivated in controlled conditions excluding contamination by metal dust deposits, average Zn value was 5060 mg $kg^{-1}$ and that for Pb 776 mg $kg^{-1}$.

Symbiotic $N_2$-fixing bacteria *Mesorhizobium metallidurans* was isolated from root nodules of *A. vulneraria* plants native to metal-contaminated soils, showing an extreme tolerance to heavy metals (Vidal et al. 2009). In addition, a second symbiont, *Rhizobium metallidurans*, was isolated from root nodules of *A. vulneraria* (Grison et al. 2015a). Consequently, symbiotic *A. vulneraria* plants can be used for soil enrichment with N, in further facilitating growth of nonsymbiotic co-occurring species at phytoremediation sites. It was calculated that due to efficient $N_2$ fixation as a result of bacterial symbiosis, *A. vulneraria* plants can supply additional 400 kg N $ha^{-1}$ to soil within 4 years both through rhizodeposition as well as decay of plant tissues (Frérot et al. 2006).

Further, *A. vulneraria* plants have been used in *Rhizobium metallidurans*-optimized phytoextraction of Zn from highly contaminated soil, with shoot Zn concentration reaching 17,428 mg $kg^{-1}$, which resulted in removal potential of 27 kg $ha^{-1}$ Zn (Grison et al. 2015b). In contrast, in *A. vulneraria* subsp. *carpatica*, both root and shoot concentration of Zn significantly decreased in nodulated plants in comparison to nitrogen-grown plants, but increasing metal tolerance (Soussou et al. 2013). These facts clearly show that different populations of *A. vulneraria* can have various degrees of both metal resistance and accumulation potential due to local genetic adaptation (Mahieu et al. 2012).

Consequently, due to ability to replenish soil nitrogen pool and possible heavy metal tolerance, *A. maritima* plants need to be further explored for their potential use in restoration of degraded or contaminated drylands.

### *8.4.5* **Alyssum montanum** ***subsp.*** **gmelinii** ***and Other Alyssum Species***

*Alyssum montanum* subsp. *gmelinii* (syn. *Alyssum gmelinii*) is a subspecies of *Alyssum gmelinii* complex with complicated taxonomy, usually occurring on sandy dunes, but recent studies using both morphological and molecular criteria do not support this type of habitat specificity (Španiel et al. 2012). However, a separate population of *Alyssum montanum* subsp. *gmelinii* with coastal-specific distribution occurs on dunes of the Eastern Baltic Sea in territory of Lithuania and Latvia. Individuals of this population are well-adapted to sand dune conditions as they are resistant to moderate sand burial (Samsone et al. 2009).

Other species of the genus are elements of European serpentine flora and can hyperaccumulate Ni, as *Alyssum murale*, *Alyssum lesbiacum*, *Alyssum bertolonii*, *Alyssum corsicum* and other species. More than 50 *Alyssum* species are Ni hyperaccumulators naturally growing on ultrabasic magmatic rocks with soil Ni concentration 2.5 to 7.0 g $kg^{-1}$ (Cecchi et al. 2010). In addition to Ni, a population of *A. montanum* from serpentine soils has been proven to exhibit high phytoextraction potential for Cd (Barzanti et al. 2011). Amino acid histidine is a main chelator of nickel, as indicated by nickel-induced 36-times higher increase in histidine concentration in hyperaccumulating *Alyssum lesbiacum* in comparison to that in *Alyssum montanum* (Krämer et al. 1996). In addition to histidine, which is the main transport ligand of Ni, malate and possibly other low molecular weight components act as complex-forming ligands for vacuolar storage of Ni in leaves of *A. murale* (McNear et al. 2010).

Using scanning electron microscopy/energy-dispersive X-ray analysis, it was shown that Ni concentration in different Ni-hyperaccumulating species of *Alyssum* in conditions of high substrate Ni reached as high as 48,400 mg $kg^{-1}$, but total metal concentration of Ni, Mn, and Ca in leaf trichomes was 15–20% from their dry mass (Broadhurst et al. 2004). This finding was supported by further studies showing that Ni is colocated with Mn only in trichome bases and in cells adjacent to trichomes, supporting a view that Ni hyperaccumulation in Alyssum species has evolved from an Mn handling system (Broadgurst et al. 2009).

We compared Ni tolerance and metal accumulation potential of coastal population of *Alyssum montanum* subsp. *gmelinii* from sand dunes of the Eastern Baltic Sea to that of *Alyssum murale* in sandy soil culture in controlled conditions. *A. montanum* subsp. *gmelinii* plants did not survive at Ni concentration

**Fig. 8.5** Growth and development of *Alyssum montanum* subsp. *gmelinii* plants from coastal sand dunes as affected by different concentrations of Ni in substrate (mg $L^{-1}$). Sixteen-week-old average plants 12 weeks after the start of the treatment are shown. Courtesy G. Ievinsh

4000 mg $L^{-1}$ substrate more than for 3 weeks, and plant shoot mass linearly decreased to 50% from control at 2000 mg $L^{-1}$ (Fig. 8.5), but growth of *A. murale* was significantly stimulated even at 2000 and 4000 mg $L^{-1}$ Ni. At 2000 mg $L^{-1}$ Ni, *A. montanum* subsp. *gmelinii* plants within 16 weeks accumulated 626 mg $kg^{-1}$ Ni on dry mass basis both in leaves and stems, and only 232 mg $kg^{-1}$ Ni in roots, in comparison to 10,000 mg $kg^{-1}$ Ni in shoots and 3869 mg $kg^{-1}$ Ni in roots of *A. murale*. At substrate Ni concentration 4000 mg $L^{-1}$, *A. murale* accumulated as much as 21,976 mg $kg^{-1}$ Ni in leaves. As stem growth of *A. montanum* subsp. *gmelinii* plants was stimulated by low Ni concentration (250 and 500 mg $L^{-1}$), but root Ni concentration was saturated at 500 mg $L^{-1}$ Ni, similar to that of *A. murale*, it is evident that the coastal population of *A. montanum* subsp. *gmelinii* shows relatively high Ni tolerance and remarkable accumulation potential in aerial parts. It is interesting to note that growth at increasing substrate Ni concentration resulted in linear increase in leaf K concentration up to 3.5-fold as compared to control in *A. montanum* subsp. *gmelinii* plants, but not in *A. murale*, indicating changes in mineral nutrition as part of defense against Ni treatment. Moreover, increasing substrate Ni concentration stimulated linear increase in Co concentration in leaves of *A. murale*, reaching 48.8 mg $kg^{-1}$ in contrast to control level of 1.4 mg $kg^{-1}$ Co. Similarly, Ni treatment induced accumulation of Mn and Ca in leaf trichomes of *A. murale* and other species, indicating that Ni-induced changes in leaf mineral balance is a common characteristic of *Alyssum* species (Broadhurst et al. 2004).

It seems that all taxonomically related *A. montanum* genotypes have relatively high level of tolerance against heavy metals, but this potential needs to be further explored. Special attention needs to be focused on possible effect of substrate moisture and edaphic conditions, as it seems that *Alyssum* species are mostly suitable for restoration of degraded drylands and soils with low fertility.

### 8.4.6 Sedum maximum *and Other Sedum Species*

Genus *Sedum* is represented exclusively by leaf succulent species from the Northern Hemisphere. Only some species are growing natively in dry coastal habitats, such as *Sedum litoreum*, *Sedum sexangulare*, and *Sedum acre*. In addition, in Northern Europe, *Sedum maximum* can be occasionally found in pine dune forests near the Baltic Sea.

Recently, several *Sedum* species from China have gained attention as heavy metal-tolerant hyperaccumulators with potential use in phytoextraction techniques. These include *Sedum alfredii* and *Sedum plumbizincicola*. Only some ecotypes of *S. alfredii* from contaminated soils are Zn and Cd hyperaccumulators (Deng et al. 2007). The main difference between the ecotypes was that only plant from the accumulating ecotype had an ability to increase shoot Zn and Cd concentration with an increase in soil contamination intensity. The ecotypes of *S. alfredii* found near Pb/Zn mines can also accumulate Pb (Huang et al. 2008). Heavy metal accumulation potential in *S. plumbizincicola* was evident at species level, reaching leaf concentration of Cd 7010 mg $kg^{-1}$ and that for Zn 18,400 mg $kg^{-1}$ (Cao et al. 2014). Other *Sedum* species have been indicated as potential accumulators of heavy metals: *Sedum spectabile* and *Sedum aizoon* for Cd (Guo et al. 2017), and *Sedum rubrotinctum* for Cu (Zhang et al. 2015).

A particular accession of *Sedum maximum* was found growing natively on a drift line vegetation on the coast of the Kalmar Strait (the Baltic Sea) near Ekerum Camping (island of Öland, Sweden) growing in saline soil with electrical conductivity 867 ± 34 mS $m^{-1}$. Part of one flowering shoot was collected and used for establishment of experimental material through vegetative propagation of nodal explants. During 5-week-long study in controlled conditions, it was established that *S. maximum* plants were highly tolerant to increased concentration of Cu, Zn, and Mn in substrate. No toxicity symptoms were found up to 250, 3000, and 4000 mg $kg^{-1}$ internal concentration of Cu, Zn, and Mn, respectively, in leaves of *S. maximum* (Fig. 8.6). Plants accumulated about 250 mg $kg^{-1}$ Cu in roots and older leaves, but both Zn (3000 mg $kg^{-1}$) and Mn (10,000 mg $kg^{-1}$) hyperaccumulation

**Fig. 8.6** Growth and development of *Sedum maximum* plants as affected by different concentration of Cu, Zn, and Mn in substrate (g $L^{-1}$) for 4 weeks. Five-week-old average plants are shown. Courtesy G. Ievinsh

threshold was exceeded in leaves of *S. maximum*, with maximum concentration found being 3200 and 15,000 mg $kg^{-1}$, for Zn and Mn, respectively. As a result, *S. maximum* accumulated as much as 110 mg Mn per plant, which is a considerable amount, and with moderate planting density of 25 plants per $m^2$, this could lead to removal of 27.5 kg $ha^{-1}$ Mn without any means of assistance by chemical agents or additional fertilization. In comparison, in field conditions without any treatment, *S. alfredii* could remove 32.7 kg $ha^{-1}$ Zn (Zhuang et al. 2007).

Due to high tolerance of *Sedum* species to dry soil conditions, other representatives of the genus from coastal dune habitats (*S. acre*, *S. sexangulare*) need to be explored for their potential use for restoration of degraded drylands. It is interesting to note that some *Sedum* species can be successfully cultivated in hydroponics when appropriate degree of aeration is provided (Chen et al. 2013). Consequently, metal-accumulating *Sedum* species and ecotypes could be useful also for different phytoextraction systems for cleanup of polluted waters.

## 8.5 Conclusions

Within this review, we tried to justify the potential of coastal plant species for practical use in different environmental phytoremediation systems. Some case species and their respective potential were described in detail as examples. It needs to be pointed out that a large number of other species of vascular plants native to coastal habitats of the temperate zone have potential for use in different phytoremediation systems, but this potential needs to be systemically and comparably explored experimentally. Important aspects to consider for further studies need to include plant responses to polymetallic substrates/wastewaters, effect of substrate moisture, and effect of edaphic conditions with emphasis on nitrogen fertilization. One of the most crucial functional aspects for successful practical implementation of phytoremediation technologies is related to possible interaction between different model species jointly used in these systems. Species with different specificity of tolerance and accumulation toward various heavy metals need to be used complimentary to provide maximum performance of the systems used in the case of polymetallic pollution. In addition, it is evident that species supporting different types of functional microbial symbiosis (mycorrhiza or N-fixing rhizobacteria) can be used as remediation assistant species, even if they do not accumulate chemicals of interest, for facilitation of tolerance or tissue concentration of the polluting components in the accumulating species.

## References

Alfadul SMS, Al-Fredan MAA (2013) Effects of Cd, Cu, Pb, and Zn combinations on *Phragmites australis* metabolism, metal accumulation and distribution. Arab J Sci Eng 38:11–19

Alhdad GM, Seal CE, Al-Azzawi MJ, Flowers TJ (2013) The effect of combined salinity and waterlogging on the halophyte *Suaeda maritima*: the role of antioxidants. Environ Exp Bot 87:120–125

Ali H, Khan E, Sajad MA (2013) Phytoremediation of heavy metals—concepts and applications. Chemosphere 91:869–881

Ampomah OY, Huss-Danell K (2011) Genetic diversity of root nodule bacteria nodulating *Lotus corniculatus* and *Anthyllis vulneraria* in Sweden. Syst Appl Microbiol 34:267–275

Andersone-Ozola U, Gaile L, Ievinsh G (2017) Physiological responses of rare coastal salt marsh species *Triglochin maritima* L. to soil chemical heterogeneity. Acta Biol Univ Daugavp 17:149–155

Andersone-Ozola U, Ievinsh G, Landorfa-Svalbe Z, Karlsons A, Osvalde A (2020a) Wetland species *Ranunculus sceleratus* from a sea coast: heavy metal tolerance and accumulation potential. Environ Exp Biol 18:45–46

Andersone-Ozola U, Samsone I, Karlsons A, Osvalde A, Ievinsh G (2020b) Heavy metal tolerance and accumulation potential of *Armeria maritima* plants from a dry coastal meadow. Environ Exp Biol 18:41–42

Anjum NA, Duarte B, Caçador I, Sleimi N, Duarte AC, Pereira E (2016) Biophysical and biochemical markers of metal/metalloid impacts in salt marsh halophytes and their implications. Front Environ Sci 4:24

Antonovics J, Bradshaw AD, Turner RG (1971) Heavy metal tolerance in plants. Adv Ecol Res 7:1–85

Baker AJM (1981) Accumulators and excluders – strategies in the response of plants to heavy metals. J Plant Nutr 3:643–654

Barrutia O, Epelde L, García-Plazaola JI, Garbisu C, Becerril JM (2009) Phytoextraction potential of two *Rumex acetosa* L. accessions collected from metalliferous and non-metalliferous sites: effect of fertilization. Chemosphere 74:259–264

Barrutia O, Garbisu C, Hernández-Allicia J, García-Plazaola JI, Becerril JM (2010) Differences in EDTA-assisted metal phytoextraction between metallicolous and non-metallicolous accessions of *Rumex acetosa* L. Environ Pollut 158:1710–1715

Barzanti R, Colzi I, Arnetoli M, Gallo A, Pignattelli S, Gabbrielli R, Gonnelli C (2011) Cadmium phytoextraction potential of different *Alyssum* species. J Hazard Mater 196:66–72

Bonanno G, Vymazal J, Cirelli GL (2018) Translocation, accumulation and bioindication of trace elements in wetland plants. Sci Total Environ 631–632:252–261

Borymski S, Piotrowska-Seget Z (2014) Rizospere of metallophytes and its role in bioremediation of heavy metals. Chemik 68:554–559

Bothe H (2011) Plants in heavy metal soils. In: Sherameti I, Varma A (eds) Detoxification of heavy metals, Soil biology, vol 30. Springer, Berlin, pp 35–57

Bothe H (2012) Arbuscular mycorrhiza and salt tolerance in plants. Symbiosis 58:7–16

Bothe H, Słomka A (2017) Divergent biology of facultative heavy metal plants. J Plant Physiol 219:45–61

Brevin LE, Mehra A, Lynch PT, Farago ME (2003) Mechanisms of copper tolerance by *Armeria maritima* in Dolfrwynog bog, North Wales – initial studies. Environ Geochem Health 25:147–156

Broadhurst CL, Chaney RL, Angle JS, Maugel TK, Erbe EF, Murphy CA (2004) Simultaneous hyperaccumulation of nickel, manganese, and calcium in *Alyssum* leaf trichomes. Environ Sci Technol 38:5797–5802

Broadhurst CL, Tappero RV, Maugel TK, Erbe EF, Sparks DL, Chaney RL (2009) Interaction of nickel and manganese in accumulation and localization in leaves of the Ni hyperaccumulators *Alyssum murale* and *Alyssum corsicum*. Plant Soil 314:35–48

Buckland SM, Price AH, Hendry GAF (1991) The role of ascorbate in drought-treated *Cochleria atlantica* pobed. and *Armeria maritima* (Mill.) Willd. New Phytol 119:155–160

Buscaroli A (2017) An overview of indexes to evaluate terrestrial plants for phytoremediation purposes (review). Ecol Indic 82:367–380

Callahan DL, Baker AJM, Kolev SD, Wedd AG (2006) Metal ion ligands in hyperaccymylating plants. J Biol Inorg Chem 11:2–12
Canalejo A, Martínez-Domínguez D, Córdoba F, Torronteras R (2014) Salt tolerance is related to a specific antioxidant response in the halophyte cordgrass, *Spartina densiflora*. Estuar Coast Shelf Sci 146:68–75
Cao D, Zhang H, Wang Y, Zheng L (2014) Accumulation and distribution characteristics of zinc and cadmium hyperaccumulator plant *Sedum plumbizincicola*. Bull Environ Contam Toxicol 93:171–176
Cecchi L, Gabbrielli R, Arnetoli M, Gonnelli C, Hasko A, Selvi F (2010) Evolutionary lineages of nickel hyperaccumulation and systematics in European Alysseae (Brassicaceae): evidence from nrDNA sequence data. Ann Bot 106:751–767
Chandra R, Yadav S, Yadav S (2017) Phytoextraction potential of heavy metals by native wetland plants growing on chlorolignin containing sludge of pulp and paper industry. Ecol Eng 98:134–145
Chen B, Ai W, Gong H, Gao X, Qiu B (2013) Cleaning up of heavy metals-polluted water by a terrestrial hyperaccumulator *Sedum alfredii* Hance. Front Biol 8:599–605
Clemens S (2006) Toxic metal accumulation, responses to exposure and mechanisms of tolerance in plants. Biochimie 88:1707–1719
Deng DM, Shu WS, Zhang J, Zou HL, Lin Z, Ye ZH, Wong MB (2007) Zinc and cadmium accumulation and tolerance in populations of *Sedum alfredii*. Environ Pollut 147:381–386
Ellouzi H, Ben Hamed K, Asensi-Fabado MA, Müller M, Abdelly C, Munné-Bosch S (2013) Drought and cadmium may be as effective as salinity in conferring subsequent salt stress tolerance in *Cakile maritima*. Planta 237:1311–1323
Ernst WHO (2006) Evolution of metal tolerance in higher plants. For Snow Landsc Res 80:251–274
Ernst WHO, Knolle F, Kratz S, Schung E (2004) Aspects of ecotoxicology of heavy metals in the Bartz region – a guided excursion. Landbaiforschung Volkenrode 54:53–71
Escarré J, Lefèbvre C, Raboyeau S, Dossantos A, Gruber W, Maret JCC, Frérot H, Noret N, Mahieu S, Collin C, van Oorts F (2010) Heavy metal concentration survey in soils and plants of the Les Malines mining district (Southern France): implications for soil restoration. Water Air Soil Pollut 216:485–504
Farahat EA, Galal TM (2018) Trace metal accumulation by *Ranunculus sceleratus*: implications for phytostabilization. Environ Sci Pollut Res 25:4214–4222
Favas PCJ, Pratas J, Varun M, D'Souza R, Paul MS (2014) Phyroemediation of soils contaminated with metals and metalloids at mining areas: potential of native flora. In: Environmental risk assessment of soil contamination. Intech Open, London, pp 485–517
Freeman JL, Persans MW, Nieman K, Albrecht C, Peer W, Pickering IJ, Salt DE (2004) Increased glutathione biosynthesis plays a role in nickel tolerance in *Thlaspi* nickel hyperaccumulators. Plant Cell 16:2176–2191
Frérot H, Lefèbvre C, Gruber W, Collin C, Dos Santos A, Escarré J (2006) Specific interactions between local metallicolous plants improve the phytostabilization of mine soils. Plant Soil 282:53–65
Gallego SM, Pena LB, Barcia RA, Azpilicueta CE, Iannone MF, Rosales EP, Zawoznik MS, Groppa MD, Benavides MP (2012) Unravelling cadmium toxicity and tolerance in plants: insight into regulatory mechanisms. Environ Exp Bot 83:33–46
Goolsby EW, Mason CM (2015) Toward a more physiologically and evolutionarily relevant definition of metal hyperaccumulation in plants. Front Plant Sci 6:33
Goolsby EW, Mason CM (2016) Response: commentary: toward a more physiologically and evolutionarily relevant definition of metal hyperaccumulation in plants. Front Plant Sci 6:1252
Grigore M-N (2019) Defining halophytes: a conceptual and historical approach in an ecological frame. In: Hasanuzzaman M, Shabala S, Fujita M (eds) Halophytes and climate change: adaptive mechanisms and potential uses. CAB International, Wallingford, pp 3–18

Grison CM, Jackson S, Merlot S, Dobson A, Grison C (2015a) *Rhizobium metallidurans* sp. nov., a symbiotic heavy metal resistant bacterium isolated from the *Anthyllis vulneraria* Zn-hyperaccumulator. Int J Syst Evol Microbiol 65:1525–2530

Grison CM, Mazel M, Sellini A, Escande V, Biton J, Grison C (2015b) The leguminous species *Anthyllis vulneraria* as a Zn-hyperaccumulator and eco-Zn catalyst resources. Environ Sci Pollut Res 22:5667–5676

Guilizzoni P (1991) The role of heavy metals and toxic materials in the physiological ecology of submersed macrophytes. Aquat Bot 41:87–109

Guo J-M, Lei M, Yang J-X, Yang J, Wan X-M, Chen T-B, Zhou X-Y, Gu S-P, Guo G-H (2017) Effect of fertilizers on the Cd uptake of two sedum species (*Sedum spectabile* Boreau and *Sedum aizoon* L.) as potential Cd accumulators. Ecol Eng 106:409–414

Haydon MJ, Cobbett CS (2007) Transporters of ligands for essential metal ions in plants. New Phytol 174:499–506

He JB, Bögemann GM, van de Steeg HM, Rijnders JGHM, Voesenek LACJ, Blom CWPM (1999) Survival tactics of *Ranunculus* species in river floodplains. Oecologia 118:1–8

Hejcman M, Křišťálová V, Červená K, Hrdličková J, Pavlů V (2012) Effect of nitrogen, phosphorus and potassium availability on mother plant size, seed production and germination ability of *Rumex crispus*. Weed Res 52:260–268

Heumann H-G (2002) Ultrastructural localization of zinc in zinc-tolerant *Armeria maritima* ssp. *halleri* by autometallography. J Plant Physiol 159:191–203

Hildebrandt U, Regvar M, Bothe H (2007) Arbuscular mycorrhiza and heavy metal tolerance. Phytochemistry 68:139–146

Huang H, Li T, Tian S, Gupta DK, Zhang X, Yang X (2008) Role of EDTA in alleviating lead toxicity in accumulator species of *Sedum alfredii* Hance. Bioresour Technol 99:6088–6096

Huang W-X, Huang Y, Ye F-Y, Shan S, Xiong Z-T (2011) Effects of copper on phenology and reproduction in *Rumex dentatus* from metalliferous and non-metalliferous sites. Ecotoxicol Environ Saf 74:1043–1049

Ievinsh G (2006) Biological basis of biological diversity: physiological adaptations of plants to heterogeneous habitats along a sea coast. Acta Univ Latv 710:53–79

Ievinsh G, Ieviņa S, Samsone I, Andersone-Ozola U (2019) Functional chemistry of coastal plant species: towards definition of Na and K metallophytes (electrolytophytes). Environ Exp Biol 17:63–64

Ievinsh G, Dišlere E, Karlsons A, Osvalde A, Vikmane M (2020a) Physiological responses of wetland species *Rumex hydrolapathum* to increased concentration of biogenous heavy metals Zn and Mn in substrate. Proc Latv Acad Sci B 74:35–47

Ievinsh G, Landorfa-Svalbe Z, Andersone-Ozola U, Bule A (2020b) Wild *Rumex* species as models in ecophysiological studies: effect of Na/K salts and nitrogen compounds on growth and electrolyte accumulation. Environ Exp Biol 18:43–44

Jones DL, Hodge A, Kuzyakov Y (2004) Plant and mycorrhizal regulation of rhizodeposition. New Phytol 163:459–480

Karlsons A, Osvalde A, Ievinsh G (2011) Growth and mineral nutrition of two *Triglochin* species from saline wetlands: adaptation strategies to conditions of heterogeneous mineral supply. Environ Exp Biol 9:83–90

Karlsons A, Druva-Lusite I, Osvalde A, Necajeva J, Andersone-Ozola U, Samsone I, Ievinsh G (2017) Adaptation strategies of rare plant species to heterogeneous soil conditions on a coast of a lagoon lake as revealed by analysis of mycorrhizal symbiosis and mineral constituent dynamics. Environ Exp Biol 15:113–126

Kavitha K, George S, Venkataraman G, Parida A (2010) A salt-inducible chloroplastic monodehydroascorbate reductase from halophyte *Avicennia marina* confers salt stress tolerance on transgenic plants. Biochimie 92:1321–1329

Kissoon LTT, Jacob DL, Otte ML (2010) Multi-element accumulation near *Rumex crispus* roots under wetland and dryland conditions. Environ Pollut 158:1834–1841

Klink A (2017) A comparison of trace metal bioaccumulation and distribution in *Typha latifolia* and *Phragmites australis*: implication for phytoremediation. Environ Sci Pollut Res 24:3834–3852
Köhl KI (1997) The effect of NaCl on growth, dry matter allocation and ion uptake in salt marsh and inland populations of *Armeria maritima*. New Phytol 135:213–225
Krämer U, CotterHowells JD, Charnock JM, Baker AJM, Smith JAC (1996) Free histidine as a metal chelator in plants that accumulate nickel. Nature 379:635–638
Landorfa-Svalbe Z, Andersone-Ozola U, Miesniece E, Ievinsh G (2019) Does *Ranunculus sceleratus* from coastal wetlands is potential electrolyte-accumulating species? Environ Exp Biol 17:65–66
Lee JA, Ignaciuk R (1985) The physiological ecology of strandline plants. Vegetatio 62:319–326
Lee JA, Woodin SJ, Press MC (1986) Nitrogen assimilation in an ecological context. In: Lambers H, Neeteson JJ, Stulen I (eds) Fundamental, ecological and agricultural aspects of nitrogen metabolism in higher plants. Martinus Nijhoff, Dordrecht, pp 331–346
Liang L, Liu W, Sun W, Huo X, Li S, Zhou Q (2017) Phytoremediation of heavy metal-contaminated saline soils using halophytes: current progress and future perspectives. Environ Rev 25:269–281
Liu J, Xionh Z, Li T, Huang H (2004) Bioaccumulation and ecophysiological responses to copper stress in two populations of *Rumex dentatus* L. from Cu contaminated and non-contaminated sites. Environ Exp Bot 52:43–51
Lutts S, Lefèvre I (2015) How can we take advantage of halophyte properties to cope with heavy metal toxicity in salt-affected areas? Ann Bot 115:509–528
Maestri E, Marmiroli M, Visiolo G, Marmiroli N (2010) Metal tolerance and hyperaccumulation: costs and trade-offs between traits and environment. Environ Exp Bot 68:1–13
Mahieu S, Soussou S, Cleyet-Marel J-C, Brunel B, Mauré L, Lefèbvre C, Escarré J (2012) Local adaptation of metallicolous and non-metallicolous *Anthyllis vulneraria* populations: their utilization in soil restoration. Restor Ecol 21:551–559
Marchand L, Mench M, Jacob DL, Otte ML (2010) Metal and metalloid removal in constructed wetlands, with emphasis on the importance of plants and standardized measurements: a review. Environ Pollut 158:3447–3461
McNaughton SJ, Folsom TC, Lee T, Park F, Price C, Roeder D, Schmitz J, Stockwell C (1974) Heavy metal tolerance in *Typha latifolia* without the evolution of tolerant races. Ecology 55:1163–1165
McNear DH Jr, Chaney RL, Sparks DL (2010) The hyperaccumulator *Alyssum murale* uses complexation with nitrogen and oxygen donor ligands for Ni transport and storage. Phytochemistry 71:188–200
Miller G, Suzuki N, Ciftci-Yilmaz S, Mittler R (2010) Reactive oxygen species homeostasis and signalling during drought and salinity stress. Plant Cell Environ 33:453–467
Mithofer A, Schulze B, Boland W (2004) Biotic and heavy metal stress response in plants: evidence for common signals. FEBS Lett 566:1–5
Moray C, Goolsby EW, Bromham L (2016) The phylogenetic association between salt tolerance and heavy metal hyperaccumulation in angiosperms. Evol Biol 43:119–130
Moreau D, Milard G, Munier-Jolain N (2013) A plant nitrophily index based on plant leaf area response to soil nitrogen availability. Agron Sustain Dev 33:809–815
Nadgórska-Socha A, Kandziora-Ciupa M, Ciepał R (2015) Element accumulation, distribution, and phytoremediation potential in selected metallophytes growing in a contaminated area. Environ Monit Assess 187:441
Nagajyoti PC, Lee KD, Sreekanth TVM (2010) Heavy metals, occurrence and toxicity for plants: a review. Environ Chem Lett 8:199–216
Nikalje GC, Suprasanna P (2018) Coping with metal toxicity – cues from halophytes. Front Plant Sci 9:777
Ovečka M, Takáč T (2014) Managing heavy metal toxicity stress in plants: biological and biotechnological tools. Biotechnol Adv 32:73–86

Padmavathiamma PK, Li LY (2007) Phytoremediation technology: hyper-accumulation metals in plants. Water Air Soil Pollut 184:105–126

Parys E, Wasilewska W, Siedlecka M, Zienkiewicz M, Drożak A, Romanovska E (2014) Metabolic response to lead of metallicolous and nonmetallicolous populations of *Armeria maritima*. Arch Environ Contam Toxicol 67:565–577

Polechońska L, Samecka-Cymerman A, Dambiec M (2017) Changes in growth rate and macroelement and trace element accumulation in *Hydrocharis morsus-ranae* L. during the growing season in relation to environmental contamination. Environ Sci Pollut Res 24:5439–5451

Puidet E, Liira J, Paal J, Pärtel M, Pihu S (2005) Morphological variation in eight taxa of *Anthyllis vulneraria s. lato* (Fabaceae). Ann Bot Fenn 42:293–304

Radić S, Štefanić PP, Lepeduš H, Roje V, Pevalek-Kozlina B (2013) Salt tolerance of *Centaurea ragusina* L. is associated with efficient osmotic adjustment and increased antioxidative capacity. Environ Exp Bot 87:39–48

Ratcliffe DA (ed) (1977) Coastlands: range of ecological variation In: A nature conservation review. Vol. 1. The selection of biological sites of national importance to nature conservation in Britain, pp. 25–48

Reeves RD, Baker AJM, Jaffre T, Erskine PD, Echevarria G, van der Ent A (2017) A global database for plants that hyperaccumulate metal and metalloid trace elements. New Phytol 218:407–411

Reeves RD, van der Ent A, Baker AJM (2018) Global distribution and ecology of hyperaccumulating plants. In: van der Ent A et al (eds) Agromining: farming for metals, Mineral resource reviews. Springer, Cham, pp 75–92

Rozema J, Gude H, Pollak G (1981) An ecophysiological study of the salt secretion of four halophytes. New Phytol 89:201–217

Sager L, Clerc C (2006) Factors influencing the distribution of *Hydrocharis morsus-ranae* L. and *Rumex hydrolapathum* Huds. In a mowed low-lying marshland, Réserve de Cheyres, lac de Neuchâtel, Switzerland. Hydrobiologia 570:223–229

Samsone I, Ievinsh G (2018) Different plant species accumulate various concentration of $Na^+$ in a sea-affected coastal wetland during a vegetation season. Environ Exp Biol 16:117–127

Samsone I, Druva-Lūsīte I, Andersone U, Nečajeva J, Karlsons A, Ievinsh G (2009) Plasticity of a dune plant *Alyssum gmelinii* in response to sand burial in natural conditions. Acta Univ Latv 763:125–136

Sharma A, Gontia I, Agarwal PK, Jha B (2010) Accumulation of heavy metals and its biochemical responses in *Salicornia brachiata*: an extreme halophyte. Mar Biol Res 6:511–518

Shiri M, Rabhi M, El Amrani A, Abdelly C (2015) Cross-tolerance to abiotic stresses in halophytes: application for phytoremediation of organic pollutants. Acta Physiol Plant 37:209

Singh A, Mishra A, Jha B (2014) Ectopic over-expression of peroxisomal ascorbate peroxidase (SbpAPX) gene confers salt stress tolerance in transgenic peanut (*Arachis hypogaea*). Gene 547:119–125

Slama I, Abdelly C, Bouchereau A, Flowers T, Savoure A (2015) Diversity, distribution and roles of osmoprotective compounds accumulated in halophytes under abiotic stress. Ann Bot 115:433–447

Smulders MJM, Horton RF (1991) Ethylene promotes elongation growth and auxin promotes radial growth in *Ranunculus sceleratus* petioles. Plant Physiol 96:806–811

Soussou S, Mahieu S, Brunel B, Escarré J, Lebrun M, Banni M, Boussetta H, Cleyet-Marel J-C (2013) Zinc accumulation patterns in four *Anthyllis vulneraria* subspecies supplemented with mineral nitrogen or grown in the presence of their symbiotic bacteria. Plant Soil 371:423–434

Španiel S, Marhold K, Thiv M, Zozomová-Lihová J (2012) A new circumscription of *Alyssum montanum* ssp. *montanum* and *A. montanum* ssp. *gmelinii* (Brassicaceae) in Central Europe: molecular and morphological evidence. Bot J Linn Soc 169:378–402

Spanò C, Malestri M, Nottega S, Grilli I, Forino LMC, Ciccarelli D (2013) *Anthemis maritima* L. in different coastal habitats: a tool to explore plant plasticity. Estuar Coast Shelf Sci 129:105–111

Sruthi P, Shackira AM, Puthur JT (2017) Heavy metal detoxification mechanisms in halophytes: an overview. Wetl Ecol Manag 25:129–148

Tang S, Fang Y (2001) Copper accumulation by *Polygonum microcephalum* D. Don and *Rumex hastatus* D. Don from copper mining spoils in Yunnan Province, P.R. China. Environ Geol 40:902–907

Teuchies J, Jacobs S, Oosterlee L, Bervoets L, Meire P (2013) Role of plants in metal cycling in a tidal wetland: implications for phytoremidiation. Sci Total Environ 445–446:146–154

Tomović G, Buzurović U, Đurović S, Vicić D, Mihailović N, Jakovljević K (2018) Strategies of heavymetal uptake by three *Armeria* species growing on different geological substrates in Serbia. Environ Sci Pollut Res 25:507–522

van der Ent A, Baker AJ, Reeves RD, Pollard AJ, Schat H (2015) Commentary: toward a more physiologically and evolutionarily relevant definition of metal hyperaccumulation in plants. Front Plant Sci 6:554

van der Sman AJM, Joosten NN, Blom CWPM (1993) Flooding regimes and life-history characteristics of short-lived species in rever forealnds. J Ecol 81:121–130

Van Oosten MJ, Maggio A (2015) Functional biology of halophytes in the phytoremediation of heavy metal contaminated soils. Environ Exp Bot 111:135–146

Vidal C, Chantreuil C, Berge O, Mauré L, Escarré J, Béna G, Brunel B, Cleyet-Marel J-C (2009) *Mesorhizobium metallidurans* sp. nov., a novel metal-resistant symbiont of *Anthyllis vulneraria* growing on metallicolous soil in Languedoc, France. Int J Syst Evol Microbiol 59:850–855

Vondráčková S, Hejcman M, Száková J, Müllerová V, Tlustoš P (2014) Soil chemical properties affect the concentration of elements (N, P, K, Ca, Mg, As, Cd, Cr, Cu, Fe, Mn, Ni, Pb, and Zn) and their distribution between organs of *Rumex obtusifolius*. Plant Soil 379:231–245

Yadav KK, Gupta N, Kumar A, Reece LM, Singh N, Rezania S, Khan SA (2018) Mechanistic understanding and holistic approach of phytoremediation: a review on application and future prospects. Ecol Eng 120:274–298

Yang J, Ye Z (2009) Metal accumulation and tolerance in wetland plants. Front Biol China 4:282–288

Yang W, Li H, Zhang T, Sen L, Ni W (2014) Classification and identification of metal-accumulating plant species by cluster analysis. Environ Sci Pollut Res 21:10626–10637

Yang J, Zheng G, Yang J, Wan X, Song B, Cai W, Guo J (2017) Phytoaccumulation of heavy metals (Pb, Zn, and Cd) by 10 wetland plant species under different hydrological regimes. Ecol Eng 107:56–64

Zhang C, Sale PWG, Clark GJ, Liu W, Doronila AI, Kolev SD, Tang C (2015) Succulent species differ substantially in their tolerance and phytoextraction potential when grown in the presence of Cd, Cr, Cu, Mn, Ni, Pb, and Zn. Environ Sci Pollut Res 22:18824–18838

Zhuang P, Wang QW, Wang HB, Shu WS (2007) Phytoxtraction of heavy metals by eight plant species in the field. Water Air Soil Pollut 184:235–242

Zuo S, Wan K, Ma S (2014) Environmental restoration effects of *Ranunculus sceleratus* L. in a eutrophic sewage system. Biochem Syst Ecol 55:34–40

# Chapter 9
# Abiotic and Biotic Factors Influencing Soil Health and/or Soil Degradation

**K. S. Anil Kumar and K. S. Karthika**

**Abstract** Soil is a highly complex system which is influenced by several factors in its existence. Soil health refers to the capacity of a soil in sustaining biological productivity by promoting plant growth while maintaining or improving environmental quality. Soil health is a prerequisite for agricultural production and thus it becomes highly essential to support crop production and in delivering ecosystem services. With increasing concerns on food security, maintenance of soil health is the foremost challenge we are posed with. It becomes difficult to understand and manage the processes occurring in soil independently as all these are interrelated and are in a unique balance. The physical, chemical and biological properties together with their interactions need to strike that perfect balance in a healthy soil, which is actually determined by several abiotic and biotic factors. When the balance between these is lost, the equilibrium is disturbed indicating the deterioration in soil quality which is defined using soil degradation. The different threats to soil by means of nutrient depletion, decline in organic matter, soil contamination, addition of toxic materials to soil and lack of proper management of soil and land use would lead to a degraded soil. In this chapter, a discussion is made on the aspects of soil health and abiotic and biotic factors that influence soil health and soil degradation.

**Keywords** Soil health · Sustainable agriculture · Abiotic and biotic factors · Soil degradation

## 9.1 Introduction

Soil is a vital natural resource, non-renewable, that performs a variety of key environmental, social and economical functions vital for life. Agriculture and forestry are dependent on soil for the basic requirements of water and nutrients.

K. S. A. Kumar (✉) · K. S. Karthika
ICAR - National Bureau of Soil Survey and Land Use Planning, Regional Centre, Bangalore, India



B. Giri, A. Varma (eds.), *Soil Health*, Soil Biology 59,
https://doi.org/10.1007/978-3-030-44364-1_9

**Table 9.1** Status of land degradation in India over years

| Year | Area (M. ha) | Organization | References |
|---|---|---|---|
| 1976 | 148.1 | National Commission on Agriculture | NCA (1976) |
| 1978 | 175.0 | Ministry of Agriculture: soil and water conservation division | MoA (1978) |
| 1980 | 95.0 | Department of Environment | Vohra (1980) |
| 1984 | 129.6 | Society for Promotion of Wastelands Development | Bhumbla and Khare (1984) |
| 1985 | 123.0 | National Wastelands Development Board | NWDB (1985) |
| 1985 | 53.3 | National Remote Sensing Agency | NRSA (2000) |
| 1985 | 173.6 | Ministry of Agriculture | MoA (1985) |
| 1994 | 107.4 | Ministry of Agriculture | MoA (1994) |
| 1994 | 187.7 | NBSSLUP | NBSSLUP (1994) |
| 2004 | 146.8 | NBSSLUP (revised) | NBSSLUP (2005) |

Source: Modified from Bhattacharyya et al. (2015)

Agricultural soil is a limited resource, the value of which was built by man over decades or even centuries. Soil has considerable storage and buffering capacity closely related to its components particularly organic matter. Storage refers to water, mineral nutrients, gases and multitude of mineral substances. Soils harbour abundant biodiversity and the biological activities are mainly contributing to the structure and fertility of the soils.

Soils possess natural and man-made contaminants, which build up in soil and released through water or air. Certain contaminants exceed the irreversibility threshold for storage and buffering capacity. Early warning systems are hence essential to prevent damage to environment and risks to public health. Soil has potentially rapid degradation rates and extremely slow formation and regeneration process. Deterioration in soil health leads to reduction in soil quality which is explained by soil degradation. Soil degradation is a major cause of concern as it has adverse effects on productivity, air and water quality and human and animal health (Manna and Sharma 2009). Available evidence points that the soil may be increasingly threatened by a range of anthropogenic activities, degrading it. Among the threats are erosion, decline in organic matter, local and diffuse contamination, compaction, acidification, salinization, loss of biodiversity and pollution (European Commission 2002). In the world, about 25% of land area has been degraded and recent estimates showed a loss of 24 billion tons of fertile soil per year indicating an alarming trend that would make 95% of Earth's land area degraded by 2050. Over the years several institutions have assessed the extent of land degradation in India. It is presented in Table 9.1. In India, around 146.8 Mha is degraded (NBSS & LUP 2004). Water erosion is the most serious land degradation problem in India, resulting in loss of topsoil and terrain deformation. Degradation of the soil resource implies ruining of the asset of not only for current farmers but also the farming options of generations to come. There should be a special focus on sustainability of farm land and soils and protection of its

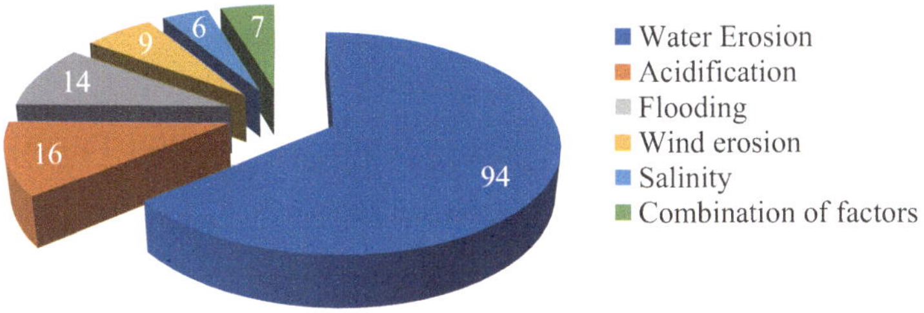


**Fig. 9.1** Extent of soil degradation in India (in Mha)

fertility and prevention of contamination. In this context, soil health gains huge importance as it is the base for supporting life by various means.

Soil health is defined as being a state of dynamic equilibrium between flora and fauna and their surrounding soil environment in which all the metabolic activities of the former proceed optimally without any hindrance, stress or impedance from the latter (Goswami and Rattan 1992). Soil health is considered as the state of a soil at a particular time, equivalent to the dynamic soil properties that change in short term (Goswami 2006). Soil degradation is a threat to agricultural productivity. It is being regarded as a global threat (Lal and Stewart 1990), which has severe impacts on environment (Lal 1997) and water quality (Lal and Stewart 1994). Soil degradation could be by processes of soil erosion, compaction, acidification, runoff, crusting, soil organic matter loss, salinization, and nutrient depletion, accumulation of heavy metals or toxins (Fig. 9.1).

In this chapter, we have attempted to understand the abiotic and biotic factors that influence soil health and soil degradation. A flowchart to understand the factors and their influence on soil health and soil degradation is represented in Fig. 9.2.

## 9.2 Abiotic Factors

Abiotic factors that are responsible for influencing soil health or soil degradation may be broadly classified under Physical factors and Chemical factors. The physical factors of texture, structure, water and temperature along with the soil reaction, acidity, alkalinity, salinity and sodicity, changes in status of soil nutrients affect soil health and quality. These factors play a vital role in the productivity of soil. It is obvious in the soil system that all the physical and chemical characteristics of the soil are interrelated and hence, it becomes a tedious task to classify each of the factors. However, an attempt has been done to classify the factors and the very important properties which are highly influenced by these factors.

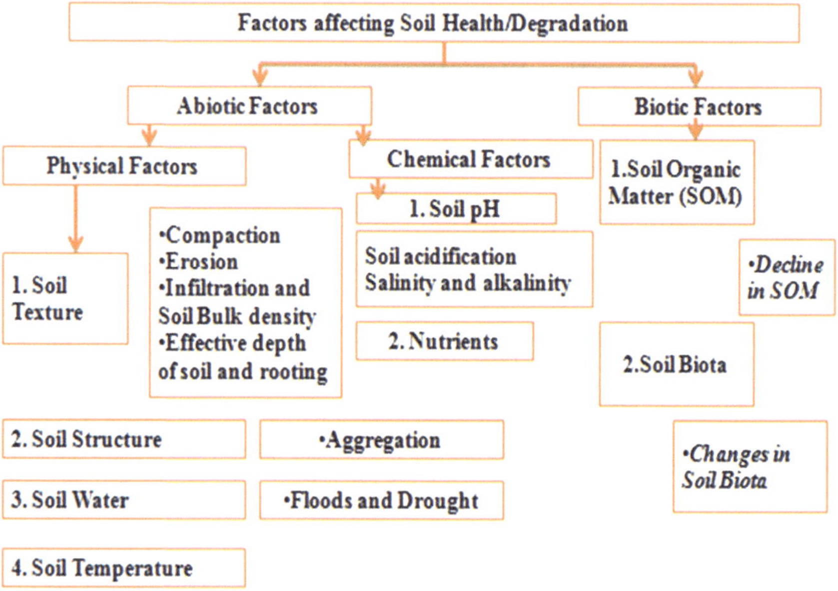


**Fig. 9.2** Factors affecting soil health and degradation

## *9.2.1 Physical Factors*

### 9.2.1.1 Soil Texture

Soil texture is defined as the relative proportions of sand, silt and clay within the fine earth fraction. The texture and coarse fragments content are very important in soils for various reasons (Shaetzl and Anderson 2005). The soil texture plays a very important role in the movement of water, and its retention in soil (Lin et al. 1999) which is dependent on the pore space. More pore space in sand and coarse textured soils allows better movement of water (Brakensiek and Rawls 1994) and the clays and fine textured soils have lower permeability due to lesser pore space restricting the movement of water indicating better water retention.

Soil surface area is also dependent on the soil texture and content of coarse fragments. The higher the soil surface area, better will be the retention of water, nutrients, cations and anions, coatings that impart colour to soil (Shaetzl and Anderson 2005). When soil texture is linked with soil quality, the shape of the particles is much important in understanding how these two are related. Sand particles are generally round, while silt and clay particles are usually thinner and flatter. This again is related with surface area. In sandy soils, dominated by round-shaped particles, more space is available for air and water, indicating better soil aeration and poor water holding capacity that could be easily prone to drought. The size distribution of these particles is equally important in the development and formation of soil aggregates and their stability. Thus, clay soils with smaller particle size could get aggregated to crumb structure with larger pores between them and smaller pores within. When considering the nutrient status as influenced by the soil texture, soils with increased clay content retain more nutrients and possess higher cation exchange capacity. These soils can sequester more organic matter.

Soil texture is important in understanding the susceptibility of soil to erosion, compaction, and understanding the effective soil depth and rooting. Soil texture is a permanent property, which does not change over time, the total porosity and the number and size of pores in a soil are influenced by the soil management practices.

#### 9.2.1.2 Soil Compaction

Soil compaction is mainly due to heavy machinery for farming operations. The compaction causes deterioration of structure and restrict the root growth, water storage capacity, fertility, biological activity and stability. If the texture is too open, it will also affect crop growth, as roots cannot hold properly on ground. Such soils are easily blown by wind, exposing the roots to air and drought. When the soil is loose aiding in unrestricted root penetration, it is described as a healthy soil.

#### 9.2.1.3 Erosion

Erosion is a natural geological phenomenon resulting from the removal of soil particles by water or wind and transporting them to elsewhere. Human activities can dramatically increase erosion rates and severe erosion is largely irreversible. More the coarse fragments, more the potential void space. This helps in resisting compaction, erosion and in maintaining good structure (Poesen et al. 1990; van Wesemael et al. 1995). The better aggregation and higher organic matter improves the capacity of soil in resisting soil erosion, which thereby positively affects the soil health.

#### 9.2.1.4 Infiltration and Soil Bulk Density

Texture of the soils determines the bulk density and infiltration capacity of soils. Many soils with more coarse fragments possess lower bulk densities (Stewart et al. 1970). Thus, these soils retain some water, affecting soil water characteristics beyond just their effect on void space (Ugolini et al. 1996). The lower the bulk density, the better the soil is. More pore space results in lowering the bulk density of the soil, which is actually advantageous. Compact soils usually have less pore space resulting in poor infiltration of water. This could lead to flood in those areas with more compact soils (Jhonson 2009). More pore space indicates better infiltration of water into soil.

#### 9.2.1.5 Effective Depth of Soil and Rooting

Soil depth is an important factor, which helps in understanding the ability of soil to support plants. The depth of soil available for roots to explore for water and nutrients is defined as the effective depth of soil. The requirement of root depth varies depending on the crops and species. Plantation crops are deep rooted, while vegetable crops are shallow rooted. A soil could be considered as good only if it has sufficient depth for extensive root development. A good arable soil should be easy to work with implements (Jhonson 2009). Usually problem soils have adverse physical properties as a result of their inherent chemical properties (Norton et al. 1999). Depth to root restricting layer along with soil texture thus plays an important role in maintaining soil health.

### *9.2.2 Soil Structure*

Soil structure refers to the arrangement of primary soil particles: sand, silt and clay. A good agricultural soil should have a good structure that allows water infiltration, with better water holding capacity, lesser problems of crusting and the ability to resist forces that results in soil erosion. These properties are linked with the soil texture as well as microbial activities. Soil organic matter influences physical properties of soil such as soil structure, water retention, surface crusting and soil erodibility (Stott et al. 1999). Knowledge on soil structure helps us to know the shape and size of pore spaces in soil, which is important in the movement of water, air and heat in soil (Jhonson 2009). Wet–dry cycles, freeze–thaw cycles, root activity and fauna all play a role in the formation of soil structure (Materechera et al. 1992). Though any structure can be formed in soils of any texture, fine-textured soils have better cohesive forces indicating lesser chances of destruction as a result of tillage, compaction etc. Soil biota influences the formation of granular and subangular blocky structure (Shaetzl and Anderson 2005). A soil is considered as healthy when the soil has a good crumb structure, without any clods and soil breaks apart easily. Maintenance of good soil structure is essential for soil health as this is very important in regulation of soil–water cycle and for providing a favourable rooting medium for plants. Soil structure could be maintained by the mechanisms of soil aggregation, formation of biostructures and pore networks. This could result from the soil biological activities mainly by the ecosystem engineers consisting of megafauna, mesofauna, fungi and bacteria (Kibblewhite et al. 2008).

#### 9.2.2.1 Aggregation

Higher the aggregation, higher will be the soil porosity. Larger pores occur usually between soil aggregates or peds (Schimel et al. 2005). When the soil gets disturbed, there will be a reduction in organic matter and aggregates turn less stable, resulting in an increased susceptibility to soil erosion (Darmody and Norton 1994). When the soil aggregates are stable, these ensure resistance to the erosive forces of water drops during irrigation and rainfall (Arshad et al. 1996). Physical degradation of soil is resisted by several microbial activities too as these biological activities help in developing and maintaining soil aggregates and soil structure (Tisdall and Oades 1982). Aggregation and the space between aggregates affects the processes of infiltration, percolation, permeability, soil aeration and nutrient leaching in soil, which are mainly related to textural characteristics. When considering the textural aspect and aggregation, larger particles have poor aggregation as evident from the beach sands as a result of poor organic matter and clay, which serve as the binding agents.

### *9.2.3 Soil Water*

Soil water content is influenced by soil texture, structure, aggregate stability, bulk density and earthworm numbers. These all are relevant in understanding a soil's ability to accommodate water entry for prolonged periods especially during high-intensity rainfall and frequent irrigation events. Soil structure plays an important role in maintaining soil water content. The granular structure is more porous indicating better infiltration and permeability. However, well-developed platy structure impedes infiltration. Similarly, prismatic structures result in slow permeability of water. Increased compaction results in reduced percolation ultimately leading to water-logged condition. Platy structure results in waterlogged soils, especially when there is a heavy downpour and intense flooding. Soil water retention is also influenced by bulk density and biological activity. Lower bulk densities are an indication of higher pore space indicating better retention of water in soil. Macropores are more in the soils with more number of earthworms as a result of the earthworm burrows and these help in easy draining of surface water (Karlen et al. 1994). A stress in soil water encompasses both salt stress and drought stress, where salt stress is dealt under soil chemical factors. Soil water controls soil biology and soil chemistry, thus indicating its inevitable role in determining the health and quality of a soil.

### *9.2.4 Soil Temperature*

Soil temperature is an abiotic factor which influences soil moisture content and biological activity. It acts as an insulator to protect soil organisms from the extremes of fluctuations in air temperature (Jhonson 2009). The chemical and biological activities in soil are affected by soil temperature as more biological activity results in warmer soils. Soil temperature affects the movement of water through soil matrix.

The soil water content has an effect on soil temperature as higher water contents result in increased absorption of sun's energy (Shaetzl and Anderson 2005).

## 9.3 Chemical Factors

### *9.3.1 Soil pH*

Soil pH indicates soil reaction. Soil acidity and alkalinity are defined based on the soil reaction. The soil reaction plays a very important role in nutrient availability, occurrence and activity of soil microbes and plant nutrition. Acidification and salinization severely impair the soil functions and reduce plant biomass productivity in the soil.

### *9.3.2 Soil Acidification*

Soil acidity could affect several factors that are related to the various physical, chemical and biological properties of soil that make it the best suited to fulfill the "desired purpose", which is an important concept in soil quality (Schjonning et al. 2004). Weathering of rocks, nature of the parent material from which the soil is formed, the interaction between these two factors, the time period involved in the process of weathering determine the degree of soil acidification and the development of different forms of acidity (McLean 1982). Soil pH affects the soil properties, and the maximum availability of nutrients occurs within pH range 5–6. The availability of most of the nutrients decreases when the pH is below 5. The ionic concentrations of Al and Mn increase with soil acidity reaching toxic levels even at times. Soils are naturally acidic in humid tropics. Agricultural practices like addition of organic matter and acid producing chemical fertilizers attenuates soil acidity. Acid soils are stressed environment for plants and microorganisms. Fine textured soils with high amounts of soil organic matter have more hydrogen ions than lighter soils with lower amounts of clay and organic matter.

### *9.3.3 Soil Salinity and Sodicity*

Soil salinity deteriorates soil health and it is one of the forms of soil degradation which is predominantly a problem in the semi-arid and arid regions of the world. Agricultural intensification, along with unfavourable natural conditions has accelerated soil salinity. Salinity develops as a result of natural and anthropogenic processes. Natural salinity could be a result of weathering or by oceanic salt deposition. Anthropogenic salinity results from an imbalance in the soil hydrologic cycle between water applied and the transpiration rates (Kumar 2013). Soil salinization is the accumulation in soils of soluble salts of sodium, magnesium and calcium to the extent that soil fertility is severely reduced. Soil salinity is measured in terms of electrical conductivity (ECe) with the exchangeable sodium percentage (ESP) of the saturated paste extract. The process is often associated with irrigation water and is common in subhumid, semi-arid and arid climates. Soil salinity results in decrease in plant available water and thereby water stress. Salinity affects soil physical properties by flocculation which is caused by binding of fine particles to form aggregates, which positively affect soil aeration, penetration of roots and root growth. Soil salinization can improve soil structure by aggregation and flocculation; however, it negatively affects crop growth and yield. Salinity is one of the serious soil degradation threats our world faces and it affects agricultural productivity. Crops grown under these conditions suffer from higher osmotic stress, poor physical conditions, nutritional disorders and toxicities and reduced crop productivity.

Sodicity is another major chemical degradation taking place in soils. Sodic soils possess unfavourable physical conditions, decreased nutrient availability, biological activity, increased soil pH, sometimes exceeding 10 and exchangeable sodium percentage generally >15%, up to 90% or so making it not suitable for cultivation of most of the crops (Kanwar and Bhumbla 1969). Sodicity could develop under certain topographic conditions and as a result of weathering of rocks (Abrol et al. 1988). Sodic soils are those that contain sodium ions on their exchangeable complex. However, a soil is considered as sodic when the destruction of soil structure initializes as a result of increased Na concentration (Rengasamy and Sumner 1998). This takes place due to the clay dispersion on addition of water to soil. Clay dispersion clogs the soil pores, develops a calcareous hard pan in the subsurface (Sharma et al. 2016) and this results in massive destruction of soil structure which negatively affects in penetration of roots to soil, establishment of seedlings and water movement in the soil (Dougherty and Anderson 2001). It also reduces water infiltration rate and permeability throughout the soil profile. The physical properties of soils are thus affected negatively by soil sodicity by developing the problems of surface crusting, poor infiltration and percolation which is hindered by the slaking associated with clay dispersion (So and Aylmore 1993). The problems of infiltration, percolation along with the degree of structural deformity of a soil are determined by two of the main factors such as soil texture and exchangeable sodium percentage (ESP) (Sharma et al. 2016). Soil sodicity induces changes in hydraulic conductivity of a soil (Levy et al. 2005). Structural stability of the soils varies according to the extent of sodicity and columnar structure dominates in sodic soils.

### *9.3.4 Nutrients*

Status of nutrients in soil and their transformations are very critical to soil health. Deficiencies of nutrients adversely affect soil health and they reduce crop productivity. Proper management of soils is very much essential in overcoming the issues related to nutrient stress, which is a very common phenomenon. Soil pH affects nutrient availability and transformations in soil and nutrient uptake by plants. An elevated soil pH affects the availability of essential nutrients for plant growth (Singh 2009; Sharma et al. 2016) and nutrient toxicities are seen associated with the development of sodicity (Sharma et al. 2016). When the soil pH is above 6, availability of P, B and Mn decreases. As and when the pH rises above 7, availability of K, Mg, Cu and Zn is likely to decrease. However, availability of Mo increases under alkaline conditions. The cations like Ca, Mg, K and Na have a direct relationship with pH. At lower pH, Fe, Mn and Al are highly soluble and attain toxic levels, whereas at higher pH these ions become deficient. Phosphorus availability is highly pH dependent and its availability is the highest when pH is neutral or slightly acidic and it decreases as the pH becomes strongly acidic or strongly alkaline. In general, with increase in soil acidity, the availability of micronutrients present in cationic forms increases and anionic forms, namely Mo and B, decreases. The activity of bacteria and actinomycetes are seen under mild acidic, neutral and high pH soils whereas fungi are active in a wide range of pH dominating in acid soils than that of soils with intermediate pH or higher pH.

### *9.3.5 Plant Nutrient Depletion*

Crop production often leads to depletion or excess of plant nutrients in soils. High input agriculture with major nutrients alone often leads to deficiency of secondary and micronutrients. Regular monitoring is necessary to forewarn such deficiencies or excesses. Nitrogen and phosphorus are the two major nutrients which are highly influential in the soil system when the soil health and ecosystem services, such as agricultural production are concerned. Soil microbes are frequently seen to be N limited though availability of C is the primary limiting factor for their activity (Schimel et al. 2005). Soil functional ability is influenced by N supply. When the N inputs are lower with slight leaching and emission losses, demand for N increases in an undisturbed natural soil system. However, in disturbed soil systems, there are evidences of increased losses of N from soil as a result of increased rates of organic matter decomposition and mineralization to inorganic forms, along with N lost in the form of agricultural produce. So an imbalance in C and N requirements of the biomass will affect the pool of N available to microorganisms to support soil functions and plant growth which is a clear indication of declining soil health. Hence, additional inputs of N are very much essential in N poor soil so as to maintain soil health in agricultural systems. Similarly, not meeting the requirement of P also causes a decline in soil health when there exists a shortage of P in its natural pool due to erosion or cropping losses.

To tackle the nutrient losses, external addition of organic manures or mineral fertilizers are essential to restore and sustain soil health. When proper management strategies are adopted for these and implemented effectively, soil health could be maintained and productivity could be achieved in well-managed agricultural systems. However, industrial agriculture results in environmental issues due to excess additions of nutrients beyond the capacity of the soil–plant system. The soil system remains unhealthy and polluted under these circumstances (Kibblewhite et al. 2008).

## 9.4 Biotic Factors

### *9.4.1 Soil Organic Matter*

Organic matter in soil is composed of organic material (plant roots, leaves and excrements), living organisms and humus, the end product of decomposition organic matter. Organic matter plays an important role in key soil functions and is essential for the soil fertility. It performs binding and buffering capacity of soils, thus limiting pollution of water. Soil organic matter is very important in maintaining soil health and it is the food source for soil microorganisms and soil fauna. It serves as soil's storehouse for nutrients viz. N, P, S and SOM improves physical and chemical properties of soil. It helps in improving physical properties like soil structure, porosity and water holding capacity. Carbon in soil plays a major role in global carbon cycle through sequestering of $CO_2$ into SOC pool. Soils higher in organic matter content have lower bulk densities.

The fungi and bacteria, primary agents of decomposition, are a food source for several microbivorous predators. Several studies have reported that these microorganisms aid in release of nutrients and stimulate microbial population which thereby regulates the rate of organic matter decomposition (Coleman and Hendrix 2000). Earthworms, termites etc. are also involved in the process of decomposition. The process of decomposition is influenced by several factors like climate and soil conditions. Soil organic matter is present in its active and passive fractions in which the active fraction contributes to nutrient cycling and passive fraction contributes to soil structural features. Many microbes such as fungi, bacteria, earthworms and termites which carry out decomposition process are also contribute towards modifications in soil structure and nutrient cycling. This indicates the importance of organic matter decomposition in improving the soil health by means of improving soil structure and nutrient cycling (Kibblewhite et al. 2008).

Conversion of forests to farming systems often results in decline of organic matter. Organic matter is the life of soil. When there is a decline in organic matter, it indicates a decline in soil health. An observation indicating a decrease in organic matter content to be seen with utmost priority and proper management strategies need to be adopted for the addition of organic manures so as to maintain the soil organic matter.

### *9.4.2 Soil Biota*

Large varieties of living organisms inhabit soil. Soil bacteria, fungus, protozoa and small organisms play essential role in maintaining the physical and biochemical properties needed for soil fertility. Larger organisms, worms, insects, snails and small arthropods break up organic matter which is further degraded by microorganisms. Soil organisms themselves serve as source of nutrients, suppress external pathogens and break down pollutants into harmless components. Soil biota helps in maintaining porosity in soil, especially macroscopic soil fauna, consisting mainly of worms and termites. Biopores are a result of their movement in soil (Dexter 1978). A fresh earthy smell of soil along with residues at various stages of decomposition on soil surface and in the topsoil along with more earthworms, many holes and casts, are indications of a healthy soil. Macropores are a result of activity of roots and soil biota, e.g. worms. Soil biological activity is influenced by the soil physical and chemical properties. Soil organisms and plant roots live and function in the pores. When the soil loses porosity, roots cannot grow as well, and many organisms have more difficulty in surviving.

The biological processes that contribute to carbon transformation are carried out by decomposers like fungi, bacteria, microbivores, detritivores and nutrient cycling by nutrient transformers like decomposers, are functional assemblages of interacting organisms which are sometimes termed as "key functional groups" (Lavelle 1997; Swift et al. 2004). Biological population is regulated by biocontrollers consisting of predators, microbivores and hyperparasites. Soil biological processes decide the ecosystem services, which along with the biotic interactions are inevitable in the concept of soil health (Young and Ritz 2005). The pores in the soil and soil biota are interrelated as the pores are where the biota exists, which in turn is related to the dynamics of water, solutes, gases and organisms in the soil matrix. The reason behind soil dynamics in soil is thereby the myriad of biotic interactions in the pore networks occurring within the soil matrix. The pore networks help in providing surfaces for colonization. The nature of pore networks defines how the organisms move through the soil volume.

Soil organisms favour the process of aggregation in soil thereby enhancing the soil structure along with improved pore networks. The process takes place by adhesion, coating, enmeshment and alignment of particles and gross movement (Lavelle et al. 1997; Ritz and Young 2004). On the other hand, all these activities and decomposition of organic matter by soil biota, may degenerate the structural integrity too indicating a two-directional interactive relationship existing between the community and the habitat. This has led to the development of a concept of "soil as a self-organizing system" encompassing the feed-forward and feedback interactions existing between the biota and architecture of the soil (Young and Crawford 2004). The self-organization capacity could be recognized as an integral component of soil health. A loss in biodiversity or a reduction in functions by soil biota along

**Table 9.2** Soil properties influenced by the abiotic and biotic factors

| Factors | Soil properties | References |
|---|---|---|
| Soil texture | Soil compaction | Shaetzl and Anderson (2005) |
| | Soil water movement | Lin et al. (1999) |
| | Soil water retention | Lin et al. (1999) |
| | Infiltration | Jhonson (2009) |
| | Bulk density | Stewart et al. (1970) |
| | Porosity | Shaetzl and Anderson (2005) |
| Soil structure | Soil water movement | Jhonson (2009) |
| | Soil air movement | Jhonson (2009) |
| | Aggregation | Shaetzl and Anderson (2005) |
| | Porosity | Shaetzl and Anderson (2005) |
| | Root penetration | Shaetzl and Anderson (2005) |
| | Infiltration | Shaetzl and Anderson (2005) |
| | Percolation | Shaetzl and Anderson (2005) |
| Soil water | Infiltration | Jhonson (2009) |
| | Percolation | Jhonson (2009) |
| | Soil chemistry | Biswas and Naher (2019) |
| | Soil biology | Biswas and Naher (2019) |
| | Soil colour | Jhonson (2009) |
| Soil temperature | Soil moisture | Jhonson (2009) |
| | Biological activity | Shaetzl and Anderson (2005) |
| Soil pH | Soil water content | Jhonson (2009), Dougherty and Anderson (2001) |
| | Nutrient availability | Singh (2009), Sharma et al. (2016) |
| | Aggregation | Jhonson (2009), So and Aylmore (1993) |
| | Soil structure | So and Aylmore (1993) |
| | Hydraulic conductivity | Levy et al. (2005) |
| Soil organic matter | Soil structure | Kibblewhite et al. (2008) |
| | Nutrient cycling | Kibblewhite et al. (2008) |
| Soil biota | Soil structure | Shaetzl and Anderson (2005) |
| | Aggregation | Shaetzl and Anderson (2005) |
| | Porosity | Dexter (1978), Shaetzl and Anderson (2005) |
| | Root penetration | Shaetzl and Anderson (2005) |
| | Nutrient transformation | Kibblewhite et al. (2008) |

with changes in soil structure and physical–chemical properties reduce the quality of delivering ecosystem services which is exhibited by a decline in soil health. Table 9.2 represents the soil properties as influenced by the abiotic and biotic factors.

## 9.5 Conclusion

In order to carry out the vital functions, maintaining soil health is essential for sustainability. However, soil is under threat from a wide range of human activities which are undermining the long-term viability of the resource. When multiple threats happen simultaneously, their effects are compounded. If not countered, they can result in soil degradation and loss of soil capacity to carry out its functions. Soil contamination may result in loss of some or several functions of soil and possible contamination of water. Contaminated farm land can lead to negative consequences on food chain and thus affect human health. Abiotic factors that affect soil health include by soil erosion resulting in loss of surface horizons alteration of soil water regime via artificial drainage, soil salinization due to poor irrigation practices, loss of natural soil organic matter caused by arable production or contamination. Soil health is therefore decided by the land use and management factors. Soil texture, soil structure, soil water and temperature decide the soil habitat conditions along with the chemical factors of soil pH, bulk density, nutrient concentration and organic matter content. These are also influenced by land use and management. These factors along with the biotic factors and their interactions determine the condition of soil system and soil health. Biotic factors include the soil organic matter, soil biota, their type and functional assemblages. Biotic factors affecting soil health indicate that all the factors are interrelated and overlapping. Soil system is a highly integrated system and hence any disturbance to any function will alter the dynamics of others. The degree of interrelatedness is higher at the higher trophic levels. This indicates that changes in the biota of the higher trophic groups influence the biota on lower trophic levels. Though organic matter decomposition occurs at the lower trophic level, it has severe implications in soil health.

## References

Abrol IP, Yadav JSP, Massoud FI (1988) Salt-affected soils and their management. FAO soils bulletin no. 39, Food and Agriculture Organization of the United Nations, Rome

Arshad MA, Lowery B, Grossman B (1996) Physical tests for monitoring soil quality. In: Doran JW, Jones AJ (eds) Methods for assessing soil quality, SSSA special pub. 49. Soil Science Society of America, Madison, WI, pp 123–141

Bhattacharyya R, Ghosh BN, Mishra PK, Mandal B, Rao Ch S, Sarkar D, Das K, Anil KS, Lalitha M, Hati KM, Franzluebbers AJ (2015) Soil degradation in India: challenges and potential solutions. Sustainability 7:3528–3570

Bhumbla DR, Khare A (1984) Estimate of wastelands in India, Society for promotion of wastelands development. Allied, New Delhi, p 18

Biswas CJ, Naher CA (2019) Soil nutrient stress and rice production in Bangladesh. In: Hasanuzzaman M, Fujita M, Nahar K, Biswas JK (eds) Advances in rice research for abiotic stress tolerance. Woodhead, Sawston, pp 431–445. https://doi.org/10.1016/B978-0-12-814332-2.00021-6

Brakensiek DL, Rawls WJ (1994) Soil containing rock fragments: effects on infiltration. Catena 23:99–110

Coleman DC, Hendrix PF (eds) (2000) Invertebrates as webmasters in ecosystems. CAB International, Wallingford, UK
Darmody RG, Norton LD (1994) Structural degradation of a prairie soil from long term management. In: Ringrose-Voase AJ, Humphreys GS (eds) Soil micromorphology: studies in management and genesis, Developments in soil science 22. Elsevier, Amsterdam, pp 641–650
Dexter AR (1978) Tunnelling of soil by earthworms. Soil Biol Biochem 10:447–449
Dougherty W, Anderson A (2001) Sodic soils: their properties and management. In: Cattle SR, George BH (eds) Describing, analysing and managing our soil. The University of Sydney and the Australian Soil Science Society, Sydney, NSW, pp 105–121
European Commission (2002) Communication of 16 April 2002 from the commission to the council, the European Parliament, the economic and social committee and the committee of the regions: towards a thematic strategy for soil protection [COM (2002) 179 final]. European Commission, Brussels
Goswami NN (2006) Soil and its quality *vis-a-vis* sustainability and society-some random thoughts. In: Proceedings of the international conference on soil, water and environmental qulaity. Indian Society of Soil Science, New Delhi, pp 43–58
Goswami NN, Rattan RK (1992) Soil health - key to sustained agricultural productivity. Fertil News 37(12):53–60
Jhonson C (2009) Biology of soil science. Oxford Book Company, Jaipur, p 298
Kanwar JS, Bhumbla DR (1969) Physico chemical characteristics of sodic soils of Punjab and Haryana and their amelioration by the use of gypsum. Agrokem Talajatan 18:315–320
Karlen DL, Wollenhaupt NC, Erbach DC, Berry EC, Swan JB, Eash NS, Jordahl JL (1994) Crop residue effects on soil quality following 10-years of no-till corn. Soil Tillage Res 31:149–167
Kibblewhite MG, Ritz K, Swift MJ (2008) Soil health in agricultural systems. Phil Trans R Soc B 363:685–701
Kumar M (2013) Crop plants and abiotic stresses. J Biomol Res Ther 3:1. https://doi.org/10.4172/2167-7956.1000e125
Lal R (1997) Degradation and resilience of soils. Phil Trans R Soc London Ser B 352:997–1010
Lal R, Stewart BA (1990) Soil degradation. A global threat. Adv Soil Sci 11:xiii–xvii
Lal R, Stewart BA (eds) (1994) Soil processes and water quality. Lewis Publishers, Boca Raton, FL, p 398
Lavelle P (1997) Faunal activities and soil processes: adaptive strategies that determine ecosystem function. Adv Ecol Res 27:93–132
Lavelle P, Bignell D, Lepage M, Wolters V, Roger P, Ineson P, Heal OW, Dhillion S (1997) Soil function in a changing world: the role of invertebrate ecosystem engineers. Eur J Soil Biol 33:159–193
Levy GJ, Goldstein D, Mamedov AI (2005) Saturated hydraulic conductivity of semiarid soils: combined effects of salinity, sodicity, and rate of wetting. Soil Sci Soc Am J 69:653–662
Lin HS, McInnes KJ, Wilding LP (1999) Effects of soil morphology on hydraulic properties. I. Quantification of soil morphology. Soil Sci Soc Am J 63:948–954
Manna MC, Sharma KL (2009) Soil and water quality. In: Goswami NN, Rattan RK, Dev G, Narayanaswamy G, Das DK, Pal DK, Rao DLN (eds) Fundamentals of soil science. Indian Society of Soil Science, New Delhi, pp 642–668
Materechera SA, Dexter AR, Alston AM (1992) Formation of aggregates by plant roots in homogenized soils. Plant Soil 142:69–79
McLean EO (1982) Soil pH and lime requirement. In: Page AL, Miller RH (eds) Methods of soil analysis part 2, Agronomy monograph no. 9, 2nd edn. ASA and SSA, Madison, WI, pp 135–151
MoA (1978). Indian Agriculture in Brief, 17th ed.; Directorate of Economics and Statistics, Ministry of Agriculture, Department of Agriculture and Cooperation, Ministry of Agriculture and irrigation, Government of India, New Delhi
MoA (1985) Indian Agriculture in Brief, 20th ed.; Directorate of Economics and Statistics, Ministry of Agriculture, Department of Agriculture and Cooperation, Ministry of Agriculture, Government of India, New Delhi

MoA (1994) Indian Agriculture in Brief, 25th ed.; Directorate of Economics and Statistics, Ministry of Agriculture, Department of Agriculture and Cooperation, Government of India, New Delhi
National Bureau of Soil Survey and Land Use Planning (NBSS & LUP) (1994) Global assessment of soil degradation (GLASOD) guidelines. NBSS & LUP, Nagpur
National Bureau of Soil Survey & Land Use Planning (NBSS & LUP) (2004) Soil map (1:1 million scale). NBSS & LUP, Nagpur
National Bureau of Soil Survey and Land Use Planning (NBSS & LUP) (2005) Annual report. NBSS & LUP, Nagpur
NCA (1976) Report of the national commission on agriculture, National commission of agriculture. Government of India, New Delhi, pp 427–472
Norton D, Shainberg I, Cihacek L, Edwards JH (1999) Erosion and soil chemical properties. In: Rattan L (ed) Soil quality and soil erosion, p 329
NRSA (2000) Waste land atlas of India. Government of India, Hyderabad
NWDB (1985) Ministry of environment and forests, national wasteland development board guidelines for action. Government of India, New Delhi
Poesen J, Ingelmo-Sanchez F, Mucher H (1990) The hydrological response of soil surfaces to rainfall as affected by cover and position of rock fragments in the top layer. Earth Surf Process Landf 15:653–671
Rengasamy P, Sumner ME (1998) Processes involved in sodic behaviour. In: Sumner ME, Naidu R (eds) Sodic soils: distribution, properties, management and environmental consequences. Oxford University Press, New York, pp 35–50
Ritz K, Young IM (2004) Interactions between soil structure and fungi. Mycologist 18:52–59. https://doi.org/10.1017/S0269915X04002010
Schimel JP, Bennett J, Fierer N (2005) Microbial community composition and soil nitrogen cycling: is there really a connection? In: Bardgett RD, Usher MB, Hopkins DW (eds) Biological diversity and function in soils. Cambridge University Press, Cambridge, UK, pp 172–188
Schjonning P, Elmholt S, Christensen BT (2004) Soil quality management-concept and terms. In: Schjonning P, Elmholt S, Christensen BT (eds) Managing soil quality challenges in modern agriculture. CABI Publishing, Wallingford, UK, p 344
Shaetzl JR, Anderson S (2005) Soils: genesis and geomorphology. Cambridge University Press, New York, p 817
Sharma PC, Thimappa K, Kaledhonkar MJ, Chaudhari SK (2016) Reclamation of alkali soils through gypsum technology. ICAR-CSSRI/Karnal/Technology Folder/2016/01, p 4
Singh G (2009) Salinity related desertification and management strategies: Indian experience. Land Degrad Dev 20:367–385
So HB, Aylmore LAG (1993) How do sodic soils behave the effects of sodicity on soil physical behavior. Soil Res 31:761–777
Stewart VI, Adams WA, Abdullah HH (1970) Quantitative pedological studies on soils derived from Silurian mudstones. II. The relationship between stone content and the apparent density of the fine earth. J Soil Sci 21:2484–2255
Stott DE, Kennedy AC, Cambardella CA (1999) Impact of soil organisms and organic matter on soil structure. In: Lal R (ed) Soil quality and soil erosion. CRC Press, Boca Raton, FL, p 329
Swift MJ, Izac AMN, Van Noorwidjk MN (2004) Biodiversity and ecosystem services in agricultural landscapes—are we asking the right questions? Agric Ecosyst Environ 104:113–134
Tisdall JM, Oades JM (1982) Organic matter and water-stable aggregates in soils. J Soil Sci 38:141–163
Ugolini FC, Corti G, Agnelli A et al (1996) Mineralogical, physical, and chemical properties of rock fragments in soil. Soil Sci 151:59–75
van Wesemael B, Poesen J, Figueiredo T (1995) Effects of rock fragments on physical degradation of cultivated soils by rainfall. Soil Tillage Res 33:229–250
Vohra BB (1980) A policy for land and water, Department of environment, vol 18. Government of India, New Delhi, pp 64–70

Young IM, Crawford JW (2004) Interactions and self-organizationin the soil-microbe complex. Science 304:1634–1637. https://doi.org/10.1126/science.1097394

Young IM, Ritz K (2005) The habitat of soil microbes. In: Bardgett RD, Usher MB, Hopkins DW (eds) Biological diversity and function in soils. Cambridge University Press, Cambridge, UK, pp 31–43

# Chapter 10
# Seaweeds: Soil Health Boosters for Sustainable Agriculture

**Inderdeep Kaur**

**Abstract** Healthy soil is a key component to growing high quality crops, and it is essential that we manage our soils well. This would be majorly possible with intensive inputs of fertilizers, and irrigation as well as integrated soil management. Fertilization of soil with chemical-based inputs which has been practiced by farmers since several decades, has challenged the very existence of humankind. Intensive eco-friendly fertilization of soil is, therefore, a major challenge in face of synthetic fertilizers opening up the pandora box of environmental degradation and health hazards. Climate change, loss of biodiversity, and urbanization have also emerged as serious challenges to farmers. New innovations in fertilizer options for soil management need to be worked out. To lift agricultural productivity and food supply, fertilizer availability and affordability is prime concern of both farmers and stakeholders. Better still would be to adopt an integrated approach to soil management. It would not only address issues of environmental quality and land degradation but would potentially improve agricultural production and crop quality. Macro algae commonly referred to as seaweeds have fast emerged as promising candidates in soil management practices and "green" agriculture. Besides eliciting a growth-promoting effect on plants, seaweeds also affect the physical, chemical, and biological properties of soil which in turn influence plant growth. Seaweeds in fresh and dried form and seaweed concentrates enhance soil health by improving moisture-holding capacity and by promoting the growth of beneficial soil microbes, besides fertilizing it.

Seaweeds offer a wide spectrum agri inputs in the form of biofertilizers, soil conditioners, amendments, enhancers and biostimulants. Seaweeds are preferred because of the high amounts of macronutrients, micronutrients, vitamins, amino acids, and growth regulators, e.g., auxins, cytokinins, and gibberellins, they contain besides phycocolloids of great commercial value. Seaweed extracts are known to enhance seed germination, increase root and plant growth, yield, protein and quality, increase resistance to insects and diseases, resistance to drought and frost and increase shelf life. They are one of the best soil supplements which improve tilth and other properties of soil. Macro algal effectiveness is due to the fact that it fulfils

I. Kaur (✉)
Sri Guru Tegh Bahadur Khalsa College, University of Delhi, Delhi, India



B. Giri, A. Varma (eds.), *Soil Health*, Soil Biology 59,
https://doi.org/10.1007/978-3-030-44364-1_10

the basic needs of both—the soil and the plants, and can efficiently alter the physical and chemical properties of soil. The potential of seaweeds as biofertilizers, biostimulants, soil conditioners, and soil additives for soil health is discussed in this chapter.

**Keywords** Seaweeds · Biostimulants · Soil supplements · Biochar · Mulch · Compost

## 10.1 Introduction

Over the last century global population has quadrupled. From 1.8 billion people around the globe in 1915, there is an increase to 7.3 billion people (UN report 2017) and in 2050, the population is expected to reach 9.7 billion (https://www.un.org/en/sections/issues-depth/population/). Accordingly, food demand is estimated to increase anywhere between 59% and 98% by 2050. Together with increase in consumption of animal products, this rise in population will drive up demand for crop species leading to pressures on agro industry. The constraints to increase crop production by various means would be transmitted to agricultural markets and farmers worldwide. Though theoretically, increasing the amount of agricultural land might appear a good option under such circumstances but practically land under agriculture would remain limited due to economic growth, urbanization, and rising affluence of developing economies. Hence, to meet the demands of about 10 billion people, enhancing productivity on existing agricultural lands would be a workable option. Nonetheless, approximately 50% of total habitable land which is under cultivation has to be "fit and healthy" in order to be productive.

Looking back in history, food demands have always remained high and new ways of farming have ensured food security from time to time. Chemical-based agricultural inputs became a popular practice in the early nineteenth century and synthetic fertilizers were frequently used to increase crop yield. Of late, however, intensive fertilization of soil with synthetic chemicals such as those with urea and potassium for crop management has resulted in significant environmental degradation. Besides environmental hazards viz., soil erosion, water contamination, pesticide poisoning, falling ground water table, water logging, and depletion of biodiversity due to overuse of chemicals (fertilizers, weedicides, and pesticides), health issues like cancer, premature ageing, abortions and many other ailments in humans are also on rise. The fallout of chemical-based agro industry has posed challenge to both, the long-run sustainability of agriculture and the survival of the farming community in developing countries.

Awareness among people on health issues and concern among environmentalists over deteriorating environment, has announced the transition of chemical-based agro industry into organic-based industry. Scientists are since long working out safe alternatives to chemicals and use of organic supplements such as green manure, vermicompost and foliar sprays along with AM fungi, nitrogen fixing bacteria is a common practice. Farming based on these inputs is considered a Good Agricultural Practice (GAP). Biofertilizers from microorganisms, such as bacteria, fungi, cyanobacteria, and macro algae and their metabolites (lipid extracted microalgal biomass residues or LMBRs) capable of enhancing soil fertility, soil health, crop growth, and/or yield, have emerged as a safe option to chemical fertilizers.

Micro algae or cyanobacteria, considered a boon for agro industry, are high value slow release organic fertilizers with established potency for soil conditioning (Metting et al.1990). They are popular amongst farmers especially for growing paddy (Tung and Shen 1985). Blue-green algae-algalization is reported as a safe agronomic practice. The ability of micro algae to produce copious amounts of mucilage that binds soil micelle, has made them popular soil conditioners. The ability of mucilage to absorb and retain water for a long time (Uysal et al. 2015), also greatly improves soil texture (Ibraheem 2007) besides stabilizing soil surfaces. Certain blue green algae (Cyanophyceae) remove sodium from saline sodic soils and increase soil fertility or reclaim damaged soil crusts.

Though micro algal biofertilizers have remained a sustainable option for long but of late micro algae are being tapped for biofuels, as phytoremediators and in maintaining ecosystem health. More so, the biggest limitation of micro algal agri inputs is the narrow range of crop species on which the effect is seen. As a result, agro industry is now flooded with algal products derived from macro algae or seaweeds.

## 10.2 What Are Seaweeds?

Algae, it is generally agreed, are simple plants that constitute a heterogenous assemblage of $O_2$-producing, photosynthetic, nonvascular organisms with unprotected reproductive structures. Regarded to be simple organisms, they are estimated to include from 30,000 to more than 1 million species (Guiry 2012). Today there is no particular classification scheme, and to both, phycologists and non-phycologists, algae are not a taxonomic category in the real sense of the term (Pereira and Neto 2014). Present molecular basis of classification which brings out inter relationships and closeness of origin between organisms, has excluded certain outlying subgroups from algae and has reserved the term for a central group naturally reconstituted within narrower limits.

In the present text, the author recognizes two major types of algae—the macro algae (commonly referred to as seaweeds) and micro algae (phytoplanktons) and in this chapter potential of the former group is unraveled as soil health boosters. Seaweeds occupy littoral zone, and are commonly referred to as greens, browns, and reds. These plants are found in all coastal areas of the world, in all climatic zones—from the warm tropics to the "icy polar regions." Generally speaking green seaweeds inhabit shallowest zones along the shore (upper intertidal), the browns are usually found in the mid-intertidal and sub-tidal zones and the reds inhabit the lower intertidal zone and deeper waters.

Seaweeds are major players in coastal ecosystems, and constitute the nutritional base for many shallow water food webs. They are architects of coastal marine meadows and underwater forests, provide homes and shelter for entire communities of associated fishes and invertebrates creating biological diversity niches. While culturing micro algae is simple and can be carried out in tanks, that of macro algae is done in the ocean and is commonly called “seaweed farming” or mariculture. It however, involves expert harvesting. Thus, obtaining large biomass of algae for farming does not require land and therefore, mariculture does not compete with crops for cultivable land.

The term seaweed was perhaps derived from the fact that several macro algae grew luxuriantly, became invasive and posed obstruction to navigation in sea and at ports. But if we exploit the invasiveness of seaweeds as feedstock for “agri inputs” and consider their ecological and economic importance, seaweeds are no longer just slimy stuff coating a seaside rock or fluttering in a tide pool but form a multibillion dollar industry (Kaur 1997). Seaweeds have dominated all spheres of our daily lives ranging from cosmeceuticals to nutraceuticals and pharmaceuticals; from fodder to food; from health supplements to esthetics and from ecological to agricultural spheres.

## 10.3 Seaweeds as Organic Fertilizers

Both types of algae have gained importance in organic farming and are rated as essential components of Good Agriculture Practices (GAP). When compared to micro algae, spectrum of macro algae appears to be broader, though no such claims have been made. The advent of synthetic fertilizers and chemical agri inputs took agro industry by storm but long-term adverse effects could not keep the promise of sustainable development. This practice emerged as one of the major causes of health risks from deadly diseases like cancer and have raised issues of environment safety as well. With organic farming picking up around the globe, seaweed resource has emerged a “safe” option. However, the combined costs of drying and transportation have confined its usage to limited places where the buyers are not too distant from the coast. The issues of overexploitation of the resource have also been raised and in several countries governments have executed plans to manage the resource while in others, regulations are still being worked out.

## 10.4 History of Seaweed as Agri Resource

According to Newton (1951), the earliest reference to seaweed manure is in the second half of the first century when the Roman Columella recommended that cabbages be transplanted at the sixth leaf stage and their roots be mulched and manured with seaweed. Since then methods for compressing seaweeds or marine plants into a compact, transportable form have appeared in literature. This is indicative of the value placed on seaweed manure and the need to transport the product over long distances. Since transportation of seaweed in wet form is uneconomical, alkaline seaweed extracts are becoming popular and the market has several such products for increasing the crop and soil health. The first attempt to produce a liquid extract was made almost a century ago (Penkala 1912). According to literature available, Dr. Reginald F. Milton, a biochemist while investigating fiber content of seaweeds worked out methods for liquefying kelp for use as a fertilizer. By 1947, he had succeeded in making a liquid product. His method, based on a hot pressurized alkaline process, was patented and has since then formed the basis for the Maxicrop process (Milton 1952) now a big name in liquid fertilizers. In the nineteenth century, coastal dwellers followed a common practice of collecting storm-cast seaweed, usually large brown seaweeds, and digging it into local soils. In the early twentieth century soil was fertilized with storm cast dried and milled macro algal material. Though in many places, the practice continued for centuries (Blunden and Gordon 1986; Metting et al. 1988; Temple and Bomke 1988) but with chemical fertilizers taking over, reports were only sporadic. However, since last decade or so seaweeds have regained global importance as eco-friendly inputs of organic farming and their use is being encouraged for sustainable crop production. Subsequent researches proved that while high fiber content of seaweed acts as a soil conditioner and assists moisture retention, the mineral content is a useful fertilizer and source of trace elements. This potential has made seaweeds farmer's first choice as soil health enhancer.

Interest in agricultural use of seaweeds is increasing rapidly as judged by the number of related publications appearing since 1950. Though the number of species with potential as fertilizer is small, the volumes of biomass they yield can be sizeable (Gibbs 1981). Studies in this field have revealed a wide range of beneficial effects of seaweed extract applications on plants, such as early seed germination and establishment, improved crop performance and yield, elevated resistance to biotic and abiotic stress, and enhanced postharvest shelf life of perishable products (Beckett and van Staden 1989; Hankins and Hockey 1990; Blunden 1991; Norrie and Keathley 2006).

## 10.5 Paradigm Shift: From Biofertilizers to Biostimulants

Seaweed extracts, concentrates, and suspensions in form of Seaweed Liquid Fertilizer (SLF) have wider use and market than seaweed mulch, seaweed meal or any other seaweed-based soil supplement. One big reason for popularity of SLF appears to be the nonavailability of the seaweed biomass in off-shore regions. The extracts are sold in concentrated form, are easy to transport and apply, and act more rapidly when given at lower concentration through aerial parts than through roots. The seaweed concentrates are applied to crops as root dips, soil drenches, soil conditioners, seed soak, or foliar sprays. Minerals and hormones in seaweed spray are absorbed through the epidermis of leaf and give resistance to numerous stresses such as frost, insect infestation, viral and fungal diseases. As the market is flooded with foliar sprays from seaweeds, one is forced to believe that extracts are more beneficial if used on aerial parts; however, it must be emphasized that the concentrates can also be used as drench and root dip which strengthen root system and healthy root system is a great binder of soil!

The seaweeds have been put to such use for centuries. Initially when researchers standardized methods, the extracts were accepted as liquid fertilizers and regarded as a tonic because of their medicine-like properties that enhanced plant growth. Subsequently, it was argued that since fertilizer had large amounts of nitrogen, phosphorus, and potassium, seaweed extracts be more correctly regarded as plant biostimulants.[1] Zhang and Schmidt (1997) from the Department of Crop and Soil Environmental Sciences, Virginia Polytechnic Institute and State University defined biostimulants as "materials that, in minute quantities, promote plant growth." By referring to biostimulants as substances or microorganisms applied to plants in "minute quantities," the authors distinguished biostimulants from nutrients. In general, biostimulants are accepted as materials, other than fertilizers, that promote plant growth when applied in low quantities.

In the revolutionized "pro organic" agro industry, demand is for broad-spectrum, organic products with dozens of nutrients as well as other benefits, and that is where biostimulants come in. Biostimulants are considered as a new generation of products for sustainable agriculture (Khan et al. 2009; Michalak et al. 2016). Biostimulants embrace the qualities possessed by the extract and therefore, the term biostimulant is generally equated with seaweed fertilizer. According to Yakhin et al. (2017) biostimulants are available in a variety of formulations and with varying ingredients but are generally classified into three major groups on the basis of their source and content. They are humic substances (HS), hormone-containing products (HCP), and amino acid-containing products (AACP). Among the biostimulants from various other sources, bioactive substances in the extract from seaweeds are the most studied. Seaweed liquid extracts which are processed from seaweed biomass using different manufacturing systems such as alkaline or acid hydrolysis or cellular disruption under pressure or fermentation, are now commercially available worldwide (Craigie 2011). They are extensively used as biostimulants by horticulturists, gardeners, farmers, and orchardists to enhance plant growth and fruit yields.

*Ascophyllum nodosum* (L.) is the most researched brown alga, and its extract has been commercialized as Acadian® for enhancing different plant growth attributes under normal and stress conditions. Kelpak® is another seaweed concentrate derived from a brown seaweed (*Ecklonia maxima*), and has been demonstrated to act as a biostimulant. A novel phlorotannin called Eckol has been isolated from Kelpak® which is found to have auxin-like activities, and its growth-promoting activity has been reported in a number of plants (Arafat Abdel Hamed Abdel Latef et al. 2017).

[1]A precise definition for biostimulants in agriculture has been proposed by the industry for consideration by the EU regulatory authority (Du Jardin 2015).

Stirk et al. (2014) reported brassinosteroids in the Kelpak™ seaweed extract. In addition to brassinosteroids, strigolactones have been found in the Seasol™ seaweed extract (from *Durvillaea potatorum* and *Ascophyllum nodosum*). The other popular products are Goemar GA 14, Seaspray, Seacrop 16, Algistim, Ujazyme and MAC.

In 1991, it was estimated that about 10,000 tons of wet seaweed were used to make 1000 tons of seaweed extracts with a value of US$ 5 million. However, the market has probably doubled in the last few decades and the global production estimate of seaweed biomass for soil and plant applications is well over 550,000 tons per annum (Nayar and Bott 2014).

Unlike, chemical fertilizers, extracts derived from seaweeds are biodegradable, nontoxic, nonpolluting, and nonhazardous to plants (Selvam and Sivakumar 2013, 2014) and contain nutrients that are essential for plant growth. In addition, seaweed extracts have many other molecules typical of plants, not yet characterized, but are thought to contribute to the efficacy of various liquid fertilizers. Bioinformatic studies have discovered plant genes that are activated when plants are treated with seaweed extracts (Nair et al. 2012; Jannin et al. 2013). There are also benefits from extracts that relate to improved soil structure, soil water holding capacity, and improved soil microbes.

## 10.6 What Makes Seaweeds a Promising Agri Resource?

Marine algae form an important component of the composting mixture and soil meal, where they contribute micro- and macroelements and chemical substances, e.g., amino acids, vitamins, plant growth hormones, and polysaccharides. The products are known to have fertilizer and/or protective role (Lacatusu et al. 2015), besides potential to improve the physical properties of soil, as amendments or supplements. They can be used in fresh, dried, powder, granular form and are also available commonly as Seaweed Liquid fertilizers. The special feature of seaweed fertilizers is the release of available nutrients into soil, which may last for several years (Eghball 2002). From among the three major types, it is the brown seaweeds—the Pheaophyceae or kelps which are preferred by the agro industry. Used directly as mulch, and "marinure," fresh and in dry form or their extracts, composts, soil conditioners, kelps (a group of browns) are known to enhance plant growth and productivity (Eryas et al. 2008; Illera-Vives et al. 2013).

Species commonly used as source of nutrients for crops are *Fucus vesiculosus, Ascophyllum nodosum (algifert), Sargassum wightii, Padina pavonica, Turbinaria, Laminaria saccharina, Padina tetrastromatica*, and *S. tenerrimum* belonging to brown algae; *Hypnea musciformis, Champia,* and *Porphyra* belonging to red algae; and *Ulva lactuca* belonging to green algae (Fig. 10.1a–d).

It must be mentioned here that red algae are widely used in cosmetics and food industry while brown seaweeds are preferred over red and green in the agro industry. The brown seaweeds not only contain vitamins common to land plants, but also vitamins which may owe their origin to bacteria that attach themselves to sea plants (in particular Vit B12). Vitamin C is present in high proportion as Lucerne, while Vit

**Fig. 10.1** Seaweeds commonly used as agri inputs. (**a**)—*Macrocystis*, (**b**)—*Fucus vesiculosus*. Both are brown algae commonly known as kelps. These are popular soil supplements. (**c**)—*Ulva* sp. a green seaweed, and (**d**)—*Porphyra*, a red seaweed. Among brown algae, *Ascophyllum*; in red, *Porphyra* and *Palmaria*, and in green seaweed species, *Ulva* has a high content of polysaccharides, up to 65% of dry weight. Green algae contain sulfuric acid polysaccharides, sulfated galactans and xylans; brown algae—alginic acid, fucoidan (sulphated fucose), laminarin (β-1, 3 glucan) and sargassan and red algae—agars, carrageenans, xylans, floridean starch (amylopectin like glucan), water-soluble sulfated galactan, as well as porphyran as mucopolysaccharides located in the intercellular spaces. Contents of both, total and species-specific polysaccharides, show seasonal variations and cold water species provide better quality phycocolloids when compared to seaweeds from warm waters

A is represented by its precursor beta carotene. B group vitamins present are B1, B2, B12 as well as pantothenic acid, folic acid, and folinic acid. Also found in the brown seaweed are Vit E, Vit K, and other PGRs besides indolyl acetic acid. Alginic acid and mannitol from brown algae are carbohydrates with chelating ability; they encircle and hold trace elements enabling plants to effectively absorb micronutrients that are generally in "unavailable" forms. These chelating agents not only make trace elements from seaweeds "available," they also make the trace elements more "available" to the plants. This may be due to the ability of soluble alginates where each metallic radical combines with one or more alginate molecules to form a polymer or large molecule with branched chains, thus leading to crumb structure in soil, an indicator of good water holding property of soil. Since alginic acid and mannitol do not immobilize available nitrogen as would cellulose, they decompose more readily than cellulose and contribute greatly to the formation of humus by stimulating microbial activity. As a result of catalytic action, alginic acid acts as a binder of soil particles and good soil texture results in a better soil aeration property with an accompanied increase in aerobic bacterial population. The aggregation of soil particles also increases the soil surface area which greatly facilitates chemical and biochemical exchange between elements thereby promoting increased productivity. According to Stephenson (1968), after the role of nutrients and trace elements, the alginic acid in brown seaweeds is important as it confers seaweed with soil conditioning properties.

In general seaweeds can contribute to the plant and soil health through four ways: (1) nutritional benefits (nitrogen, phosphorous, potassium, trace elements), (2) disease resistance (sulphated polysaccharides), (3) endocrine effects (cytokinins, auxins, gibberellins), and (4) soil conditioning (water holding capacity, beneficial soil biota) (Winberg et al. 2011).

## 10.7 Tapping Drift and Invasive Seaweeds: As Soil Amendments

Waste algae—the drifts and the casted seaweeds are composted and directly applied as soil conditioners and/or fertilizers in many coastal regions of the world (Castlehouse et al. 2003). The biomass is subjected to different composting technologies and is stabilized. The obtained compost quality varies with the feedstocks that have been used to produce it, the methods applied for pretreatment of biomass before composting, the composting time, and the method used to process the compost (Vendrame and Moore 2005). Seaweed compost treated as organic fertilizer has several advantages over regular plant compost, especially in the content of micro- (Mn, Zn, I) and macroelements (P, K, Ca, Mg), as well as the content of plant hormones such as total auxins and cytokinins (adenine) (~5 times more) and total amino acids (~7 times more) it contains (Abou El-Yazied et al. 2012).

### *10.7.1 Drift Seaweeds*

Since a large biomass of dry seaweeds is required for application in fields as soil amendments, it would mean harvesting tons of wet algal biomass and ultimately disappearance of species. This limitation can be overcome if we exploit drift and invasive seaweeds as soil amendments, and which are otherwise a nuisance on seacoasts (drift plants). The large biomass invites bugs and emanates unpleasant smell, and may also pose problems in navigation (invasive plants).

Drift seaweed is known to accumulate in lines left behind by the receding tide. As the height of the high tide recedes during the spring–neap cycle (which happens twice a month) successive lines are left behind. Such drifts are seen in exposed coasts where drift lines are made generally of brown seaweeds (Fig. 10.2a–c). In more sheltered localities, however, green algae may be more prevalent. In certain regions like south coast of England, huge amounts of such material accumulate, often necessitating removal, else rotting of biomass leads smell bad.

**Fig. 10.2** Drift seaweeds provide a large biomass which is processed as fertilizer. (**a**)—Giant kelp, (**b**)—*Sargassum muticum* thrown on the beach is frequently collected and used to amend soil. By Graça Gaspar—uploaded with the author's permission, CC BY-SA 3.0, https://commons.wikimedia.org/w/index.php?curid (**c**)—Strandlines formed by the beach-cast seaweed

The drift seaweed in olden days was occasionally dumped inland without any treatment, constituting a new source of pollution. But gradually people became aware of the potential of this biomass and experimented with it as soil amendment where the detached and broken seaweed thalli were used to fertilize soil. In one such example, in Aran Islands off the west coast of Ireland, the drift seaweed along with shore sand was transferred to bare limestone to make "buaile" or small pastures. Of late, such practices have become popular as eco-friendly and sustainable options. Nowadays part of this algal mass is composted and then used for growing crops and vegetables in various types of "seaweed amended" soils. In several parts of the world, drift seaweed or beach-washed seaweed is collected and dried for use (McHugh 2003; Michalak and Chojnacka 2013; Cole et al. 2016) while the leftover biomass portion is designated for the landfill disposal. In many places like Breton, farmers transport large quantities of a brown seaweed, *Himanthalia elongata* for raising the artichoke crops (The seaweed site: information on algae). In Cornwall (United Kingdom), the practice is to mix seaweed with sand, then allow it to rot and later dig it in soil. This checks the unpleasant sight of bugs and smell emanating due to decomposition of the biomass. Detached seaweeds sometimes called "total drift" have been used to raise crops like potato. In Brittany (France), for over a few hundred kilometers of the coast line around the beach-cast, brown seaweed is regularly collected by farmers and used on fields. In Scotland farmers use

*Ascophyllum* which after composting is obtained as a dried powder. Similar practices are carried out in countries with a vast coastline and more tropical climate like the Philippines, where not only large quantities of *Sargassum* have been collected, used wet locally, but sun-dried biomass is also transported to other areas. "Afrikelp" is another example of a commercially available dried seaweed sold as a fertilizer and soil conditioner. It is produced from brown seaweed *Ecklonia maxima* that is washed up on the beaches of west coast of Africa and Namibia. In Puerto Madryn (Argentina), large quantities, about 8000 tons of green seaweeds are cast ashore every summer which interfere with recreational uses of beaches besides unpleasantly rotting in situ (Eyras and Rostagno 1995). In all the cases, addition of compost is reported to result in improving water holding capacity and plant growth. Such practice is being encouraged elsewhere also where drift seaweed provides compost, thus reducing problems of environmental pollution.

Processing of drifted biomass is easily carried out by mixing with a lignocellulosic substratum (80:20 seaweed: sawdust proportion, in dry weight, DW). There are evidences that *Ascophyllum* drying under controlled conditions for 11–12 days, results in breaking of alginate chains into smaller chains which retain the property of forming gels with calcium but are weaker. The composted product is dark brown, granular with 20–25% water and can be easily stored and used in this form. The product has yielded good results in areas with steep slopes which are difficult to cultivate with conventional equipment and are likely to suffer soil loss by runoff. Spraying such slopes with composted *Ascophyllum*, clay, fertilizer, seed, mulch, and water has given good results, even on bare rock. The spray is thixotropic, i.e., it is fluid when a force is applied to spread it but it sets to a weak gel when standing for a time and sticks to the sloping surface.

Since algal biomass has become a popular soil supplement, different composting technologies are applied for algal biomass stabilization. They may be divided into three groups: (1) passive piles or windrows—A method in which material is left undisturbed, relies on natural convective air flows for aeration; (2) turned or aerated piles or windrows—Here, the air is provided by mechanically turning and mixing the material; and (3) in-vessel systems—optimum environmental conditions of aeration, moisture, and temperature for the quick decomposition of algal biomass are maintained (Michalak and Chojnacka 2013).

Besides drift seaweed, "mid-beach" plants can also be gathered. These are often found scattered on the beach from the water's edge to the highest point of recent high tides. The seaweed "mid-beach" is drier than seaweed at the tide line and therefore lighter to carry. It also has fewer bugs than the seaweed high up on the beach, making it a little more pleasant to gather. For a sustainable approach, fine broken up seaweed patches smaller than leaf size are collected and are applied as mulch. To get best results, seaweed mulch is normally not used on heavy soils but is preferred as a surface dressing on light sandy soils. This is because seaweeds have some insoluble fibers in their cell wall and if the alga is dug into such soils, the fibers form an impermeable layer. However, in light soils, the seaweed (kelp) biomass provides nutrients and other growth promoting substances.

Co-composting of drifting and beach-cast seaweeds is recommended for utilization of the waste biomass (Illera-Vives et al. 2013). If care is taken to avoid mixing of treated straw or urban and industrial sludge in the compost as co-composting materials, the produced seaweed compost is not only of good quality but is also with very low content of pesticides, such as organochlorine compounds and cereal growth regulators (Morand 1990).

### *10.7.2 Invasive Seaweeds*

Some seaweed species like *Sargassum muticum, Undaria pinnatifida, Caulerpa taxifolia*, and *Enteromorpha* have invaded ecosystems along the coasts of many countries displacing native algae and seagrasses, reducing biodiversity and impairing habitats of fish and invertebrates. Eradicating these invasive species has so far not only been a costly affair but has also met with little success as the biomass is huge. Recently such invasive seaweeds have been used for high value products like biofuel and antioxidants. If biomass of such species is also exploited for fertilizing soil, it could help in making the "undesirable" bioresource available and more so make its removal from the site justified. This would cater to the market demand for biomass. A few reports from various countries on use of invasive seaweeds as agricultural inputs have appeared indicating the huge potential of such biomass in agriculture. *Gracilaria salicornia*, a common invasive red algal species found in Kāne'ohe Bay and around the world, is a potential potassium fertilizer source. Analysis of invasive brown algae, *Turbinaria ornata* and *Sargassum mangarevense* indicate them to be rich in potassium, nitrates, calcium, iron, and polyunsaturated amino acids, with a very low level of any heavy metals (Zubia et al. 2003). In another study, soils amended with *Eucheuma* spp. (invasive) used to raise sweet potato resulted in better plant growth when compared to the control (https://scholarspace.manoa.hawaii.edu/handle/10125/101058) while *Enteromorpha* used in China has increased crop yield in several cases. Realizing their importance, commercial products from invasive seaweeds have appeared in the market, e.g., *NZBioActive*, a biostimulant is extracted exclusively from brown alga *Undaria pinnatifida,* an invasive seaweed.

However, the entire process has limitations of the type—high costs incurred in processing and transporting biomass to the fields especially if it is to be sent to off-shore sites. Moreover, the drift algal biomass invites bugs besides consuming area suitable otherwise for waste settlement, and issues related to leachate and biogas production during decomposition of algae (Vallini et al. 1993).

## 10.8 The New Biomass Service: Biochar

Seaweed biomass besides liquid fertilizer or manure also offers a feedstock for the production of nutrient-rich biochar which can be used as soil ameliorant. The product is obtained from intensely cultivated seaweeds such as *Saccharina*, *Undaria*, and *Sargassum*—brown seaweeds, and *Gracillaria*, *Kappaphycus* and *Euchema*—red seaweeds. These products have introduced the concept of fertigation in raising kitchen gardens and on commercial farms (Fig. 10.3a–g).

Biochar, a C-rich "biological charcoal," is a solid material obtained by pyrolysis which is the decomposition of organic material under oxygen-limited conditions and at a temperature ranging from 350 °C to 900 °C (http://www.biochar-international.org/). When compared to raw biomass, biochar contains a higher amount of carbon (C) when compared to cow-derived manure, besides other elements such as hydrogen (H), oxygen (O), nitrogen (N), and sulfur (S). Biochar is known for its high stability and surface properties, such as large surface area, porous structure, and surface functional groups, though the properties may show variations based on the source of raw material (i.e., biomass), and production conditions (Roberts et al. 2015).

**Fig. 10.3** Various forms of seaweed fertilizers. (**a**)—The dried form can be added to soil to improve tilth while (**b**)—the crushed granular form can be added as supplement. The dried seaweed is packed (**c**) and sent to factories where liquid fertilizers are prepared (**d**). Biochar is the new product (**e**) where seaweed is occasionally mixed with pine wood to render a balanced fertilizer. For large-scale farming, the plants are fertilized and irrigated (fertigation) with one application (**f**, **g**). Source: wikivisually and creative commons

Biochar is used as soil amendment and fertilizer (Bird et al. 2011); as a reducer of greenhouse gas emissions; as adsorbent of wide range of pollutants in the air, soil, and water; and as an energy source, among other applications. Application of biochar to the soil allows fixing carbon, improving soil quality by neutralizing acidic soil, promoting cationic exchange capacity, and increasing the activities of microorganisms. When biochar is applied, the basic cations of biochar are discharged into the soil, with aluminum (Al) and $H^+$ being replaced. As a consequence, the cationic exchange capacity of the soil increases (Chan et al. 2007). In addition, biochar has high N, phosphorus (P), calcium (Ca), and potassium (K) concentrations, which directly provide nutrients to the soil or associated microorganisms.

While there is some variability in biochar properties as a function of origin of seaweed, there are several defining and consistent characteristics of seaweed biochar. In particular a relatively low C content and surface area but high yield, essential trace elements (N, P, and K) and exchangeable cations (particularly K) define biochar. The pH range of biochar is from neutral (7) to alkaline (11), making it work for broad-spectrum applications in diverse soil types. Blending of seaweed biochar with rice husk and pine saw dust is known to act as a value-added soil ameliorant that combines a high fixed C content with a mineral-rich substrate to enhance crop productivity.

## 10.9 Limitations

Interest in seaweeds as "safe" or eco-friendly fertilizers has increased over the years. The most intriguing feature of this resource is the availability of biomass which when wet appears bulky but the dried form produced is quiet less. To date, seaweeds only produce a small fraction of the global supply of biomass with below 30 × 106 fresh weight (FW) tons of seaweed, in comparison to 16 × 1011 tons of terrestrial crops, grasses, and forests. The problem gets compounded when the effect of climate change on the growth of seaweeds is considered. The seaweed biomass, which is known to have multiple benefits, contains vitamins that get decomposed when the biomass is exposed to intense heat outside the sea. For using the dry biomass therefore, one important step would be quick harvest and efficient drying process. This would also be applicable to drift seaweeds which need to be removed quickly from coast and processed for use. The second limitation is the amount of biomass required for compost or mulch is too high and if seaweeds are harvested unmindfully, environmental concerns are bound to be raised. It is therefore, recommended that only species that can be cultivated through mariculture be used. If at all harvesting from sea is required, it should be done under strict surveillance. The harsh reality is that if the waters are polluted especially with oil spilling and nuclear run off, the biomass becomes unfit for use as it gets contaminated with toxins including heavy metals. According to Smith et al. (2010), algae can contain high levels of organic arsenic, which could be toxic if mineralized. Large amounts of cadmium have also been measured in different kinds of seaweeds (Besada et al.

2009). However, Verkleij (1992) stated that only in heavily and chronically polluted waters, problems are to be expected regarding seaweed quality (for consumption and agri input). As long as seaweeds are collected in clean areas, no toxicity is expected, but still monitoring water quality would be necessary when large areas of seaweeds are harvested for soil fertilization purposes.

The other constituents in biomass, like polysaccharides especially alginic acid are known to undergo seasonal variation and may be extremely low at some point in life cycle of the seaweed. Harvesting plants during such stages would yield a poor-quality soil amendment. Hence, basic knowledge of the life cycle of these plants is necessary before they are harvested, discouraging collection of poor-quality biomass and wasting it in the process. The economics of the technology is also lopsided—The seaweed decomposition in most cases is slow and transporting wet biomass inland becomes not only costly but also wasteful if the plants start to rot. This would mean that the resource is more available to coastlands while basic agriculture is carried out in inland areas. Similarly, the drift seaweed biomass needs to be cleared from the site before it gets infested with flies. Awareness among local populations about importance of seaweeds is required. Local population should be educated that howsoever abundant and vast seaweeds may appear in sea, they are not quick to replenish especially the seaweeds that produce maerl. They have to be harvested scientifically and crude methods would be damaging to the resource.

For agri industry to derive maximum benefits from seaweeds, a major research effort is needed to elucidate the complex modes of action and forms of seaweed applications on diverse crops. Each plant has special requirements which are generally not met with by soil unless it is amended. Overdose of seaweed fertilizers/biochar/manure may result in inhibitory effect and the results may not be convincing. In addition, we need to recognize that seaweed extracts are inherently different since their source and processing is different which confers them with specific stability properties (Stirk et al. 2014). Furthermore, their capacity to elicit plant responses also depends in part upon the application usage rates, application frequency and the timing of applications in relation to plant development life cycle. For best results seaweed products can be applied in combination with other organic products. Therefore, we need to standardize methods for determining the appropriate time and plant stages where the results would be encouraging in terms of crop yield. We also need to define the optimal dosages required to maximize the yield besides working out combinations to obtain a value-added product. Some soils may also have special requirements where amendment has to be worked out for specifications. For example, in dry areas or in areas with poor water retention quality of soil, adding seaweed in mulch form or granular form is effective in improving water retention of soil but it may also increase the salt content of soil. It is therefore, important that soil testing is carried out. In several cases, the soil becomes "hot" upon amending with solid form of seaweed and seedlings get "burnt." Whole seaweeds or seaweed meals are reported to inhibit seed germination and plant growth, to reduce N availability on the short term and possibly to release toxic sulfhydryl compounds (Craigie 2011). In such cases, extracts from seaweed are recommended.

Putatively, seaweed biomass may be a useful amendment for crop production due to provision of primary plant nutrients and micronutrients (e.g., N, P, K, Ca), effects on soil water holding capacity, and promotion of microbial activity, among other plant production benefits, but may be limited by high sulfur, salt, and heavy metal content. The amount of literature on detrimental effects of seaweed application to crops is relatively scarce compared to the beneficial effects; these detrimental effects disappear after a few weeks. The biggest limitation is that the seaweed agri products are slow to show results. The bottom line, however, is that in the long run these would prove healthy, economical, and beneficial—only we need to be patient with results.

## 10.10 Conservation

Popularity of seaweed agri inputs has put pressures on marine resources and overharvesting in several instances, has resulted in dwindling biomass and disappearance of a few species while others face dangers of extinction. It is possible that overexploitation of natural seaweed resources leads to significant ecological, economic, and social consequences at local, regional, and even global scales (Graham et al. 2007; Rebours and Karlsen 2007). This has forced local governments to develop stringent regulations and directives for sustainable exploitation of seaweed resources. In many regions seaweed harvesting is under stringent regulation by the local governments. Countries like Chile, Norway, Portugal, and Canada have developed and implemented coastal management plans including well-established and sustainable exploitation of their natural seaweed resources (Rebours et al. 2014). The State of Hawaii prohibits collection of *Gracilaria* with "bumps" on it. These bumps are reproductive bodies on the female plants, that allow the plants to multiply themselves if left in the water.

Resource scientists, managers, conservationists, governments, and other stakeholders need to be proactive in the sustainable management of these valuable resources. Each country is, however, in need of long-term and ecosystem-based management plans to ensure that exploitation is sustainable.

Besides making rules for seaweed harvesting, "seaweed farming" needs to be carried out on mass scale for biomass availability, raw material for agri products (McHugh 2002). The seaweed aquaculture industry still requires technological and management improvements, institutional changes, and appropriate environmental and social frameworks, especially in developing countries which have vast resources but less of know-how in this field (Valenti 2008; Oliveira 2009; Abreu et al. 2011; Marroni and Asmus 2013).

## 10.11 Concluding Remarks

The seaweed agri products have since long been cited as promising soil amendments and fertilizers. A large variety of products are available in market, but little do we realize that a large biomass is harvested to bring these products to the market. If harvesting of seaweeds is carried out unmindfully, most seaweed resources will get depleted. The solutions lie in improving the strains and mass cultivation of commercial species. With climate change adversely affecting the crop yield, it is important that integrated methods of agriculture are adopted. Seaweed resources can be used in several combinations and in forms that do not require much expertise. They enhance the effect of other forms of manure when used in combinations. Though biotechnology has provided us with better quality GM crops, but some genetically engineered products such as brinjal have generated controversies, focus is therefore, on natural means of increasing crop yield. The first step in such an endeavor is to improve soil health. Seaweeds if used judiciously have a great deal to offer. It may be interpreted as role reversal—from our dining table as food they have moved to fields as food providers! To derive maximum benefit from seaweeds, limitations of the usage must be worked out.

## References

Abou El-Yazied A, El-Gizawy AM, Ragab MI, Hamed ES (2012) Effect of seaweed extract and compost treatments on growth, yield and quality of snap bean. J Am Sci 8:1–20. http://www.sciencepub.net/american

Abreu MH, Pereira R, Yarish C, Buschmann AH, Sousa-Pinto I (2011) IMTA with *Gracilaria vermiculophylla*: productivity and nutrient removal performance of the seaweed in a land-based pilot scale system. Aquaculture 312:77–87

Beckett RP, van Staden J (1989) The effect of seaweed concentrate on the growth and yield of potassium stressed wheat. Plant Soil 116:29–36

Besada JM, Andrade JM, Schultze F, Gonzalez JJ (2009) Heavy metals in edible seaweeds commercialised for human consumption. J Mar Syst 75:305–313

Bird MI, Wurster CM, de Paula Silva PH, Bass AM, de Nys R (2011) Algal biochar – production and properties. Bioresour Technol 102:1886–1891

Blunden G (1991) Agricultural uses of seaweeds and seaweed extracts. In: Guiry MD, Blunden G (eds) Seaweed resources in Europe: uses and potential. Wiley, Chichester, NH, pp 65–68

Blunden G, Gordon SM (1986) Betaines and their sulphono analogues in marine algae. In: Round FE, Chapman DJ (eds) Progress in phycological research, vol 4. Biopress, Bristol, pp 39–80

Castlehouse H, Smith C, Raab A, Deacon C, Meharg AA, Feldmann J (2003) Biotransformation and accumulation of arsenic in soil amended with seaweed. Environ Sci Technol 37:951–957

Chan KY, Van Zwieten L, Meszaros I, Downie A, Joseph S (2007) Agronomic values of green waste biochar as a soil amendment. Soil Res 45:629–634

Cole AJ, Roberts DA, Garside AL, de Nys R, Paul NA (2016) Seaweed compost for agricultural crop production. J Appl Phycol 28:629–642

Craigie JS (2011) Seaweed extract stimuli in plant science and agriculture. J Appl Phycol 23:371–393 du Jardin P (2015) Plant biostimulants: definition, concept, main categories and regulation. Sci Hortic 196:3–14

Eghball B (2002) Soil properties as influenced by phosphorus-and nitrogen-based manure and compost applications. Agron J 94:128–135

Eyras M, Rostagno C (1995) Bioconversión de algas marinas de arribazón: experiencias en Puerto Madryn, Chubut, Argentina. Naturalia Patagónica 3:25–39

Eyras MC, Defossé GE, Dellatorre F (2008) Seaweed compost as an amendment for horticultural soils in Patagonia, Argentina. Compost Sci Util 16:119–124

Gibbs SP (1981) The chloroplasts of some algal groups may have evolved from endosymbiotic eukaryotic algae. Ann N Y Acad Sci 361(1):193–208

Graham MH, Vásquez JA, Buschmann AH (2007) Global ecology of the giant kelp *Macrocystis*: from ecotypes to ecosystems. Oceanogr Mar Biol Annu Rev 45:39–88

Guiry MD (2012) A catalogue of Irish seaweeds. A.R.G. Gantner, Ruggell

Hankins SD, Hockey HP (1990) The effect of a liquid seaweed extract from *Ascophyllum nodosum* (Fucales, Phaeophyta) on the two spotted red spider mite *Tetranychus urticae*. Hydrobiologia 204(205):555–559

Ibraheem IBM (2007) Cyanobacteria as alternative biological conditioners for bioremediation of barren soil. Egypt J Phycol 8:99–116

Illera-Vives M, Seoane Labandeira S, Lo'pez-Mosquera ME (2013) Production of compost from marine waste: evaluation of the product for use in ecological agriculture. J Appl Phycol 25:1395–1403

Jannin L, Arkoun M, Etienne P, Laine P (2013) *Brassica napus* growth is promoted by *Ascophyllum nodosum* (L.) Le Jol seaweed extract: microarray analysis and physiological characterization of N, C, and S metabolisms. J Plant Growth Regul 32(1):31–52

Kaur I (1997) Potential and future prospects. In: Vijayaraghavan MR, Kaur I (eds) Brown algae: structure, ultrastructure and reproduction. APH Publishing, New Delhi. ISBN 81-7024-879-5

Khan W, Rayirath UP, Subramanian S, Jithesh MN, Rayorath P, Hodges DM, Critchley AT, Craigie JS, Norrie J, Prithiviraj B (2009) Seaweed extracts as biostimulants of plant growth and development. J Plant Growth Regul 28:386–399. https://doi.org/10.1007/s00344-009-9103-x

Lacatusu R, Capatana R, Lacatusu AR, Meghea A (2015) A new compost for organic farming. Testing by plants. SGEM2015 conference proceedings. 15th international multidisciplinary scientific geoconference SGEM 2015 www.sgem.org, June 18–24, 2015 Book 3(2):11–18

Latef AAHA, Srivastava AK, Saber H, Alwaleed EA, Tran L-SP (2017) *Sargassum muticum* and *Jania rubens* regulate amino acid metabolism to improve growth and alleviate salinity in chickpea. Sci Rep 7:10537

Marroni EV, Asmus ML (2013) Historical antecedents and local governance in the process of public policies building for coastal zone of Brazil. Coast Ocean Manag 76:30–37

McHugh DJ (2002) Prospect for seaweed production in developing countries, FAO fisheries circular no 968 FIIU/C968. FAO, Rome

McHugh DJ (2003) A guide to seaweeds industry, FAO fisheries technical paper, no. 441. Rome, FAO, p 105

Metting B, Rayburn WR, Reynaud PA (1988) Algae and agriculture. In: Lembi CA, Waaland RA (eds) Algae and human affairs. Cambridge University Press, Cambridge, pp 335–370

Metting B, Zimmerman WJ, Crouch I, Van Staden J (1990) Agronomic uses of seaweed and microalgae. In: Akatsuka I (ed) Introduction to applied phycology. SPB Academic Publishing, The Hague, pp 269–307

Michalak I, Chojnacka K (2013) Algal compost - toward sustainable fertilization. Rev Inorg Chem 33:161–172

Michalak I, Górka B, Wieczorek PP, Rój E, Lipok J, Łęska B et al (2016) Supercritical fluid extraction of algae enhances levels of biologically active compounds promoting plant growth. Eur J Phycol 51:243–252. https://doi.org/10.1080/09670262.2015.1134813

Milton RF (1952) Improvements in or relating to horticultural and agricultural fertilizers. UK Patent 664:989
Morand P, Carpentier B, Charlier RH, Maze J, Orlandini M, Plunkett BA, de Waart J (1990) Bioconversion. In: Guiry MD, Blunden G (eds) Seaweed resources in Europe – uses and potential. Heydon & Son, London
Nair P, Kandasamy S, Zhang J, Ji X, Kirby C, Benkel B, Hodges MD, Critchley AT, Hiltz D, Prithiviraj B (2012) Transcriptional and metabolomic analysis of *Ascophyllum nodosum* mediated freezing tolerance in *Arabidopsis thaliana*. BMC Genet 13:643
Nayar S, Bott K (2014) Current status of global cultivated seaweed production and markets. World Aquacult 45:32–37
Newton GW (1951) Seaweed manure for perfect soil and smiling fields. Sampson Low, London, p 188
Norrie J, Keathley JP (2006) Benefits of *Ascophyllum nodosum* marine-plant extract applications to 'Thompson seedless' grape production. Proceedings of the Xth international symposium on plant bioregulators in fruit production. Acta Hortic 727:243–224
Oliveira RC (2009) O panorama da aquacultura no Brasil: a prática com foco na sustentabilidade. Rev Intertox Toxicol Risco Ambient Soc 2(1):71–89
Penkala L (1912) Method of treating seaweed. British Patent 27:257
Pereira L, Neto JM (2014) Marine algae: biodiversity, taxonomy, environmental assessment, and biotechnology. CRC Press, Boca Raton, FL, p 398. ISBN 9781138582088
Rebours C, Karlsen Å (2007) Seaweeds in the north: new scopes for coastal farming. Bioforsk Fokus 2(13):73–76
Rebours C, Marinho-Soriano E, Zertuche-Gonzalez JA, Hayashi L, Vásquez JA, Kradolfer P, Soriano G, Ugarte R, Abreu MH, Bay-Larsen I, Hovelsrud G, Rodven R, Robledo D (2014) Seaweeds: an opportunity for wealth and sustainable livelihood for coastal communities. J Appl Phycol 26:1939–1951
Roberts DA, Paul NA, Dworjanyn SA, Bird MI, de Nys R (2015) Biochar from commercially cultivated seaweed for soil amelioration. Sci Rep 5:9665
Selvam GG, Sivakumar K (2013) Effect of foliar spray form seaweed liquid fertilizer of *Ulva reticulata* (Forsk) on *Vigna mungo* L. and their elemental composition using SEM-energy dispersive spectroscopic analysis. Asian Pac J Reprod 2(2):119–125
Selvam GG, Sivakumar K (2014) Influence of seaweed extract as an organic fertilizer on the growth and yield of *Arachis hypogea* L. and their elemental composition using SEM–energy dispersive spectroscopic analysis. Asian Pac J Reprod 3(1):18–22
Smith J, Summers G, Wong R (2010) Nutrient and heavy metal content of edible seaweeds in New Zealand. N Z J Crop Hortic Sci 38(1):19–28. https://doi.org/10.1080/01140671003619290
Stephenson WA (1968) Seaweed in agriculture and horticulture. Faber & Faber, London
Stirk WA, Tarkowská D, Turečová V, Strnad M, van Staden J (2014) Abscisic acid, gibberellins and brassinosteroids in Kelpak®, a commercial seaweed extract made from *Ecklonia maxima*. J Appl Phycol 26:561–567. https://doi.org/10.1007/s10811-013-006
Temple WD, Bomke AA (1988) Effects of kelp (*Macrocystis integrifolia*) on soil chemical properties and crop responses. Plant Soil 105:213–222
The seaweed site: information on algae. UN report 2017
Tung HF, Shen TC (1985) Studies of the *Agolla pinnata-Anabaena azollae* symbiosis: concurrent growth of *Azolla* with rice. Aquat Bot 22:145–152
Uysal O, Uysal FO, Ekinci K (2015) Evaluation of microalgae as microbial fertilizer. Eur J Sustain Dev 4(2):77–82
Valenti WC (2008) Aqüicultura Brasileira é sustentável? Palestra apresentada durante o IV Seminário Internacional de Aqüicultura, Maricultura e Pesca, Aquafair 2008, Florianópolis, 13–15 de maio de 2008. pp 1–11
Vallini G, Pera A, Cecchi F, Valdrighi MM, Sicurani MA (1993) Compost stabilization of algal biomass drawn in eutrophic lagoon ecosystems. Compost Sci Util 1:49–53

Vendrame W, Moore KK (2005) Comparison of herbaceous perennial plant growth in seaweed compost and biosolids compost. Compost Sci Util 13:122–126

Verkleij FN (1992) Seaweed extracts in agriculture and horticulture: a review. Biol Agric Hortic 8 (4):309–324. https://doi.org/10.1080/01448765.1992.9754608

Winberg PC, Skropeta D, Ullrich A (2011) Seaweed cultivation pilot trials – towards culture systems and marketable products. Australian Government Rural Industries Research and Development Corporation, RIRDC publication no. 10/184. PRJ - 000162. rirdc.infoservices.com.auitems/10-184

Yakhin OI, Lubyanov AA, Yakhin IA (2017) Biostimulants in plant science: a global perspective. Front Plant Sci 7:2049

Zhang X, Schmidt RE (1997) The impact of growth regulators on the α-tocopherol status in water-stressed *Poa pratensis*. Int Turfgrass Soc Res J 8:1364–1137

Zubia M, Payri CE, Deslandes E, Guezennee J (2003) Chemical composition of attached and drift specimens of *Sargassum mangarevense* and *Turbinaria ornate* (Phaeophyta: Fucales) from Tahiti, French Polynesia. Bot Mar 46:562–571

# Chapter 11
# Arbuscular Mycorrhizal Fungi: The Potential Soil Health Indicators

**Manju M. Gupta**

**Abstract** Over the years the concept and understanding of the importance of the soil health have been well accepted in the agricultural community and there has been a greater focus on standardizing newer parameters as soil health indicators. In the present chapter scope of arbuscular mycorrhizal fungal (AMF) symbioses not only as an indicator but also key determinants of soil health is discussed. Role of AMF in soil health development is discussed via three main mechanisms: influences on plant physiology, soil ecological interactions, and soil structural engineering. Their potential to serve as a soil health indicator is explored with reference to their role in soil aggregation and land or ecological restoration.

**Keywords** Soil aggregation · Glomalin · Ecological restoration · AMF

## 11.1 Introduction

Soil health is commonly defined as the ability of soil to function as a living ecosystem that sustains biological productivity, maintain air or water and promote plant, animal, and human health (Doran 2002; Bonfante et al. 2019). Conceptually it is different from soil quality which focuses on maintaining soil functions (Apfelbaum et al. 2019). Soil health presents the soil as a finite nonrenewable and dynamic living resource (Laisram et al. 2012) and is an integrated concept that deals with management and optimization of the chemical, physical, and biological processes of soil that are important for sustained productivity and environmental quality (Kibblewhite et al. 2008; van Es and Karlen 2019). The term soil health originated from the observation that soil quality influences the health of animals and humans via the quality of crops (e.g., Warkentin 1995). Soil health has also been explained

M. M. Gupta (✉)
Sri Aurobindo College, University of Delhi, Delhi, India
e-mail: mmgupta@aurobindo.du.ac.in



B. Giri, A. Varma (eds.), *Soil Health*, Soil Biology 59,
https://doi.org/10.1007/978-3-030-44364-1_11

by several authors via the analogy to the health of an organism or a community (Doran and Parkin 1996; Bünemann et al. 2018; Bonfante et al. 2019).

Over the years the concepts and understanding of the importance of the soil health have been well accepted in the agricultural community and have moved beyond a few innovative producers and scientists, to become a focus in broader circles. Predetermination of soil health through several measurable parameters/indicators is the mainstay to turn the phenomenon of soil health determination to one's advantage. Compilations of soil health indicators, measurements, and monitoring protocols have been or are being proposed for diverse purposes (Apfelbaum et al. 2019; van Es and Karlen 2019). The documents known as soil health cards have many applications including guidance fertilizer recommendations driven by crop yield, declining water supplies, declining food nutritional density, and soil loss allowance guidance (Amezaga et al. 2016; Purakayastha et al. 2019).

Indicators of soil health are measurable parameters of soil or plant properties that provide clues about how well the soil will function in any agroecosystem. Scientific relevance and practical adoption of an indicator of soil health depends on its sensitivity to variations in soil management, good correlation with the beneficial soil functions and other variables which are difficult to access or measure, helpfulness in revealing ecosystem processes and comprehensibility and utility for land managers, costs and ease of measurement (Parisi et al. 2005; Congreves et al. 2015). The common soil health indicators are classified into physical, chemical, or biological. There exist several exhaustive lists and catalogs on type and methodology followed for these indicators (Moebius-Clune 2016). Numerous reviews in recent times highlight the need for standardizing newer and innovative methods of prediction of soil health (Cardoso et al. 2013; Bünemann et al. 2018; Apfelbaum et al. 2019). In the present chapter potential of arbuscular mycorrhizal fungal (AMF) symbiosis not only as indicators but also key determinants of soil health are discussed.

AMF symbiosis refers to symbiotic associations of crop roots with fungi of subphylum Glomeromycotina of phylum Mucoromycota (Spatafora et al. 2016) and are found in most of the crop plants except few members belonging to Brassicaceae and certain species such as *Arabidopsis* which form rudimentary arbuscular mycorrhiza (Cosme et al. 2018). Their potential to act as a soil health indicator originated from the fact that they are ubiquitous in occurrence, (Davison et al. 2015) they have an important role in nutrient both macro and micro acquisition, and thus have important consequences for crop nutritional value. Additionally, AMF enhance plant water relations through several mechanisms, potentially contributing to increased crop drought resistance (Augé 2001; Sendek et al. 2019). AMF have been known to be keystone organisms with myriads of ecosystem roles (Powell and Rillig 2018). They inhabit simultaneously two habitats: the host plant root and the soil (Powell and Rillig 2018). Structures produced in soil includes a network of thicker hyphae that function as channels for the supply of nutrients absorbed through thin highly branched hyphae called absorptive hyphae (Fig. 11.1) (Brundrett et al. 1996). Spores are large asexual spherical structures (20–1000 μm or more in diameter) formed on hyphae in soil, and/or in roots depending on the species.

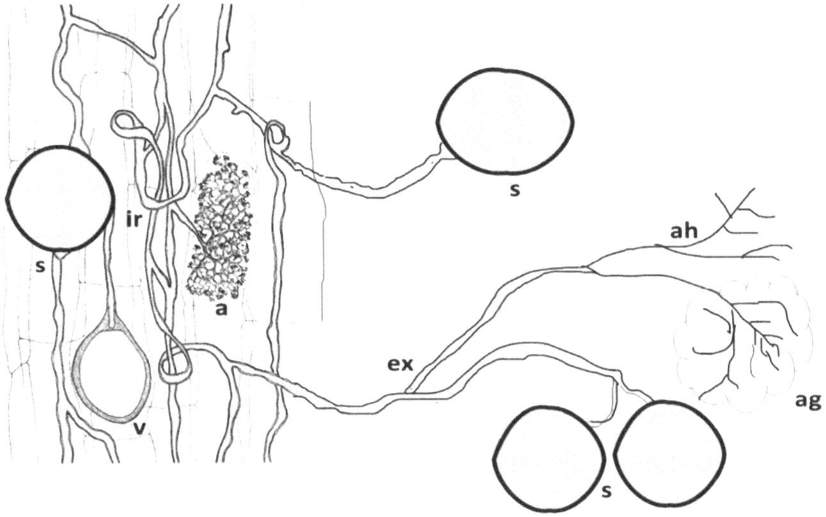


**Fig. 11.1** Diagram of structures produced by AMF inside roots and soil (from left to right *s* spore, *v* vesicle, *ir* intraradical hypha, *a* arbuscule, *ex* extraradical hypha, *ah* absorptive hypha, *ag* soil aggregates)

Structures in roots include nonseptate hyphae which ramify within the cortex, arbuscules, intricately branched haustoria in cortex cells and vesicles the storage structures formed by many fungi.

The present chapter will examine the prospects of AMF to act as a soil health indicator. Different aspects of the functioning of AMF are discussed in relation to soil health, which includes their role in creating a healthy soil environment. Their potential to serve as a soil health indicator is explored with reference to their role in soil aggregation and land or ecological restoration.

## 11.2 Role of AMF in Creating and Predicting a Healthy Soil Environment

AMF has been known to be important for soil health via three main mechanisms, namely, influences on plant physiology, soil ecological interactions, and soil structural engineering (Fig. 11.2) (Powell and Rillig 2018). They affect almost all-important aspects of the physiology of plant partner especially during conditions of biotic and abiotic stress. Nutrients uptake and transfer (Smith and Gianinazzi-Pearson 1988; Bago et al. 2000), water status (Zhu et al. 2014), and photosynthesis (Zhu et al. 2014) of plants when exposed to stress, the effect of AMF on lipid peroxidation, osmotic adjustment (Chen 2014; Wipf et al. 2019) and root exudation (Bansal and Mukerji 1994; Gupta et al. 2018a) are some of the well-studied aspects.

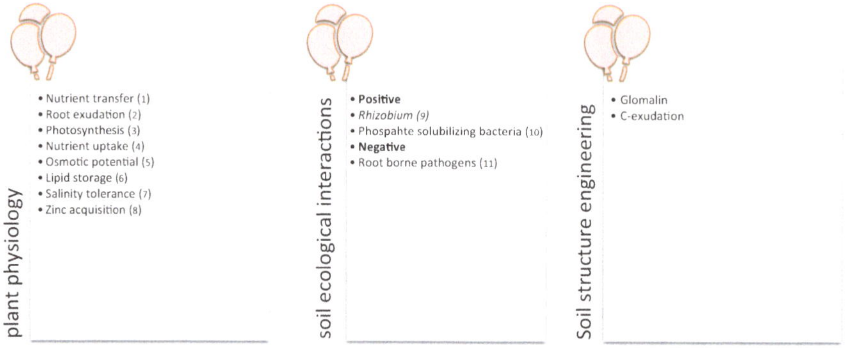


**Fig. 11.2** Three domains of influence of AMF on soil health (1—Azcón-Aguilar and Barea 1997; 2—Gupta et al. 2018a; 3—Zhu et al. 2014; 4—Smith and Gianinazzi-Pearson 1988; Bago et al. 2000; 5—Chen 2014; 6—Wipf et al. 2019; 7—Giri et al. 2007; 8—Watts-Williams and Cavagnaro 2012; 9—Gupta et al. 2018a; 10—Toro et al. 1997; 11—Jacott et al. 2017)

AMF increase root absorbing surface 100 or even 1000 times through external hyphal network that permeates into the microsites of rocks and soils surrounding the plant roots (Larcher 1995), thus increasing plants' nutrient and water availability and absorption (Banerjee et al. 2013; Sendek et al. 2019). The symbiosis is especially beneficial to plants that are situated in an environment containing heavy metals. The hyphae of fungi can accumulate these toxic elements (Cu, Pb, Zn, etc.) in their bodies and thus protect the root of the host plant (Hildebrandt et al. 2007). This improves plants' soil water relations, tolerance to biotic and abiotic stresses, increase nutrient supply, growth, yield and reproductive success and reduce fertilizer requirement (Chen et al. 2018; Gupta et al. 2019; Wipf et al. 2019). The potential application of managing AMF occurrence and diversity to improve soil quality and health to increase the productivity of agricultural ecosystems has gradually evolved as a new and very promising technology known as next-generation mycorrhizal technology (Rillig et al. 2016; Bagyaraj and Ashwin 2017; Gupta et al. 2019) (Fig. 11.2).

AMF influence ecological interactions of an individual and thus influence plant community structure (Van der Heijden et al. 1998) and are considered to have a pivotal role in plant community assembly and succession (Janos 1980; Kikvidze et al. 2010). They have an important role in determining soil microbial interactions in soil, for example, evidence for AMF protection against fungal root pathogens (Azcón-Aguilar and Barea 1997; Jacott et al. 2017). AMF could interact positively with beneficial microbes, such as phosphate solubilizing bacteria, with potential beneficial contributions to nutrient cycling and plant nutrition. Interaction of AMF and other soil microbes are driving many fundamental nutrient cycling processes, soil structural dynamics, degradation of pollutants, various other ecosystem services (Powell and Rillig 2018) and respond quickly to natural perturbations and environmental stress. This allows microbial analyses to discriminate soil quality status and shifts in microbial population and activity could be used as an indicator of changes in soil quality.

Recent studies have established AMF to be important in determining the soil structure through their direct effects on soil aggregation in soils and through secretion soil proteins and exudates (Gupta et al. 2018a; Lehmann et al. 2019). AMF produce glomalin, a soil protein named after *Glomus*, the most common AMF, to coat hyphae to keep water and nutrients from getting lost on the way to and from the plant. When a hypha stops transporting nutrients, glomalin sloughs off onto surrounding soil particles (Fig. 11.1). Hundreds of meters of hyphae can grow throughout a small sample of soil resulting in the production of large amounts of glomalin. High glomalin concentrations are related to the formation and stabilization of aggregates in undisturbed and no-till systems compared to nearby conventionally tilled sites (Nichols and Wright 2006). Given the importance of soil aggregation to the functioning of ecosystems and the role played by AMF in this context, they have a pivotal role in determining soil health (Pankhurst et al. 1995).

AMF like other microbial soil health indicators could be used to measure or monitor the impact management decisions selected toward expected ecological intensification (Li et al. 2018). A list of criteria, considered to be critical for soil health indicators according to USDA (Moore-Kucera et al. 2014), includes most of the criteria which are satisfied by AMF, such as they cover a diverse set of soil biological, physical, or chemical functions or processes that are relevant to agriculture or ecological systems; they must be sensitive to changes in soil and crop management systems and should reflect these changes within 1–3 years; the indicator chosen and the method used should be able to be adopted by commercial laboratories; they should be relatively easy to sample and measure for they should be cost-effective and repeatable. However, they need to be specifically standardized for these fungi.

## 11.3 AMF, Glomalin, or GRSP as an Indicator of Soil Aggregation

Soil aggregation refers to the formation of aggregates and the resulting matrix of pore spaces. It is a key ecosystem process resulting in the formation and stabilization of soil structure. The aggregation of soil is a complex process, regulated by a range of abiotic factors (e.g., texture) and mediated by plants and multiple biota groups and their interactions (Rillig 2004). Soil aggregation is important for root growth, plant emergence, water filtration and for a wide range of soil features and ecosystem process rates, such as carbon storage and resistance to erosion (Lehmann et al. 2019). Prior information on the aggregation of soil is very important in soil erosion and land restoration studies.

Plant roots and their mycorrhizal symbionts have been consistently reported to play a crucial force in driving soil aggregation (Leifheit et al. 2014; Rillig et al.

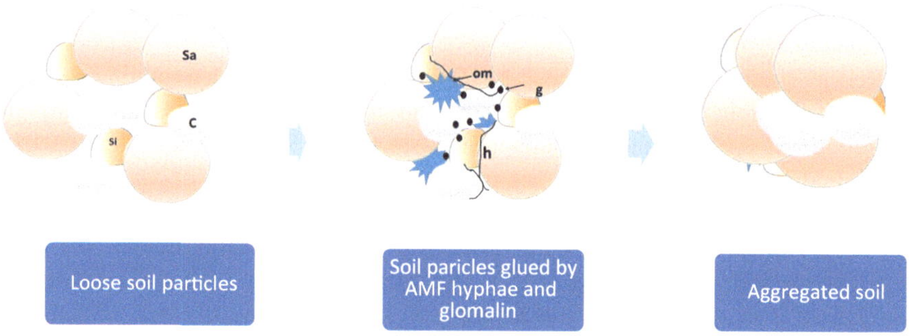


**Fig. 11.3** Role of AMF and glomalin in soil aggregation. The loose soil particles (minerals, organic matter, etc.) are glued together by AMF hyphae and glomalin molecules into pellets. These pellets are rich in nutrients and resist erosion (Abbreviations in the figure are *sa* sand, *si* silt, *c* clay, *om* organic matter, *h* fungal hypha, *g* glomalin molecule)

2015). AMF are known to contribute to soil aggregate stability directly by a physical effect of a hyphal network around soil particles, and indirectly by the hyphal exudation of an iron-containing, heat stable glycoprotein (extracted at 121 °C) glomalin as an aggregate binding agent (Gonzalez-Chavez et al. 2004; Rillig 2004). Glomalin occurs in very large amounts in soil, typically in the range of several to 10 mg $g^{-1}$ soil (Rillig 2005). Glomalin decomposes slower than the AMF hyphae producing it, and the turnover time for glomalin in the soil is estimated to be of the order of several decades (Godbold et al. 2006). Most of the work on glomalin has been done in pot cultures or soil; however, it has also been reported to be produced in vitro vultures of AMF (Rillig 2005). Wright et al. (1996) microscopically observed immunofluorescence of glomalin on the surface of AMF colonized roots, and first suggested that glomalin may be useful as an indicator of AMF colonization. Driver et al. (2005) showed that AMF extraradical fungal mycelium produces glomalin under sterile conditions and could be used as a biomarker for determination of AMF biomass in soil (Rosier et al. 2008) (Fig. 11.3).

Glomalin is quantified from the soil as glomalin-related soil protein (GRSP) and is estimated by the indirect method of Bradford protein analysis (Wright and Upadhyaya 1999). GRSP estimation by this method includes other proteins in addition to glomalin. Other methods, such as immunoreactive total glomalin, have also been used to reduce the interference of other substances (Wright and Upadhyaya 1999). GRSP has been related to carbon content, mycorrhization, root exudates and enzyme activities and soil management (Fokom et al. 2012). Reports have emerged that GRSP may also have a non-AMF origin (Gillespie et al. 2011). Several studies have successfully used GRSP as a suitable proxy for estimation of AMF extraradical hypha length or biomass in soil (Lovelock et al. 2004). It is also used for the quantification of AMF in roots (Rosier et al. 2008). However, different AMF species may produce different amounts of glomalin per unit of hyphae (Lovelock et al. 2004; Rosier et al. 2008). Also, the two most common ways of quantifying the amount of GRSP in a soil extract, ie the Bradford protein assay and the ELISA with the monoclonal antibody, (MAb 32B11), overestimate and underestimate GRSP respectively (Janos et al. 2008). Thus, there needs to be a calibration done for glomalin concentration in soil and the amount of external hyphae of these fungi.

AMF and GRSP have been suggested and reviewed by several researchers to be used as biological indicators of soil quality or health (Rillig et al. 2003; Fokom et al. 2012). The GRSP glue helps to bind soil tiny particles into small aggregates of different sizes. Well-aggregated soil is favorable for plant and microbial growth because it is stable enough to resist wind and water erosion and has better air and water infiltration rates (Lehmann et al. 2019). Additionally, GRSP is known to have a longer residence time in soils and it plays a pivotal role in long-term carbon/nitrogen storage and heavy metal sequestration (Yang et al. 2017). According to Cornejo et al. (2008) and Chern et al. (2007), 1 g of glomalin can bind 4.8 mg of Cu and 188 mg of Pb. In addition to heavy metals, the accumulation was also observed in organic pollutants, e.g., phenanthrene (Gao et al. 2017). Thus, the release and accumulation of GRSP in soils can be a very important mechanism for the ecological restoration of degraded soils, which has the potential to work as a soil health indicator at different stages of restoration.

Few studies have examined the effects of site age, land management, and soil attributes on GRSP in natural and planted forests (Rillig et al. 2003; Fokom et al. 2012). For example, in Brazil, de Souza et al. (2013) suggested that integrating the GRSP content with soil attributes in a recovery process with contrasting areas of different ages will help to understand its behavior in soil recovery processes and to find relationships with other biological, physical, and chemical soil properties and to improve our knowledge about GRSP as biological indicator of soil quality.

## 11.4 AMF Abundance and Diversity as an Indicator of Land Restoration

Land or ecological restoration refers to the improvement of degraded land on a large scale that rebuilds ecological integrity and enhances people's lives. It is the process of ecological restoration of a site to a natural landscape and habitat, safe for humans, wildlife, and plant communities. The potential role of AMF in ecological restoration has been well recognized even before restoration ecology emerged as a scientific field of study (Janos 1980; Brundrett and Abbott 2002; Jeffries et al. 2002). Assessment of measurable ecological restoration parameters and ecological processes in relation to AMF diversity and abundance need to be investigated to be used as indicators.

Degraded lands are characterized by low levels of AMF abundance and diversity (Cardozo-Junior et al. 2012; Gupta et al. 2018b). An extensive study was carried out in Brazil that compared the abundance and diversity of AMF on lands of differing degradation levels and also compared this with a young restoring site. This revealed that as the scale of degradation increases, the abundance and diversity of AMF reduces and when restoration presumes both AMF abundance and diversity increase (Cardozo-Junior et al. 2012). Elsewhere also, it was reported that an effective exclosure increased AMF abundance (Birhane et al. 2010; Gupta et al. 2018b). Further, a study carried out in a greenhouse with simulated erosion was able to demonstrate that erosion of soil beyond 7.5 cm could make soil loose AMF completely (Habte 1989). In another study, it was demonstrated that while AMF could maintain their infective potential in extremely dry soil conditions, but their infective potential was significantly lowered when the soil was disturbed (Jasper et al. 1989). External hyphae of AMF are the important source of inoculum but are highly susceptible to disturbance and hence, disturbance leads to lowered propagation (Brundrett and Abbott 2002). Similarly, Gupta et al. (2018b) measured the AM fungal diversity at nine sites located in Delhi forests, which had different types of urban usage in terms of heavy vehicular traffic pollution, littering, defecation, and recreational activities. This study revealed a significant decrease in alpha diversity of AMF and their abundance (measured as spore density, biovolume, mean infection percentage in roots), soil hyphal length and easily extractable glomalin-related soluble proteins (EE-GRSP) at polluted sites.

Furthermore, several studies conducted in agricultural fields have shown that disturbance not only reduces AMF abundance, diversity, and infectivity but also results in a drastic shift in the AMF community structure (Brundrett and Ashwath 2013; Gupta et al. 2018b). Most species of the most common AMF families (Glomeraceae, Acaulosporaceae, and Gigasporaceae) have distinctive biomass allocation strategies whereby species of the Glomeraceae allocate most of their biomass in the intraradical hyphae while species of the Gigasporaceae allocate most of their biomass in the extraradical hyphae and species of the Acaulosporaceae produce low biomass both intra- and extra-radically (Maherali and Klironomos 2007; Gupta et al. 2018b). According to Brundrett and Ashwath (2013), difference in spore size in different AMF families may also be related to their strategy of biomass allocation.

Distinctive AMF fungal groups have different life-history strategy. Some studies revealed that most species of the Glomeraceae are ruderals while that of Gigasporaceae and Acaulosporaceae are competitors and stress tolerators, respectively (Chagnon et al. 2013). AMF species growing on disturbed soil or wasteland are disturbance tolerant since they have an adjusted life history strategy viz., grow faster, have short life cycle and invest earlier and more abundantly in spore formation, fuse fragmented hyphae more readily, and form cross-walls that enable infected root pieces and severed hyphal fragments to recolonize host roots and have shorter extra-radical mycelium (Chagnon et al. 2013). AMF communities of disturbed sites are characteristically dominated by disturbance tolerant species most of the family Glomeraceae (Chagnon et al. 2013) or *Glomus constrictum* in another study (Gupta et al. 2018b). The number of surviving propagules of AMF in soils also declines with time in the absence of host plants (Brundrett and Abbott 2002). Another study in a Malaysian forest reported that heavy logging significantly reduced (75% reductions) the abundance and infectivity of AMF propagules (Alexander et al. 1992). Therefore, because land degradation significantly reduces plant cover, increases soil disturbance and erosion, low levels of AMF abundance, diversity and infective potential can be considered as a peculiar feature of degraded lands. Degraded lands are also prone to invasion by exotic alien species.

Determination of soil microbial activity and use of microbial soil health indicators is especially important phenomenon for determination heavy metals or pesticides because their impact is dependent on the presence of microbes (Hildebrandt et al. 2007). For example, the concentration of heavy metals in the soil will not change over small time periods, but their bioavailability may change (Yang et al. 2016). Similarly, microbial degradation of Polycyclic aromatic hydrocarbons (PAHs) depends on various environmental conditions, such as nutrients, number and kind of the microorganisms, nature as well as the chemical property of the PAH being degraded (Ghosal et al. 2016). Therefore, the total content of chemicals in soil is not a reliable indicator of its bioavailability (Logan 2000) and thereby soil health. Instead, bioavailability must be measured in relation to bioassays and specific microbial processes. In context of this, microbial responses also integrate the effect of chemical mixtures, the information not obtained by studying the chemical mixtures themselves.

## 11.5 Approaches and Way Forward

AMF are promising soil microorganisms that improve soil health as a tradeoff of nutrient uptake, disease control, and phytoremediation (Jacott et al. 2017; Yang et al. 2017). Besides the need for constant monitoring and evaluations of physical, chemical, and biological processes to achieve better soil health, it is imperious to keep in mind that soil microorganisms and their interactions are the main agents of nutrient cycling and also have a complex interaction with plants (Jacott et al. 2017). Any land-use strategy that contributes to a better equilibrium of soil microorganisms can result in greater crop productivity, at low cost, and contributes to minimizing the use of mineral fertilizers or pesticides, favoring high sustainability (Gupta et al. 2018a; Chen et al. 2018). Rillig et al. (2015) proposed there is a need for dedicated experiments for example for separation of formation and stabilization of aggregates or for comparing different root systems or fungal isolates under otherwise identical conditions. Also, there is a need to build mycorrhizal fungal strain database since Lehmann et al. (2019) found soil aggregate formation capability ranging from neutral to positive and 39 large differences in trait expression among different AMF strains.

## References

Alexander I, Norani A, Lee SS (1992) The role of mycorrhizas in the regeneration of some Malaysian forest trees. Phil Trans R Soc Lond Ser B 335:379–388

Amezaga IM, Elustondo EM, Crespo CG, Hortala MA, Sierra LE (2016) Health cards for the evaluation of agricultural sustainability. Span J Soil Sci 6(1):1–20

Apfelbaum SI, McElligott K, Thompson R, Tiller E (2019) AES white paper: defining soil health within the context of ecosystem health—a framework. Appl Ecol Serv:1–16

Augé RM (2001) Water relations, drought and vesicular-arbuscular mycorrhizal symbiosis. Mycorrhiza 11:3–42

Azcón-Aguilar C, Barea JM (1997) Arbuscular mycorrhizas and biological control of soil-borne plant pathogens—an overview of the mechanisms involved. Mycorrhiza 6(6):457–464

Bago B, Pfeffer PE, Shachar-Hill Y (2000) Carbon metabolism and transport in arbuscular mycorrhizas. Plant Physiol 24(3):949–958

Bagyaraj DJ, Ashwin R (2017) Can mycorrhizal fungi influence plant diversity and production in an ecosystem. In: Bagyaraj DJ, Jamaluddin D (eds) Microbes for restoration of degraded ecosystems. NIPA, New Delhi, pp 1–7

Banerjee K, Gadani MH, Srivastava KK, Verma N, Jasrai YT, Jain NK (2013) Screening of efficient arbuscular mycorrhizal fungi for *Azadirachta indica* under nursery condition: a step towards afforestation of semi-arid region of Western India. Braz J Microbiol 44(2):587–594

Bansal M, Mukerji KG (1994) Positive correlation between root exudation and VAM induced changes in *Rhizosphere mycoflora*. Mycorhiza 5:39–44

Birhane E, Kuyper TW, Sterck FJ, Bongers F (2010) Arbuscular mycorrhizal associations in *Boswellia papyrifera* (frankincense-tree) dominated dry deciduous woodlands of Northern Ethiopia. For Ecol Manag 260(12):2160–2169

Bonfante A, Terribile F, Bouma J (2019) Refining physical aspects of soil quality and soil health when exploring the effects of soil degradation and climate change on biomass production: an Italian case study. Soil 5(1):1–4

Brundrett MC, Abbott LK (2002) Arbuscular mycorrhizas in plant communities. In: Microorganisms in plant conservation and biodiversity. Springer, Dordrecht, pp 151–193

Brundrett MC, Ashwath N (2013) Glomeromycotan mycorrhizal fungi from tropical Australia III. Measuring diversity in natural and disturbed habitats. Plant Soil 370(1–2):419–433

Brundrett M, Bougher N, Dell B, Grove T, Malajczuk N (1996) Working with mycorrhizas in forestry and agriculture (No. 435-2016-33680). Australian Centre for International Agricultural Research, Canberra

Bünemann EK, Bongiorno G, Bai Z, Creamer RE, De Deyn G, de Goede R, Fleskens L, Geissen V, Kuyper TW, Mäder P, Pulleman M (2018) Soil quality–a critical review. Soil Biol Biochem 120:105–125

Cardoso EJ, Vasconcellos RL, Bini D, Miyauchi MY, Santos CA, Alves PR, Paula AM, Nakatani AS, Pereira JD, Nogueira MA (2013) Soil health: looking for suitable indicators. What should be considered to assess the effects of use and management on soil health? Sci Agric 70 (4):274–289

Cardozo-Junior FM, Carneiro RFV, Goto BT, Bezerra AAC, Araújo ASF, Nunes LAPL (2012) Arbuscular mycorrhizal fungi in degraded lands in Northeast Brazil. Afr J Microbiol Res 6:7198–7205

Chagnon PL, Bradley RL, Maherali H, Klironomos JN (2013) A trait-based framework to understand life history of mycorrhizal fungi. Trends Plant Sci 18:484–449

Chen LQ (2014) Sweet sugar transporters for phloem transport and pathogen nutrition. New Phytol 201:1150–1155

Chen M, Arato M, Borghi L, Nouri E, Reinhardt D (2018) Beneficial services of arbuscular mycorrhizal fungi–from ecology to application. Front Plant Sci 9:1270. https://doi.org/10.3389/fpls.2018.01270

Chern EC, Tsai DW, Ogunseitan OA (2007) Deposition of glomalin-related soil protein and sequestered toxic metals into watersheds. Environ Sci Technol 41:3566–3572

Congreves KA, Hayes A, Verhallen EA, Van Eerd LL (2015) Long-term impact of tillage and crop rotation on soil health at four temperate agroecosystems. Soil Tillage Res 152:17–28

Cornejo P, Meier S, Borie G, Rillig MC, Borie F (2008) Glomalin-related soil protein in a Mediterranean ecosystem affected by a copper smelter and its contribution to Cu and Zn sequestration. Sci Total Environ 406:154–160
Cosme M, Fernández I, Van der Heijden MG, Pieterse CM (2018) Non-mycorrhizal plants: the exceptions that prove the rule. Trends Plant Sci 23(7):577–587
Davison J, Moora M, Opik M, Adholeya A, Ainsaar L, Ba A, Johnson NC (2015) Global assessment of arbuscular mycorrhizal fungal diversity reveals very low endemism. Science 349(6251):970–973
de Souza RG, da Silva DKA, de Mello CMA, Goto BT, da Silva FSB, Sampaio EVSB, Maia LC (2013) Arbuscular mycorrhizal fungi in revegetated mined dunes. Land Degrad Dev 24:147–155
Doran JW (2002) Soil health and global sustainability: translating science into practice. Agric Ecosyst Environ 88:119–127
Doran JW, Parkin TB (1996) Quantitative indicators of soil quality: a minimum data set. In: Doran JW, Jones AJ (eds) Methods for assessing soil quality. Soil Science Society of America, Madison, pp 25–37
Driver JD, Holben WE, Rillig MC (2005) Characterization of glomalin as a hyphal wall component of arbuscular mycorrhizal fungi. Soil Biol Biochem 37:101–106
Fokom R, Adamou S, Teugwa MC, Begoude Boyogueno AD, Nana WL, Ngonkeu MEL, Tchameni NS, Nwaga D, Tsala NG, Amvam Zollo PH (2012) Glomalin related soil protein, carbon, nitrogen and soil aggregate stability as affected by land use variation in the humid forest zone of South Cameroon. Soil Tillage Res 120:69–75
Gao Y, Zhou Z, Ling W, Hu X, Chen S (2017) Glomalin related soil protein enhances the availability of polycyclic aromatic hydrocarbons in soil. Soil Biol Biochem 107:129–132
Ghosal D, Ghosh S, Dutta TK, Ahn Y (2016) Current state of knowledge in microbial degradation of polycyclic aromatic hydrocarbons (PAHs): a review. Front Microbiol 7:1369
Gillespie AW, Farrell RE, Walley FL, Ross AR, Leinweber P, Eckhardt KU, Regier TZ, Blyth RI (2011) Glomalin-related soil protein contains non-mycorrhizal-related heat-stable proteins, lipids and humic materials. Soil Biol Biochem 43:766–777
Giri B, Kapoor R, Mukerji KG (2007) Improved tolerance of *Acacia nilotica* to salt stress by arbuscular mycorrhiza, Glomus fasciculatum may be partly related to elevated K/Na ratios in root and shoot tissues. Microb Ecol 54:753–760
Godbold DL, Hoosbeek MR, Lukac M, Cotrufo MF, Janssens IA, Ceulemans R, Polle A, Velthorst EJ, Scarascia-Mugnozza G, De Angelis P, Miglietta F (2006) Mycorrhizal hyphal turnover as a dominant process for carbon input into soil organic matter. Plant Soil 281(1–2):15–24
Gonzalez-Chavez MC, Carrillo-Gonzalez R, Wright SF, Nichols KA (2004) The role of glomalin, a protein produced by arbuscular mycorrhizal fungi, in sequestering potentially toxic elements. Environ Pollut 130(3):317–323
Gupta MM, Aggarwal A, Asha (2018a) From mycorrhizosphere to rhizosphere microbiome: the paradigm shift. In: Giri B, Parasad R, Varma A (eds) Root biology. Springer, Cham, pp 487–500
Gupta MM, Gupta A, Kumar P (2018b) Urbanization and biodiversity of arbuscular mycorrhizal fungi: the case study of Delhi, India. Rev Biol Trop 66(4):1563–1574
Gupta MM, Chourasiya D, Sharma MP (2019) Diversity of arbuscular mycorrhizal fungi in relation to sustainable plant production systems. In: Microbial diversity in ecosystem sustainability and biotechnological applications. Springer, Singapore, pp 167–186
Habte M (1989) Impact of simulated erosion on the abundance and activity of indigenous vesicular-arbuscular mycorrhizal endophytes in an Oxisol. Biol Fertil Soils 7:164–167
Hildebrandt U, Regvar M, Bothe H (2007) Arbuscular mycorrhiza and heavy metal tolerance. Phytochemistry 68(1):139–146
Jacott CN, Murray JD, Ridout CJ (2017) Trade-offs in arbuscular mycorrhizal symbiosis: disease resistance, growth responses and perspectives for crop breeding. Agronomy 7(4):75–80
Janos DP (1980) Mycorrhizae influence tropical succession. Biotropica 12:56–64

Janos DP, Garamszegi S, Beltran B (2008) Glomalin extraction and measurement. Soil Biol Biochem 40:728–739
Jasper DA, Abbott LK, Robson AD (1989) Hyphae of a vesicular-arbuscular mycorrhizal fungus maintain infectivity in dry soil, except when the soil is disturbed. New Phytol 112:101–107
Jeffries P, Craven-Griffiths A, Barea JM, Levy Y, Dodd JC (2002) Application of arbuscular mycorrhizal fungi in the revegetation of decertified Mediterranean ecosystems. In: Gianinazzi S, Schuepp H, Barea JM, Haselwandter K (eds) Mycorrhizal technology in agriculture. Birkhauser, Basel, pp 151–174
Kibblewhite MG, Ritz K, Swift MJ (2008) Soil health in agricultural systems. Phil Trans R Soc Lond Ser B Biol Sci 363(1492):685–701
Kikvidze Z, Armas C, Fukuda K, Martínez-García LB, Miyata M, Oda-Tanaka A (2010) The role of arbuscular mycorrhizae in primary succession: differences and similarities across habitats. Web Ecol 10:50–57
Laisram J, Maikhuri RK, Rao KS (2012) Soil quality and soil health: a review. Int J Ecol Environ Sci 38(1):19–37
Larcher W (1995) Physiological plant ecology, 4th edn. Springer, Berlin
Lehmann A, Zheng W, Ryo M, Soutschek K, Rongstock R, Maass S, Rillig MC (2019) Fungal traits important for soil aggregation. BioRxiv 1:732628
Leifheit EF, Verbruggen E, Rillig MC (2014) Arbuscular mycorrhizal fungi reduce decomposition of woody plant litter while increasing soil aggregation. Soil Biol Biochem 81:323–328
Li S, Tian Y, Wu K, Ye Y, Yu J, Zhang J, Liu Q, Hu M, Li H, Tong Y (2018) Modulating plant growth–metabolism coordination for sustainable agriculture. Nature 560:595–600
Logan TJ (2000) Soils and environmental quality. In: Sumner ME (ed) Handbook of soil science. CRC Press, Boca Raton, pp G155–G169
Lovelock CE, Wright SF, Nichols KA (2004) Using glomalin as an indicator for arbuscular mycorrhizal hyphal growth: an example from a tropical rain forest soil. Soil Biol Biochem 36 (6):1009–1012
Maherali H, Klironomos JN (2007) Influence of phylogeny on fungal community assembly and ecosystem functioning. Science 316:1746–1748
Moebius-Clune BN (2016) Comprehensive assessment of soil health: the Cornell framework manual, 3.1 edn. Cornell University, Ithaca, NY
Moore-Kucera J, Cox SB, Peyron M, Bailes G, Kinloch K, Karich K, Miles C, Inglis DA, Brodhagen M (2014) Native soil fungi associated with compostable plastics in three contrasting agricultural settings. Appl Microbiol Biotechnol 98:6467–6485
Nichols KA, Wright SF (2006) Carbon and nitrogen in operationally-defined soil organic matter pools. Biol Fertil Soils 43:215–220
Pankhurst CE, Hawke BG, MacDonald HJ, Kirkby CA, Buckerfield JC, Michelsen P, O'Brien KA, Gupta VVSR, Doube BM (1995) Evaluation of soil biological properties as potential bioindicators of soil health. Aust J Exp Agric 35:1015–1028
Parisi V, Menta C, Gardi C, Jacomini C, Mozzanica E (2005) Microarthropod communities as a tool to assess soil quality and biodiversity: a new approach in Italy. Agric Ecosyst Environ 105:323–333
Powell JR, Rillig MC (2018) Biodiversity of arbuscular mycorrhizal fungi and ecosystem function. New Phytol 220(4):1059–1057
Purakayastha TJ, Pathak H, Kumari S, Biswas S, Chakrabarty B, Padaria RN, Kamble K, Pandey M, Sasmal S, Singh A (2019) Soil health card development for efficient soil management in Haryana, India. Soil Tillage Res 191:294–305
Rillig MC (2004) Arbuscular mycorrhizae, glomalin, and soil aggregation. Can J Soil Sci 84:355–363
Rillig MC (2005) Polymers and microorganisms. In: Hillel D (ed) Encyclopedia of soils in the environment. Oxford, UK, Elsevier, pp 287–294
Rillig MC, Ramsey PW, Morris S, Paul EA (2003) Glomalin, an arbuscular-mycorrhizal fungal soil protein, responds to land-use change. Plant Soil 253(2):293–299

Rillig MC, Aguilar-Trigueros CA, Bergmann J, Verbruggen E, Veresoglou SD, Lehmann A (2015) Plant root and mycorrhizal fungal traits for understanding soil aggregation. New Phytol 205:1385–1388

Rillig MC, Sosa-Hernández MA, Roy J, Aguilar-Trigueros CA, Vályi K, Lehmann A (2016) Towards an integrated mycorrhizal technology: harnessing mycorrhiza for sustainable intensification in agriculture. Front Plant Sci 7:1625. https://doi.org/10.3389/fpls.2016.01625

Rosier CL, Piotrowski JS, Hoye AT, Rillig MC (2008) Intraradical protein and glomalin as a tool for quantifying arbuscular mycorrhizal root colonization. Pedobiologia 52:41–50

Sendek A, Karakoç C, Wagg C, Domínguez-Begines J, do Couto GM, van der Heijden MG, Naz AA, Lochner A, Chatzinotas A, Klotz S, Gómez-Aparicio L (2019) Drought modulates interactions between arbuscular mycorrhizal fungal diversity and barley genotype diversity. Sci Rep 9(1):9650

Smith SE, Gianinazzi-Pearson V (1988) Physiological interactions between symbionts in vesicular-arbuscular mycorrhizal plants. Annu Rev Plant Physiol Plant Mol Biol 39(1):221–244

Spatafora JW, Chang Y, Benny GL, Lazarus K, Smith ME, Berbee ML, Bonito G, Corradi N, Grigoriev I, Gryganskyi A, James TY (2016) A phylum-level phylogenetic classification of zygomycete fungi based on genome-scale data. Mycologia 108(5):1028–1046

Toro M, Azcon R, Barea J (1997) Improvement of arbuscular mycorrhiza development by inoculation of soil with phosphate-solubilizing rhizobacteria to improve rock phosphate bioavailability ((sup32) P) and nutrient cycling. Appl Environ Microbiol 63(11):4408–4412

Van der Heijden MGA, Klironomos JN, Ursic M, Moutoglis P, Streitwolf-Engel R, Boller T et al (1998) Mycorrhizal fungal diversity determines plant biodiversity, ecosystem variability and productivity. Nature 396:69–72

van Es HM, Karlen DL (2019) Reanalysis validates soil health indicator sensitivity and correlation with long-term crop yields. Soil Sci Soc Am J 83(3):721–732. https://doi.org/10.2136/sssaj2018.09.0338

Warkentin BP (1995) The changing concept of soil quality. J Soil Water Conserv 50:226–228

Watts-Williams SJ, Cavagnaro TR (2012) Arbuscular mycorrhizas modify tomato responses to soil zinc and phosphorus addition. Biol Fertil Soils 48:285–294

Wipf D, Krajinski F, van Tuinen D, Recorbet G, Courty PE (2019) Trading on the arbuscular mycorrhiza market: from arbuscules to common mycorrhizal networks. New Phytol 223 (3):1127–1142

Wright SF, Upadhyaya A (1999) Quantification of arbuscular mycorrhizal fungi activity by the glomalin concentration on hyphal traps. Mycorrhiza 8:283–285

Wright SF, Franke-Synder M, Morton JB, Upadhyaya A (1996) Time course study and partial characterization of a protein on arbuscular mycorrhizal hyphae during active colonization of roots. Plant Soil 181:193–203

Yang X, Yu L, Chen Z, Xu M (2016) Bioavailability of polycyclic aromatic hydrocarbons and their potential application in eco-risk assessment and source apportionment in urban river sediment. Sci Rep 6:23134

Yang Y, He C, Huang L, Ban Y, Tang M (2017) The effects of arbuscular mycorrhizal fungi on glomalin-related soil protein distribution, aggregate stability and their relationships with soil properties at different soil depths in lead-zinc contaminated area. PLoS One 12(8):e0182264

Zhu XQ, Wang CY, Chen H, Tang M (2014) Effects of arbuscular mycorrhizal fungi on photosynthesis, carbon content, and calorific value of black locust seedlings. Photosynthetica 52:247–252

# Chapter 12
# Significance and Management of Green Manures

**K. Das, N. Biswakarma, R. Zhiipao, A. Kumar, P. C. Ghasal, and V. Pooniya**

**Abstract** Green manuring is an arable-farming practice in which undecomposed green material is incorporated (in situ/harvested elsewhere) into soil in order to increase productivity of subsequent crops. Green manure crop is to be turned into the soil at the point of flowering, i.e., about 7–8 weeks from sowing in most crops. The continuous use of green manures enhances the organic matter content and also supplements the nutrient pool of the soil which ultimately improves the soil physical, chemical and biological properties and also suppresses the weeds. It provides nutrient-rich organic matter for the soil microorganisms which easily converts organically bound nutrients in plant residues to easily available nutrient form to the crop. The portion of green manure nitrogen available to a succeeding crop is usually about 40–60% of the total amount contained in the legume and large amounts of legume N retained in soil mostly in organic forms. However, beneficial effects of green manure on succeeding crops depend largely on residue quantity and quality, soil type, soil fertility, soil acidity, biological activity, soil moisture, and temperature.

**Keywords** Green manure · Nutrient availability · Productivity · Soil fertility

K. Das (✉)
ICAR-Central Research Institute for Jute and Allied Fibres, Barrackpore, India

N. Biswakarma · R. Zhiipao · V. Pooniya
ICAR-Indian Agricultural Research Institute, New Delhi, India

A. Kumar
ICAR-Indian Institute of Maize Research, Ludhiana, Punjab, India

P. C. Ghasal
ICAR-Indian Institute of Farming Systems Research, Meerut, India

 
B. Giri, A. Varma (eds.), *Soil Health*, Soil Biology 59,
https://doi.org/10.1007/978-3-030-44364-1_12

## 12.1 Introduction

Maintaining healthy soils is a huge task due to high input–cost relationships, costly fertilizers, and issue of soil degradation. Under such situation green/organic manures are considered as the cheap and good source for plant nutrients and soil amendment. Therefore, integration of green manuring crops into cropping systems brings not only positive influence on the soil properties but also enhances crop productivity, conserve soil moisture, and check weed infestations. In this chapter, we describe the work on green manuring crops that has been carried out during the past few decennium and present days with respect to pros and cons of adopting such practices for making agriculture more attractive and remunerative to the farmers. Degradation of soil properties is the foremost reason for the continuous decline in soil fertility and productivity, consequently leading to reduced economic output. Green manure, also referred to as fertility building crops, constitutes a technology that has been proven by researchers and as efficient and economically viable practice to achieve sustainable agricultural production. Although, it has many roles and been used in traditional agriculture for thousands of years, these are still often underutilized by today's organic farmers.

Green manuring is an arable-farming practice in which undecomposed green material is incorporated into soil in order to increase productivity of subsequent crops. This material may either be obtained from quick-growing green manure crops (GMs) grown in situ or harvested elsewhere, i.e. green leaf manuring usually from a perennial crop, and brought in field (Meelu et al. 1994a). Green manures have been tested by the authors of this chapter in cereal crops. Published reports showed that sesbania added the highest crop residue, i.e. 38.56 t $ha^{-1}$ leading to recycling of major and micronutrients followed by cowpea and mungbean green manures (Pooniya et al. 2012; Pooniya and Shivay 2013). Presently more emphasis on mitigating the environmental impact of all farming systems has led to a growing interest on green manuring due to amelioration of soil physical, chemical and biological properties. Furthermore, this practice increases biological security by suppressing weeds and soil-borne diseases and has the potential to disrupt the life cycle of agriculture pest (Kumar et al. 2014; Varma et al. 2017). In addition, it caters ground cover and, as regard with legume, fixes nitrogen too. Green manure is considered as the complete manure because it contains most of the essential nutrients and releases them slowly during decomposition processes. According to Agricultural Statistics (2005), the area under green manures in India was ~67 lakh ha and subsequently supplemented 7.55, 1.24 and 4.19 lakh tonnes of NPK into the soil. Green manuring is widely practised in Indian states like Andhra Pradesh, Uttar Pradesh, Karnataka, Punjab, Orissa, Gujarat, Madhya Pradesh, Himachal Pradesh and Haryana, while it is practised on a limited scale in other states (Table 12.1).

Table 12.1 Green manuring scenario in different states of India

| State | Area (lakh ha) | Expected green and dry matter yield (lakh tonnes) | Nutrients contribution (lakh tonnes) | | | |
|---|---|---|---|---|---|---|
| | | | N | P | K | Total |
| Andhra Pradesh | 27.20 | 544 | 3.02 | 0.49 | 1.67 | 5.18 |
| Uttar Pradesh | 11.00 | 220 | 1.23 | 0.20 | 0.69 | 2.12 |
| Karnataka | 7.30 | 146 | 0.82 | 0.14 | 0.46 | 1.42 |
| Punjab | 4.25 | 85 | 0.48 | 0.08 | 0.27 | 0.83 |
| Orissa | 3.55 | 70 | 0.40 | 0.07 | 0.22 | 0.69 |
| Gujarat | 2.10 | 42 | 0.24 | 0.04 | 0.13 | 0.41 |
| Madhya Pradesh | 1.94 | 38.8 | 0.22 | 0.03 | 0.12 | 0.37 |
| Himachal Pradesh | 1.30 | 26 | 0.15 | 0.02 | 0.08 | 0.25 |
| Haryana | 1.13 | 22.6 | 0.13 | 0.02 | 0.07 | 0.22 |
| Others | 7.32 | 146.6 | 0.82 | 0.14 | 0.46 | 1.42 |
| Total | 67.09 | 1341.8 | 7.55 | 1.24 | 4.19 | 12.98 |

Source: Agricultural Statistics (2005)

## 12.2 General Requirements That Should Be Met by Green Manure/Cover Crops

To actualize the benefits of green manure/cover crops, it is necessary to know all aspects related to them, i.e. family to which they belong, growth habitat, developmental cycle, nature of weed competition, effects on soil environment, nutrient recycling pattern, N-fixing capacity, seed production performance, time of pests and diseases infestation and ways to manage, etc. Furthermore, it is necessary to know the goal intended to be attained by including them as well as aspects concerning the production systems in which they will be included, i.e. climate, soil type, soil fertility, crops with which they will be integrated in the system, available machinery and/or implements, etc. There exist some desirable features to be possessed by green manure/cover crops to make them more favourable for incorporation into an agricultural production system, the most important being (Florentín et al. 2010): (1) have low establishment and management costs, and easy to sow or manage; (2) faster growth and weed suppression habit; (3) having residual effect on succeeding crops, require few crop management practices; (4) present good conservation characteristics, avoid the proliferation of pests and diseases; (5) avoid competition for land, labour, time, and space with cash or subsistence crops.

The major criteria and characteristics for selection of green manure/cover crops are the following (Monegat 1991): (1) crops belonging to leguminaceae family and quick initial growth to suppress the weeds and to produce huge dry matter in shorter period; (2) deeper and stronger root system for more nutrient and water acquisition and also improves the soil structural properties; (3) green manures have the capacity to tolerate biotic and abiotic stresses with wider range of ecological adaptability.

## 12.3 Types of Green Manuring

### *12.3.1 Green Manuring In Situ*

In this system, short-duration (~45 to 60 days) green manure crops are grown and incorporated into the soil at same site, either as a pure crop or as an intercrop with the main crop. In later case, the green manure is sown between the rows and is particularly well adapted to situations where the soil has to be used as intensively as possible. This system of generating nutrient resources on-site is popular in cropping systems with rice as the main crop and predominant in northern and southern parts of India. This type of green manure should be treated with caution in order to avoid competition with the main crop and consequently lower yields. The main advantages of the system are intensive use of the soil, the efficient erosion control and the reduction in weed population. Crops and tree species suitable for green manuring in different agro-climatic zones are mentioned in Tables 12.2 and 12.3.

### *12.3.2 Green Leaf and Brown Manuring*

It refers to turning into the soil green leaves and tender green twigs collected from shrubs and trees grown on bunds, wastelands and nearby forest areas. They are usually turned down or mixed into the soil 15–20 days before sowing of the principal/main crops based on the tenderness of the foliage or plant parts. *Gliricidia maculata*, *Pongamia pinnata*, *Sesbania speciosa*, *Leucaena leucocephala*, *Azadirachta indica*, *Delonix regia* and *Peltophorum pterocarpum* are used for green leaf manuring. Brown manuring is a technique to grow sesbania in standing rice crop and kill them with the help of herbicide like 2,4-D ester or bispyribac-Na for manuring. After killing, the colour of the sesbania residue becomes brown and so it is called brown manuring.

**Table 12.2** Green manure crops suitable for different agro-climatic zones

| Tropical areas | | Temperate areas | |
|---|---|---|---|
| Common name | Scientific name | Common name | Scientific name |
| Pueraria | *Pueraria phaseoloides* | Faba bean | *Vicia faba* |
| Cowpea | *Vigna unguiculata* | Ladino clover | *Trifolium repens* |
| Green gram | *Vigna radiata* | Crimson clover | *Trifolium incarnatum* |
| Cluster bean | *Cyamopsis tetragonoloba* | Subterranean clover | *Trifolium subterraneum* |
| Lablab | *Lablab purpureus* | Soybean | *Glycine max* |
| White lupin | *Lupinus albus* | Black lentil | *Lens culinaris* |
| Sunn hemp | *Crotalaria breviflora* | Red clover | *Trifolium pratense* |
| Gray bean | *Mucuna cinerecum* | Alfalfa | *Medicago sativa* |
| Pigeon pea | *Cajanus cajan* | Hairy vetch | *Vicia villosa* |
| Jack bean | *Canavalia ensiformis* | Barrel medic | *Medicago truncatula* |
| Buffalo bean | *Mucuna aterrima* | Milk vetch | *Astragalus sinicus* |
| Stylo | *Stylosanthes guianensis* | Sweet clover | *Melilotus officinalis* |
| Velvet bean | *Mucuna deeringiana* | Winter pea | *Pisum sativum* |
| Milk vetch | *Astragalus sinicus* | Cura clover | *Trifolium ambiguum* |
| Desmodium | *Desmodium ovalifolium* | Common vetch | *Vicia sativa* |
| Dhaincha | *Sesbania aculeata, S. rostrata* | Purple vetch | *Vicia benghalensis* |
| Zornia | *Zornia latifolia* | (B) Ex situ green leaf manuring plants | |
| Black gram | *Phaseolus mungo, P. trilobus* | Common name | Scientific name |
| Kudzu | *Pueraria phaseoloides* | Wild indigo | *Indigofera teysmannii* |
| Soybean | *Glycine max* | Gliricidia | *Gliricidia sepium* |
| Adzuki bean | *Vigna angularis* | Subabul | *Leucaena leucocephala* |
| Wild indigo | *Indigofera tinctoria* | Kassod | *Cassia tora* |
| Alfalfa | *Medicago sativa* | Tephrosia | *Tephrosia purpurea* |
| Jumby bean | *Leucaena leucocephala* | Karanj | *Pongamia glabra* |
| Berseem | *Trifolium alexandrinum* | Sesbania | *Sesbania rostrata* |
| Sunn hemp | *Crotalaria juncea, C. striata* | Milkweed | *Calotropis gigantea* |

Source: Meena et al. (2018)

## 12.4 Benefits of Green Manuring

### *12.4.1 Building of Organic Matter and Improved Soil Structure*

A major benefit from green manures is the addition of organic matter to the soil. Biomass and N accumulation of green manure legumes are influenced by species grown, water regime, nutrient supply, soil type, photoperiod and age of incorporation. The continuous use of green manures enhances the organic matter content and also supplements the nutrient pool of the soil (Kumar et al. 2008), which ultimately improves the soil physical properties and quantities of organic acid, amino acids,

**Table 12.3** In situ green manuring weeds and their mineral composition on dryweight basis

| Weed | Nutrient content (%) | | |
|---|---|---|---|
| | N | $P_2O_5$ | $K_2O$ |
| *Amaranthus viridis* | 3.16 | 0.06 | 4.51 |
| *Cassia occidentalis* | 3.08 | 1.56 | 2.31 |
| *Chenopodium album* | 2.59 | 0.37 | 4.34 |
| *Cleome viscosa* | 1.96 | 1.53 | 5.81 |
| *Dactyloctenium aegyptium* | 2.78 | 0.24 | 1.65 |
| *Digitaria sanguinallis* | 2.00 | 0.36 | 3.48 |
| *Echinochloa crus-galli* | 2.98 | 0.40 | 2.96 |
| *Portulaca quadrifida* | 2.40 | 0.09 | 5.57 |
| *Solanum xanthocarpum* | 2.56 | 1.63 | 2.12 |
| *Trianthema portulacastrum* | 2.34 | 0.30 | 1.15 |
| *Eupatorium* spp. | 2.93 | 0.49 | 1.47 |
| *Parthenium hysterophorus* | 2.66 | 0.88 | 1.29 |
| *Eichhornia crassipes* | 2.83 | 0.90 | 1.79 |

Source: Kumar and Pawar (2018)

sugars, vitamins and mucilage (Shukla et al. 2011). These organic substances are capable of binding the soil particles and provide better soil aggregation and help in improvement of soil structure (MacRae and Mehuys 1985; Schutter and Dick 2001). The increase in soil organic matter (SOM) following GMs often ranges from 0 to 1% of total soil mass (Utomo et al. 1990; Reddy et al. 2003). Compounds resistant to decomposition such as gums, waxes and resins are formed while organic matter are broken down by microorganisms. These compounds and the mycelia, mucus, and slime produced by the microorganisms bind soil particles as granules or aggregates. Soils with better aggregates are easy to till apace with enhanced aeration and higher water infiltration rate. Soil humus is positively correlated with the amount of organic matter content in it. Humus is the end product of the decay of plant and animal materials in the soil which provides a wide range of benefits to crop production. Pooniya et al. (2012) reported that sesbania and cowpea green manuring accumulated higher fresh and dry weight than mungbean. The fast and determinate growth habit of sesbania can lead to enhanced organic C content of soil and nutrient availability. In another long-term study conducted at IARI indicated that maize-chickpea-sesbania (MCS) system resulted in 18.9% and 20.4% higher soil organic carbon (SOC) than maize–mustard–mungbean and maize–maize–sesbania cropping system (Parihar et al. 2016). This could be a result of differential quantity and chemical constituent in root exudates and/or crop residue biomass among the crop rotations (Congreves et al. 2015). The narrow C:N ratio of legume residue causes rapid decomposition and hence higher SOC compared to other cropping sequences. Further, tap roots of the legume result in higher SOC content in deeper layers.

### 12.4.2 Enhanced Soil Microbial Activities

The soil microorganisms multiply promptly and attack the freshly green manure crop (young, relatively lush green) once it is incorporated into the soil. Further decomposition of the incorporated green manure serves major functions for microflora providing both C and energy for growth and formation of new cell material, which further multiplies its colony saprophytically on the decomposing organic matter (Ye et al. 2014). The process of decomposition has an immense significance for longer existence of primary products, through the release of energy and nutrients by microbial activity (Kumar et al. 2014). During microbial breakdown, nutrients held within the plant tissues are released and made available to the succeeding crop. The breaking down of organic matter by microorganisms is influenced by soil moisture and temperature as well as carbon-to-nitrogen (C:N) ratio of the plant material. The C:N ratio of plant tissue reflects the kind and age of the plants from which it was derived. The C:N ratio between 15:1 and 25:1 is favourable for accelerated decomposition of organic matter. The incorporation of green manures particularly legume green manures (LGM) has two main positive outcomes from the micro-biological point of view: (1) primarily it provides nutrient-rich OM for the microbial community which easily converts organically bound nutrients in plant residues to easily available nutrient form to the crops; and (2) it enhances the microbial diversity of soil microorganisms (Schutter and Dick 2001; Eriksen 2005; Kumar et al. 2016).

### 12.4.3 Improvement in Soil Physical Properties

Green manuring has favourable effects on soil physical properties. The soil physical properties that are affected by the incorporation of the green manures include the structure, moisture retention capacity, consistency and density. Other properties such as the porosity, aeration, hydraulic conductivity and infiltration are related to the modifications in the soil structure. The legume green manuring exaggerates total pore space by decreasing the soil bulk density leading to enhanced root development, soil water content and nutrient use efficiency (Anderson et al. 1997; Parihar et al. 2016). The penetration of green manure roots, especially dhaincha and sunn hemp prevents soil erosion by holding the soil in its place (Schumann et al. 2000).

### 12.4.4 Weed Suppression

Weeds flourish on bare soil and green manure crops take up space and light, thereby shading the soil and reducing the opportunity for weeds to establish themselves. The effects of incorporating green manures on weeds are often variable and have multiple mechanisms, whereby weeds may be suppressed by either direct competition or allelopathy or through phytotoxic effects on germinating weeds (Creamer et al. 1996; Hoffman et al. 1996). While, on the other hand, it may promote weed infestation by hindering chemical or mechanical weed control (Thorup-Kristensen et al. 2003). The prime intent of a non-legume green manure such as rye, millet, or sudan grass is to provide weed control, add organic matter, and improve soil tilth. Thus, whenever possible, annual grain or vegetable crops should follow a legume green manure to derive the benefit of farm-produced nitrogen. The use of green manure/cover crops lowers the weed population in the cotton field, besides favouring the crop by reducing competition for water, light and nutrients. Furthermore, it could provide additional economic advantage for the farmer by saving the labour (hoeing) for weed control (Florentín et al. 2010). The use of sesbania alone as green manure has an identical effect on reducing the population and dry weight of weeds as that of integrated use of urea + sesbania (Khankahre et al. 2002). Employing allelopathic cover crops and live mulches in suppressing weeds has become an integral method of weed control in sustainable agriculture. Allelopathic plants are those that inhibit or slow down the growth of other nearby plants by releasing natural toxins, or 'allelochemicals'. Cover crops exhibiting allelopathy include the small grains like rye and summer annual forages related to sorghum and sudan grass.

### *12.4.5 Nitrogen Fixation*

The current environment has been threatened due to overuse of nitrogenous fertilizer to enhance agricultural production. The use of chemical fertilizers in agriculture and different industrial units has resulted in enhanced concentrations of reactive forms (e.g. NOx, $N_2O$, $NO_3^-$, $NH_3$) of N to around 120% in the atmosphere (Meena et al. 2018). Therefore, the scenario compels us to rethink about the role of biological nitrogen fixation (BNF) whereby green manuring with inclusion of legumes appears to be the most feasible option. The legume green manuring may have a realistic and applicable potential in enhancing the agricultural sustainability through improved nutrient retention (Dinnes et al. 2002), enhancing the soil fertility status by decreasing soil erosion (Fageria and Baligar 2005). There are certain cultural and environmental conditions that limit legume crop growth such as delayed planting, poor stand establishment and drought which in turn reduce the amount of nitrogen fixation and its production. While conditions favouring good nitrogen fixation include proper crop establishment, optimum soil nutrient levels and soil pH, good nodulation, and adequate soil moisture. Estimates of N accumulation for leguminous GMs and the relative contribution of biological N fixation in this process range broadly depending on soil fertility, water availability and GM crops species.

Generally, most legumes accumulate N from biological fixation when their demand cannot be met by uptake of N from the soil (Gardner 1985). For example, sunn hemp (*Crotalaria juncea* L.) has been estimated to fix 27–39% (Ramos et al. 2001), 72–81% (Ladha and Kundu 1997), and 91% (Seneratne and Ratnasinghe 1995) of its total N in different study locations and conditions. Reduction of soil N

**Table 12.4** Potential N contributions of N-fixing legumes to succeeding crops in Indian soils

| Name | Botanical name | Sowing season | An average yield of green matter (t/ha) | N (% on a green weight basis) | N added (kg/ha) |
|---|---|---|---|---|---|
| Sunn hemp | *Crotalaria juncea* | Kharif | 15.2 | 0.43 | 84.0 |
| Dhaincha | *Sesbania aculeata* | Kharif | 14.4 | 0.42 | 77.1 |
| Mungbean | *Vigna radiate* | Kharif | 5.7 | 0.53 | 38.6 |
| Cowpea | *Vigna ungiculata* | Kharif | 10.8 | 0.49 | 56.3 |
| Guar | *Cyamopsis tetragonoloba* | Kharif | 14.4 | 0.34 | 62.3 |
| Senji | *Melilotus alba* | Rabi | 20.6 | 0.51 | 134.4 |
| Khesari | *Lathyrus sativus* | Rabi | 8.8 | 0.54 | 61.4 |
| Berseem | *Trifolium alexandrinum* | Rabi | 11.1 | 0.43 | 60.7 |

Source: Yawalkar et al. (1996)

through competition generally increases rates of biological N fixation by legumes. Water stress and deficiency of nutrients other than N may significantly reduce N fixation (Gardner 1985). The portion of green manure nitrogen available to a succeeding crop is usually about 40–60% of the total amount contained in the legume. However, beneficial effects of green manure on succeeding crops depend largely on residue quantity and quality, soil type, soil fertility, soil acidity, biological activity, soil moisture and temperature (Mary and Recous 1994; Thonnissen et al. 2000). Ladd et al. (1983) and Harris et al. (1994) reported that less than 30% of legume N was recovered by succeeding non-legume crops, and large amounts of legume N retained in soil is mostly in organic forms. The potentials of different green manure crops for atmospheric $N_2$ fixation are shown in Table 12.4.

### 12.4.6 Enhanced Crop Yields

The positive impact of green manuring particularly legume green manuring is well reflected on grain yield through enhanced soil organic matter and increased nutrition to growing crops. The application of synthetic N along with green manure enhances the N use efficiency (NUE). Furthermore, green manure constantly supplies N and releases slowly matching the plants' requirement resulting in improved crop performance (Westcott and Mikkelson 1987; IRRI 1990). The resultant effect of green manuring on the growth and yield of various succeeding crops through field experiments by a number of researchers are summarized in Table 12.5.

**Table 12.5** Percentage increase in crop yield by adopting green manure practice

| Crop | State | Green manure crop | Increase in yield (%) |
|---|---|---|---|
| Rice | Tamil Nadu | Sunn hemp | 24 |
| | West Bengal | Sunn hemp | 20 |
| | | Dhaincha | 21 |
| | Orissa | Dhaincha | 24 |
| | Uttar Pradesh | Dhaincha | 51 |
| | Bihar | Dhaincha | 60 |
| | | Sunn hemp | 63 |
| | Andhra Pradesh | Sunn hemp | 114 |
| | | Dhaincha | 80 |
| | Delhi | Dhaincha | 16 |
| | | Cowpea | 15 |
| Wheat | Uttar Pradesh | Cowpea | 21 |
| | | Sunn hemp | 45 |
| | | Dhaincha | 16 |
| | Bihar | Sunn hemp | 106 |
| | Madhya Pradesh | Sunn hemp | 13 |
| | Delhi | Sunn hemp | 35 |
| Sugarcane | Uttar Pradesh | Sunn hemp | 30 |
| | Assam | Sunn hemp | 9 |
| | | Dhaincha | 8 |
| Cotton | Gujarat | Dhaincha | 21 |
| | Maharashtra | Dhaincha | 12 |
| | Tamil Nadu | Sunn hemp | 21 |

Sunnhemp—*Crotalaria juncea*; Dhaincha—*Sesbania aculeata*; Cowpea: *Vignaunguiculata*
Source: Yawalkar et al. (1996) and Mishra et al. (2004)

## 12.4.7 Disease Suppression/Control

Growing non-host species of green manure crops have a significant potential to break the life cycle of crop-specific pathogens, though some important soil-borne diseases such as rhizoctonia spp. and fusarium spp. have a wide host range (Wallwork 2000). Even so, the potential for green manure crop bestowing disease reduction should not be overlooked; integrating a specific green manure crop in sequence with another non-host crop could provide the extended disease break necessary for the susceptible crop (Roper et al. 2012). The use of green manure crops, especially legumes, promotes the decomposition of cereal stubbles by balancing C:N ratio and could help in reducing disease pressure where stubbles are left on the soil surface (minimum or no-tillage) (Roper and Gupta 1995; Felton et al. 1998).

### 12.4.8 Soil and Water Conservation

The soil conservation benefits provided by green manure crops extend beyond protection of bare soil during fallow periods. The mulch that results from a chemically or mechanically incorporated GM crop increases water infiltration and reduces water evaporation from the soil surface. The inclusion of green manure crop in rotation improved water availability and enhanced soil water storage carried through to the following season in contrast to that of continuous cereal crops (López-Bellido et al. 2007; Blackshaw et al. 2001). It also reduces soil crusting and subsequent surface water runoff during rainy periods.

### 12.4.9 Enhanced Availability of Native Soil Nutrients

Legumes not only add nitrogen to the soil but also help in recycling of other nutrients on the farm. Nitrogen (N), phosphorous (P), potassium (K), calcium (Ca), magnesium (Mg), sulphur (S), and micronutrients are accumulated by green manure crops during a growing season. When the green manure is incorporated or laid down as no-till mulch, these nutrients become slowly available during decomposition. Legume green manures lower soil pH, lower C:N ratio in soil organic matter, scavenge nutrients from the subsoil and translocate them upwards to the surface rooting zone, where they become available to the succeeding crop (Hargrove 1986). During decomposition of organic matter carbonic and other organic acids are formed as a by-product of microbial activity. These organic acids react with insoluble mineral rocks and phosphate precipitates, releasing phosphates and exchangeable nutrients. Green manures may enhance P nutrition of the succeeding crops via a number of mechanisms. Green manure crops may convert relatively unavailable native and residual fertilizer P to chemical forms more available to succeeding crops. The availability of phosphorus (P) is enhanced through green manuring (Cavigelli and Thien 2003) as well as the utilization of phosphorous fertilizers markedly enhanced from 3% to 39% (Hundal et al. 1992; Bah et al. 2006).

Green manures also influence the availability of micronutrients in soil, especially through changes in oxidation reduction capacity and releasing micronutrients during decomposition. Increases in soil solution concentrations of $Fe^{2+}$ and $Mn^{2+}$ with submergence have been found to be accelerated when soils are green manured. Flooding generally causes a decline in the availability of Zn in soils, which is accentuated by green manuring. However, in sodic soils, green manuring increases the availability of Zn because of its favourable effect on soil pH. Under flooded soil conditions, soluble and exchangeable Fe and Mn increase at the expense of organic or oxide-held metals, and this process provides surfaces of high absorptive capacity on which Zn and Cu may be bounded.

## 12.5 Agronomy of Green Manure Crops

### *12.5.1 Method of Sowing*

Green manures are usually broadcasted (Raheja 1952; Chela and Gill 1973). The seed of sesbania may be soaked overnight in water to hasten germination. The close contact between seed and soil in moist condition is necessary for satisfactory germination. In rainfed areas of eastern India, rice together with 15–20 kg/ha sesbania seed are dry-seeded by broadcasting (Garrity and Flinn 1988). Seed can also be sown by manual drilling behind a plough in rows about 20–30 cm apart followed by planking, or by bullock/tractor-drawn seed drill (Chela and Gill 1973; Meelu et al. 1992). Arakeri and Patil (1957) and Tirol-padre and Ladha (1990) suggested that drilling is better than broadcasting.

### *12.5.2 Number of Cultivations*

In rainfed areas, where a green manure crop is raised in fallow fields during the rainy season, two cultivations are usually enough (Raheja 1952; Chela and Gill 1973). However, in Philippines, the conventional method of land preparation for growing a green manure crop is one ploughing and one or two harrowings of the flooded soil (Ventura and Watanabe 1991). In irrigated areas in India, green manures can be grown in the summer season without preparatory tillage. In salt-affected soils, the land is usually ploughed two or three times followed by levelling (Dargan et al. 1982).

### *12.5.3 Time of Sowing*

Green manure is grown in the summer dry season for use in the wet-season crop (Singh et al. 1982; Meelu et al. 1992), or in the rainy season for the benefit of the dry-season crop (Becker et al. 1990; Manguiat et al. 1989). There can be no rigid time schedule for sowing rainfed green manure crops as the commencement of the monsoon rains varies from region to region. In northern India, rainfed green manure crops are sown in early July or as early as possible after the break of the monsoon (Mirchandani and Khan 1952). Under irrigated conditions, a green manure crop can be sown from mid-April to mid-May depending upon the harvesting of the dry-season crop. It may be concluded that a growing period for the green manure crop of 6–8 weeks before transplanting/planting the main crop is necessary.

### 12.5.4 Seed Rate

A high seed rate is recommended for green manuring in order to delay the development of woodiness. For sesbania and sunn hemp a seed rate of 50 kg $ha^{-1}$ is normal. However, in salt affected soils where germination is likely to be low, an even higher seed rate may be used. The seed rates for different green manure crops in normal and salt affected soils are given in Table 12.6.

### 12.5.5 Inoculation

Most leguminous crops can form root nodules and fix atmospheric N without prior inoculation with rhizobium. In a field trial, Singh (1990) observed no significant effect of inoculation on nodulation and dry matter and N yield of sesbania. In the regions where such a crop is introduced for the first time, there may be difficulty in obtaining nodulation without inoculation, which enhances the onset and number of effective nodules and hence the amount of $N_2$ fixed by a legume. In China, milk vetch was found to respond to inoculation (Chen 1988). Jia (1986) found that, during the first year of cultivation, inoculation of milk vetch seed with Rhizobium gave a fourfold increase in green matter production. In Bangladesh, Rhizobium inoculation increased nodulation and dry matter yield of Sesbania (Subba Rao 1988). In Pakistan, Siddiqui et al. (1985) reported that inoculation increased the number of nodules/plant and height of Sesbania only at low levels of P and K application. Gaur

**Table 12.6** Seeding rate (kg/ha) and special features of different green manure crops

| Species | Seed rate (kg $ha^{-1}$) | Special features |
|---|---|---|
| *Sesbania aculeate* | | A quick growing succulent green manure crop and could correct alkalinity |
| –Normal soils | 50–60 | |
| –Salt-affected soils | 60–70 | |
| *Sesbania rostrata* | 30–40 | Nodules both on stem and roots, but photosensitivity nature restrict its use in winter |
| *Sesbania speciosa* | 50 | Adaptive to different soil conditions |
| Cowpea | 30–35 | Drought hardy and has wide and droopy leaves—conserved soil and moisture |
| Cluster bean | 50 | The deep tap root system and high adaptability to water stress condition |
| Soybean | 50 | |
| Mungbean/pigeonpea/Indigofera/lablab | 30 | |
| Tephrosia purpurea | 45 | Hardy, drought-resistant and suited for summer |
| Crotolaria juncea | 25–40 | Quick growing green manure-cum-fibre crop |
| Phaseolus trilobus | 20–25 | Dual-purpose crop, yielding good fodder for cattle and green manure for land |

(1978) reported that Rhizobium inoculation increased the yield of Sesbania by about 19%. Bhardwaj (1974) considered that Rhizobium was abundantly distributed in salt-affected soils of India as good nodulation of Sesbania plants was noted. However, a considerable number of nodules were ineffective. Application of gypsum for soil reclamation stimulated growth and multiplication of Rhizobium which in turn gave a better start to the leguminous crops (National Academy of Sciences 1979).

### 12.5.6 Fertilizer Application

In conventional agriculture, GM crops respond to basal N (15–25 kg N $ha^{-1}$) and phosphate fertilization (30–45 kg $P_2O_5$ $ha^{-1}$) which enhanced the biomass production and also its availability to the succeeding crop. Furthermore, a fertilizer containing Mo and B could be applied to soils showing their deficiency for proper N fixation. In contrast, fertilizer application cannot be practised in organic farming and therefore is not recommended.

### 12.5.7 Irrigation

Dry matter and N yields of green manure crops are affected by soil moisture supply, but their deep rooting conveys an advantage over other non-leguminous crops under conditions of soil moisture stress. Singh and Lamba (1971) recommended that cowpeas should be irrigated when the available water in the 180-cm profile was depleted to 35%. Gaul et al. (1976) reported that during summer in northern India, about 600–650 mm of irrigation water was required for raising a 74-day-old green manure crop of sesbania on sodic soils. Singh et al. (1991) studied water expense and water-use efficiency of sesbania, cowpea and cluster bean green manure crops. The water-use efficiency was the highest for cowpea under all irrigation schedules. Total water expense of 7-week-old green manures was 36–38 cm under favourable irrigation schedules (IW/PAN-E = 1.0).

### 12.5.8 Stage of Incorporation

The age at which a green manure crop should be ploughed-in is important for maximum benefit. In India, Mirchandani and Khan (1952) concluded that a green manure crop should be turned into the soil at the point of flowering, i.e. about 8 weeks from sowing in most crops. Chela and Gill (1973) considered that sesbania attained maximum growth about 8 weeks after sowing, sunn hemp was ready for ploughing in 60–70 days after sowing when flowering began all over the crop, and cluster bean reached flowering 7–8 weeks after sowing. Panse et al. (1965) reported a crop of 7–8 weeks old sunn hemp/sesbania produced the best responses with rice and wheat. Vachhani and Murty (1964) found that 8-week-old sesbania was tender and succulent and should be turned-in at the time of transplanting of rice for maximum response.

### 12.5.9 Method of Incorporation

Proper incorporation of the green manure into soil containing sufficient moisture is important for rapid decomposition. Raheja (1952) reported that green manure crops as high as 2 m can be readily buried by a soil-inverting plough after the standing crop has been planked down. The soil-inverting plough is run in the direction the crop has been laid flat in the field and again planked to compact the soil. Standard tractor-drawn disc harrows can also be used. In Philippines, Meelu et al. (1992) used a power tiller-drawn mouldboard plough for in situ incorporation of eight 60-days-old green manure crops. To help cut and incorporate a green manure crop, the IRRI developed an attachment of ring-knives to the cage wheels of a power tiller-drawn mouldboard plough. A local animal-drawn implement used by farmers in the Philippines has also shown promise for green manure incorporation. The implement flattens a sesbania crop on the first pass while five sharp blades slice the branches. On the second pass at right angles, plant stems are cut into 25 cm sections and driven down into the mud; conventional ploughing and harrowing follow with no additional effort (Garrity and Flinn 1988). In the rainfed areas of eastern India where rice and sesbania seeds are broadcasted dry, the green manure crop is incorporated by an operation known as beushening. The field is ploughed and cross-ploughed in standing water after 4–6 weeks and the sesbania plants are trampled into the mud. This operation also thins the young rice crop and controls weeds.

### 12.5.10 Depth of Incorporation

The depth of incorporation influences the susceptibility of the green manure N to loss and thus determines the efficiency of its use by the crop. In Indonesia, Staker (1958) reported an increase in rice yield with incorporation of green manure rather than its surface application. From a 3-year study, Williams and Finfrock (1962) showed that rice responded significantly better to deep (10–15 cm) incorporation of vetch green manure than too shallow incorporation. As with fertilizer N, deep placement of green manure N reduces ammonia volatilization. A green manure crop should, therefore, be incorporated into the soil for full effectiveness.

### 12.5.11 Interval Between Incorporation and Crop Establishment

The interval between turning-in the green manure crop must accommodate with the transplanting/sowing of the succeeding crop, particularly under intensive cropping where time available for growing a green manure crop is relatively short. Hence the rate at which green manures decompose and mineralize in both upland and wetland ecosystems is important. Meelu et al. (1992) studied the response of rice to eight tropical green manure crops and found that succulence and tissue N, which are expected to be related to decomposition rate, were determined more by plant age than by species. On average, they also reported a 29% increase in grain yield of rice by green manuring of eight different legumes over control (no green manuring). In rainfed areas, green manure is generally grown in the rainy season (early July in northern India) for the benefit of a dry-season crop about 8 weeks old GM crop is turned-under at the end of August before the monsoon ends. Wheat is sown at the end of October, giving a decomposition period of about 2 months. In the pre-Green Revolution era, Khan and Mathur (1957) studied the effect of age at incorporation (4–10 weeks) of sunn hemp and time-interval between burial and sowing of a succeeding wheat crop. They found that 8 weeks was the correct timing for both periods and made optimum use of plant nutrients. Mirchandani and Khan (1952) also found that a 2-month interval for sunn hemp green manuring gave the best results. Anant Rao et al. (1957) used an interval of 50–60 days between incorporating 49 to 56 days-old sunn hemp green manure and sowing of wheat. Singh and Sinha (1964) applied green matter in early September and allowed it to decompose in the soil until sowing wheat in mid-November.

In irrigated areas under intensive cropping in India, only about 8 weeks are available for growing a green manure crop after the harvest of a dry-season crop and transplanting/sowing of a succeeding wet-season crop; thus, a long decomposition period is not feasible. Results of experiments have shown that long decomposition periods for green manure are not necessary before rice transplanting (Singh et al. 1990; Meelu et al. 1994b). Adequate moisture content in the soil and high temperature enhance the rapidity of mineralization, and longer aerobic decomposition may result in N loss on flooding at rice transplanting.

## 12.6 Constraints in the Adoption of Green Manures

Despite high $N_2$ fixation, reduced N losses, increased grain yields, and various positive effects on soil physical and chemical parameters, the use of green manure legumes in crop production systems has been dramatically declining over the last 30 years (Roger and Watanabe 1986), and its applicability still remains in the research farms (Meena et al. 2018). In India, green manuring is a simple yet viable technology, which will bring both short- and long-term benefits. Meelu et al.

(1994b), Becker et al. (1990), Meena et al. (2018) have identified the following constraints and the possible reasons behind it in the adoption and spread of green manuring. They are listed as under:

1. No obvious or immediate return in cash or kind, except for dual-purpose cropping, and hence labour input considered unproductive.
2. Narrow window period between the two crops for growing and incorporating green manure crops during most of the cropping season.
3. Green manure crop, if not incorporated at proper growth stage and time, may lead to immobilization of N on a temporary basis.
4. Cost-effectiveness unattractive compared with fertilizer, particularly when high-yielding varieties of rice are grown.
5. Being high water requiring crop, it may not be suitable for dryland agriculture.
6. Difficulty and/or cost of seed supply.
7. Control of the quantity and timing of nutrients applied more complex than with fertilizers.
8. When intercropped, possible competition for soil moisture, nutrients and space with the main crop.
9. Problems of decomposition of green manuring in the sowing of the following crop if proper moisture is not available, particularly in semiarid regions.
10. When grown on rice bunds may reduce crop yield by shading and root competition.
11. Residual effect not always obvious, particularly in the short term.
12. Low emphasis on organic manures by research and extension workers, particularly where demonstrations on farmers' fields are concerned.

## 12.7 Economics of Green Manuring

The most obvious direct economic benefit derived from legume green manure or cover crops is nitrogenous fertilizer savings. In most cases, these savings can offset crop establishment costs. Indirectly, allelopathic cover crop reduces herbicide application and control costs of nematode and insect apart from conserving water infer through no-till mulch and scavenging residual nitrate of ground water. Longer term benefits are derived from the build-up of organic matter resulting in increased soil health. Healthy soil lessened the erosivity of soil, resulting in better absorption of rain water and nutrient recycling, ultimately leading to healthy crops with good yields. Many studies have shown that legume crops can replace a portion of the fertilizer nitrogen requirements for the succeeding crop. The economic value of these nitrogen replacements can be calculated by using a local nitrogen price. These costs can then be compared to green manure crop seed and planting costs. These simple nitrogen cost comparisons do not take into account the benefits of improved soil tilth and increased water infiltration resulting from these crops. Water consumption by green manure crops is a concern and pronounced in areas with less than 30 inches of precipitation per year. There is always additional management required when GM crops of any sort are added to the rotation. Turning green manures requires additional time and expense, compared to having no crop at all. It is also worth accounting that legume green manures have many additional effects like the supply of additional on-farm fodder, which could be included in economic analysis, or an impact on soil erosion which is difficult to quantify economically. The economic benefits of green manuring need to be quantified and clearly demonstrate them to be effectively adopted by the farming community.

Overall, green manure practice is an efficient management approach for fertility restoration, and sustainability as it has favourable influences on soil properties and consequently increased crop yields.

## References


Agricultural Statistics (2005) Agricultural statistics, Ministry of Agriculture and Coopration, GOI, New Delhi

Anant Rao NK, Agarwal JP, Pawar KS, Yadav NR (1957) Phosphate manuring of sunhemp green manure. Indian J Agron 1:215–219

Anderson I, Buxton DR, Karlen DL, Cambardella C (1997) Cropping system effects on nitrogen removal, soil nitrogen, aggregate stability, and subsequent corn grain yield. Agron J 89:881–886

Arakeri HR, Patil SV (1957) The why and how of green manuring. Indian Farming 7(1):24–26

Bah AR, Zaharah AR, Hussin A (2006) Phosphorus uptake from green manures and phosphate fertilizers applied in an acid tropical soil. Commun Soil Sci Plant Anal 37:2077–2093

Becker M, Ladha JK, Ottow JCG (1990) Growth and $N_2$ fixation of two stem-nodulating legumes and their effect as green manure on lowland rice. Soil Biol Biochem 22:1109–1119

Bhardwaj KKR (1974) Note on the distribution of effectiveness of *Sesbania aculeata* (Poir.) in saline-alkali soils. Indian J Agric Sci 44:683–684

Blackshaw RE, Moyer JR, Doram RC, Boswall AL, Smith EG (2001) Suitability of undersownsweetclover as a fallow replacement in semiarid cropping systems. Agron J 93:863–868

Cavigelli MA, Thien SJ (2003) Phosphorus bioavailability following incorporation of green manure crops. Soil Sci Soc Am J 67:1186–1194

Chela KS, Gill GS (1973) Green manuring. Punjab Agricultural University, Ludhiana, p 34

Chen L (1988) Green manuring cultivation and use for rice in China. In: Green manure in rice farming. International Rice Research Institute, Los Banos, pp 63–70

Congreves KA, Hayes A, Verhallen EA, Van Eerd LL (2015) Long-term impact of tillage and crop rotation on soil health at four temperate agroecosystems. Soil Tillage Res 152:17–28

Creamer NG, Bennett MA, Stinner BR, Cardina J, Regnier EE (1996) Mechanisms of weed suppression in cover crop-based production systems. Hortic Sci 31(3):410–413

Dargan KS, Singh OP, Gupta IC (1982) Forages and green manure crops. In: Crop production in salt affected soils. Oxford and IBH Publishing, New Delhi, p 237

Dinnes DL, Karlen DL, Jaynes DB, Kaspar TC, Hatfield JL, Colvin TS, Cambardella CA (2002) Nitrogen management strategies to reduce nitrate leaching in tile-drained Midwestern soils. Agron J 94:153–171

Eriksen J (2005) Gross sulphur mineralization-immobilization turnover in soil amended with plant residues. Soil Biol Biochem 37:2216–2224

Fageria NK, Baligar VC (2005) Role of cover crops in improving soil and row crop productivity. Commun Soil Sci Plant Anal 36:2733–2757

Felton WL, Marcellos H, Alston C, Martin RJ, Backhouse D, Burgess LW, Herridge DF (1998) Chickpea in wheat-based cropping systems of northern New South Wales II. Influence on biomass, grain yield, and crown rot in the following wheat crop. Aust J Agric Res 49:401–407
Florentín MA, Penalva M, Calegari A, Derpsch R, McDonald MJ (2010) Green manure/cover crops and crop rotation in conservation agriculture on small farms. Integrated Crop Management 12
Gardner FP (1985) Physiology of crop plants. Iowa State Univ, Press, Ames
Garrity DP, Flinn JC (1988) Farm level management systems for green manure crops in Asian rice environments. In: Green manuring in rice farming. International Rice Research Institute, Los Banos, pp 111–119
Gaul B, Abrol IP, Dargan KS (1976) Notes on the irrigation needs of *Sesbania aculeata* Poir. for green manuring during summer. Indian J Agric Sci 46:434–436
Gaur AC (1978) Recycling and utilization of organic wastes as fertilizers. In: India/FAO/Norway seminar on the development of the complementary use of mineral fertilizers and organic materials in India. Ministry of Agriculture and Irrigation, KrishiBhawan, New Delhi, pp 109–126
Hargrove WL (1986) Winter legumes as a nitrogen source for no-till grain sorghum. Agron J 78:70–74
Harris GH, Hesterman OB, Paul EA, Peters SE, Janke RR (1994) Fate of legume and fertilizer nitrogen$^{15}$ in a long-term cropping systems experiment. Agron J 86:910–915
Hoffman ML, Weston LA, Snyder JC, Regnier EE (1996) Allelopathic influence of germinating seeds and seedlings of cover crops on weed species. Weed Sci 44(3):579–584
Hundal HS, Dhillon NS, Dev G (1992) Contribution of different green manures to P nutrition of rice. Indian J Soil Sci Soc 40:76–81
IRRI (1990) World rice statistics. The International Rice Research Institute, Los Bafios, p 320p
Jia Z (1986) Green manure and soil microorganisms. In: Bin J (ed) Green manure in China. Publishing House of Agriculture, Beijing, pp 113–120
Khan AR, Mathur BP (1957) Effect of sannhemp as green manure on the yield of wheat. Indian J Agric Sci 27:171–176
Khankahre PJ, Barman KK, Yaduraju NT (2002) Effect of dhaincha green manure on weed infestation and grain yield of transplanted rice. Proc. National Seminar on Developments in soil science. JNKVV, Jabalpur, 11–15
Kumar V, Pawar K (2018) A review on soil health and fertility management in organic agriculture through green manuring. J Pharmacogn Phytochem SP1:3213–3217
Kumar B, Gupta RK, Bhandari AL (2008) Soil fertility changes after long-term application of organic manures and crop residues under rice-wheat system. J Indian Soc Soil Sci 56(1):80–85
Kumar R, Mahajan G, Srivastava S, Sinha A (2014) Green manuring: a boon for sustainable agriculture and pest management – a review. Agric Rev 35(3):196–206
Kumar S, Sheoran S, Kumar SK, Kumar P, Meena RS (2016) Drought: a challenge for Indian farmers in context to climate change and variability. Prog Res Int J 11:6243–6246
Ladd JN, Amato M, Jadickckson RB, Butler JH (1983) Utilization by wheat crops of nitrogen from legume residues decomposing in soils in the field. Soil Biol Biochem 15:231–238
Ladha JK, Kundu DK (1997) Legumes for sustaining soil fertility in lowland rice. In: Rupela OP, Johansen C, Herridge DF (eds) Extending nitrogen fixation research to farmers' fields. International Crops Research Institute for the Semi-Arid Tropics, Patancheru, Andra Pradesh, pp 76–102
López-Bellido RJ, López-Bellido L, Benítez-Vega J, López-Bellido FJ (2007) Tillage system, preceding crop, and nitrogen fertilizer in wheat crop: II. Water utilization. Agron J 99:66–72
MacRae RJ, Mehuys RG (1985) The effect of green manuring on the physical properties of temperate area soils. Adv Soil Sci 3:71–94
Manguiat IJ, Perez AS, Jalalon AT (1989) Green manuring of corn with inoculated *Sesbania rostrata*. Philipp J Crop Sci 14:15–19
Mary B, Recous S (1994) Measurement of nitrogen mineralization and immobilization fluxes in soil as a means of predicting net mineralization. Eur J Agron 2:291–300

Meelu OP, Morris RA, Furoc RE, Dizon MA (1992) Grain yield responses in rice to eight tropical green manures. Trop Agric Trinidad 66:133–136

Meelu OP, Singh B, Singh Y (1994a) Effect of green manuring and crop residue recycling on N economy, organic matter and physical properties of soil in rice-wheat cropping system. In: Dhiman SD, Chaudhary MK, Panwar DVS (eds) Proceedings of rice-wheat cropping systems in India. CCS Haryana Agricultural University, Hisar, pp 115–124

Meelu OP, Singh Y, Singh B (1994b) Green manuring for soil productivity improvement. World Soil Resources Report 76. Food and Agriculture Organization of the United Nations, Rome. pp 123

Meena BL, Fagodiya RK, Prajapat K, Dotaniya ML, Kaledhonkar MJ, Sharma PC, Meena RS, Mitran T, Kumar S (2018) Legume green manuring: an option for soil sustainability. In: Legumes for soil health and sustainable management. Springer, Singapore, pp 387–408

Mirchandani TJ, Khan AR (1952) Effect of age of sunnhemp on the succeeding wheat crop. Green Manuring. ICAR Review Series 6, pp 37

Mishra BN, Prasad R, Gangaiah B (2004) Organic manures for increased rice productivity and sustained supply of Fe to rice. Acta Agron Hungarica 52:391–397

Monegat C (1991) Plantas de cobertura del suelo: características y manejo en pequeñas propiedades. Chapeco, p 337

National Academy of Sciences (1979) Tropical legumes: resources for the future. Washington, p 331

Panse VG, Abraham TP, Leelavathi CR (1965) Green manuring of crops (Review of experimental results in India), Indian Council of Agricultural Research Technical Bulletin No. 2, p 84

Parihar CM, Jat SL, Singh AK, Kumar B, Singh-Yadvinder, Pradhan S, Pooniya V, Dhauja A, Chaudhary V, Jat ML, Jat RK, Yadav OP (2016) Conservation agriculture in irrigated intensive maize-based systems of north-western India: effects on crop yields, water productivity and economic profitability. Field Crop Res 193:104–116

Pooniya V, Shivay YS (2013) Enrichment of Basmati rice grain and straw with zinc and nitrogen through ferti-fortification and summer green manuring crops under Indo-Gangetic Plains of India. J Plant Nutr 36(1):91–117

Pooniya V, Shivay YS, Rana A, Nain L, Prasanna R (2012) Enhancing soil nutrient dynamics and productivity of Basmati rice through residue incorporation and zinc fertilization. Eur J Agron 41:28–37

Raheja PC (1952) Why, where, when and how to green manure? Indian Farming 2(2):28–29

Ramos MG, Villatoro MAA, Urquiaga S, Alves BJR, Boddey RM (2001) Quantification of the contribution of biological nitrogen fixation to tropical green manure crops and the residual benefit to a subsequent maize crop using $^{15}$N-isotope techniques. J Biotechnol 91:105–115

Reddy K, Zablotowicz RM, Locke MA, Koger CA (2003) Cover crop, tillage, and herbicide effects on weeds, soil properties, microbial populations, and soybean yields. Weed Sci 51:987–994

Roger PA, Watanabe I (1986) Technologies for utilizing biological nitrogen fixation in wetland rice: Potentialities, current usage and limiting factors. Fertil Res 9:39–77

Roper MM, Gupta VVSR (1995) Management practices and soil biota. Aust J Soil Res 33:321–339

Roper MM, Milroy SP, Poole ML (2012) Green and brown manures in dryland wheat production systems in Mediterranean-type environments. Adv Agron 117:275–313

Schumann RA, Meyer JH, Antwerpen RV (2000) A review of green manuring practices in sugarcane production. Proc S Afr Sugar Technol Assess 74:93–100

Schutter M, Dick R (2001) Shifts in substrate utilization potential and structure of soil microbial communities in response to carbon substrates. Soil Biol Biochem 33(11):1481–1491

Seneratne R, Ratnasinghe DS (1995) Nitrogen fixation and beneficial effects of some grain legumes and green manure crops on rice. Biol Fertil Soils 19:49–54

Shukla KP, Sharma S, Singh NK, Singh V, Tiwari K, Singh S (2011) Nature and role of root exudates: efficacy in bioremediation. Afr J Biotechnol 10(48):9717–9724

Siddiqui MA, Aslam M, Hayat MY, Sandhu GR (1985) Nodulation studies on *Sesbania sesban* (L.) Merr. 2. Green manuring for wheat. Pak J Sci Ind Res 28:407–411

Singh B (1990) Quantification of dinitrogen fixed by dhaincha (*Sesbania* spp.). M.Sc. Thesis, Punjab Agricultural University, Ludhiana, India
Singh S, Lamba PS (1971) Agronomic studies on cowpea FS-68. I. Effect of soil moisture regimes, seed rates and levels of phosphorus on growth characters and yield. J Res Haryana Agric Univ Hisar 3:1–7
Singh MP, Sinha SK (1964) Studies on green manuring of wheat in Bihar. III. Optimum dose of green manure. Indian J Agron 9:138–143
Singh R, Vig AC, Singh NT (1982) Nitrogen substitution with green manures in maize-wheat rotation. Indian J Agron 27:371–376
Singh Y, Singh B, Khind CS, Meelu OP (1990) Nitrogen equivalence of green manure for wetland rice on coarse textured soil. Int Rice Newsl 15(1):23
Singh NT, Singh R, Vig AC (1991) Yield and water expenses of cowpea, clusterbean and Sesbania as summer green manures in semi-arid regions of Punjab. Indian J Agric Sci 56:417–421
Staker EV (1958) Green manure crops in rotation to paddy production South-East Asia. Int Rice Commun Newsl (Bangkok) 7(4):1–20
Subba Rao NS (1988) Microbiological aspects of green manure in lowland rice soils. In: Green manure in rice farming. International Rice Research Institute, Los Banos
Thonnissen C, Midmore DJ, Ladha JK, Olk DC, Schmidhalter U (2000) Legume decomposition and nitrogen release when applied as green manures to tropical vegetable production systems. Agron J 92:253–260
Thorup-Kristensen K, Magid J, Jensen LS (2003) Catch crops and green manures as biological tools in nitrogen management in temperate zones. Adv Agron 79:227–302
Tirol-Padre A, Ladha JK (1990) Effect of planting method and optimum seeding rate on biomass production and nitrogen fixation in *Sesbania rostrata*. Int Rice Commun Newsl 15(6):15
Utomo M, Frye WW, Blevins RL (1990) Sustaining soil nitrogen for corn using hairy vetch cover crop. Agron J 82:979–983
Vachhani MV, Murty KS (1964) Green manuring for rice. Indian Council of Agricultural Research, New Delhi, Res. Rep. Ser. No. 17, p 50
Varma D, Meena RS, Kumar S (2017) Response of mungbean to fertility and lime levels under soil acidity in an alley cropping system of Vindhyan Region, India. Int J Chem Stud 5(2):384–389
Ventura W, Watanabe I (1991) Azolla and Sesbania: organic fertilizers. In: The Philippine environment: opportunities in conservation and rehabilitation. The Philippines Futuristic Society, Manila, pp 171–178
Wallwork H (2000) Cereal root and crown diseases. Australian 2000 Edition. Canberra: Grains Research and Development Corporation
Westcott MP, Mikkelson DS (1987) Comparison of organic and inorganic nitrogen sources for rice. Agron J 79:937–943
Williams WA, Finfrock DC (1962) Effect of placement and time of incorporation of vetch on rice yields. Agron J 54:547–549
Yawalkar KS, Agarwal JP, Bokde S (1996) Manures and fertilisers, 8th edn. Agri-Horticultural Publishing House, Nagpur, p 331
Ye X, Liu H, Li Z, Wang Y, Wang Y, Wang H, Liu G (2014) Effects of green manure continuous application on soil microbial biomass and enzyme activity. J Plant Nutr 37(4):498–508

# Chapter 13
# Green Manuring and Its Role in Soil Health Management

**Sanjeev Kumar, Samiksha, and Premasis Sukul**

**Abstract** The recent advancement in technology has undoubtedly improved the agriculture in overcoming many constraints which earlier led to loss in the crop yields. But many of these improvements include use of chemicals in the form of fertilizers and pesticides and therefore have led to further problems of reduced C/N ratio, low microbial diversity and less organic matter per unit area. These all have resulted in degradation of soil quality and fertility. Therefore, present chapter discusses the role of green manure (GM) on the expansion of the soil nitrogen source, the mobilization of soil phosphorus and the improvement of soil organic matter contain in soil. Another part of the chapter deals with the kinds of green manure crops cultivated and its propter utilization for sustainable agricultural practices. In future, we propose holistic system and precise agricultural practices to better deal with constrain encountered by farmer during adaptation of GM-based cropping system.

**Keywords** Sustainable agriculture · Microbial diversity · Organic content · Mineralization

---

S. Kumar
Department of Genetics and Plant Breeding, School of Agriculture Lovely Professional University, Jalandhar, India

Samiksha
Department of Biotechnology, School of Biosciences, Lovely Professional University, Jalandhar, India

P. Sukul (✉)
Department of Soil Science, School of Agriculture, Lovely Professional University, Jalandhar, India
e-mail: premasis.20644@lpu.co.in



B. Giri, A. Varma (eds.), *Soil Health*, Soil Biology 59,
https://doi.org/10.1007/978-3-030-44364-1_13

## 13.1 Introduction

Green manuring has become a very important agricultural practice since last three decades, but the interest gradually has declined as the fertilizers were available readily in the market. Reliance on synthetic fertilizers has proved to be detrimental for human health as well as to the environment. A dead zone has been created in the Gulf of Mexico due to the run off of soils and such fertilizers (Pimentel et al. 2005). Therefore, sufficient production of quality food grains and to feed ever growing population, green manuring practices are promising tools for the better sustainability of agriculture in the longer run and with almost no hazards to the environment (Kumar et al. 2014). Moreover, due to the continuous exertion of soil and incompetent methods of soil management like excessive tillage and burning of vegetative residues, exposure to climatic changes, the accelerated soil degradation has led to reduced crop production (Florentín et al. 2010). The soil management practices to increase fertility and productivity should include an increase in biomass along with reducing its decomposition (Bunch 2012). Green manuring, on the other hand, being a practice in which the undecomposed plant material is turned under the soil to provide organic material and nutrients to the soil, can be considered as a suitable alternative to positively affect the physical, chemical and biological properties of the soil (Fageria 2007). Giving proper attention and gaining knowledge regarding the composition, rates and placement of GM crops, the application of fertilizers can be reduced drastically (Schröder 2005). The potential of green manures is not limited to only improvement of soil quality and increasing the yields of subsequent crops, rather pest management and weed control has also been reported when green manuring practices are followed appropriately (Boydston and Hang 1995; Al-Khatib et al. 1997). However, the increased production cost of planting green manure led to minimize farmers' acceptance to this approach, which is showing a lack of support for its sustainability (Ntakirutimana et al. 2019).

## 13.2 Effect of Green Manure in Soil Health and Fertility Management

In the modern days of agricultural science, crop rotation and green manuring (GM) offer a technology to achieve sustainable production efficiently. Since continuous and conventional cultivation leads to a decline in soil OM content, soil nutrients holding capacity are affected severely, resulting in a loss of soil sustainability. One of the options to maintain sustainability in agriculture by restoring soil quality (especially in tropical soils) and reclaiming degraded soil is to increase soil OM content by GM (Kumar et al. 2010, 2014; Chimouriya et al. 2018) because this practice is eco-friendly, non-polluting and non-hazardous to soil, water and air (Yang et al. 2018). Moreover, like chemical fertilizers they do not exhibit any adverse effect on food commodities.GM with high nutrient concentrations and low

C/N ratios have a great impact on their value as organic fertilizer in crop production (Talgre et al. 2012). Use of GM together with adequate residue management and crop rotation could be useful to conserve or increase soil fertility, promote nutrient cycling at farm scale, bring crop nutrients up from lower soil profiles, smother weeds and prevent weed seedling growth, and reduce the external nutrient inputs (Drinkwater et al. 1998; Melero et al. 2006; Yadav et al. 2019). GM plants are mainly grown for the benefit of the soil and are very commonly referred as soil fertility building crops. In general, green manure enriches soil with organic matter and nutrients, improves soil physical, chemical and biological properties, causes nutrient and soil conservation, increases the biochemical activity in soil, reduces soil compaction, increases soil porosity, water infiltration and rooting depth and finally enhances crop health and yield (Schutter and Dick 2001; Golec et al. 2007; Bhattarai et al. 2012). Although main objectives in using GM are to increase organic matter content and replenish nutrients in soil, potential benefits from GM include reduced nutrient, particularly nitrate, leaching and lowering N fertilizer application to the succeeding crops (Fageria 2007). However, lack of synchrony between nutrient mineralization from GM and subsequent crop requirements for its nutrients becomes a major challenge (Mafongoya et al. 1998). GM plants are raised between the main crops to provide a shield to soils from erosion, restore productivity of exhausted land, protect fallow land from nitrate leaching and substitute the chemical fertilizers (Xie et al. 2016; Yang et al. 2018). It has been observed that during maize cultivation nearly 15–30% of external inputs in the form of chemical fertilizers were reduced, when GM was used (Yang et al. 2018). However, the correct selection of GM plant species to precede a main crop is considered critical to maximize crop yield associated with greater economic return.

## 13.3 Decomposition and Mineralization

Primarily, decomposition of organic compounds involves two phases: rapid and slow. In rapid phase complex organic molecules are transformed to water soluble compounds that on further decomposition gives rise to energy, carbon dioxide, water and others under aerobic condition, and energy, methane, carbon dioxide and others under anaerobic condition. Released energy level is more in aerobic condition than anaerobic condition. Slow phase of decomposition of organic compounds involves the degradation of lignin and other resistant substrates. They need longer time to be decomposed. Decomposition with subsequent nutrient release occurs at a quicker rate in GMs with lower C/N ratios, lignin and polyphenol contents (Palm et al. 2001; Parton et al. 2007). These parameters are considered as critical to determine the speed of the decomposition and thus they govern the dynamics of nutrient release from GM to the soil. Rate of decomposition and mineralization of GM are influenced by several factors such as type of plants used as GM crop and the growth stages during their incorporation in soil, climatic conditions and factors related to soil (texture, structure, pH, soil microbial status, etc.) (Dinnes et al. 2002; Dhakal et al.

**Table 13.1** Classification of green manures based on their functions

| Types | Function | Examples |
|---|---|---|
| Cover crop | Prevent erosion | Oats, winter rye, Sirius peas, lentils, clovers, vetch |
| Break crops | Interrupt the lifecycle of pests or diseases | *Mustard, rye, brassica, alfalfa* |
| Leguminous crops (N-fixer) | Enrich soils of available nitrogen | *Clovers, lupins, vetches, alalfa, peas, beans, soybeans* |
| Nutrient conserving crops | Minimize nutrient leaching, and further enrich soil with more nutrient's addition | *Oil radish, red clover, buck wheat, rye grass* |
| Smother crops: | Smother weeds by outcompeting them in growth | *Buckwheat, oilradish, winter rye, yellow sweet clover* |

2015; Reddy 2016). Fast rate of mineralization is observed in moist and warm soil. Both legumes (clovers, lupins, vetches, alfalfa, peas, beans, soybeans) and non-legumes (phacelia, buckwheat, chicory, mustard, turnips, ryegrass, oats, barley, rye) are used as GM crops. Leguminous green manure is more effective than other non-legumes (Bhattarai et al. 2012). However, rate of decomposition of plant materials and thus nutrient discharging ability may differ within legumes; such as Mucuna (*Mucuna pruriens*) exhibited faster decomposition as compared to green gram (*Vigna radiate L.*), implying that green gram possesses relatively more resistant materials to decompose than mucuna (Saria et al. 2018). Thus, chemical composition and the quality of the incorporated materials influence their decomposition and mineralization. It is not always the C:N ratio to determine the occurrence of mineralization or immobilization. Chemical composition of GM plant materials also plays a significant role in their decomposition. Carbon may exist in plant materials in various forms. Cellulose is fast degradable, while lignin is relatively more resistant to decomposition. Organic compounds containing phenols inhibit microbial action. Therefore, GM may be selected depending on their specific functions (Table 13.1) as well as their nutrient contents (Table 13.2). However, according to the need of the requirement, two different plants may also be considered to achieve best outcome. Planting together barley and white clover in the fall ensures soil enrichment with organic carbon and nitrogen for the main crop in spring.

Decomposition of fresh GM by several microorganisms leads to the release of nutrients and relatively resistant organic residues (humus) that improve soil fertility status and soil physical conditions, respectively. By decomposing organic matter, soil microbes obtain nutrients and energy for their growth and multiplications (Akpor et al. 2006) and at the same time nutrients are released for plant uptake, thus helping in nutrient recycling. GMs have the high potentials for biomass production. Soil organic matter (SOM) level is improved after incorporation and subsequent decomposition and mineralization of GM in soil (Meena et al. 2018a). SOM provides energy and carbon source to the soil-inhabiting microorganisms which further multiply saprophytically on the decomposing SOM (Ye et al. 2014). Eventually soil microbes convert organic forms of nutrients that are unavailable to plants, to their inorganic but plant available forms. GM is also responsible for

**Table 13.2** Nutrient contents of green manuring and green leaf manuring plants

| Plants | Nutrient content (%) on air dry basis | | |
|---|---|---|---|
| | N | $P_2O_5$ | $K_2O$ |
| Green manure | | | |
| *Sunnhemp (Crotalaria juncea)* | 2.3 | 0.5 | 1.8 |
| Dhaincha *(Sesbaniaaculeata)* | 3.5 | 0.6 | 1.2 |
| Sesbania *(Sesbania speciosa)* | 2.7 | 0.5 | 2.2 |
| Kolinji (*Tephrosia purpurea*) | 3.1 | 0.5 | 1.2 |
| Green leaf manure | | | |
| Neem (*Azadirachta indica*) | 2.8 | 0.3 | 0.4 |
| Subabul (*Leucaena leucocephala*) | 3.5 | 0.5 | 0.8 |
| Pongamia (*Pongamia glabra*) | 3.3 | 0.4 | 2.4 |
| Gliricidia (*Gliricidia sepium*) | 2.8 | 0.3 | 4.6 |
| Gulmohur *(Delonix regia)* | 2.8 | 0.5 | 0.5 |
| Parthenium(*Parthenium hysterophorus*) | 2.7 | 0.7 | 1.5 |
| Water hyacinth (*Eichhornia crassipes*) | 3.0 | 0.9 | 0.2 |
| Ipomoea | 2.0 | 0. | 0.4 |
| Cassia (*Cassia fistula*) | 1.6 | 0.2 | 1.2 |

microbial species richness in soil causing biodiversity as well as microbial species abundance leading to an increase in soil microbial population (Kumar and Adholeya 2016).

## 13.4 Soil Fertility in Relation to Physical, Chemical and Biological Properties

Incorporation of GM plants in soil increases soil organic matter levels and thus improves soil physical conditions in terms of increasing water and nutrient holding capacity, enhancing soil aggregation, decreasing soil bulk density and thus decreasing soil compactness, increasing infiltration rate, increasing air and water movement in soil etc. (Badanur et al. 1990; Chimouriya et al. 2018). Interestingly, these soil physical characteristics encourage more microbial proliferation which facilitates the major conversion of soil available nutrients from their unavailable forms and thus soil fertility (Pandey and Singh 2016; Zaccheo et al. 2016; Nayak and Vaidya 2018). However, it is also true that organics present in leguminous and non-leguminous GM behave differently. Legumes decompose quickly than non-legumes. Legumes exhibit little effect on long-term soil organic matter building, but they have profound effect on microbial growth and their dynamics during initial few months after their mixing in soil. On the other hand, non-legumes decompose slowly in comparison to legumes. However, in general, effects of GM on soil physical properties may only become significant after growing several GM crops over a period of perhaps five to ten years. GM plant roots entangle soil particles, thus encouraging aggregate stabilization and consequently influencing the infiltration rate by increasing pore size. Water holding capacity of sandy soil is increased when organic matter is added to it due to a dramatic change in its physical properties such as decreasing pore sizes and thus reducing the percolation rate (Selvi and Kalpana 2009; Yadav et al. 2019). Besides, GM increases microbial growth and their dynamics in soil (Eriksen 2005) by releasing nutrients and energy materials as root exudates and eventually enhances soil fertility and soil health (Doran et al. 1988). Additionally, microorganisms synthesize polysaccharide gums that bind the soil particles together to form soil aggregates and thus soil structure is maintained properly. Except Brassicas and lupins, most of the green manures are known to maintain the population of soil mycorrhiza which is normally associated with phosphorus nutrition in soil. Nevertheless, they are also responsible in maintaining soil structure (Dubey et al. 2015) by entrapping soil aggregates. Additionally, GM prevents water and wind erosion (Logsdon et al. 1993) due to the root–soil binding effect (Schumann et al. 2000) and due to the crop canopy, which lessens the beating effect of water and wind on soil surface so that soil particles are not detached and transported quickly. It has been found that green manuring by lucerne, chicory and red clover facilitates to break up compacted soil as they possess deep tap root system (Rayns and Rosenfeld 2010).

GM between crop sequence prevents nutrients being lost from the soil. GM draws nutrients from the soil and stores them in their bodies. The nutrients held in GM plants are again released in the soil, which become available to the succeeding crop when the plants are decomposed after their mixing in soil, and thus help in nutrient recycling. GM serves as a source of nutrients and energy for innumerable number of soil organisms which are essentially helpful to maintain soil health. Release of nitrogen, phosphorus and potassium in soil from decomposing *Crotalaria juncea* L. was demonstrated earlier (Sinha et al. 2009); however, the rate of nutrient release was found to be influenced significantly by climatic factors such as temperature and moisture (Sinha et al. 2009) as well as plant types (Saria et al. 2018). Buckwheat, lupin, oil radish etc. are known to enrich soils with phosphorus. Lupin demonstrated more phosphorus uptake and their utilization than grain crops. Therefore, once GMs are incorporated in soil, phosphorus from their body will be released and will become available to the subsequent crop. GM has also been found to increase phosphorus availability from rock phosphate in rice (Cavigelli and Thien 2003). In general, they increase fertilizer phosphorus utilization (Bah et al. 2004, 2006). Bah et al. (2004) demonstrated the benefit of integrating GM and inorganic phosphatic fertilizers to overcome acid soil infertility. Legume GM shifted the soil acidic pH to over 6.5 for several weeks (8–16 weeks) that enhanced nitrogen (N), phosphorus (P), potassium (K), calcium (Ca) and magnesium (Mg) status of soil. Irrespective of their quality, GM significantly improved the availability of fertilizer P by as much as 40 to over 80%. The increase in pH may be due to abstraction of protons by organic anions to form water, ammonification, ligand exchange reactions between hydroxyl groups on soil surfaces and organic anions, as well as dissociation/association reactions of various organic acids and decarboxylation of organic anions by microbial action (Helyar and Porter 1989; Tang et al. 1999; Haynes and Mokolobate 2001). However, soil pH may be reduced due to organic acids and carbon dioxides released by GM through their roots and during their overall decomposition. Thus, GM affects soil pH significantly (Buragohain et al. 2018). GM, more specifically the leguminous one is primarily considered as a source of nitrogen, since they contain low C:N ratio (Bhuiyan and Zaman 1996). Additionally, leguminous type of GM in association with specific type of microorganisms can fix atmospheric N and increase soil N level after their incorporation in soil, while non-leguminous GMs increase the SOM content and do not fix atmospheric N (Tejada et al. 2008). Therefore, application of GM reduces the use of chemical N fertilizers and additionally, improves soil physical and biological properties by increasing soil organic carbon (SOC) content (Pung et al. 2004). Soil microbes decompose organic substances present in soil and become instrumental to transform unavailable form of nutrients to their available form for crops. During decomposition of plant residues in soil, the distribution and population dynamics of soil microorganisms are significantly influenced (Akpor et al. 2006; Bokhtiar et al. 2003) reported that the GM with dhaincha and sunn hemp supplemented with urea at different levels exhibited an increase in sugarcane yield up to 57% along with the significant increase in SOM, total N, available P and S in the soil. N loss from soil amended with GM is considered significantly low compared to chemical nitrogenous fertilizers. Thus, GM application to soil minimizes air (acid rain, global warming etc.) and water pollution (eutrophication, nitrate toxicity etc.). About 14% and 35% loss of nitrogen was demonstrated in flooded soil with applied GM and split dose of urea, respectively (Becker et al. 1995). This might be due to synchronization of demand and supply of nitrogen as GM acts as slow release nitrogen source.

The importance of GMs in the improvement of soil CEC has been demonstrated by many researchers (Kimetu et al. 2008; Saria et al. 2018). Therefore, it may be concluded that GMs protect cation from leaching out of the plant root zone and makes them available to plant roots.

Use of GM has potential to improve crop growth accompanied with both quantitative and qualitative yield as compared to N, P and K fertilizers. It is proved that the nutritional effects of GMs on soil and crop plants depends on their residue quality (Agbede 2018; Adekiya et al. 2019). Therefore, choice and selection of GMs also play a pivotal role for the beneficial discharge to the succeeding crop, considering both qualitative and quantitative aspect of the crop yield. Out of four GMs Moringa (*Moringa oleifera* Lam.), Pawpaw (*Carica papaya* L.), Neem (*Azadirachta indica* A. Juss.) and Mesquite (*Prosopis Africana* Guill., Perr. & A. Rich)], Moringa leaves were found to be the best green manure improving the fruit quality of Okra in terms of K, Ca, Fe, Zn, Cu and vitamin C contents compared with other green manures (Adekiya et al. 2019). They recommended Moringa for obtaining quality of okra fruits and Mesquite for quantity. Likewise, cassava productivity was increased by Gliricidia, while Moringa improved root quality (Agbede 2018). Overall, GMs such as Neem (*Azadirachta indica* A. Juss.), Moringa (*Moringa oleifera* Lam.), Gliricidia (*Gliricidia sepium* (Jacq.) Kunth ex Walp.) and Leucaena (*Leucaena leucocephala* (Lam.) de Wit) were found to increase mineral and starch contents and reduced HCN content in the cassava tuber roots compared with the control (Agbede 2018). Such variations might be due to the varied potential of GM plant materials to improve soil quality because of the difference in their chemical composition, rate of degradation, extent of released nutrient elements in soil and in turn influencing the crop uptake.

It has been observed that microbial population, their growth and diversity are significantly influenced with GM and soil admixture. This might be attributed to the fact that easily available energy and nutrient source through GM was delivered to the soil microbial community, which stimulate their activity and growth. More microbial proliferation leads to more extra cellular enzymes and their favourable abundance causes major transformations of nutrients from plant unavailable forms to their available forms. During 4 years of continuous experiment with ryegrass (*Loliummul tiflorum* L.) application to soil, soil microbial biomass carbon ($C_{mic}$) and nitrogen ($N_{mic}$), soil respiration, soil enzymatic activities such as urease, invertase, and catalase were increased as compared to the control treatment (Ye et al. 2014). GM, *Trifolium pratense* L. applied @ 25 t/ha increased $C_{mic}$ by 79.2% (Tejada et al. 2008). Increase in microbial $C_{mic}$ and $N_{mic}$, their size and activity were also observed by several other researchers (Elfstrand et al. 2007; Ochiai et al. 2008; Balota and Chaves 2011).

To increase agricultural production, in conventional agriculture chemical N fertilizers are often overused to such an extent that environment is adversely affected. Concentrations of several reactive oxidized and reduced forms of N such as NOx, $N_2O$, $NO_3^-$, $NH_3$ are reported to exhibit an increase in their concentrations in the environment (Fagodiya et al. 2017; Meena et al. 2018b). Obviously, this alarming hike in N species in the environment is not only due to chemical fertilizers, but also due to other agricultural practices such as crop stubble burning after harvesting, and indeed due to emissions from various industrial units. Nevertheless, biological N fixation (BNF) from the atmosphere should be a better option to reduce using chemical fertilizers that cause water pollution ($NO_3^-$), air pollution (NOx), and climate change ($N_2O$) (Sulieman and Tran 2016). Under practical situations in organic systems, the leguminous GM acts as the main source of N. Nowadays in conventional system also legume GMs are used extensively primarily to minimize chemical fertilizer application and to build up organic C pool in soil. Improved organic matter level in soil helps to increase soil N as there exists a positive correlation between them. Soil N level can be improved with improving levels of SOM. Application of legume GM is an important option to optimize the BNF and to ensure soil sustainability (Meena et al. 2018b).

Legumes in mutualistic symbiotic association with soil bacteria, called N-fixers, (*Rhizobia*) can fix nitrogen from atmosphere and enrich soil with nitrogen through mineralization, once they are decomposed after their incorporation in soil (Fig. 13.1). In this process of BNF, the plants act as the C and energy source for the bacteria which reside in the roots, forming nodules and in turn, supply N to the plant. The available N present in the soil is critical to determine the extent of N fixation. If there is abundance of available N in soil during plantation of legume, GM prefers to use available form of N from soil instead of fixing them from atmosphere. Therefore, to maximize the soil fertility gain through leguminous GM, one should be careful during its placement in crop sequence and current soil fertility status.

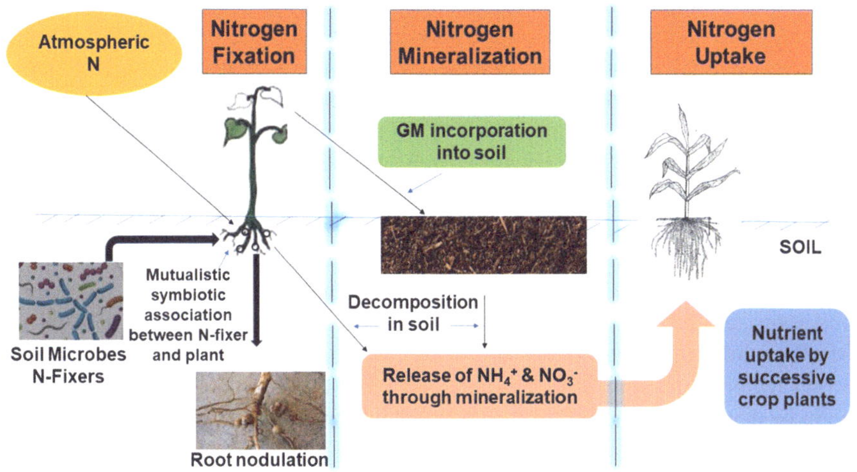


**Fig. 13.1** Soil enrichment with N by fixation and mineralization

## 13.5 Nutrient Availability

It has been observed that N mineralization is normally very fast during the initial days after incorporation followed by flooding, afterward release of N decreases. *Sesbania* released 31% N in 20 days, which went up to 39% in 40 days (Singh et al. 1992). In a separate experiment also maximum effect on soil nutrient content was found in sixth and seventh weeks after incorporation of GM, Mucuna (*Mucunapruriens*) and green gram (*Vigna radiate L.*) into the soil (Saria et al. 2018). Green gram and mucuna caused an increase in soil available P content from 0.03 to 0.39 and 0.37 mg/kg, respectively. They observed a hike in total organic C in soils by a factor of 2.3 to 3.2. Total N increased significantly from 1.28% to 2.64% at sixth week in soil with green gram and 2.83% at seventh week in soil with mucuna. However, N mineralization from GM plants is significantly influenced by soil characteristics such as pH and texture, management practices, climate, age of the plants and their C:N ratio (Nagarajah et al. 1989; Singh et al. 1992; Kumar and Power 2018).

GM addition in soil may increase P availability to succeeding crops (Vanlauwe et al. 2000; Cavigelli and Thien 2003; Chimouriya et al. 2018) by several mechanisms such as (1) GM transforms unavailable P arising from native and residual fertilizer P to some other chemical forms that become available to succeeding crops; (2) There are some GMs, such as alfalfa, red clover, sweet clover, lupine etc., which show more ability to absorb available P from soil as compared to other crops (Braum and Helmke 1995). Their further decomposition and mineralization increase available P in soil; (3) Carbon dioxide, released during GM decomposition after their incorporation in soil or metabolically released from root and microbial respiration, forms carbonic acid in soil solution, that facilitates dissolution of P containing primary minerals like apatite. Additionally, organic acids, generated from the decaying GM during their decomposition, are responsible for dissolution of P from rocks and minerals; (4) P availability may be increased in soils of high P fixing capacity because P adsorption sites may be masked by organic compounds that are synthesized during GM decomposition (Easterwood and Sartain 1990). Moreover, organic compounds may also release P from adsorbed sites through anion exchange phenKangomenon (Kafkaf et al. 1988). Therefore, in general GM decreases soil P sorption (Singh and Jones 1976; Bumaya and Naylor 1988); and (5) Since repeated application of GM improves soil physical conditions by decreasing bulk density, increasing moisture holding capacity and improving soil aggregation, succeeding crop roots and mycorrhizal growth are improved facilitating more P availability (MacRae and Mehuys 1985). But this is also true that increased P availability is not always reciprocated with GM incorporation in soil (Bumaya and Naylor 1988). This might be due to subsequent P sorption on soil and more utilization of available P by the soil microorganisms (White and Ayoub 1983).

It has been established that organic matter amendments to soil improves utilization of naturally stocked soil micronutrients, and thereby reduce any need for major external inputs (Aghili et al. 2014). They applied $ZnSO_4$ alone as fertilizer, $^{65}Zn$ radiolabelled green manure (sunflower) and their combined application. $ZnSO_4$ application to soil led to an increase in bread-wheat grain Zn concentration from 20 to 39 mg Zn/kg and sole application of green manure of sunflower to soil raised bread-wheat grain Zn concentration to 31 mg Zn/kg. Their combined application to soil increased grain Zn concentration even further to 54 mg Zn/kg.

## 13.6 Reclamation of Saline and Sodic Soil

N, organo-mineral, organic C and total carbohydrate levels in salt-affected soil are improved to a great extent, when soils were amended with GM (Zubair et al. 2012). However, this essentially improved the impaired soil physical conditions too. In sodic soil, green manuring is a common practice to reclaim the soil as well as to improve the soil fertility. GM lowers down high pH of the soil by adding acids to the soil and further replaces the exchangeable sodium ions by other favourable ions such as calcium. Sodic soil is low in its organic contents because sodium carbonate and sodium bicarbonate salts in the solution dissolve humus, particularly fulvic and humic acid fractions. Hence, mixing of GM in sodic soil enhances SOM content (Khan et al. 2000) and eventually improves the soil physical conditions. Additionally, they help to replace sodium ions at the exchange sites by calcium ions. Lime always exists in sodic soil as insoluble calcium carbonate and thus sodic soil is always with less calcium ionic activity. As GM addition to soil releases more carbon dioxide during its decomposition, soil solution is enriched with more soluble calcium that replaces sodium ions at the exchange site and sodic soil is eventually reclaimed (Maitra et al. 2018). To reclaim sodic soils, *Sesbania aculeata* and *Delonix elata*, which are resistant to sodicity stress, are found very effective GM and green leaf manures, respectively (Baig and Zia 2006; Chandrasekaran et al. 2010). *Sesbania aculeata* contains 34% Ca on dry weight basis which helps to replace sodium from sodic soils. Partially decayed tamarind (*Tamarindus indicus*) leaves were also reported to ameliorate soil salinity (Vakeesan et al. 2008).

## 13.7 Nutrient and Soil Conservation

Nutrient and soil conservation are the most important two benefits which can be achieved from GM, as the topsoil is covered by the crop canopy (Larson and Pierce 1996; Maitra et al. 2018). It protects the soil from the direct impact of rain and heavy wind which may loosen the soil particles and transport them to other places. Apart from soil conservation, GM consumes excess nutrients present during the fallow period and releases them back to the soil during the main crop cultivation. Large quantities of nitrates are lost from soil by leaching and surface run-off as nitrates are not strongly attracted by soil particles. This is more pronounced during the gap period between two main crops. GM crops act as cover crops, protecting the soil from erosion and nutrient loss (Florentin et al. 2011). Winter GMs can be very effective crops for 'mopping up' excess nitrate in the soil in the autumn. Reduction in soil nutrient loss is due to the uptake of soluble nutrients by GMs, that might have been otherwise lost in drainage water or due to erosion. After subsequent mixing with soil, GMs make C, N, P, S and other nutrients available to the succeeding crops.

Similar to nitrate trapping (Tonitto et al. 2006; Constantin et al. 2011; Tribouillois et al. 2015), S catching by GM was also reported (Eriksen and Thorup-Kristensen 2002; Eriksen et al. 2004; Couëdel et al. 2018). Crucifers grown as cover crops minimizes sulphate leaching (S catch-crop service) and subsequently when they are incorporated into the soil, give back significant quantity of sulphate to the soil, which are utilized by the successive main crop. These S catch-crop and green manure services were validated for a wide variety of crucifer and legume species that differ in architecture, precocity and C:S ratio (Couëdel et al. 2018). In N green manure service, however, legume is the sole cover crop providing a lower S green manure service than crucifer sole cover crops because of their lower S uptake and higher C:S ratio, which may lead to net S immobilization instead of net S release (Eriksen 2005). Hence, cruciferous crops such as winter rape or fodder radish may be effective to prevent sulphur leaching. GM such as chicory shows its mining effect with its deep tap root. It may accumulate large amounts of sulphur, boron, manganese, molybdenum and zinc from rich subsoil (Rayns and Rosenfeld 2010).

## 13.8 Pest Management

GM controls insect-pests, weeds and soil-borne diseases and disrupts the life cycles of several crop pests by creating a space in the crop rotation sequence where suitable hosts for cereal or oilseed insect pests or diseases are not available (Kumar et al. 2014; Varma et al. 2017). Thus, pest incidence may be reduced by providing a break from cereals with a GM legume.

GM controls weed population (Blackshaw et al. 2001) and inhibits weed seed germination by their allelopathic effect (Boydston and Hang 1995). Population and dry weight of weeds were found to be reduced by dhaincha alone or integrated use of 60 kg N/ha through urea +60 kg N/ha through dhaincha.

GM plants encourage proliferation of microbial growth and their activities which may consequently suppress pathogen. Additionally, they may exhibit direct biocidal effect on the pathogen. Williams-Woodward et al. (1997) reported a control of common root rot of pea (*Aphanomyces eutieches*) bylucerne residues. Similarly, buckwheat controlled common scab (*Streptomyces scabies*) and verticillium wilt of potatoes (Wiggins and Kinkel 2005). In a separate experiment verticillium wilt of potatoes was found to be suppressed by Austrian winter pea, *Pisum sativum* L.; Broccoli, *Brassica oleracea* L. or Sudan grass, *Sorghum vulgare* (Ochiai et al. 2007). GM plants (such as Brassica, Crotalaria etc.) exhibited positive effect on soil-borne disease management (Pung et al. 2004; Larkin and Griffin 2007; Kamil et al. 2009; Kumar et al. 2010). Brassica GM crops produce isothiocyanates type of biofumigant to inhibit density of *Sclerotinia* spp. which causes various soil-borne plant diseases (Pung et al. 2004). Crucifer type of GM plants contains glucosinolates that control *Pythium* spp. and in general suppress total fungal population (Lazzeri and Manici 2001). There are also reports of releasing fungitoxic compounds such as avenacin, saponins (Deacon and Mitchell 1985; Engelkes and Windels 1994) from GM plants. Soil-borne diseases caused by pathogens such as *Rhizoctonia solani*, *Phytophthora erythroseptica*, *Pythium ultimum*, *Sclerotinia sclerotiorum*, *Fusarium sambucinum* and *Sclerotium rolfsii* were found to be effectively controlled by *Crotalaria juncea* L (Kamil et al. 2009) and Brassica crops such as canola, rapeseed, radish, turnip, yellow mustard, and Indian mustard (Larkin and Griffin 2007). Suppression of nematode attack to agricultural crops by GM was also reported (Pakeerathan et al. 2009; Agbenin 2012). Incorporation of green leaf manuring with Potria (*Thespesia amnea*), Calotropis (*Calotropis gigantia*), Neem (*Azadiracta indica*), Gliricidia (*Gliricidia maculata*) and Glycosmis (*Glycosmis pentaphylla*) reduced the incidence of *Meloidogyne incognita*, a root-knot nematode, in tomato (Pakeerathan et al. 2009). All in all, GM promotes beneficial microbial activity in soil (Manici et al. 2004).

## 13.9 Types of Green Manuring and Its Application in Sustainable Agriculture

Green manuring is practiced in different ways depending upon the soil and the climatic conditions of the area. On the basis of how biomass is added to the soil, the green manuring practices can be considered of two types.

### *13.9.1 In Situ Green Manuring*

In the northern parts of India, green manuring is done by growing green manure crop as an intercrop with the main crops (Singh 1984). The crops like sunn hemp, black gram and green gram can be used as green manure crops. These crops are grown along the main crop and after reaching flowering stage are cut and buried into the soil and allowed to decompose. *Sesbania aculeata*, *Crotolaria juncea* and *Aeschynomene americana* are some of the most commonly used green manures (Wilson et al. 1986). *Sesbania aculeate* (GM crops) also known as Dhaincha, is a fast growing plant which can be incorporated after 8–10 weeks of sowing. It can be grown as green manure crop and helps in improving physical properties of soil including infiltration rate and soil organic carbon (Boparai et al. 1992). It is known for root nodules, fixing nitrogen and growth in adverse conditions like water logging and drought; it has gained momentum in context of being a sustainable agricultural crop (Ambika et al. 2015). *Sesbania* ranks first among green manures in contributing biomass and nitrogen to the fields (Dubey et al. 2015). Furthermore, *Crotalaria juncea*, also known as sunn hemp or Tropic sun, is generally considered as a native to India and is grown on a wide scale in other countries. Sunn hemp is grown to serve the purpose of green manure; it should be ploughed down before the full flowering stage. Application of *Crotalaria* in long term in the fields improves the soil quality and promotes shoot growth in *Zea mays* (Sangakkara et al. 2004). Being a leguminous plant, Crotalaria is known to provide nitrogen content to the subsequent plant but experiments have shown that input of nitrogen by adding either shoot or root or both shoot or root showed different results. The subsequent plant in pot with only root showed that the N recovery rate was significantly higher than the one with both root and shoot (Choi et al. 2008). *Aeschynomene americana* is another type of leguminous annual herb which is used as green manure in rice fields in southern India. It is grown during the fallow season and ploughed down before the rice crop has to be sown (Rao et al. 1980). It can be used as GM crops to improve yield, fix large amount of nitrogen and may help in providing sufficient nitrogen content to rice plants. Therefore, it can be used as a substitute by resource poor rice farmers (Dickmann et al. 1996).

### 13.9.2 Green Leaf Manuring

Leaf of GM crops is practiced by incorporation of the leaves which are brought to the fields in bundles and kept covered for the process of decomposition (Vakeesan et al. 2008). In that context, green leaves are harvested from the plants which are pruned three times in a year, i.e. in June, November and March (Rao et al. 1980). This practice is followed mostly in central and eastern India. The crops which are known to be suitable for this purpose include *Gliricidia maculate*, *Leucaena leucocephala*, *Azadirachta indica*, etc. (Ambika et al. 2015; Lokanadhan et al. 2012). *Gliricidia* being thornless, is a fast growing leguminous tree that has the potential to play a multipurpose role in the agricultural system. It can be used as a GM crop as well as a fodder to the cattle and can improve the soil nitrogen content and organic content. Hence it can be used as a fertilizer for rice paddies (Falvey 1982). Experiments by Bah and Rahman (2001) have shown that the application of *Gliricidia* leaves to the surface increased the uptake of nitrogen by maize, thus proving that it potentially can improve the productivity of soils. Application of *Gliricidia* pruning combined with inorganic fertilizers in maize fields helps in enhancing the availability of phosphorus to the soil which can be utilized by maize plants (Mweta et al. 2007). *Leucaena leucocephala* is a deep-rooted GM crop which is able to survive in various soil types and is known for its resistance to drought conditions (Aganga and Tshwenyane 2003). Studies and experiments have shown that incorporation of leaves and prunings of *Leucaena* in maize field may serve as a source of nitrogen for the subsequent crop as effectively as some inorganic nitrogen source (Kang et al. 1984). *Azadirachta indica* (neem) is commercially exploitable and useful; the leaves of tree have gained popularity as being potential manures (Biswas et al. 2007; Lokanadhan et al. 2012). Because neem is rich in potassium, sulphur, nitrogen, calcium, etc., it can nurture the plants and soil by providing the required nutrients and help in increasing the crop yield (Lokanadhan et al. 2012).

### 13.9.3 Impact of Green Manure on Microbial Diversity

The organic matter provided by green manure serves as a substrate for microorganisms and thus leads to an increase in microbial populations (Feng et al. 2003; Elfstrand et al. 2007; Mohammadi 2011). The structure and function of these microbial communities depend upon the soil and the plant species in the field (Berg and Smalla 2009). The *Denaturing gradient gel electrophoresis* (*DGGE*) analysis has also confirmed the effect of soil and plant type on the microbial communities (Wieland et al. 2001). These microorganisms like bacteria, fungi, actinomycetes etc. play a very important role in maintaining fertility of the soil. The rhizobacteria help in fixing nitrogen whereas many fungi like Vesicular Arbuscular Mycorrhiza (VAM) can function to convert insoluble phosphates to soluble forms, accumulate plant nutrients and reduce the severity of plant pathogenic diseases (Mohammadi and Sohrabi 2012). However, long-term application of inorganic fertilizers in soil led to deficiency in organic content and microbial activity (Fauci and Dick 1994). This reduction in organic content can lead to a decreased microbial activity and diversity (Kumar and Adholeya 2016). Mineral fertilizers, like nitrogen fertilizers, if applied in high amounts can lead to recession in number of bacteria *Arthrobacter* and *Streptomyces* by 50%. Not only this, the fertilizers can lead to a complete extinction of genera like *Azotobacter*, *Rhizobium* and *Bradyrhzobium* (Barabasz et al. 2002). Many approaches have been developed to estimate the microbial communities and biomass in the soil. Soil microbial biomass can be assessed either directly by microscopic counting or indirectly by estimation of any constituent of microbial cell like C, P, S, N, ATP and phospholipids. The comparison between fields receiving no organic matter and field receiving organic inputs over years, showed that soils receiving organic inputs showed an increase in the Phospholipid fatty acid (PPLA) content indicating increases in the microbial biomass (Elfstrand et al. 2007). Another promising approach to estimate the microbial communities is metagenomic analysis. This approach is based on collection of samples from the environment and amplifying the 16S rDNA from the DNA present in collected samples (Đokić et al. 2010). The comparative studies carried out by researchers in organic and conventionally farmed soils showed a significant difference in the taxonomic composition of the microbes as the organically farmed soils were found to be dominated by plant growth-promoting bacteria (Pershina et al. 2016). The analyses also show that manuring has proved to be very effective in improving the microbial community in the soils, where on the other hand, mineral fertilization was not found very effective in inducing changes in community structure (Lentendu et al. 2014).

## 13.10 Management Practices and Conservation of Green Manure Crops

The management of green crops is very important for improving productivity and organic content of soil and for ensuring proper usage of resources. These practices may vary depending upon the type of green manure used in terms of sowing season, duration of crops and the farming system used. Other factors that are considered include fertility management, seed bed preparation, agronomic feasibility etc. Every crop requires sufficient nutrients, adequate moisture levels and balanced pH for proper growth. These factors if not available in appropriate quantity to the green manure might influence the performance of green manures as their nitrogen content and biomass production may vary according to the field conditions (Ladha and Garrity 1994). Moreover for proper production of the crops, the depth at which seed is sown has to be taken care of. According to the size of seed to be sown, the depth can be determined. The seed bed should be prepared in such a way that the soil should be loosened leaving no compaction at the surface or in the depth, keeping the track of available moisture content. Most of the green manure crops that are used in cropping systems are leguminous, nitrogen fixing plants. But these plants might also serve as food sources for various other microorganisms which can cause diseases in plants and therefore, farmers seem reluctant while growing such plants. Therefore, to serve as a potential green manure crop, the crops are either needed to be disease resistant or other or to be managed by other disease resistance strategies (Ladha and Garrity 1994). The amount of biomass and nitrogen green manures contribute to the soil is variable in different plant species. Besides this there are other important factors like sensitivity to photoperiods, ability to withstand harsh conditions and yields which are needed to be in a plant to serve as green manure crop. Except these the overall costs are considered to count a plant as a feasible agronomic green manure crop (Ladha and Garrity 1994).

## 13.11 New Insight of Green Manure for Crop Improvement

Green manure crops have proved to be very beneficial to farmers as it provides fixed nitrogen to the subsequent plants improving soil health and significantly reducing the pest infestation and thus the diseases (Berry and Rhodes 2006). Thus, for them, to efficiently work as GM crops, they need to be good in biomass, have high nitrogen fixation rate, resistant to stresses like drought and salinity. Since genetic modification of plant uses highly heritable traits, it can be considered as an initiation step in optimizing the plant (Huyghe 1998). Genetic transformation and regeneration using *A. tumefaciens*-mediated transformation were confirmed to create stable herbicide (Basta) resistant cultivars in Legumes (Atkins and Smith 1997; Fageria 2007). Moreover, Biswas et al. (2007) reported an increase nodulation efficiency in Sesbania by two to four-fold using mutant strain of Sinorhizobium saheli, originated from the root nodule of Sesbania *aculeate*. Additionally, many literatures manipulated agronomical important gene in cover crops for better biomass, photoperiod sensitivity, vigorous shoot type to suppress weeds, minimize pod shattering and create strong resistance against pathogen (such as *Fusarium* wilt) (Tani et al. 2017). Recently, a major quantitative trait locus against root-knot nematodes disease in legumes were identified and mapped by Huynh et al. (2016). Moreover, Pottorff et al. (2012) mapped *Fusarium* resistance loci of 85 diverse line of legumes against *Fusarium oxysporum* f.sp. *tracheiphilum. Besides, introduction of novel gene pyramiding technology transfer Fusarium wilt and Root knot nematodes resistance genes in single breeding line of leguminosae crops.* wilt into single breeding line of leguminous crops. New modifications based on increase in plant biomass of legume cover crop by inhibiting reproductive development was also described by Harfouche et al. (2011). Similarly, Alfalfa is another crop which is routinely being used as a green manure crop since past few years. However, crop is susceptible to alfalfa weevil (*Hypera postica*). To improve the variety of alfalfa, new transgenic plants were developed which possess resistance against alfalfa weevil. Additionally, new transgenic cultivar of Alfalfa was developed using Agrobacterium mediated transformation which showed resistance against insect (Alfalfa weevil) (Tohidfar et al. 2012).

**Acknowledgements** The authors wish to thank financial contributions from Lovely Professional University India. We thank Chancellor, Shri. Ashok Mital, for providing the necessary infrastructure for carrying out the research work.

## References

Adekiya AO, Agbede TM, Aboyeji CM, Dunsin O, Ugbe JO (2019) Green manures and NPK fertilizer effects on soil properties, growth, yield, mineral and vitamin C composition of okra (*Abelmoschus esculentus* (L.) Moench). J Saudi Soc Agric Sci 18:218–223

Aganga AA, Tshwenyane SO (2003) Lucerne, Lablab and *Leucaena leucocephala* forages: production and utilization for livestock production. Pak J Nutr 2(2):46–53

Agbede TM (2018) Effect of green manure application on cassava (*Manihot esculenta* Crantz) growth, yield quantity and quality in degraded alfisols. Pertanika J Trop Agric Sci 41 (4):1757–1777

Agbenin NO (2012) Biological control of plant parasitic nematodes: prospects and challenges for the poor Africa farmer. Plant Prot Sci 47(2):62–67

Aghili F, Gamper HA, Eikenberg J, Khoshgoftarmanesh AH, Afyuni M, Schulin R, Frossard E (2014) Green manure addition to soil increases grain zinc concentration in bread wheat. PLoS One 9(7):e101487

Akpor OB, Okoh AI, Babalola GO (2006) Culturable microbial population dynamics during decomposition of *Theobroma cacao* leaf litters in a tropical soil setting. J Biol Sci 6(4):768–774

Al-Khatib K, Libbey C, Boydston R (1997) Weed suppression with Brassica green manure crops in green pea. Weed Sci 45(3):439–445

Ambika S, Somasundaram G, Sundralingam K (2015) Importance of green and green leaf manures in sustainable agriculture. Popular kheti 3(1):62–65

Atkins CA, Smith PMC (1997) Genetic transformation and regeneration of legumes. In: Legocki A, Bothe H, Puhler A (eds) Biological fixation of nitrogen for ecology and sustainable agriculture. Springer, Berlin, pp 283–304

Badanur VP, Poleshi CM, Naik BK (1990) Effect of organic matter on crop yield and physical and chemical properties of a vertisol. J Indian Soc Soil Sci 38(3):426–429

Bah AR, Rahman ZA (2001) Gliricidia (*Gliricidia sepium*) green manures as a potential source of N for maize production in the tropics. Sci World J 1:90–95

Bah AR, Zaharah AR, Hussin A (2004) Effect of interactions on the nutrient status of a tropical soil treated with green manures and inorganic phosphate fertilizers. Sci World J 4:393–414

Bah AR, Zaharah AR, Hussin A (2006) Phosphorus uptake from green manures and phosphate fertilizers applied in an acid tropical soil. Commun Soil Sci Plant Anal 37:2077–2093

Baig MB, Zia MS (2006) Rehabilitation of problem soils through environmental friendly technologies-II: role of sesbania (*Sesbania aculeata*) and gypsum. Agric Trop Subtrop 39:1

Balota EL, Chaves JCD (2011) Microbial activity in soil cultivated with different summer legumes in coffee crop. Braz Arch Biol Technol 54(1):35–44

Barabasz W, Albinska D, Jaskowska M, Lipiec J (2002) Biological effects of mineral nitrogen fertilization on soil microorganisms. Pol J Environ Stud 11(3):193–198

Becker M, Ladha JK, Ali M (1995) Green manure technology: potential, usage, and limitations. A case study for lowland rice. Plant Soil 174:181–194

Berg G, Smalla K (2009) Plant species and soil type cooperatively shape the structure and function of microbial communities in the rhizosphere. FEMS Microbiol Ecol 68(1):1–13

Berry S, Rhodes R (2006) Green manure crops: agronomic characteristics and effect on nematodes. Proc S Afr Sug Technol Ass 80:269–273
Bhattarai N, Vaidya GS, Baral B (2012) Effect of mycorrhizal soil and green manures on growth of Ipil Ipil (*Leucaena diversifolia* L.). Scientific World 10(10):66–69
Bhuiyan NI, Zaman SK (1996) In: Rahman M et al (eds) Biological nitrogen fixation associated with rice production. Kluwer Academic, Dordrecht, pp 51–64
Biswas S, Das RH, Sharma GL, Das HR (2007) Isolation and characterization of a novel cross-infective rhizobia from *Sesbania aculeata* (Dhaincha). Curr Microbiol 56(1):48–54
Blackshaw RE, Moyer JR, Doram RC, Boswell AL (2001) Yellow sweet clover, green manure, and its residues effectively suppress weeds during fallow. Weed Sci 49(3):406–413
Bokhtiar SM, Gafur MA, Rahman ABMM (2003) Effects of Crotalaria and *Sesbania aculeata* green manures and N fertilizer on soil fertility and the productivity of sugarcane. J Agric Sci 140:305–309
Boparai BS, Singh Y, Sharma BD (1992) Effect of green manuring with *Sesbania aculeata* on physical properties of soil and on growth of wheat in rice-wheat and maize-wheat cropping systems in a semiarid region of India. Arid Land Res Manag 6(2):135–143
Boydston RA, Hang A (1995) Rapeseed (*Brassica napus*) green manure crop suppresses weeds in potato (*Solanum tuberosum*). Weed Technol 9:669–675
Braum SM, Helmke PA (1995) White lupin utilizes soil phosphorus that is unavailable to soybean. Plant Soil 176:95–100
Bumaya AH, Naylor DV (1988) Phosphorus sorption and extractability in Andic soil incubated with plant residues of variable P content. Plant Soil 112(1):77–81
Bunch R (2012) Restoring the soil. A guide for using green manure/cover crops to improve the food security of smallholder farmers. Canadian Foodgrains Bank. Winnipeg, Canada
Buragohain S, Sarma B, Nath DJ, Gogoi N, Meena RS, Lal R (2018) Effect of 10 years of biofertiliser use on soil quality and rice yield on an Inceptisol in Assam, India. Soil Res 56 (1):49–58
Cavigelli MA, Thien SJ (2003) Phosphorus bioavailability following incorporation of green manure crops. Soil Sci Soc Am J 67(4):1186–1194
Chandrasekaran B, Annadurai K, Somasundaram E (2010) A textbook of agronomy. New Age International, New Delhi, p 442
Chimouriya S, Lamichhane J, Gauchan DP (2018) Green manure for restoring and improving the soil nutrients quality. Int J Res 5(20):1064–1074
Choi B, Ohe M, Harada J, Daimon H (2008) Role of belowground parts of green manure legumes, *Crotalaria spectabilis* and *Sesbania rostrata*, in N uptake by the succeeding tendergreen mustard plant. Plant Prod Sci 11(1):116–123
Constantin J, Beaudoin N, Laurent F, Cohan JP, Duyme F, Mary B (2011) Cumulative effects of catch crops on nitrogen uptake, leaching and net mineralization. Plant Soil 341:137–154
Couëdel A, Alletto L, Justes E (2018) Crucifer-legume cover crop mixtures provide effective sulphate catch crop and sulphur green manure services. Plant Soil 426(1–2):61–76
Deacon JW, Mitchell RT (1985) Toxicity of oat roots, oat root extracts, and saponins to zoospores of *Pythium* spp. and other fungi. Trans Br Mycol Soc 84:479–487
Dhakal Y, Meena RS, De N, Verma SK, Singh A (2015) Growth, yield and nutrient content of mungbean (*Vigna radiata* L.) in response to INM in eastern Uttar Pradesh, India. Bangladesh J Bot 44(3):479–482
Dickmann KH, Ottow JC, Diekmann KH, De Datta SK (1996) Yield and nitrogen response of lowland rice (*Oryza sativa* L.) to *Sesbania rostrata* and *Aeschynomene afraspera* green manure in different marginally productive soils in the Philippines. Biol Fertil Soils 21(1–2):103–108
Dinnes DL, Karlen DL, Jaynes DB, Kaspar TC, Hatfield JL, Colvin TS, Cambardella CA (2002) Nitrogen management strategies to reduce nitrate leaching in tile-drained Midwestern soils. Agron J 94:153–171
Đokić L, Savić M, Narančić T, Vasiljević B (2010) Metagenomic analysis of soil microbial communities. Arch Biol Sci (Serbia) 62(3):559–564

Doran JW, Fraser DG, Culik MN, Liebhardt WC (1988) Influence of alternative and conventional agricultural management on soil microbial process and nitrogen availability. Am J Altern Agric 2:99–106

Drinkwater LE, Wagoner P, Sarrantonio M (1998) Legume based cropping systems have reduced carbon and nitrogen losses. Nature 396:262–265

Dubey L, Dubey M, Jain P (2015) Role of green manuring in organic farming. Plant Arch 15 (1):23–26

Easterwood GW, Sartain JB (1990) Clover residue effectiveness in reducing orthophosphate sorption on ferric hydroxide coated soil. Soil Sci Soc Am J 54:1345–1350

Elfstrand S, Hedlund K, Mårtensson A (2007) Soil enzyme activities, microbial community composition and function after 47 years of continuous green manuring. Appl Soil Ecol 35 (3):610–621

Engelkes CA, Windels CE (1994) B-escin (saponin) oat seedlings and oat residue in soil affects growth of Aphanomyces cochlioides hyphae zoospores and oogonia (Abstr.). Phytopathology 84:1158

Eriksen J (2005) Gross sulphur mineralization-immobilization turnover in soil amended with plant residues. Soil Biol Biochem 37:2216–2224

Eriksen J, Thorup-Kristensen K (2002) The effect of catch crops on sulphate leaching and availability of S in the succeeding crop on sandy loam soil in Denmark. Agric Ecosyst Environ 90(3):247–254

Eriksen J, Thorup-Kristensen K, Askegaard M (2004) Plant availability of catch crop sulfur following spring incorporation. J Plant Nutr Soil Sci 167:609–615

Fageria NK (2007) Green manuring in crop production. J Plant Nutr 30(5):691–719

Fagodiya RK, Pathak H, Kumar A, Bhatia A, Jain N (2017) Global temperature change potential of nitrogen use in agriculture: a 50-year assessment. Sci Report 7:44928

Falvey JL (1982) *Gliricidia maculate* – a review. Int Tree Crops J 2(1):1–14

Fauci MF, Dick RP (1994) Soil microbial dynamics: short- and long-term effects of inorganic and organic nitrogen. Soil Sci Soc Am J 58:801–806

Feng Y, Motta A, Reeves D, Burmester C, Santen EV, Osborne J (2003) Soil microbial communities under conventional-till and no-till continuous cotton systems. Soil Biol Biochem 35 (12):1693–1703

Florentín MA, Peñalva M, Calegari A, Derpsch R, McDonald MJ (2010) Green manure/cover crops and crop rotation in conservation agriculture on small farms. Integr Crop Manag 12:7

Florentín MA, Peñalva M, Calegari A, Derpsch A (2011) Green manure/cover crops and crop rotation in conservation agriculture on small farms. FAO Integr Crop Manag 12:97

Golec AFC, Pérez PG, Lokare C (2007) Effective microorganisms: myth or reality? Rev Peru Biol 14(2):315–319

Harfouche A, Meilan R, Altman A (2011) Tree genetic engineering and applications to sustainable forestry and biomass production. Trends Biotechnol 29(1):9–17

Haynes RJ, Mokolobate MS (2001) Amelioration of Al toxicity and P deficiency in acid soils by additions of organic residues: a critical review of the phenomenon and the mechanisms involved. Nutr Cycl Agroecosyst 59:47–63

Helyar KR, Porter WM (1989) Soil acidification, its measurement and the processes involved. In: Robson AD (ed) Soil acidity and plant growth. Academic Press, Sydney, pp 61–101

Huyghe C (1998) Genetics and genetic modifications of plant architecture in grain legumes: a review. Agronomie EDP Sci 18(5–6):383–411

Huynh BL, Matthews WC, Ehlers JD, Lucas MR, Santos JRP, Ndeve A, Close TJ, Roberts PA (2016) A major QTL corresponding to the Rk locus for resistance to root-knot nematodes in cowpea (*Vigna unguiculata* L. Walp.). Theor Appl Genet 129:87–95

Kafkaf U, Bar-Yosef B, Rosenberg R, Sposito G (1988) Phosphorus adsorption by kaolinite and montmorillonite II Organic anion competition. Soil Sci Soc Am J 52:1585–1589

Kamil D, Kumar R, Sinha A (2009) Effect of green manuring of *Crotalaria juncea* L. on some soil-borne pathogens. Indian Phytopath 62(3):304–309

Kang BT, Wilson GF, Lawson TL (1984) Alley cropping: a stable alternative to shifting cultivation. Adv Agron 43:315–359

Khan AR, Ghorai AK, Ingh SR (2000) Improvement of crop and soil sustainability through green manuring in rainfed lowland rice ecosystem. Agrochimica 44(1–2):21–29

Kimetu JM, Lehmann J, Ngoze SO, Mugendi DN, Kinyangi JM, Riha S, Pell AN (2008) Reversibility of soil productivity decline with organic matter of differing quality along a degradation gradient. Ecosystems 11(5):726

Kumar S, Adholeya A (2016) Impact of land use and soil types on Arbuscular mycorrhizal fungal diversity in tropical soil of India. AJMR 10(38):1595–1606

Kumar V, Power K (2018) A review on soil health and fertility management in organic agriculture through green manuring. J Pharmacogn Phytochem SP1:3213–3217

Kumar R, Srivastava S, Srivastava M, Sinha A (2010) Effects of organic amendments on soil mycoflora. Asian J Plant Path 4(2):73–81

Kumar R, Mahajan G, Srivastava S, Sinha A (2014) Green manuring: a boon for sustainable agriculture and pest management – a review. Agric Rev 35(3):196–206

Kumar S, Sheoran S, Kumar SK, Kumar P, Meena RS (2016) Drought: a challenge for Indian farmers in context to climate change and variability. Prog Res An Int J 11:6243–6246

Ladha JK, Garrity DP (eds) (1994) Green manure production systems for Asian ricelands: selected papers from the international rice research conference. International Rice Research Institute, Manila

Larkin RP, Griffin TS (2007) Control of soil borne potato diseases using Brassica green manures. Crop Prot 26(7):1067–1077

Larson W, Pierce FJ (1996) Conservation and enhancement of soil quality, the soil quality concept. Edited by The Soil Quality Institute, United States Department of Agriculture and Natural Resources Conservation Service 11–38

Lazzeri L, Manici LM (2001) Allelopathic effect of Glucosinolate containing plant green manure on *Pythium* sp. and total fungal population in soil hort. Science 36(7):1283–1289

Lentendu G, Wubet T, Chatzinotas A, Wilhelm C, Buscot F, Schlegel M (2014) Effects of long term differential fertilization on eukaryotic microbial communities in an arable soil: a multiple barcoding approach. Mol Ecol 23(13):3341–3355

Logsdon SD, Radke JK, Karlen DL (1993) Comparison of alternative farming system infiltration techniques. Am J Altern Agric 8:15–20

Lokanadhan S, Muthukrishnan P, Jeyaraman S (2012) Neem products and their agricultural applications. J Biopest 5:72

MacRae RJ, Mehuys GR (1985) The effect of green manuring on the physical properties of temperate-area soils. Adv Soil Sci 3:71–94

Mafongoya PL, Giller KE, Palm CA (1998) Decomposition and nitrogen release patterns of tree prunings and litter. Agrofor Syst 38:77–97

Maitra S, Zaman A, Mandal TK, Palai JB (2018) Green manures in agriculture: A review. J Pharmacogn Phytochem 7(5):1319–1327

Manici LM, Caputo F, Babini V (2004) Effect of green manure on *Pythium* spp. population and microbial communities in intensive cropping systems. Plant Soil 263(1/2):133–142

Meena BL, Fagodiya RK, Prajapat K, Dotaniya ML, Kaledhonkar MJ, Sharma PC, Kumar S (2018a) Legume green manuring: an option for soil sustainability. In: Legumes for soil health and sustainable management. Springer, Singapore, pp 387–408

Meena RS, Vijayakumar V, Yadav GS, Mitran T (2018b) Response and interaction of *Bradyrhizobium japonicum* and arbuscular mycorrhizal fungi in the soybean rhizosphere. Plant Growth Regul 84(2):207–223

Melero S, Riuz Porrai JC, Herencia JF, Madejon E (2006) Chemical and biochemical properties in a sylty loam soil under conventional and organic management. Soil Tillage Res 90:162–170

Mohammadi K (2011) Soil management, microorganisms and organic matter interactions: a review. Afr J Biotechnol 10(86):19840–19849

Mohammadi K, Sohrabi Y (2012) Bacterial biofertilizers for sustainable crop production: a review. J Agric Biol Sci 7:307–316

Mweta DE, Akinnifesi FK, Saka JDK, Makumba W, Chokotho N (2007) Green manure from prunings and mineral fertilizer affect phosphorus adsorption and uptake by maize crop in a gliricidia-maize intercropping. Sci Res Essays 2(10):446–453

Nagarajah S, Neue HU, Alberto MCR (1989) Effect of Sesbania, Azolla and rice straw incorporation on the kinetics of NH 4, K, Fe, Mn, Zn and P in some flooded rice soils. Plant Soil 116 (1):37–48

Nayak JJ, Vaidya AK (2018) Green manure in crop production and soil health. Int J Innov Res Sci Eng Technol 7(6):7378–7381

Ntakirutimana L, Li F, Huang X, Wang S, Yin C (2019) Green manure planting incentive measures of local authorities and farmers' perceptions of the utilization of rotation fallow for sustainable agriculture in Guangxi, China. Sustainability 11(10):2723

Ochiai N, Powelson ML, Dick RP, Crowe FJ (2007) Effects of green manure type and amendment rate on *Verticillium* wilt severity and yield of Russet Burbank potato. Plant Dis 91(4):400–406

Ochiai N, Powelson ML, Crowe FJ, Dick RP (2008) Green manure effects on soil quality in relation to suppression of *Verticillium* wilt of potatoes. Biol Fertil Soils 44(8):1013–1023

Pakeerathan K, Mikunthan G, Tharshani N (2009) Eco-friendly management of root-knot nematode *Meloidogyne incognita* (Kofid and White) chitwood using different green leaf manures on tomato under field conditions. Am Eurasian J Agric Environ Sci 6(5):494–497

Palm CA, Gachengo CN, Delve RJ, Cadisch G, Giller KE (2001) Organic inputs for soil fertility management in tropical agroecosystems: application of an organic resource database. Agric Ecosyst Environ 83(1–2):27–42

Pandey AK, Singh MK (2016) Importance and uses of green manuring in field crops. Rashtriyakrishi 11(2):35–35

Parton W, Silver WL, Burke IC, Grassens L, Harmon ME, Currie WS, King JY, Adair EC, Brandt LA, Hart SC, Fasth B (2007) Global-scale similarities in nitrogen release patterns during long-term decomposition. Science 315:361–364

Pershina EV, Valkonen JP, Kurki P, Ivanova EA, Chirak EL, Korvigo IO, Andronov EE (2016) Correction comparative analysis of prokaryotic communities associated with organic and conventional farming systems. PLoS One 11(5):e0155155

Pimentel D, Hepperly P, Hanson J, Douds D, Seidel R (2005) Environmental, energetic, and economic comparisons of organic and conventional farming systems. Bioscience 55 (7):573–582

Pottorff M, Wanamaker S, Ma YQ, Ehlers JD, Roberts PA, Close TJ (2012) Genetic and physical mapping of candidate genes for resistance to *Fusarium oxysporum* f. sp. tracheiphilum race 3 in cowpea [*Vigna unguiculata* (L.) Walp]. PLoS One 7(7):e41600

Pung H, Aird PL, Cross S (2004) The use of Brassica green manure crops for soil improvement and soil borne disease management. 3rd Australian soil borne diseases symposium 8–11 February 2004, pp 1–2

Rao NS, Tilak KVBR, Singh CS (1980) Root nodulation studies in *Aeschynomene aspera*. Plant Soil 56(3):491–494

Rayns F, Rosenfeld A (2010) Green manures – effects on soil nutrient management and soil physical and biological properties HDC, Factsheet 24/10, Soil grown crops Projects FV 299 and 299a

Reddy PP (2016) Sustainable intensification of crop production. Springer, Singapore, pp 241–252

Sangakkara UR, Liedgens M, Soldati A, Stamp P (2004) Root and shoot growth of maize (*Zea mays*) as affected by incorporation of *Crotalaria juncea* and *Tithonia diversifolia* as green manures. J Agron Crop Sci 190(5):339–346

Saria AG, Sibuga KP, Semu E, Jensen HH (2018) Soil fertility dynamics of ultisol as influenced by greengram and mucuna green manures. J Plant Sci Agric Res 2(2):14

Schröder J (2005) Revisiting the agronomic benefits of manure: a correct assessment and exploitation of its fertilizer value spares the environment. Bioresour Technol 96(2):253–261

Schumann RA, Meyer JH, Van Antwerpen R (2000) A review of green manuring practices in sugarcane production. Proc S Afr Sug Technol Ass 74:93–100

Schutter M, Dick R (2001) Shifts in substrate utilization potential and structure of soil microbial communities in response to carbon substrates. Soil Biol Biochem 33(11):1481–1491

Selvi RV, Kalpana R (2009) Potentials of green manure in integrated nutrient management for rice: a review. Agric Rev 30(1):40–47

Singh NT (1984) Green manures as sources of nutrients in rice production. Organic matter and rice, pp 217–228

Singh BB, Jones JP (1976) Phosphorus sorption and desorption characteristics of soil as affected by organic residues. Soil Sci Soc Am J 40:389–394

Singh Y, Singh B, Khind CS (1992) Nutrient transformations in soils amended with green manures. Adv Soil Sci 20:238–298

Sinha A, Kumar R, Kamil D, Kapur P (2009) Release of nitrogen, phosphorus and potassium from decomposing *Crotalaria juncea* L. in relation to different climatic factors. Environ Ecol 27 (4B):2077–2081

Sulieman S, Tran L (2016) Legume nitrogen fixation in a changing environment. Springer, Cham

Talgre L, Lauringson E, Roostalu H, Astover A, Makke A (2012) Green manure as a nutrient source for succeeding crops. Plant Soil Environ 58(6):275–281

Tang C, Sparling GP, McLay CDA, Raphael C (1999) Effect of short-term legume residue decomposition on soil acidity. Aust J Soil Res 37:561–573

Tani E, Abraham E, Chachalis D, Travlos I (2017) Molecular, genetic and agronomic approaches to utilizing pulses as cover crops and green manure into cropping systems. Int J Mol Sci 18 (6):1202

Tejada M, Gonzalez JL, García-Martínez AM, Parrado J (2008) Application of a green manure and green manure composted with beet vinasse on soil restoration: effects on soil properties. Bioresour Technol 99(11):4949–4957

Tohidfar M, Zare N, Jouzani GS, Eftekhari SM (2012) Agrobacterium-mediated transformation of alfalfa (*Medicago sativa*) using a synthetic cry3a gene to enhance resistance against alfalfa weevil. Plant Cell Tiss Org Cult 113(2):227–235

Tonitto C, David MB, Drinkwater LE (2006) Replacing bare fallows with cover crops in fertilizer-intensive cropping systems: a meta-analysis of crop yield and N dynamics. Agric Ecosyst Environ 112:58–72

Tribouillois H, Fort F, Cruz P, Charles R, Flores O, Garnier E, Justes E (2015) A functional characterisation of a wide range of cover crop species: growth and nitrogen acquisition rates, leaf traits and ecological strategies. PLoS One 10:1–17

Vakeesan A, Nishanthan T, Mikunthan G (2008) Green manures: nature's gift to improve soil fertility. LEISA Magazine 24(2):16–17

Vanlauwe B, Diels J, Sanginga N, Carsky RJ, Deckers J, Merckx R (2000) Utilization of rock phosphate by crops on a representative toposequence in the Northern Guinea savanna zone of Nigeria: response by maize to previous herbaceous legume cropping and rock phosphate treatments. Soil Biol Biochem 32:2079–2090

Varma D, Meena RS, Kumar S (2017) Response of mungbean to fertility and lime levels under soil acidity in an alley cropping system of Vindhyan Region, India. Int J Chem Stud 5(4):1558–1560

White RE, Ayoub AT (1983) Decomposition of plant residues of variable C/P ratio and the effect on soil phosphate availability. Plant Soil 74:163–173

Wieland G, Neumann R, Backhaus H (2001) Variation of microbial communities in soil rhizosphere and rhizoplane in response to crop species soil type and crop development. Appl Environ Microbiol 67(12):5849–5854

Wiggins BE, Kinkel LL (2005) Green manures and crop sequences influence potato diseases and pathogen inhibitory activity of indigenous streptomycetes. Phytopathology 95:178–185

Williams-Woodward JL, Pfleger F, Fritz VA, Allmaras RR (1997) Green manures of oat, rape and sweet corn for reducing common root rot in pea (*Pisum sativum*) caused by *Aphanomyces euteiches*. Plant Soil 188(1):43

Wilson GF, Kang BT, Mulongoy K (1986) Alley cropping: trees as sources of green-manure and mulch in the tropics. Biol Agric Hortic 3(2–3):251–267

Xie Z, Tu S, Shah F, Xu C, Chen J, Han D, Liu G, Li H, Muhammad I, Cao W (2016) Substitution of fertilizer-N by green manure improves the sustainability of yield in double-rice cropping system in south China. Field Crop Res 188:142–149

Yadav GS, Lal R, Meena R, Babu S, Das A, Bhoumik SN, Datta M, Layak J, Saha P (2019) Conservation tillage and nutrient management effects on productivity and soil carbon sequestration under double cropping of rice in North Eastern Region of India. Ecol Indic 105:303–315

Yang L, Bai J, Liu J, Zeng N, Cao W (2018) Green manuring effect on changes of soil nitrogen fractions, maize growth, and nutrient uptake. Agronomy 8:261

Ye X, Liu H, Li Z, Wang Y, Wang Y, Wang H, Liu G (2014) Effects of green manure continuous application on soil microbial biomass and enzyme activity. J Plant Nutr 37(4):498–508

Zaccheo PVC, Neves CSVJ, de Cinque Mariano D, Zorzenoni TO, Higashibara LR, Piccinin GG, Okumura RS (2016) Green manure in fruit culture: aspects on soil quality and use in agriculture. Afr J Agric Res 11(17):146

Zubair M, Anwar F, Ashraf M, Ashraf A, Chatha SAS (2012) Effect of green and farmyard manure on carbohydrates dynamics of salt affected soil. J Soil Sci Plant Nutr 12(123):497–510

# Chapter 14
# Mighty Microbes: Plant Growth Promoting Microbes in Soil Health and Sustainable Agriculture

**Imtiaz Ahmad and Sania Zaib**

**Abstract** Plants undergo a variety of biotic and abiotic stresses and to cope with such stresses, they have developed different direct and indirect mechanisms. Soil contains a diverse microbial community that have a diverse functional niche. For maximum growth and protection benefits, plants establish partnership with a variety of different beneficial microbes such as bacteria and fungi. These microbes can live inside and outside of plant tissue and have a broad host range. Microbes can modulate plant defense signaling pathways in the presence of stress conditions. Beneficial microbes can antagonize stresses and improve plant growth and fitness. We need to better understand the ecology and biology of the microbes associated with plants to exploit maximum services associated with beneficial microbes. For environmentally friendly and sustainable agriculture, we should reduce input of chemical fertilizers, pesticides, and fungicides and in this regard, use of beneficial plant-associated microbes is a promising alternative.

**Keywords** Beneficial microbes · Sustainable agriculture · Soil health · Symbiosis · PGPR · Fungi · Biotic and abiotic stresses

## 14.1 Introduction

In the past 50 years, a tremendous increase in the food production has been observed which is usually the outcome of widespread agriculture system called conventional or industrial farming system. According to an estimation of World Bank this system has risen the food production between 70% and 90%. However, for the enhancement

---

I. Ahmad (✉)
Department of Entomology, Pennsylvania State University, University Park, PA, USA
e-mail: ixa12@psu.edu

S. Zaib
Department of Biochemistry, Faculty of Biological Sciences, Quaid-i-Azam University, Islamabad, Pakistan



B. Giri, A. Varma (eds.), *Soil Health*, Soil Biology 59,
https://doi.org/10.1007/978-3-030-44364-1_14

of food production, this agriculture system relies on various chemicals and artificial enhancements such as fertilizers, pesticides, genetically modified organisms and external energy inputs like fossil fuels, large-scale farms and use of large machine for their management. Although intensive agricultural production is the need of the time, the massive utilization of the natural resources that are nonrenewable such as soil, water, fossil fuels and rock mineral reserves etc. and energy-intensive industrial processes for the fertilizers production and the runoff of the soluble nutrients into the aquatic systems are the sources of environmental contamination (Browne et al. 2013). In addition, intensive agriculture is also affecting the stability of biosphere by increasing the production of greenhouse gases and therefore rising the earth's temperature. This is ultimately generating various types of stress conditions such as salinity, drought, nutrient deficiency, soil erosion, contamination, pest and disease etc. which are affecting both agricultural system and natural ecosystems. Despite enhanced food production by the industrialized agriculture system, due to the negative aspects of this technology, many scientists and farmers are shifting toward sustainable agriculture which is environmentally, economically, and socially more sustainable.

## 14.2 Sustainable Agriculture

The word "sustain" comes from the Latin word sustinere (sus means "from below" and tinere means "to hold"), implying for long-term support. With respect to agriculture, the term sustainable means a farming system that for an indefinite time can maintain its productivity and efficacy to a society that is socially supportive, commercially competitive, resource conserving, and environmentally sound. So sustainable agriculture mainly emphasizes on long-term production of crops with minimum effects on the environment. In addition to food, it also helps in lessening the utilization of fertilizers and pesticides, water conservation and promoting crops biodiversity. In addition, this system assists in the maintenance of economic stability and assists the farmers in improving their quality of life by using better techniques. In accordance to Food, Agriculture, Conservation and Trade Act (FACTA) of 1990, sustainable agriculture is an integrated system that in the long run will satisfy the needs of human food and fiber, promote agricultural economy by enhancing the quality of natural resources and environment, efficiently utilize nonrenewable resources, and improve the life quality of farmers and society as a whole. In agriculture, environmental sustainability means good stewardship of the natural system and the resources on which farms depend upon. This includes the maintenance of healthy soil, management of water, reduction of pollution, and enhancement of plant biodiversity.

## 14.3 Benefits of Sustainable Agriculture

The various benefits of sustainable agriculture can be divided into human health benefits and environmental benefits. With respect to human health, in sustainable agricultural crops, people are not being exposed to chemicals, i.e., fertilizers and pesticides, so these crops ensure consumer safety as it reduces their chances of illness after exposure to these chemicals. Additionally, these crops could be more nutritious being more natural and healthier than the crops grown through industrialized agriculture. It also benefits the environment through saving soil water, decreasing soil erosion, and by the maintenance of soil quality. Moreover, it also contributes for enhancing the biodiversity by the provision of natural and healthy environment to a variety of the organisms. One of the major benefits for the environment is the 30% less energy consumption for one unit of crop yield. This reduces the dependence on fossil fuels, and consequently environment is less polluted.

## 14.4 Sustainable Agricultural Practices

World population is expected to reach 10 billion in 2050 and almost all this growth increase is expected to take place in the developing countries. This will generate many challenges, increasing the demand of more food, fiber and bioenergy and the demand for the preservation of the biosphere. Researchers are aware to address their research efforts for satisfying the food demands of a growing and urbanized world population. In this regard, agricultural practices are basic to fulfill the future agricultural demands of the world (Altieri 2004). Some of the most common techniques include crops rotation, planting cover crops, reduction or elimination of tillage, application of integrated pest management, integration of livestock and crops and adoption of agroforestry practices. Besides traditional solutions, a recommended approach is the use of beneficial native soil microbes as biocontrol agents, which is a strategic technique for the achievement of sustainable agricultural production. These microbes play fundamental role to produce healthy and sustainable crops and preserve the biosphere. They improve the plant nutrition and health and improve the quality of soil (Lugtenberg 2015).

## 14.5 Plant-Associated Soil Microbiome and Sustainable Agriculture

A goal of sustainability is to look for competent methods of nutrient recycling, pathogens and pest control and the alleviation of negative effects of abiotic stress factors. Microbes offer many services to plants and soil (Ray et al. 2016; Zolla et al.

2013). Various types of organisms are found in the soil microbiome, but among those bacteria and fungi received much attention (Spence and Bais 2013). Several factors affect the recruitment and maintenance of the microbiome in the rhizosphere (a thin layer of soil encircling the plant roots). Plant morphology as well as the rhizodeposition of the plant give rise to distinct microbial population (Rosier et al. 2016). Along with other environmental factors, the chief traits of soil such as micro- and macronutrients, pH, redox and water potential, play a vital role in shaping composition and activity of microbiome. Most of these microbes are found in the rhizosphere and few of them reside inside the plant tissues, known as endophytes (Mercado-Blanco 2015). About 90–100 million bacteria and 200,000 fungi are usually found in a gram of soil. The exudates of plant roots containing a large amount of photosynthetically fixed carbon is the one of the major sources of attraction for these microbes. The relationship between plants and microbes may be beneficial, harmful, or neutral for the plants. Rhizobacteria and fungi that play beneficial role for plant growth and development are known as plant growth promoting rhizobacteria (PGPR) and plant growth promoting fungi (PGPF), respectively. Presently, based on successful interaction of PGPR and PGPF with plants, scientists are now utilizing these in agriculture, horticulture and for other environment services (Gamalero and Glick 2011). Hence understanding the diversity in structure and key function of the microbiome can play important role in sustainability of agriculture.

## 14.6 Role of Soil Microbes for Stress Tolerance in Crops

Meta-analysis of the plants under various stresses simultaneously show a complex regulation of plant growth and immunity. It is fundamental to understand the phytohormonal interaction in the signaling network, in order to understand the survival of plant microbiome systems under stressed conditions, which will be useful for designing biotechnological strategies to optimize the plants stress adaptive mechanisms and to improve microbes' ability to ameliorate stress situations in crops (Pozo et al. 2015). Although under stress situations the mechanisms involved in plant–microbe interactions are not well understood. However, various ongoing researches are showing evidences of the microbes induced alterations in the plant excreted root exudates, activities of transporters and in the plant morphology and physiology, changes that enable plants to recruit stress alleviating microbes, a strategy that helps in enhancing crop yield under stress environments (Zolla et al. 2013). As the environmental stress factors badly affect the agriculture productions, so rhizosphere microbes play a vital role in assisting plants to survive under such harmful conditions (Barea et al. 2013). Applications of the stress-tolerant PGPR and Arbuscular mycorrhiza (AM) fungi help in plant growth and survival under severe conditions (Nadeem et al. 2014). During stress, various direct and indirect mechanisms are utilized by these microbes to combat the adverse effects (Ahmad et al. 2019). Under unfavorable circumstances, microbes use various biochemical and molecular mechanisms to promote the plant growth and development. For instance, PGPR regulate hormonal and nutritional balance, produce phytohormones, induce systemic resistance against phytopathogens, produce certain metabolites like siderophores that help in reducing the phytopathogens population around the plants (Spence and Bais 2015; Złoch et al. 2016). Similarly, they can also fix atmospheric nitrogen and solubilize phosphate that facilitate plant growth (Ahmad et al. 2011). In addition, mobilization of nutrients and production of exopolysaccharide and rhizobitoxine, etc. are other mechanisms used by the PGPR (Vardharajula et al. 2011). Rhizobitoxine by inhibiting the ethylene production helps the plant to cope stress situation (Kumar et al. 2009). Besides certain key enzymes such as ACC deaminase, glucanase and chitinase are also produced by these microbes to overcome adverse conditions (Farooq et al. 2009b). Some bacteria also possess sigma factors to alter the gene expression in order to overcome negative effects of stress situations (Gupta et al. 2013). Besides PGPR, fungi also play an important role in the growth and development of plants. Arbuscular mycorrhiza is the most widespread mycorrhizal association present in the agriculture field, which plays an important role in nutrient cycling. So, by utilizing these various mechanisms, microbes assist the plants in maintaining their original growth under the stress conditions as shown in Fig. 14.1. As PGPR are the potential alternate of fertilizers and pesticides, so they are essential for sustainable agriculture production and to deal with future food security concerns. Use of the stress-tolerant microbes in the sustainable agriculture practices can help in enhancing the nutritional values and yields of food grains under the changing environment and can save 20–25% cost of the chemical fertilizers and pesticides. Further utilization of these practices can also assist the farmers in enhancing their financial income by producing organic foods and vegetables. We will focus on the role of PGPR and PGPF in plant growth promotion and protection (Fig. 14.1).

## 14.7 Plant Growth Promoting Rhizobacteria

Among the plant-associated microbiome, PGPR are the potential microbes having ability to colonize the plant roots and stimulate plant growth (Goswami et al. 2016). These soil bacteria belong to various genera such as *Pseudomonas*, *Bacillus*, *Rhizobium*, *Azospirillum Azotobacter*, *Azoarcus*, *Clostridium*, *Enterobacter*, and *Serratia* (Arora et al. 2015). PGPR are also potential microbes that provide plant protection against various environmental stresses (Kang et al. 2014). The effective role of PGPR in sustainable agriculture has been described by an ample literature (Ahemad and Kibret 2014; Arora and Mishra 2016). Initially PGPR's role was explored only for the enhancement of crop productivity, but according to several recent studies they are also important for the proper functioning of agroecosystems (Cheng 2009). PGPR can assist in restoring the degraded land, reducing the environmental pollutants from the soil and improving the soil quality and combating changes in the climate (Kuiper et al. 2004) by a variety of mechanisms.

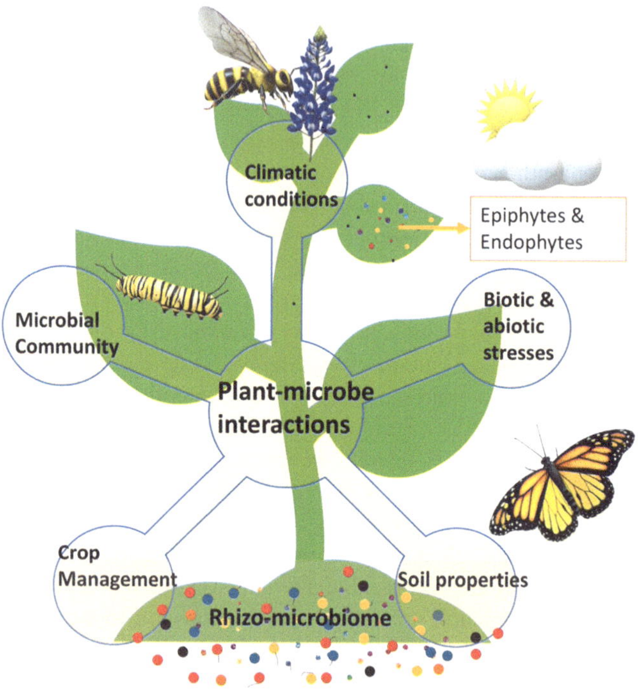


**Fig. 14.1** Role of soilborne microbes in plant growth promotion and protection against stresses. PGPR, PGPF and other rhizospheric and endophytic soilborne microbes with a diverse host range help plants in combating several environmental stresses

### 14.7.1 Mechanisms of Plant–PGPR Interactions

The in-depth understanding of the interactions at the soil–root nexus will help in unlocking the potential of the microbes for agriculture. During recent past, extensive research has been conducted to reveal the mechanisms of plant–microbe interactions (Beneduzi et al. 2012). PGPR usually facilitate the plant growth either directly or indirectly as shown in Fig. 14.2.

### 14.7.2 Direct Mechanisms

As the availability of the mineral nutrients in the rhizosphere is important for the plant growth, the improvement of mineral nutrition in plants by PGPR is one of the best mechanisms studied among the direct mechanisms (Pii et al. 2015b). Biological nitrogen fixation, siderophores production, phosphorus solubilization, and the manipulation of biochemical and molecular pathways for nutrient acquisition are

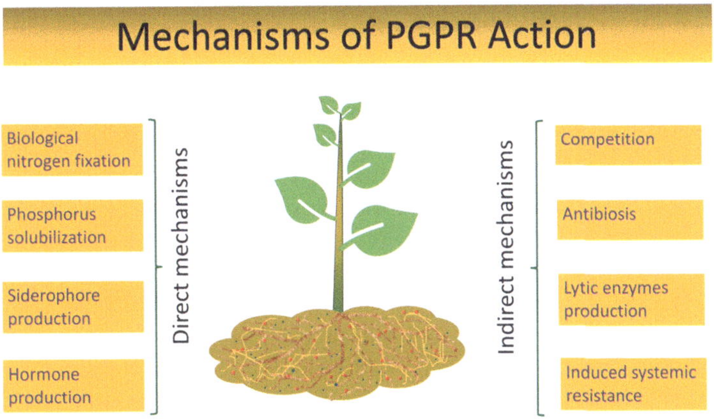


**Fig. 14.2** Mechanisms of plant growth promotion by PGPR. PGPR can promote growth and their mutualistic or symbiotic plant hosts from different biotic and abiotic stresses by multimechanistic approach

the various ways utilized by PGPR for mineral nutrients acquisition (Terrazas et al. 2016).

#### 14.7.2.1 Biological Nitrogen Fixation

Nitrogen (N) is the most vital element for plant growth and development. Although 78% of N is available in atmosphere, plants are unable to use this molecular nitrogen directly. Biological nitrogen fixation helps in providing this important nutrient to plants in a usable form. In this process, N is converted into ammonia ($NH_3$) using ATP, which is catalyzed by the nitrogenase enzyme. Various microbial genera such as *Psudomonas*, *Bacillus*, *Azospirillum*, *Erwinia*, and *Rhizobium* can fix atmospheric nitrogen (Silva et al. 2016). Both symbiotic and nonsymbiotic bacteria can fix the atmospheric nitrogen (Bhat et al. 2015). The nonsymbiotic bacteria include *diazotrophicus*, *Azospirillum*, *Azotobacter*, and *cyanobacteria*. The symbiotic nitrogen fixation is carried out by the members of *Rhizobia* genera through forming symbiotic association with leguminous plants and most of the atmospheric nitrogen is fixed through this process. However free-living *diazotrophs* contribute very less for the bioavailability of N to the plants (Jones et al. 2007). Biological nitrogen fixation is economically viable and environmentally safe alternative of chemical fertilizers. Some of these bacteria-forming root nodules also play an important role in interaction between plants and bacteria (Yan et al. 2008). Different abiotic stresses like salinity and drought inhibit the root nodulation. Nitrogenase enzyme is very sensitive to salt stress and its activity is severely hampered under salt stress condition, which ultimately inhibits nitrogen fixation. PGPR help plants in combating these stress conditions and offer great services to promote and protect plant growth.

#### 14.7.2.2 Solubilization of Phosphorus

Like nitrogen, phosphorus (P) is also an essential nutrient important for plant growth and development. It is important for root growth and for various physiological processes of plants. Plants usually acquire P from the soil in the form of phosphate ions. Phosphorus in the soil is present in excess both in the organic and inorganic forms but less than 1% is readily available for plants (Terrazas et al. 2016). So, in order to ensure the availability of immobilized P to plants, solubilization of inorganic forms and the enzymatic mineralization of organic P is required (Gerke 2015). PGPR make it available to plants through mineralization and solubilization. The plant–microbe association in soil helps in enhancing the mobility of phosphorus and its availability to plants. The phosphate solubilizing rhizobacteria are very common in nature and their numbers differ from soil to soil. They govern biogeochemical cycle in natural agriculture system and usually mobilize phosphate by producing organic acids and phosphatase. The phosphate solubilization may result either due to decrease in pH or due to cation chelation. Various genera of PGPR and fungi are able to solubilize phosphate (Yadav et al. 2014). Most of the plant-associated PGPR improve phosphorus uptake in plants from soil under the P deficient condition possibly through production of organic acids and phosphatases and by lowering pH. These organic acids compete for binding sites on soil and make the availability of P to plants. The organic acids decrease the surrounding pH and release phosphate from the $H^+$ ions. Among the various organic acids, 2-ketogluconic acid is the effective acid. Moreover, cotransport of protons along with exudation process acidify the external medium that promotes acid solubilization of the immobilized P (Terrazas et al. 2016). The phosphatase enzymes of PGPR assist in mineralizing the organic forms of P and releasing orthophosphate groups in soil (Azeem et al. 2015). Several PGPR are reported to have ability of phosphate solubilization, so their use in agriculture may be cost-effective and sustainable.

#### 14.7.2.3 Production of Siderophores

Iron is a micronutrient vital for plant growth playing crucial role in photosynthesis, respiration, DNA synthesis, and nitrogen fixation. Iron in nature exists in ferrous Fe (II) and ferric Fe (III) oxidation states. Under aerobic environment, iron is mostly present as ferric ($Fe^{3+}$) form which has very less solubility due to formation of insoluble hydroxides and oxyhydroxides that are not available to both plants and microbes (Mimmo et al. 2014). Microbes usually obtain iron by synthesizing and secreting low molecular weight organic compounds usually known as microbial siderophores (Lemanceau et al. 2009), that have high affinity for ferric form of iron (Mishra et al. 2011). These siderophores form stable complexes with $Fe^{3+}$ which through ABC transporters are transported into the microbial cell (Braun and Hantke 2011; Hider and Kong 2010) where they are reduced from $Fe^{3+}$ to soluble $Fe^{2+}$ form (Rajkumar et al. 2010). Siderophores are water soluble and are usually divided into intracellular and extracellular siderophores. Plants mostly acquire iron through microbial siderophores using various means either by direct uptake or by ligand exchange reaction that helps in enhancing chlorophyll contents of plant leaves. Application of the siderophores producing PGPR in iron deficient plants helps in alleviating the Fe deficiency, showing the possible role of these microbes in assisting the plant roots for nutrients acquisition by making them available to soil (Pii et al. 2015a).

#### 14.7.2.4 Modulation of Biochemical and Molecular Mechanisms

Although plants may have ability of enhancing the mineral uptake by themselves but PGPR basically enhance the availability of the key nutrients in the rhizosphere, which improves the mineral nutrition in the PGPR-associated plants. ATPases in plasma membrane are responsible for the generation and maintenance of $H^+$ electrochemical gradients across the transmembrane, that is necessary for the transport of numerous nutrients such as phosphate $PO_4^{2-}$, sulfate $SO_4^{2-}$ and nitrate $NO_3^-$. Hence PGPR improve the mineral uptake by enhancing $H^+$ extrusion across the membrane. Concerning the N nutrition, PGPR application induces the expression of various nitrate transporter genes (Kechid et al. 2013). Similarly, PGPR application also enhance the root iron chelate reductase activity that is involved in reducing $Fe^{3+}$ to soluble $Fe^{2+}$ prior to uptake by transporters (Zhao et al. 2014).

#### 14.7.2.5 Production of Phytohormones

Phytohormones play a key role in plant development and growth (Glick et al. 2010) and their activity is mainly associated with the plasticity of the root system (Kloepper et al. 2007). PGPR can also produce auxin, cytokinin, and gibberellins that influence the root architecture (Vacheron et al. 2013). Many PGPR have the ability to synthesize and secrete indole-3-acetic acid (IAA, auxin) as secondary metabolites (Scagliola et al. 2016). Bacterial- and plant-derived IAA play an additive role by enhancing the root length and surface area to provide better access to soil nutrients (Glick 2012). Additionally, IAA have been found to induce the expression of the key genes both in plants ad microbes which are required for the better establishment of efficient plant–PGPR interactions (Spaepen and Vanderleyden 2011). PGPR producing 1-aminocyclopropane-1-carboxylate (ACC) deaminase are able to cleave the ethylene precursor (ACC) into 2-oxobutanoate and ammonia and hence facilitate the plant growth improvement by reducing the stress ethylene level (Glick 2014). Under stress, ethylene is produced in two peaks, the first one being much smaller is believed to activate the defense mechanisms whereas the second peak has adverse effects on plant growth. PGPR having ACC deaminase activity act as a sink for ACC, thus mitigating the negative effects of ethylene on plant growth (Glick 2014). As a result, plants applied with ACC deaminase PGPR show better associations with microbes, enhance nutrients uptake and ultimately improve the plant growth (Nadeem et al. 2014).

### 14.7.3 *Indirect Mechanisms*

Indirect mechanisms have basic role in biocontrol and based on the features of the mechanisms, can be further be distinguished into three groups: (1) competition, (2) antibiotic and lytic enzyme production, and (3) induced systemic resistance (Beneduzi et al. 2012).

#### 14.7.3.1 Competition

PGPR compete with pathogenic microbes at two different levels: (1) competition for essential nutrients and (2) competition for niches. Direct competition among PGPR and pathogens help in reducing the incidence and severity of the diseases. Similarly, production of microbial siderophores represents an effective biocontrol mechanism, which due to higher affinity for Fe, reduces its availability to phytopathogens for their proliferation (Lugtenberg and Kamilova 2009).

#### 14.7.3.2 Antibiotic and Lytic Enzymes Production

2,4-Diacetylphloroglucinol, pyrrolnotrin, zwittermicin A, phenazine, tensin, and xanthobaccin are some of the antibiotics produced by the biocontrol PGPR, especially preventing the proliferation of pathogenic fungi (Mazurier et al. 2009). Depending on the induction by various environmental stimuli, more than one kind of antibiotics might be produced by each PGPR strain (Duffy and Défago 1999). Many PGPR acting as biocontrol agents have been commercialized in order to avoid the use of pesticides. Similarly, many biocontrol PGPR strains are able to produce lytic enzymes such as proteases, lipases, chitinases, glucanases and cellulases that help in preventing the infection of various fungi by compromising the integrity of their cell walls (Kim et al. 2008).

#### 14.7.3.3 Induced Systemic Resistance

Although plants have many chemical and physical methods to defend against pathogens. Under such circumstances, PGPR colonizing the root surface help plants by activating their defense response against pathogens (Salas-Marina et al. 2011). Plants are usually protected from pathogens by preventing the growth of pathogens through competition for nutrition and space and by reducing symptoms of diseases (Ghazalibiglar et al. 2016). Various mechanisms of defense against biotic stresses are being used by the PGPR. Systemic acquired resistance (SAR) and induced systemic resistance (ISR) are the two kinds of resistances induced against pathogens. It is basically a physiological state of enhanced defensive capacity elicited by various environmental stimuli, priming innate defenses of the plants which react faster against the future pathogenic attack (Nawrocka and Małolepsza 2013). For the induction of resistance, both SAR and ISR use different signaling mechanisms. ISR consists of jasmonic acid (JA) and ethylene pathways while in SAR, salicylic acid (SA) and pathogenesis-related (PR) proteins accumulate (Dimkpa et al. 2009). PR proteins act synergistically with antioxidant enzymes and provide higher degree of protection against pathogens. SA plays a role in systemic development which acquire resistance. In addition, nonpathogenic bacteria also stimulate ISR in plants (Wani et al. 2016). They enhance the level of peroxidase and chitinase enzymes in plant leaves and roots. Moreover, induced resistance is systemic because in addition to primary infection site, defensive capacity also spread the whole plant, which is mostly due to JA and ET pathways, stimulating the plant defense mechanisms against variety of pathogens (Verhagen et al. 2004). The efficacy of SAR and ISR is widespread and includes bacterial, viral, and fungal pathogens. Their effects vary from plant to plant and also depend on the level of PGPR–plant interactions. Activation of defense system enhances the activity of antioxidant enzymes such as superoxide dismutase, peroxidase, catalase and guaiacol etc. which provide protection against oxidative damage caused by reactive oxygen species (ROS) and pathogen (Ghazalibiglar et al. 2016). Various bacterial components such as flagellar proteins, b-glucans, chitin and lipopeptides also stimulate ISR inside the plants (Annapurna et al. 2013). Various changes in plants are induced due to ISR like alteration in the cell wall, change in the level of chitinase, phenylalanine ammonia lyase and expression of stress-related genes (Choudhary and Johri 2009). *Pseudomonas* and *Bacillus* are well known for their induced systemic and antagonistic effects. Although plants have many chemical and physical methods, use of these PGPR is very cost-effective, beneficial, and sustainable approach for providing protection to plants against pathogens.

## 14.8 Plant Growth Promoting Fungi

### *14.8.1 Arbuscular Mycorrhizal Fungi*

Arbuscular Mycorrhizal Fungi (AMF) establish symbiotic relationship with a variety of plant species by forming fungal hyphal network and specialized structures called arbuscules. AMF improve plant growth by accruing nutrient absorption that significantly affect the soil health status specifically under stress conditions. AMF can help plants to withstand a variety of biotic and abiotic stresses. One of the important factors for a better crop production is the availability and absorption of nutrients. Under limited nutrients conditions, plants establish symbiotic relationship with AMF to access and utilize the available nutrient resources efficiently for better growth, photosynthetic efficiency, and enzyme activity. AMF-inoculated plants show increased accumulation and absorption of mineral nutrients by improved nutrient availability, uptake, and assimilation (Hashem et al. 2015). This effect may be due to the hyphal networks that interact with roots under the soil (Muthukumar et al. 2014) and may change the root architecture to ameliorate nutrient deficiency. One of the important and growth limiting environmental stress is water. It can hamper various metabolic functions such as nutrient uptake, nutrient assimilation, protein synthesis, and photosynthesis (Ahmad et al. 2019; Nazar et al. 2015). Water stress accompanied with heat stress can lead to extreme drought stress thereby limiting the plant's ability to grow (Ahanger and Agarwal 2017). Under drought stress, plants produce reactive oxygen species (ROS) that are scavenged by different antioxidant enzymes to minimize the stress symptoms (Hashem et al. 2015; Wu and Zou 2017). However, symbiotic relationship of AMF with roots can offer great services to withstand the water stress by a better uptake of nutrients, osmolyte production, and scavenging of ROS (Hameed et al. 2014; Yooyongwech et al. 2013). AMF can activate the antioxidant system to scavenge free radicals to maintain cellular redox levels (Wu and Zou 2017). Temperature stress (high and low) can halt various plant developmental and metabolic processes thus causing adverse effects on global agricultural productivity as it is often accompanied with drought, salinity, and mineral stresses (Jedmowski et al. 2015; Machado and Paulsen 2001). Temperature stress can damage fruits, cause delays in germination, reduce growth, biomass accumulation and photosynthetic efficiency (Adam and Murthy 2014; Farooq et al. 2009a; Jedmowski et al. 2015; Liu et al. 2013; Paredes and Quiles 2015). AMF inoculation increased the plant growth and productivity by increasing photosynthetic efficiency (Xu et al. 2016; Zhu et al. 2012), accumulation of proline and sugars (Evelin et al. 2009; Xu et al. 2016) and improved water-use efficiency (Elhindi et al. 2017). Water and temperature stresses are usually accompanied with salinity stress that is also a major abiotic stress that can severely affect plant growth and development (Khan et al. 2014). High salinity can affect ionic homeostasis, enzyme activity and nutrient acquisition by disrupting the cellular redox level (Iqbal et al. 2015; Khan et al. 2014). AMF play an important role in improving plant growth, biomass, productivity, and photosynthetic activity under saline conditions (Alqarawi et al. 2014; Aroca et al. 2013; Elhindi et al. 2017; Evelin et al. 2009; Latef and Chaoxing 2014). More interestingly, AMF-inoculated plants showed reduced salt-triggered oxidative damage by activating cellular antioxidant and osmoprotectant defense system (Alqarawi et al. 2014; Scagel et al. 2017; Yang et al. 2014). Soil accumulation of toxic heavy metals can affect the crop growth and productivity. Once present in soil, these heavy metals such as lead, arsenic, cadmium, and mercury can transfer easily from soil into plant tissues and can potentially cause several health risks in humans and animals. However, certain plant species can accumulate heavy metals and expel them by volatilization. AMF are found under very high saline conditions in a variety of soils (Ahanger et al. 2014; Hameed et al. 2014). Some plants in association with AMF can help in the phytoremediation of heavy metals from agricultural soil thus can reduce the associated health risks by phytoextraction and phytostabilization (Latef et al. 2016). AMF have very beneficial effects in stress amelioration and activation of key defense mechanisms. However, extensive research is needed to understand the exact mechanism of amelioration by AMF–plant interactions.

### *14.8.2 Entomopathogenic Fungi*

Entomopathogenic fungi or insect pathogenic fungi have been studied very well as microbial control agents in the field and greenhouse experiments (Ahmad et al. 2020; Vega 2018; Vega et al. 2009). When a spore encounters the insect body, it develops infection structures that penetrate the cuticle, cause intoxication and ultimately kill the insect. The killed insect sporulates and produces millions of spores that can start a new infection cycle in a host. Scientists have focused more on insect–fungal interactions than the biology and ecology of plant–fungal interactions. In addition to their negative effects on insect pests, various entomopathogenic fungi have been reported as plant growth promotors and protection agents through rhizospheric and endophytic colonization (Jaber and Enkerli 2016, 2017; Lacey et al. 2015; Vega 2018). Endophytes are the beneficial or mutualistic microbes that reside within plant tissues without causing any harmful symptoms. Research has mainly focused on the rhizospheric colonizers and have not tapped much about the endophytes that is emerging as a great area of research. More than 170 bioformulations have been developed from almost 12 fungal species as inundative biopesticides against insects, mites, and ticks (de Faria and Wraight 2007). Relatively, less research has focused on the potential of fungal entomopathogens such as *Verticillium*, *Metarhizium*, *Beauveria*, *Paecilomyces* etc. as endophytic agents for plant protection and growth promotion (Akutse et al. 2013; Gurulingappa et al. 2010; Lopez et al. 2014). *Beauveria* belongs to the group of entomopathogens that can cause infection in a wide range of insect and plant hosts (Arnold and Lewis 2005; de Faria and Wraight 2007). *Metarhizium* spp. colonize many plant species, including switchgrass, haricot bean, tomato, wheat, and soybean (Akello and Sikora 2012; Behie et al. 2012, 2015; de Faria and Wraight 2007; Elena et al. 2011; Greenfield et al. 2016; Sasan and Bidochka 2012). Another important aspect of fungal entomopathogens, such as *Metarhizium* spp., is their role as nutrient translocator from a killed insect pest to the host plant (Behie et al. 2012, 2017). Moreover, fungal entomopathogens can confer plant protection against plant pathogens (Akello and Sikora 2012; Akutse et al. 2013; Jaber 2015; Lopez et al. 2014; Mantzoukas et al. 2015; Vidal and Jaber 2015; Zhou et al. 2016). Microbial consortia can be facilitated in nature to achieve long-term ecosystem services to benefit plant productivity and sustainable agriculture. In order to achieve better biological control and plant growth promotion, the use of compatible microbes as consortia has gained a great attention. When used as consortia, fungal entomopathogens and other beneficial microbes, the bioformulations helped plants synergistically and effectively in stress management (Shrivastava et al. 2015). These ecological roles provide opportunities for the multiple use of entomopathogenic fungi in integrated pest management (IPM) strategies for sustainability in agriculture. We need more research to unravel the mechanisms of action of the plant growth promoting and pest suppressive effects of entomopathogenic fungi in soil and their short- and long-term multitrophic interactions. Understanding such interactions will provide answers to complex but important ecological links and molecular cross-talks and may facilitate more exploration and exploitation of biocontrol agents to achieve our sustainability goals in natural and managed agroecosystems. We also need a concerted approach to explore the potential of entomopathogenic fungi in establishing systemic or localized endophytic colonization in naturally occurring biota for improved plant productivity and protection. Future research on such entomopathogenic fungi as endophytes, biocontrol agents, plant growth promotors, and disease antagonists could lead to understand in depth their ecological niches, improved production and formulations and better pest management.

### 14.8.3 *Trichoderma*

*Trichoderma* spp. have been used to control and manage plant diseases and plant growth. Historically, *Trichoderma* spp. have been known as an effective biocontrol agent of several phytopathogens. It is ubiquitously available in soil and can be easily isolated and cultured to be used for sustainable agriculture. *Trichoderma* spp. have the potential to control many plant diseases including sheath blight (de França et al. 2015), Fusarium wilt (Al-Ani 2017), bacterial wilt (Yuan et al. 2016), downy mildew (Perazzolli et al. 2012), root-knot nematode (Al-Hazmi and Tariq 2016) etc. *Trichoderma* spp. employ an array of mechanisms that are very useful to improve plant growth, mineral nutrients assimilation, secondary metabolites production, plant defense modulation, and siderophore production. Such mechanisms of action establish *Trichoderma* spp. as suitable candidates for biocontrol of many phytopathogens and can be formulated as biopesticides. The mode of action of *Trichoderma* can be either mycoparasitism or non-mycoparasitism. *Trichoderma* spp. act against fungal phytopathogens mycoparasitically (Elad et al. 1982) that involves development of infection structures, penetration and subsequently killing the opponent fungus thus offering direct biocontrol. The mycoparasitic mechanism of *Trichoderma* spp. is species dependent and is due to the production of cell wall degrading enzymes (Harman et al. 2004; Sivan and Chet 1989). Some *Trichoderma* spp. are not mycoparasites of other fungi and the mechanism of action can be antibiosis, competition and by mediating plant defenses against phytopathogen (Howell 2003).

### 14.8.4 Other Plant Growth Promoting Fungi

Endophytes from the genus *Epichloë* (Clavicipitaceae: Ascomycota) can establish symbiotic relationships with above ground parts of certain grass species and can be transmitted vertically (Gundel et al. 2011; Schardl 2010). The proportion or prevalence of endophytic colonization depends upon the fitness of host plant and potential of vertical transmission to progeny (Gundel et al. 2011). *Epichloë* spp. protect plants from herbivores by producing defensive compounds such as alkaloids (Saikkonen et al. 2013; Schardl 2010) and get photosynthetic products, protection, nutrients, and reproduction mechanisms (vertical transmission) in return. They can also enhance plant growth, photosynthetic efficiency and tolerance to biotic and abiotic stresses in host (Bao et al. 2015; Saikkonen et al. 2013). The profile of alkaloids depends upon the fungal species and strains, plant species and the environmental conditions (Ryan et al. 2015; Saikkonen et al. 2013). *Epichloë* spp. have been found to modulate SA signaling pathway in plants and protect them from pathogens (Wiewióra et al. 2015).

*Fusarium* is a genus that is composed of cosmopolitan species with a wide range of host plants. Many *Fusarium* spp. are the causative agents of various diseases in plants whereas some species have been reported as plant growth promoting agents with a wide range of host plants to colonize endophytically (Imazaki and Kadota 2015). They can produce secondary metabolites that are the drivers of steering their biology and ecology (Bills and Gloer 2016; Kaul et al. 2016; Stępień et al. 2018). *Fusarium* spp. can assist the host plant from insects and pathogens (Ji et al. 2009). The potential of *Fusarium* spp. as a symbiont needs a plenty of research to fill the knowledge gaps. There are many other plant-associated epiphytic and endophytic fungi that have the potential to affect the herbivores negatively and plant growth positively and can be promising microbial agents for soil health and sustainable agriculture.

## 14.9 Concluding Remarks and Future Perspective

Plants have coevolved multifaceted direct (production of toxins) and indirect mechanisms to defend themselves against a multitude of biotic and abiotic stresses. To achieve maximum benefits, plants do partnership with a variety of different beneficial microbes (e.g., bacteria and fungi) through mutualism or symbiosis. Microbes mediate plant defense signaling by modulating and cross-talking and fine-tuning their phytohormone (JA, SA, etc.) signaling pathways in the presence of stresses (Pieterse et al. 2012). These defense mechanisms are mostly associated with costs that can compromise the growth and reproduction. Beneficial microbes can efficiently antagonize stresses and improve plant growth and fitness (Pineda et al. 2017). In order to exploit the maximum benefits associated with beneficial microbes, we need to understand the ecology and biology of the microbes associated with plants for a sustainable agriculture. The ecological roles of plant-associated microbes remains elusive and needs deeper digging. We should promote the strategies to conserve the beneficial microbes among growers for environmentally friendly and sustainable agriculture while reducing input of chemical fertilizers, pesticides, and fungicides.

**Conflict of Interests** Authors declare no conflict of interests.

## References

Adam S, Murthy SDS (2014) Effect of cold stress on photosynthesis of plants and possible protection mechanisms. In: Gaur R, Sharma P (eds) Approaches to plant stress and their anagement. Springer, New Delhi, pp 219–226

Ahanger MA, Agarwal R (2017) Potassium up-regulates antioxidant metabolism and alleviates growth inhibition under water and osmotic stress in wheat (*Triticum aestivum* L). Protoplasma 254:1471–1486

Ahanger MA, Hashem A, Abd-Allah EF et al (2014) Arbuscular mycorrhiza in crop improvement under environmental stress. In: Ahmad P, Rasool S (eds) Emerging technologies and management of crop stress tolerance. Elsevier, New York, pp 69–95

Ahemad M, Kibret M (2014) Mechanisms and applications of plant growth promoting rhizobacteria: current perspective. J King Saud Univ Sci 26:1–20

Ahmad M, Zahir ZA, Asghar HN et al (2011) Inducing salt tolerance in mung bean through coinoculation with rhizobia and plant-growth-promoting rhizobacteria containing 1-aminocyclopropane-1-carboxylate deaminase. Can J Microbiol 57:578–589

Ahmad I, Zaib S, Alves PCMS et al (2019) Molecular and physiological analysis of drought stress responses in *Zea mays* treated with plant growth promoting rhizobacteria. Biol Plant 63:536–547

Ahmad I, del Mar Jiménez-Gasco M, Luthe DS, Shakeel SN, Barbercheck ME (2020) Endophytic Metarhizium robertsii promotes maize growth, suppresses insect growth, and alters plant defense gene expression. Biol Control, 104167

Akello J, Sikora R (2012) Systemic acropedal influence of endophyte seed treatment on *Acyrthosiphon pisum* and *Aphis fabae* offspring development and reproductive fitness. Biol Control 61:215–221

Akutse KS, Maniania NK, Fiaboe KKM et al (2013) Endophytic colonization of *Vicia faba* and *Phaseolus vulgaris* (Fabaceae) by fungal pathogens and their effects on the life-history parameters of *Liriomyza huidobrensis* (Diptera: Agromyzidae). Fungal Ecol 6:293–301

Al-Ani L (2017) Potential of utilizing biological and chemical agents in the control of Fusarium wilt of banana. PhD School of Biology Science, Universiti Sains Malaysia Pulau, Pinang, Malaysia 259

Al-Hazmi AS, Tariq JM (2016) Effects of different inoculum densities of *Trichoderma harzianum* and *Trichoderma viride* against *Meloidogyne javanica* on tomato. Saudi J Biol Sci 23:288–292

Alqarawi A, Hashem A, Abd-Allah E et al (2014) Effect of salinity on moisture content, pigment system, and lipid composition in *Ephedra alata Decne*. Acta Biol Hung 65:61–71

Altieri MA (2004) Linking ecologists and traditional farmers in the search for sustainable agriculture. Front Ecol Environ 2:35–42

Annapurna K, Kumar A, Kumar LV et al (2013) PGPR-induced systemic resistance (ISR) in plant disease management. In: Bacteria in agrobiology: disease management. Springer, Berlin, pp 405–425

Arnold AE, Lewis LC (2005) Ecology and evolution of fungal endophytes, and their roles against insects. In: Vega FE, Blackwell M (eds) Insect-fungal associations: ecology and evolution. Oxford University Press, New York, pp 74–96

Aroca R, Ruiz-Lozano JM, Zamarreño ÁM et al (2013) Arbuscular mycorrhizal symbiosis influences strigolactone production under salinity and alleviates salt stress in lettuce plants. J Plant Physiol 170:47–55

Arora NK, Mishra J (2016) Prospecting the roles of metabolites and additives in future bioformulations for sustainable agriculture. Appl Soil Ecol 107:405–407

Arora R, Behera S, Kumar S (2015) Bioprospecting thermophilic/thermotolerant microbes for production of lignocellulosic ethanol: a future perspective. Renew Sust Energ Rev 51:699–717

Azeem M, Riaz A, Chaudhary AN et al (2015) Microbial phytase activity and their role in organic P mineralization. Arch Agron Soil Sci 61:751–766

Bao G, Saikkonen K, Wang H et al (2015) Does endophyte symbiosis resist allelopathic effects of an invasive plant in degraded grassland? Fungal Ecol 17:114–125

Barea J, Pozo M, López-Ráez J et al (2013) Arbuscular mycorrhizas and their significance in promoting soil-plant systems sustainability against environmental stresses. In: González MB, Gonzalez-López J (eds) Beneficial plant-microbial interactions: ecology and applications. CRC Press, New York, pp 353–387

Behie SW, Zelisko PM, Bidochka MJ (2012) Endophytic insect-parasitic fungi translocate nitrogen directly from insects to plants. Science 336:1576–1577

Behie SW, Jones SJ, Bidochka MJ (2015) Plant tissue localization of the endophytic insect pathogenic fungi *Metarhizium* and *Beauveria*. Fungal Ecol 13:112–119

Behie SW, Moreira CC, Sementchoukova I et al (2017) Carbon translocation from a plant to an insect-pathogenic endophytic fungus. Nat Commun 8:14245

Beneduzi A, Ambrosini A, Passaglia LM (2012) Plant growth-promoting rhizobacteria (PGPR): their potential as antagonists and biocontrol agents. Genet Mol Biol 35:1044–1051

Bhat TA, Ahmad L, Ganai MA et al (2015) Nitrogen fixing biofertilizers; mechanism and growth promotion: a review. J Pure Appl Microbiol 9:1675–1690

Bills GF, Gloer JB (2016) Biologically active secondary metabolites from the fungi. Microbiol Spectr 4:6

Braun V, Hantke K (2011) Recent insights into iron import by bacteria. Curr Opin Chem Biol 15:328–334

Browne P, Barret M, Morrissey JP et al (2013) Molecular based strategies to exploit the inorganic phosphate solubilization ability of *Pseudomonas* in sustainable agriculture. Mol Microbial Ecol Rhizosphere 1:615–628

Cheng W (2009) Rhizosphere priming effect: its functional relationships with microbial turnover, evapotranspiration, and C–N budgets. Soil Biol Biochem 41:1795–1801

Choudhary DK, Johri BN (2009) Interactions of *Bacillus* spp. and plants–with special reference to induced systemic resistance (ISR). Microbiol Res 164:493–513

de Faria MR, Wraight SP (2007) Mycoinsecticides and mycoacaricides: a comprehensive list with worldwide coverage and international classification of formulation types. Biol Control 43:237–256

de França SKS, Cardoso AF, Lustosa DC et al (2015) Biocontrol of sheath blight by *Trichoderma asperellum* in tropical lowland rice. Agron Sustain Dev 35:317–324

Dimkpa C, Weinand T, Asch F (2009) Plant–rhizobacteria interactions alleviate abiotic stress conditions. Plant Cell Environ 32:1682–1694

Duffy BK, Défago G (1999) Environmental factors modulating antibiotic and siderophore biosynthesis by *Pseudomonas fluorescens* biocontrol strains. Appl Environ Microbiol 65:2429–2438

Elad Y, Chet I, Henis Y (1982) Degradation of plant pathogenic fungi by *Trichoderma harzianum*. Can J Microbiol 28:719–725

Elena GJ, Beatriz PJ, Alejandro P et al (2011) *Metarhizium anisopliae* (Metschnikoff) Sorokin promotes growth and has endophytic activity in tomato plants. Adv Biol Res 5:22–27

Elhindi KM, El-Din AS, Elgorban AM (2017) The impact of arbuscular mycorrhizal fungi in mitigating salt-induced adverse effects in sweet basil (*Ocimum basilicum* L.). Saudi J Biol Sci 24:170–179

Evelin H, Kapoor R, Giri B (2009) Arbuscular mycorrhizal fungi in alleviation of salt stress: a review. Ann Bot 104:1263–1280

Farooq M, Aziz T, Wahid A et al (2009a) Chilling tolerance in maize: agronomic and physiological approaches. Crop Pasture Sci 60:501–516

Farooq M, Wahid A, Kobayashi N et al (2009b) Plant drought stress: effects, mechanisms and management. Agron Sustain Dev 29:185–212

Gamalero E, Glick BR (2011) Mechanisms used by plant growth-promoting bacteria. In: Maheshwari D (ed) Bacteria in agrobiology: plant nutrient management. Springer, Berlin, pp 17–46

Gerke J (2015) The acquisition of phosphate by higher plants: effect of carboxylate release by the roots. A critical review. J Plant Nutr Soil Sci 178:351–364

Ghazalibiglar H, Hampton JG, de Jong EZ et al (2016) Is induced systemic resistance the mechanism for control of black rot in *Brassica oleracea* by a *Paenibacillus sp.*? Biol Control 92:195–201

Glick BR (2012) Plant growth-promoting bacteria: mechanisms and applications. Scientifica 2012:1–15

Glick BR (2014) Bacteria with ACC deaminase can promote plant growth and help to feed the world. Microbiol Res 169:30–39

Glick BR, Cheng Z, Czarny J et al (2010) Promotion of plant growth by ACC deaminase-producing soil bacteria. In: Bakker PA, Raaijmakers JM, Bloemberg G et al (eds) New perspectives and approaches in plant growth-promoting Rhizobacteria research. Springer, New York, pp 329–339

Goswami D, Thakker JN, Dhandhukia PC (2016) Portraying mechanics of plant growth promoting rhizobacteria (PGPR): a review. Cogent Food Agric 2:1127500

Greenfield M, Gomez-Jimenez MI, Ortiz V et al (2016) *Beauveria bassiana* and *Metarhizium anisopliae* endophytically colonize cassava roots following soil drench inoculation. Biol Control 95:40–48

Gundel P, Rudgers J, Ghersa C (2011) Incorporating the process of vertical transmission into understanding of host–symbiont dynamics. Oikos 120:1121–1128

Gupta G, Panwar J, Jha PN (2013) Natural occurrence of *Pseudomonas aeruginosa*, a dominant cultivable diazotrophic endophytic bacterium colonizing *Pennisetum glaucum* (L.) R. Br. Appl Soil Ecol 64:252–261

Gurulingappa P, Sword GA, Murdoch G et al (2010) Colonization of crop plants by fungal entomopathogens and their effects on two insect pests when in planta. Biol Control 55:34–41

Hameed A, Wu Q-S, Abd-Allah EF et al (2014) Role of AM fungi in alleviating drought stress in plants. In: Miransari M (ed) Use of microbes for the alleviation of soil stresses. Springer, Cham, pp 55–75

Harman GE, Howell CR, Viterbo A et al (2004) *Trichoderma* species—opportunistic, avirulent plant symbionts. Nat Rev Microbiol 2:43–56

Hashem A, Abd-Allah E, Ahmad P (2015) Effect of AM fungi on growth, physio-biochemical attributes, lipid peroxidation, antioxidant enzymes and plant growth regulators in *Lycopersicon esculentum* mill. subjected to different concentration of NaCl. Pak J Bot 47:327–340

Hider RC, Kong X (2010) Chemistry and biology of siderophores. Nat Prod Rep 27:637–657

Howell C (2003) Mechanisms employed by *Trichoderma* species in the biological control of plant diseases: the history and evolution of current concepts. Plant Dis 87:4–10

Imazaki I, Kadota I (2015) Molecular phylogeny and diversity of *Fusarium* endophytes isolated from tomato stems. FEMS Microbiol Ecol 91:fiv098

Iqbal N, Umar S, Khan NA (2015) Nitrogen availability regulates proline and ethylene production and alleviates salinity stress in mustard (*Brassica juncea*). J Plant Physiol 178:84–91

Jaber LR (2015) Grapevine leaf tissue colonization by the fungal entomopathogen *Beauveria bassiana* s.l. and its effect against downy mildew. Biol Control 60:103–112

Jaber LR, Enkerli J (2016) Effect of seed treatment duration on growth and colonization of *Vicia faba* by endophytic *Beauveria bassiana* and *Metarhizium brunneum*. Biol Control 103:187–195

Jaber LR, Enkerli J (2017) Fungal entomopathogens as endophytes: can they promote plant growth? Biocontrol Sci Tech 27:28–41

Jedmowski C, Ashoub A, Momtaz O et al (2015) Impact of drought, heat, and their combination on chlorophyll fluorescence and yield of wild barley (*Hordeum spontaneum*). J Bot 2015:1–9

Ji HF, Li XJ, Zhang HY (2009) Natural products and drug discovery. EMBO Rep 10:194–200

Jones KM, Kobayashi H, Davies BW et al (2007) How rhizobial symbionts invade plants: the Sinorhizobium–Medicago model. Nat Rev Microbiol 5:619–633

Kang S-M, Radhakrishnan R, Khan AL et al (2014) Gibberellin secreting rhizobacterium, *Pseudomonas putida* H-2-3 modulates the hormonal and stress physiology of soybean to improve the plant growth under saline and drought conditions. Plant Physiol Biochem 84:115–124

Kaul S, Sharma T, Dhar KM (2016) "Omics" tools for better understanding the plant–endophyte interactions. Front Plant Sci 7:955

Kechid M, Desbrosses G, Rokhsi W et al (2013) The NRT 2.5 and NRT 2.6 genes are involved in growth promotion of *Arabidopsis* by the plant growth-promoting rhizobacterium (PGPR) strain *Phyllobacterium brassicacearum* STM 196. New Phytol 198:514–524

Khan MIR, Asgher M, Khan NA (2014) Alleviation of salt-induced photosynthesis and growth inhibition by salicylic acid involves glycinebetaine and ethylene in mungbean (*Vigna radiata* L.). Plant Physiol Biochem 80:67–74

Kim YC, Jung H, Kim KY et al (2008) An effective biocontrol bioformulation against *Phytophthora* blight of pepper using growth mixtures of combined chitinolytic bacteria under different field conditions. Eur J Plant Pathol 120:373–382

Kloepper J, Gutierrez-Estrada A, McInroy J (2007) Photoperiod regulates elicitation of growth promotion but not induced resistance by plant growth-promoting rhizobacteria. Can J Microbiol 53:159–167

Kuiper I, Lagendijk EL, Bloemberg GV et al (2004) Rhizoremediation: a beneficial plant-microbe interaction. Mol Plant-Microbe Interact 17:6–15

Kumar S, Pandey P, Maheshwari D (2009) Reduction in dose of chemical fertilizers and growth enhancement of sesame (*Sesamum indicum* L.) with application of rhizospheric competent *Pseudomonas aeruginosa* LES4. Eur J Soil Biol 45:334–340

Lacey L, Grzywacz D, Shapiro-Ilan D et al (2015) Insect pathogens as biological control agents: back to the future. J Invertebr Pathol 132:1–41

Latef AAHA, Chaoxing H (2014) Does inoculation with *Glomus mosseae* improve salt tolerance in pepper plants? J Plant Growth Regul 33:644–653

Latef AAHA, Hashem A, Rasool S et al (2016) Arbuscular mycorrhizal symbiosis and abiotic stress in plants: a review. J Plant Biol 59:407–426

Lemanceau P, Bauer P, Kraemer S et al (2009) Iron dynamics in the rhizosphere as a case study for analyzing interactions between soils, plants and microbes. Plant Soil 321:513–535

Liu W, Yu K, He T et al (2013) The low temperature induced physiological responses of *Avena nuda* L., a cold-tolerant plant species. Sci World J 2013:1–7

Lopez DC, Zhu-Salzman K, Ek-Ramos MJ et al (2014) The entomopathogenic fungal endophytes *Purpureocillium lilacinum* (formerly *Paecilomyces lilacinus*) and *Beauveria bassiana* negatively affect cotton aphid reproduction under both greenhouse and field conditions. Pone 9: e103891

Lugtenberg B (2015) Life of microbes in the rhizosphere. In: Lugtenberg B (ed) Principles of plant-microbe interactions. Springer, Cham, pp 7–15

Lugtenberg B, Kamilova F (2009) Plant-growth-promoting rhizobacteria. Annu Rev Microbiol 63:541–556

Machado S, Paulsen GM (2001) Combined effects of drought and high temperature on water relations of wheat and sorghum. Plant Soil 233:179–187

Mantzoukas S, Chondrogiannis C, Grammatikopoulos G (2015) Effects of three endophytic entomopathogens on sweet sorghum and on the larvae of the stalk borer *Sesamia nonagrioides*. Entomol Exp Appl 154:78–87

Mazurier S, Corberand T, Lemanceau P et al (2009) Phenazine antibiotics produced by fluorescent pseudomonads contribute to natural soil suppressiveness to Fusarium wilt. ISME J 3:977

Mercado-Blanco J (2015) Life of microbes inside the plant. In: Lugtenberg B (ed) Principles of plant-microbe interactions. Springer, Cham, pp 25–32

Mimmo T, Del Buono D, Terzano R et al (2014) Rhizospheric organic compounds in the soil–microorganism–plant system: their role in iron availability. Eur J Soil Sci 65:629–642

Mishra PK, Bisht SC, Ruwari P et al (2011) Alleviation of cold stress in inoculated wheat (*Triticum aestivum* L.) seedlings with psychrotolerant Pseudomonads from NW Himalayas. Arch Microbiol 193:497–513

Muthukumar T, Priyadharsini P, Uma E et al (2014) Role of arbuscular mycorrhizal fungi in alleviation of acidity stress on plant growth. In: Miransari M (ed) Use of Microbes for the Alleviation of Soil Stresses. Springer, Berlin, pp 43–71

Nadeem SM, Ahmad M, Zahir ZA et al (2014) The role of mycorrhizae and plant growth promoting rhizobacteria (PGPR) in improving crop productivity under stressful environments. Biotechnol Adv 32:429–448

Nawrocka J, Małolepsza U (2013) Diversity in plant systemic resistance induced by *Trichoderma*. Biol Control 67:149–156

Nazar R, Umar S, Khan N et al (2015) Salicylic acid supplementation improves photosynthesis and growth in mustard through changes in proline accumulation and ethylene formation under drought stress. S Afr J Bot 98:84–94

Paredes M, Quiles MJ (2015) The effects of cold stress on photosynthesis in Hibiscus plants. PLoS One 10:e0137472

Perazzolli M, Moretto M, Fontana P et al (2012) Downy mildew resistance induced by *Trichoderma harzianum* T39 in susceptible grapevines partially mimics transcriptional changes of resistant genotypes. BMC Genomics 13:660

Pieterse CM, Van der Does D, Zamioudis C et al (2012) Hormonal modulation of plant immunity. Annu Rev Cell Dev Biol 28:489–521

Pii Y, Mimmo T, Tomasi N et al (2015a) Microbial interactions in the rhizosphere: beneficial influences of plant growth-promoting rhizobacteria on nutrient acquisition process. A review. Biol Fertil Soils 51:403–415

Pii Y, Penn A, Terzano R et al (2015b) Plant-microorganism-soil interactions influence the Fe availability in the rhizosphere of cucumber plants. Plant Physiol Biochem 87:45–52

Pineda A, Kaplan I, Bezemer TM (2017) Steering soil microbiomes to suppress sboveground insect pests. Trends Plant Sci 22:770–778

Pozo MJ, López-Ráez JA, Azcón-Aguilar C et al (2015) Phytohormones as integrators of environmental signals in the regulation of mycorrhizal symbioses. New Phytol 205:1431–1436

Rajkumar M, Ae N, Prasad MNV et al (2010) Potential of siderophore-producing bacteria for improving heavy metal phytoextraction. Trends Biotechnol 28:142–149

Ray S, Alves PC, Ahmad I et al (2016) Turnabout is fair play: herbivory-induced plant chitinases excreted in fall armyworm frass suppress herbivore defenses in maize. Plant Physiol 171:694–706

Rosier A, Bishnoi U, Lakshmanan V et al (2016) A perspective on inter-kingdom signaling in plant–beneficial microbe interactions. Plant Mol Biol 90:537–548

Ryan G, Rasmussen S, Parsons A et al (2015) The effects of carbohydrate supply and host genetic background on *Epichloë* endophyte and alkaloid concentrations in perennial ryegrass. Fungal Ecol 18:115–125

Saikkonen K, Gundel PE, Helander M (2013) Chemical ecology mediated by fungal endophytes in grasses. J Chem Ecol 39:962–968

Salas-Marina MA, Silva-Flores MA, Uresti-Rivera EE et al (2011) Colonization of Arabidopsis roots by *Trichoderma atroviride* promotes growth and enhances systemic disease resistance through jasmonic acid/ethylene and salicylic acid pathways. Eur J Plant Pathol 131:15–26

Sasan RK, Bidochka MJ (2012) The insect-pathogenic fungus *Metarhizium robertsii* (Clavicipitaceae) is also an endophyte that stimulates plant root development. Am J Bot 99:101–107

Scagel CF, Bryla DR, Lee J (2017) Salt exclusion and mycorrhizal symbiosis increase tolerance to NaCl and $CaCl_2$ salinity in 'Siam Queen' basil. Hort Sci 52:278–287

Scagliola M, Pii Y, Mimmo T et al (2016) Characterization of plant growth promoting traits of bacterial isolates from the rhizosphere of barley (*Hordeum vulgare* L.) and tomato (*Solanum lycopersicon* L.) grown under Fe sufficiency and deficiency. Plant Physiol Biochem 107:187–196

Schardl CL (2010) The *epichloë*, symbionts of the grass subfamily Poöideae. Ann Missouri Bot Gard 97:646–665

Shrivastava G, Ownley BH, Augé RM et al (2015) Colonization by arbuscular mycorrhizal and endophytic fungi enhanced terpene production in tomato plants and their defense against a herbivorous insect. Symbiosis 65:65–74

Silva K, Perin L, Gomes ML et al (2016) Diversity and capacity to promote maize growth of bacteria isolated from the Amazon region. Acta Amazon 46:111–118

Sivan A, Chet I (1989) Degradation of fungal cell walls by lytic enzymes of *Trichoderma harzianum*. Microbiology 135:675–682

Spaepen S, Vanderleyden J (2011) Auxin and plant-microbe interactions. Cold Spring Harb Perspect Biol 3(4):a001438

Spence C, Bais H (2013) Probiotics for plants: rhizospheric microbiome and plant fitness. In: Frans JDB (ed) Molecular microbial ecology of the rhizosphere. Wiley, New York, pp 713–721

Spence C, Bais H (2015) Role of plant growth regulators as chemical signals in plant–microbe interactions: a double edged sword. Curr Opin Plant Biol 27:52–58

Stępień Ł, Lalak-Kańczugowska J, Witaszak N et al (2018) *Fusarium* secondary metabolism biosynthetic pathways: so close but so far away. In: Merillon JM, Ramawat K (eds) Co-evolution of secondary metabolites. Springer, Cham, pp 1–37

Terrazas RA, Giles C, Paterson E et al (2016) Plant–microbiota interactions as a driver of the mineral turnover in the rhizosphere. Adv Appl Microbiol 95:1–67

Vacheron J, Desbrosses G, Bouffaud M-L et al (2013) Plant growth-promoting rhizobacteria and root system functioning. Front Plant Sci 4:356

Vardharajula S, Zulfikar Ali S, Grover M et al (2011) Drought-tolerant plant growth promoting *Bacillus* spp.: effect on growth, osmolytes, and antioxidant status of maize under drought stress. J Plant Interact 6:1–14

Vega FE (2018) The use of fungal entomopathogens as endophytes in biological control: a review. Mycologia 110:4–30

Vega FE, Goettel MS, Blackwell M et al (2009) Fungal entomopathogens: new insights on their ecology. Fungal Ecol 2:149–159

Verhagen BW, Glazebrook J, Zhu T et al (2004) The transcriptome of rhizobacteria-induced systemic resistance in *arabidopsis*. Mol Plant-Microbe Interact 17:895–908

Vidal S, Jaber LR (2015) Entomopathogenic fungi as endophytes: plant–endophyte–herbivore interactions and prospects for use in biological control. Curr Sci 109:46–54

Wani SH, Kumar V, Shriram V et al (2016) Phytohormones and their metabolic engineering for abiotic stress tolerance in crop plants. Crop J 4:162–176

Wiewióra B, Żurek G, Żurek M (2015) Endophyte-mediated disease resistance in wild populations of perennial ryegrass (*Lolium perenne*). Fungal Ecol 15:1–8

Wu Q-S, Zou Y-N (2017) Arbuscular mycorrhizal fungi and tolerance of drought stress in plants. In: Wu QS (ed) Arbuscular mycorrhizas and stress tolerance of plants. Springer, New York, pp 25–41

Xu H, Lu Y, Zhu X (2016) Effects of arbuscular mycorrhiza on osmotic adjustment and photosynthetic physiology of maize seedlings in black soils region of northeast China. Braz Arch Biol Technol 59:e16160392

Yadav J, Verma JP, Jaiswal DK et al (2014) Evaluation of PGPR and different concentration of phosphorus level on plant growth, yield and nutrient content of rice (*Oryza sativa*). Ecol Eng 62:123–128

Yan Y, Yang J, Dou Y et al (2008) Nitrogen fixation island and rhizosphere competence traits in the genome of root-associated *Pseudomonas stutzeri* A1501. Proc Natl Acad Sci 105:7564–7569

Yang Y, Tang M, Sulpice R et al (2014) Arbuscular mycorrhizal fungi alter fractal dimension characteristics of *Robinia pseudoacacia* L. seedlings through regulating plant growth, leaf water status, photosynthesis, and nutrient concentration under drought stress. J Plant Growth Regul 33:612–625

Yooyongwech S, Phaukinsang N, Cha-um S et al (2013) Arbuscular mycorrhiza improved growth performance in *Macadamia tetraphylla* L. grown under water deficit stress involves soluble sugar and proline accumulation. Plant Growth Regul 69:285–293

Yuan S, Li M, Fang Z et al (2016) Biological control of tobacco bacterial wilt using *Trichoderma harzianum* amended bioorganic fertilizer and the arbuscular mycorrhizal fungi *Glomus mosseae*. Biol Control 92:164–171

Zhao L, Wang F, Zhang Y et al (2014) Involvement of *Trichoderma asperellum* strain T6 in regulating iron acquisition in plants. J Basic Microbiol 54:S115–S124

Zhou W, Jia CG, Wu X et al (2016) ZmDBF3, a novel transcription factor from maize (*Zea mays* L.), is involved in multiple abiotic stress tolerance. Plant Mol Biol Rep 34:353–364

Zhu X, Song F, Liu S et al (2012) Arbuscular mycorrhizae improves photosynthesis and water status of *Zea mays* L. under drought stress. Plant Soil Environ 58:186–191

Złoch M, Thiem D, Gadzała-Kopciuch R et al (2016) Synthesis of siderophores by plant-associated metallotolerant bacteria under exposure to $Cd^{2+}$. Chemosphere 156:312–325

Zolla G, Bakker MG, Badri DV et al (2013) Understanding root–microbiome interactions. Mol Microbial Ecol Rhizosphere 1:743–754

# Chapter 15
# Fertilizers and Pesticides: Their Impact on Soil Health and Environment

**Pooja Baweja, Savindra Kumar, and Gaurav Kumar**

**Abstract** The agricultural practices around the world are dependent upon extensive use of fertilizers and pesticides. These chemical formulations are being added to improve crop quality and meet the global food demand. Fertilizers and pesticides are also considered as critical farmland tools for food security. On the other hand, the inorganic fertilizers and pesticides have many undesirable aspects which cannot be overlooked. They have properties to remain in soil and environment for a long time and affect various biotic and abiotic factors. They have adverse effects on soil, microflora, other organisms, environment, and human health. These undesirable properties of fertilizers and pesticides have led to the search of another option, i.e., sustainable agriculture, which is attracting the farmers and gaining the attention. In this system, the use of harsh chemicals is avoided and other methods such as organic farming, biofertilizers, composting, and use of bio control agents etc. are adopted and that is sustainable agriculture. Keeping all these aspects in view, this chapter aims at discussing various impacts of fertilizers and pesticides on soil structure, composition and environment along with the various alternatives to inorganic fertilizers and pesticides, so that preventive measures can be taken to conserve the nature.

**Keywords** Organic and inorganic fertilizers · Sustainable agriculture and environment · Chemical fertilizers and pesticides

P. Baweja (✉)
Department of Botany, Maitreyi College, University of Delhi, New Delhi, India

S. Kumar
Department of Botany, Zakir Husain Delhi College, University of Delhi, New Delhi, India

G. Kumar
Department of Environmental Studies, PGDAV College, University of Delhi, New Delhi, India



B. Giri, A. Varma (eds.), *Soil Health*, Soil Biology 59,
https://doi.org/10.1007/978-3-030-44364-1_15

## 15.1 Introduction

All around the world the governments are struggling to meet the demand of ever increasing population. There is high demand for agricultural produce as well as the industrial products. To feed all the people on the earth is a biggest challenge and all around the world efforts are being done to increase the agricultural production. But, these efforts are costing too much, as due to overuse of fertilizers and pesticides there is not only pollution and contamination of soil but also it is severely affecting human health. Also, availability of nutrients in sufficient and balanced quantities for optimum plant growth is also a challenge. Natural reserves of plant nutrients in soil are largely unavailable to plants. Only a minor portion of plant nutrients is released through biological activity or chemical processes which is too slow to compensate for the removal of nutrients by agricultural production and to meet crop requirements (Chen 2006). Therefore, fertilizers are designed in a way so that they can supplement the nutrients already present in the soil. Industrially manufactured or synthetic fertilizers are being added enormously in crop fields, which are composed of known quantities of macronutrients like nitrogen, potassium, and phosphorous along with micronutrients like zinc, boron, iron, etc. Approximately, 300 million pounds of different categories of fertilizers and pesticides are being added in the soils under different brand names (Tomkins and Bird 2002). In general, out of the total fertilizers or pesticides applied to crops, 60–90% is lost and the remaining 10–40% is taken up by plants (Bhardwaj et al. 2014). A lot of benefits are associated with fertilizers and pesticides. They are considered as most crucial element in farming and agriculture. Both, fertilizers and pesticides have a strong bonding with each other. They stimulate crop's growth with greater yield, help in irrigation, crop protection, control pests, kill weeds, prevent diseases, and protect food and much more.

Although these chemicals are being added to increase the fertility of soil and crop productivity, they are also impacting the health and environment. These chemicals tend to stay in the soil, change its physical properties, disrupt the ecological balance of soil microflora and environment, disturbing many activities of soil such as nutrient balance and availability, reduction in rate of decomposition of organic and inorganic matter. Apart from changing soil characteristics the fertilizers and pesticides when used in excess quantity release greenhouse gases (N – fertilizers); develop algal blooms and pest resistance (Prashar and Shah 2016). It is well understood that fertilizers and pesticides are integral part of agricultural system, but now as we know their impacts; it is desirable to evaluate their role and effects on soil, environment and crop productivity besides considering the importance of sustainable agriculture.

## 15.2 Classifications of Fertilizers

Fertilizers are compounds that are applied to agricultural lands either on the soil or directly on crops to increase the yield as well as the quality. Fertilizers are rich in plant nutrients which are required for the growth and development of plants. Based on origin or composition fertilizers are classified as natural or synthetic/industrial fertilizers and organic or inorganic fertilizers, respectively. Based on application fertilizers are also classified as direct or indirect fertilizers. The direct fertilizer provides essential nutrients such as nitrogen, phosphorus and potassium (NPK), compound or microelements directly to the crops. Indirect fertilizers such as gypsum, lime, and bacterial fertilizers are used for the improvement of soil quality (https://www.fertilizer-machine.net/solution_and_market/types-of-fertilizer.html).

### *15.2.1 Organic Fertilizers*

These are the fertilizers which are natural and are composed of farmyard manure and residues of animal or plant products. They are rich in carbonic materials, organic acids, and nutrients. Examples of organic fertilizers are agricultural waste, livestock manure, industrial waste, or municipal sludge.

### *15.2.2 Inorganic Fertilizers*

Inorganic fertilizers are made chemically in industries such as nitrogen fertilizers, urea, phosphate fertilizers, potash fertilizers, macro- or micronutrients etc. On the basis of composition inorganic fertilizers can also be classified as straight, complex, and mixed fertilizers.

The straight fertilizers are those which supply only one primary nutrient, for example sodium, potassium, or phosphorous. The complex fertilizers are composite forms of 2–3 primary nutrients, for example nitrophosphate, ammonium phosphate, diammonium phosphate (DAP) etc. and mixed fertilizers are normally the mixture of straight fertilizers. On the basis of physical form they are classified as solid or liquid (Fig. 15.1) and on the basis of application, the fertilizers are either direct fertilizers (applied directly on plant) or indirect fertilizers (which are mixed in soil) (https://www.sciencedirect.com/topics/agricultural-and-biological-sciences/fertilizers) (Fig. 15.1).

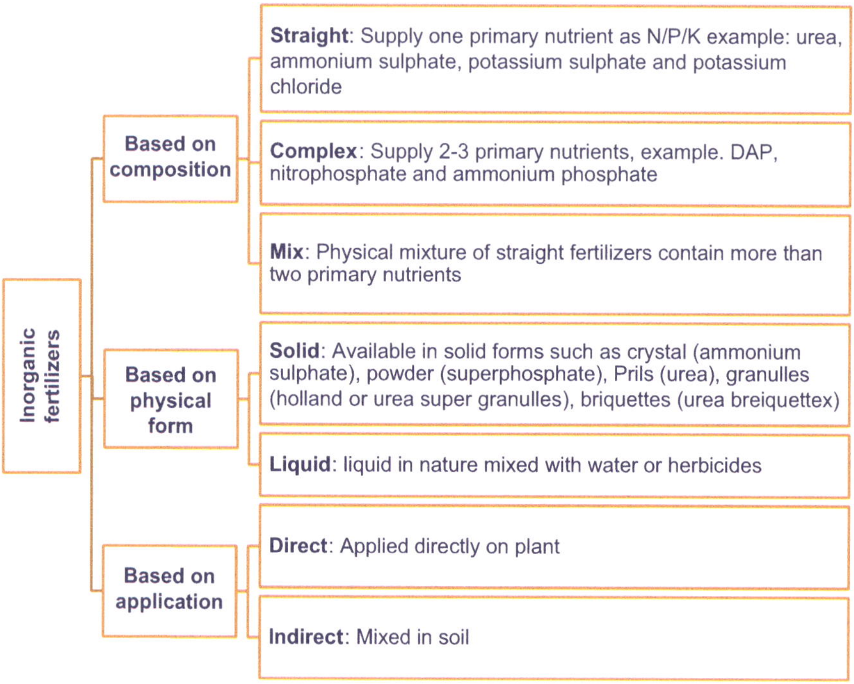


**Fig. 15.1** Classification of inorganic fertilizers

## 15.3 Classification of Pesticides

Pesticides are bioactive toxic substances that are meant to control pests, including weeds. Pesticides are a group of chemicals which includes various algicides, avicide, bactericide, fungicide herbicides, insecticides, molluscicide, nematicide, piscicide, and rodenticide. Out of these various pesticides herbicides are commonly used, which account for approximately 80% of all. In general, we can say that a pesticide is a chemical or biological agent that may be a virus, bacterium, or a fungus which can deter, incapacitate, kill, or otherwise discourage pests (Aktar et al. 2009). Generally, the pesticides are considered to have a low persistence and biodegradable properties, which do not affect the nontarget organisms, but the pesticides do not have any of these properties. They are basically the BIOCIDES, which can harm or kill the nontarget organism (Prashar and Shah 2016). The pesticides have chronic and acute toxicity. The pesticides may be classified into various categories based on target organism as listed in Table 15.1.

Based on chemical composition the pesticides are classified as organochlorines (DDT), organophosphates (Malathion), carbamates (Mexacarbate), and Pyrethroids (Permethrin). On the basis of toxicity level four classes have been assigned to pesticides such as class Ia, Ib, II, III. The pesticides of class Ia are extremely hazardous (Parathion), class Ib are highly hazardous (Eldrin), class II is composed

**Table 15.1** Various types of pesticides, their target organisms and examples

| S. No. | Types | Target organism | Examples |
|---|---|---|---|
| 1 | Acaricides | Arachnid (includes ticks and mites) | Carbamate; Dicofol; Dienochlor; Ivermectin; Permethrin |
| 2 | Algicides | Algae | Benzalkonium Chloride; Dichlone; Dichlorophen; Diuron; Endothal; Fentin |
| 3 | Avicides | Birds | Avitrol; Chloralose; CPTH; DRC-1339; Strychnine |
| 4 | Bactericides | Bacteria | Cephalosporins; Daptomycin; Hypochlorites; Potassium Persulfate; Vancomycin |
| 5 | Fungicides | Fungi | Acibenzolar; Benomyl; Famoxadone Monocerin; Nimbin |
| 6 | Herbicides | Unwanted Plants | 2,4-D; Aminopyralid; Dicamba; Fluroxypyr; Metolachlor; Picloram |
| 7 | Insecticides | Insects | Allethrin; Chlordane; Fenitrothion; Lindane; Resmethrin |
| 8 | Molluscicides | Molluscs | Allicin; Ferric Sodium EDTA; Metaldehyde; Methiocarb; Trifenmorph |
| 9 | Nematicides | Nematodes | Abamectin; Benclothiaz; Carvacrol Diamidafos; Imicyafos |
| 10 | Piscicides | Fish | Fintrol (Antimycin A); Niclosamide Rotenone; TFM (3-Trifluoromethyl-4-nitropheno) |
| 11 | Rodenticides | Rodents | Bromethalin; Difethialone; Flupropadine; Norbormide; Pindone; Pyrinuron |
| 12 | Virucides | Viruses | Cyanovirin-N; Scytovirin; Urumin |

**Table 15.2** Classification of pesticides based on chemical composition and level of toxicity

| S. No. | Chemical class | Example | S. No. | Toxicity class | Toxicity level (mg/kg) (LD50 determination in rats) | Example |
|---|---|---|---|---|---|---|
| 1 | Organochlorines | DDT | 1 | Class Ia | Extremely hazardous | Parathion |
| 2 | Organophosphates | Malathion | 2 | Class Ib | Highly hazardous | Eldrin |
| 3 | Carbamates | Mexacarbate | 3 | Class II | Moderately hazardous | DDT |
| 4 | Pyrethroids | Permethrin | 4 | Class III | Slightly hazardous | Malathion |

of pesticides which are moderately hazardous (DDT) and class III pesticides are slightly hazardous (Malation) (Table 15.2).

## 15.4 Impacts of Chemical Fertilizers and Pesticides

The fertilizers and pesticide usage is not new in world and they are being used in fields since the early 1900s to increase the crop production. Initially people were not aware of usage and problems associated with the use of such chemical, but today the awareness is getting spread across the globe. Although chemical fertilizers and pesticides are used to improve the growth of plants and increase the yields of fruits and vegetables in relatively shorter period, the over usage of fertilizers and pesticides is imposing possible risks and adverse effects on the soil health, crop productivity, environment, and human health which are discussed in the following paragraphs.

### *15.4.1 Soil Health and Properties*

A soil is a living and dynamic system. It is considered as healthy, when it has capacity to sustain biological productivity by maintaining its environment, is able to promote and support air, water, humans, animals and plant's health and habitat. Also, soil is healthy if it can sustain itself and without undergoing degradation it is able to support plant growth. The three parameters which define a soil healthy include biological (microbial activities, respiration); chemical (pH, salinity and soil organic matter), and physical parameters (soil texture and water holding capacity of soil). All these three parameters are interrelated and influence each other.

The overuse of chemical fertilizers and pesticides have various effects on soil health. The chemical fertilizers are rich in N, P, K which are highly water soluble and change biochemical properties including organic carbon content, nitrogen content, pH, moisture, altered enzymatic activities etc. The change in biochemical properties of soil leads to variable nutrient availability. The inorganic content of fertilizers decreases the organic and nitrogen content of soil thus decreasing the microflora and alter the properties of various soil organisms. These inorganic fertilizers are also highly water soluble and with irrigation they leach down to subsoil layers and alter its characteristic properties. Also through surface runoffs or leaching they reach the adjoining water bodies and contaminate them. The pesticides are known to remain in soil for longer period, form transformation products having toxic and harmful effects, and are retained in soil. This retention is directly proportional to the organic matter content of soil. The fertilizers and pesticides also cause the acidification of soil by lowering its pH. The acids in soil fasten the process of dissolving of soil crumbs, which are rich in minerals and are important for oil drainage, resulting in highly compacted soil with reduced drainage and air circulation.

### 15.4.2 Microorganisms

The overuse of inorganic fertilizers or pesticides leads to decline in the population of beneficial soil microorganisms. The fertilizer controls the functional diversity of the soil microbial community. The fertilizers and pesticides applied to soil directly affect the microbial community present in agricultural fields. As stated earlier inorganic fertilizers tend to decrease the microbial population, for example *Pseudomonas* sp. population decreases tremendously with the application of inorganic fertilizers. As the microbial population decreases, the activity of various soil-based enzymes such as dehydrogenase, catalase, invertase, urease, casein protease, and arylsulphatase has also been found less (Prashar and Shah 2016).The pesticides which are present in soil for longer periods have toxic and harmful effects. The more pesticides stay in soil the more is the shift in microbial population. With the increase in inorganic content of soil the microbes in soil decrease. The chemical compounds present in fertilizers and pesticides lower the pH of the soil. The altered pH makes the soil acidic and eliminates the beneficial microorganisms. The sensitive microbes tend to die immediately on application of long and persistent pesticides. These microorganisms are an integral component of healthy soil. They maintain soil structure, transformation, and mineralization of organic matter. They are also capable of degrading various pollutants and pesticides. Various plants such as legumes depend on soil bacteria for conversion of atmospheric nitrogen to nitrates. The mycorrhizal fungi present in soil also aid in nutrient uptake. The pesticides like oryzalin, trifluralin, triclopyr are found to be toxic to mycorrhizal fungi and their spores (Aktar et al. 2009); on the other hand there are some pesticides like carbofuran (insecticide), iprodione (fungicide), simazine (herbicide) which do not have any negative impact on soil microflora (Prashar and Shah 2016).

### 15.4.3 Crop Productivity

The disproportionate use of fertilizers causes chemical burns to crops. The use of synthetic chemical fertilizers leads to imperfectly synthesized protein in leaves, which is responsible for poor crops and in turn for pathological conditions in humans and animals fed with such deficient food (Talukdar et al. 2003). The undue application may cause damage to leaves (chemical leaf scorch) and lead to less crop productivity. The herbicide glyphosate causes reduced seed quality. Exposure to clopyralid decreases fruit production in tomato plants (Aktar et al. 2009).

### 15.4.4 Nontarget Organisms

The pesticide sprays can directly affect the nontarget vegetation and either kill them or have sublethal action on them. Fertilizers and pesticides present as contaminants can also harm nontarget plants and animals such as fish. Chlorpyrifos and trifluralin present in pesticides are highly toxic to fish. 2,4-D herbicides affect salmon and trout, and also it reduces the percentage of successful hatching of chicken eggs. Pesticide poisoning is also common in dolphins and has been reported worldwide. In addition, there is also adverse effect on reproductive and immunological activities of aquatic animals such as shellfish, shrimp, mussels, otters, and mink by DDT and PCBs. A herbicide oxadiazon is found to be toxic to bees, which are the potent pollinators. Also many insect and spider population also get affected and get declined when such plants and foliage are destroyed which are needed for their food and shelter. Brodifacoum (a rodenticide) can be lethal to many birds when they eat. Various herbicides which tend to kill unwanted plants may have adverse effect on nontarget aquatic plant, for example oxadiazon, atrazine, and alachlor reduce algal growth and diatoms by damaging cells, blocking photosynthesis, and stunted growth (Aktar et al. 2009).

### 15.4.5 Environment and Human Health

Excessive use of synthetic fertilizers and pesticides has caused tremendous harm to the environment as well as human population indirectly. Continuous use of these two has also resulted in the development of resistance against the pests, which become difficult to control by other means. They are sprayed onto crops, food, especially fruits and vegetables, and they may leach into soils and groundwater ending up in drinking water and may drift and pollute the air or soil. The fertilizers and pesticides may enter the human body through inhalation of polluted air and dust or inhalation of vapors that contain pesticides; by consumption of contaminated food and water; and through dermal exposure by direct contact (Yadav and Devi 2017). Chemical fertilizers and pesticides not only cause environmental pollution at the application site but also the adjoining sites get contaminated. The leftover from manufacturing sites, when released without any treatment, pollutes the nearby streams and water bodies. When water soluble nitrogen fertilizers are applied to the soil, a good portion of added nutrients does not become available to the plants, but is lost to the groundwater through leaching or runoff. The excess nitrate leached into rivers or ponds causes blooms, encourages the growth of organisms, and thus produces a lot of organic matter which on decomposition lead to foul smell, having an adverse effect on health. The contaminants also reach sites far from the application site and pollute the soil and water bodies of that area (Fig. 15.2).

Chemicals such as DDT, polychlorinated biphenyls, or organo chemicals used in fertilizers and pesticides undergo biomagnification. These contaminants are taken by

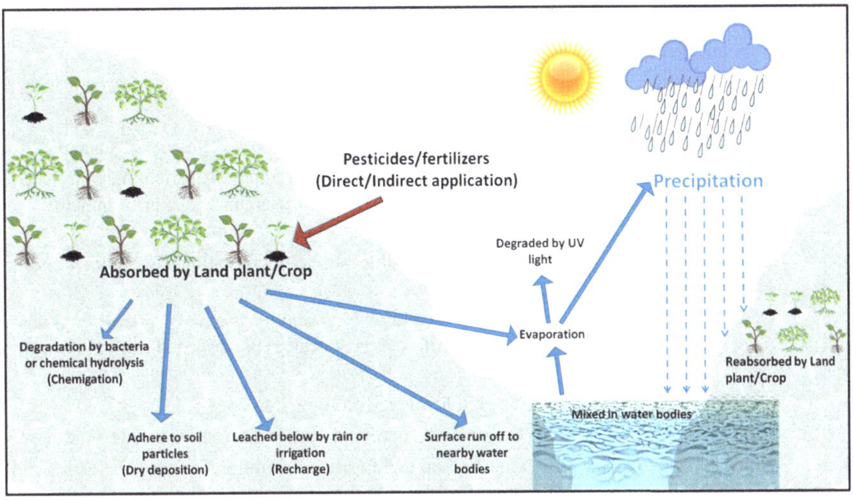


**Fig. 15.2** Various mechanisms involved in the movement of fertilizers and pesticides

organisms through the food they consume, and when other organisms at the higher level in food chain feed on such organisms the toxins get accumulated in the organism at higher trophic level. The pesticides, insecticides, fungicides, and fertilizers when added in the soil undergo leaching, deposition, chemigation, or evaporation etc. and enter the food chain. Approximately, 60% of all herbicides (weed killers), 90% of all fungicides (mold killers), and 30% of all insecticides (insect killers) are potentially cancer causing. Residues of these chemicals affect the central nervous system, respiratory and gastro intestinal system. They may cause depression, insomnia, hyperreflexia etc. Some pesticides can also cause wheezing and nausea by irritating the lungs if large amounts are inhaled. Foods grown with overapplication of chemical fertilizers or pesticides cause various deteriorating health hazards in animals as well as human beings as listed in Table 15.3.

## 15.5 Alternatives to the Chemical Fertilizers and Pesticides

> Sustainable development is the development that meets the need of the present generation, without compromising the need of the future generations (WCED 1987).

In recent years, because of the concerns related to chemical fertilizers and pesticides, the concepts of sustainable agricultural development have been much debated. Sustainable agriculture is the need of the hour, as it provides food security to the growing human population. Also, there are a number of ecological, economic, social, and philosophical issues being addressed by sustainable agriculture and is thus essential for the overall development of any nation. The different methods of sustainable agricultural are organic farming, biofertilizers, and biocontrol agents. In

**Table 15.3** Effects of various fertilizers on humans and animals

| S. No. | Type of fertilizer | Effects |
|---|---|---|
| 1 | DDT (Dichloro diphenyl trichloro ethane) | Affects central nervous system, carcinogen |
| 2 | Potassium fertilizer | Decreases vitamin C and carotene content in vegetables |
| 3 | Nitrogen fertilizers | Infant disease, methemoglobinemia, also amines produced from the nitrogenous fertilizer cause cancer in human beings |
| 4 | Aluminum based fertilizers | High levels lead to birth defects, asthma, Alzheimer and bone diseases |
| 5 | Calcium rich fertilizers | Developmental and neurological toxicity, growth retardation, cognitive delay, kidney, nervous and immune system damage |
| 6 | Cobalt | At high level leads to lung damage |
| 7 | Boron | Causes low sperm count, nose, throat and eye irritation |
| 8 | Chloropyripos | Causes fetal malnutrition, pneumonia, muscle paralysis and even death to respiratory failure |
| 9 | Malathion | Damages nervous system, if it enters the body |

the present scenario, these strategies are gaining the attention of scientists, farmers, and the general public worldwide. After green, white, and blue revolution it is a new revolution in the field of agriculture. This revolution has come in the wake of a realization of the ill effects of the Green Revolution such as public health, loss of biodiversity, and environmental pollution as people are now health conscious and are ready to pay for clean, healthy, and naturally produced food. The concept of sustainable agriculture has given a sense of direction, and an urgency, that has sparked innovative thinking in the agricultural world.

### *15.5.1 Organic Farming: A Sustainable Option*

Organic farming is one of the sustainable agricultural methods, which not only ensures food safety but also is environment friendly. According to Sir Albert Howard (Father of Organic Farming), "The maintenance of the fertility of the soil is the first condition of any permanent system of agriculture" (Howard 1943). Organic farming can broadly be described as a form of special cultivation technique on the concept of working "with nature" instead of against it. Organic farming relies on components such as green manure, compost, crop rotations, and biological pest control as well as it may involve the use of biofertilizers and pesticides which are obtained from natural sources to maintain soil productivity (Epule 2019). It is an age-old chemical-free farming system followed in India since ancient time and practiced centuries before the advent of the Green Revolution. Natural microflora of the soil such as useful bacteria, algae, and fungi including the Plant Growth Promoting Rhizobacteria (PGPR) are the main components of organic farming.

**Table 15.4** Advantages and disadvantages of organic fertilizers

| Advantages |
|---|
| 1. Balanced nutrient supply to keep plants healthy |
| 2. Enhanced soil biological activity to improve nutrient mobilization from organic and chemical sources and decomposition of toxic substances |
| 3. Enhanced colonization of mycorrhizae to improve P supply |
| 4. Better root growth due to better soil structure |
| 5. Increased organic matter content of the soil to improve the exchange capacity of nutrients, increased soil water retention, promotes soil aggregates and buffers the soil against acidity, alkalinity, salinity, pesticides, and toxic heavy metals |
| 6. Slow release of nutrients and contribution toward the residual pool of organic N and P in the soil, reducing the loss of N through leaching and P fixation; they can also supply micronutrients |
| 7. Better growth of beneficial microorganisms and earthworms |
| 8. Suppresses certain plant diseases, soilborne diseases and parasites |
| Disadvantages |
| 1. Too slow nutrient release rate to meet crop requirements in a short time, hence some nutrient deficiency may occur |
| 2. The major plant nutrients may not exist in organic fertilizer in sufficient quantity to sustain maximum crop growth |
| 3. Variable nutrient composition of compost and high cost |
| 4. Long-term or heavy application to agricultural soils may result in salt, nutrient or heavy metal accumulation and may adversely affect plant growth, soil organisms, water quality, and animal and human health |
| 5. Method of application and storage of the product are all critical to the success of a biological product. Short shelf life, lack of suitable carrier materials, susceptibility to high temperature, problems in transportation and storage are bottlenecks that still need to be solved in order to obtain effective inoculation |

According to various theoreticians, organic farming is a holistic practice which promotes soil health, biodiversity of soil microorganisms, insects and microbes which are part of biogeochemical cycles.

Although, organic farming represents only 1% of world agricultural area, it is one of the fastest growing sectors of world agriculture.

The various advantages of organic farming are: Organic fertilizers provide nutrients to crops directly and also by microbial actions. The biocontrol agents are an environmental friendly and cheap alternative to chemical pesticides. The resources used in organic farming are natural, nonpolluting, and inexpensive. Besides improving fertility of soil, the organic fertilizers enhance water uptake, provide disease resistance etc. Organic farming also protects water quality and helps in pest and disease control. However, compared with conventional farming practices, organic farming reportedly has lower productivity. Although there are some advantages associated with organic fertilizers, some disadvantages are also associated with it as mentioned in Table 15.4.

#### 15.5.1.1 Biofertilizers for Organic Farming

Any microbial or organic substances when applied to soil, seeds, or plant surface help in increasing the availability of nutrients to crop plants, which are known as biofertilizers. Biofertilizers offer an eco-friendly and cheap alternative technology over chemical fertilizers to replenish the required deficient crop nutrients. Different kinds of microorganisms play a vital role in fixing/solubilizing/mobilizing/recycling nutrients in the agricultural ecosystem. Even though all these microorganisms are naturally present in soil and play crucial role in various crops, their population is often scanty. Therefore, these microorganisms have to be artificially supplemented in the form of biofertilizers or microbial inoculants. The different sources of biofertilizers are microbial, algal, fungal, and agro-waste. Recently seaweeds are gaining recognition for their use as biofertilizers.

The rhizosphere (narrow zone surrounding plant roots) is a hot spot for numerous organisms and is considered as one of the most complex ecosystems on Earth containing ~1011 microbial cells per gram of root and >30,000 prokaryotic species that in general improve plant growth and productivity (Bhowmik and Das 2018; Mendes et al. 2013). These rhizosphere microbial communities or biofertilizers have been identified as an alternative to chemical fertilizers to increase soil fertility and crop production in sustainable agriculture and biosafety program (Wu et al. 2005). In a broader perspective, the biofertilizers may include all organic resources (manure) for plant growth which are concentrated in an available form for plant absorption through microorganisms or plant associations or interactions. They have their own advantages and disadvantages in the context of nutrient supply, crop growth, and environmental quality.

#### 15.5.1.2 Bacterial Biofertilizers

The knowledge of applied microbial inoculums is long history which passes from generation to generation of farmers. Nitrogen, phosphorous, and potassium are available in our environment in abundance but they are not freely available for plants. Most of the soil bacteria have close relationship with plant roots. In place of N fertilizer, use of rhizobium inoculation is a well-known practice to ensure adequate N supply to legumes. Heavier application of inoculums mixed into peat granules trickled into soil as the seeds are planted is an alternative technique to encourage nodulation (Khosla 2017). Examples of free living nitrogen fixing bacteria are obligate anaerobes (*Clostridium* sp., *Desulfovibrio* sp., *Rhodospirillum* sp., *Rhodopseudomonas* sp., *Desulfotomaculum* sp., *Desulfovibrio* sp., *Chromatium* sp., *Chlorobium* sp.); obligate aerobes (*Azotobacter* sp., *Beijerinckia* sp., *Tolypothrix* sp., *Aulosira* sp.); facultative anaerobes (*Klebsiella pneumonia*, *Bacillus polymyxa*); photosynthetic bacteria (*Rhodobacter*); and some methanogens. *Azotobacter* sp. and *Azospirillium* sp. inhabit root surface or rhizosphere soil. These bacteria are not host specific but can fix nitrogen. The bacteria, for example

*Azotobacter* sp. can also produce antifungal compounds which can fight against many plant pathogens, increase germination and vigor in young plants (Khosla 2017). *Azospirillum* sp., *Herbaspirillum* sp., *Gluconobacter diazotrophicus*, *Azoarcus* sp. are examples of associative nitrogen fixing bacteria. Apart from nitrogen fixation many bacterial species also help plant in solubilizing potassium (e.g., *Bacillus mucilaginous*) or phosphorous (e.g., *Bacillus megaterium*, *Bacillus circulans*, *Bacillus subtilis* and *Pseudomonas straita*). Phosphobacterin, mainly bacteria and fungi, can make insoluble phosphorus available to the plant because under acidic or calcareous soil conditions, large amounts of phosphorus are fixed in the soil but are unavailable to the plants. The phosphobacterins have solubilization effect as they produce organic acids that can lower the soil pH and bring about the dissolution of bound forms of phosphate (Khosla 2017).

#### 15.5.1.3 Blue-Green Algae (Cyanobacteria)

The cyanobacteria or blue-green algae are an ancient group of Gram-negative prokaryotes with a very old and diverse evolutionary history. Some of them are able to fix atmospheric nitrogen due to the presence of heterocysts which is often referred to as nitrogen fixing factories because of the presence of nitrogenase enzyme. Cyanobacteria play a potential role in the enhancement of agriculture productivity and mitigation of GHG emissions. Furthermore cyanobacteria naturally occur in several agroecosystems like paddy fields around the globe, and in Antarctica to Arctic poles because they can easily survive on bare minimum requirement of light, carbon dioxide ($CO_2$), and water (Singh et al. 2016). Important nitrogen fixing cyanobacterial genera can be classified as unicellular (*Aphanothece* sp., *Dermocapsa* sp., *Synechococcus* sp., *Gloeocapsa* sp., *Myxosarcina* sp., *Pleurocapsa* sp., *Xenococcus* sp. etc.); filamentous heterocystous (*Anabaena* sp., *Anabaenopsis* sp., *Aulosira* sp., *Calothrix* sp., *Nodularia* sp., *Nostoc* sp., *Rivularia* sp., *Scytonema* sp., *Stigonema* sp., *Tolypothrix* sp. etc.); and filamentous non-heterocystous (*Lyngbya* sp., *Oscillatoria* sp., *Pseudanabaena* sp., *Schizothrix* sp., *Trichodesmium* sp. etc.). Many nitrogen fixing genera also live symbiotically with other plants such as Anabaena sp., *Nostoc* sp., *Scytonema* sp., *Pleurocapsa* sp. etc. in *Azolla* sp., *Anthoceros* sp., cycads, *Gunnera* sp. etc. Permanent perpetual symbiosis of cyanobacteria *Anabaena azollae* on dorsal leaf cavities of the host plant *Azolla*, naturally found in the rice paddy fields is capable of high rates of nitrogen fixation. In addition to the property of nitrogen fixation they are also able to release plant growth promoting substances such as sugars, vitamins, and growth hormones which may be used by other microorganisms leading to better colonization, enrichment of microflora, and the maintenance of microbial diversity. The reported N content in maximum standing crops of *Azolla* ranged from 20 to 146 kg $ha^{-1}$ and averaged 70 kg $ha^{-1}$, and $N_2$-flxing rate ranged from 0.4 to 3.6 kg N $ha^{-1}$ $d^{-1}$ and averaged 2 kg N $ha^{-1}$ $d^{-1}$ in a growing cycle of approximately 40 days. The N balance in the 0–50-cm soil profile after 27 cropping cycles confirmed an average annual gain of 76 kg N $ha^{-1}$ under *Azolla* treatment (Reddy et al. 2002). The system is known as "green gold mine" owing to its multifaceted uses. Reddy et al. (2002) highlighted the following benefits of cyanobacteria to the agroecosystem:

1. Cyanobacteria enhance solubilization and mobilization of nutrients of limited supply.
2. They make complexes of heavy metals and xenobiotics to limit their mobility and transport in plants.
3. Mineralization of simpler organic molecules such as amino acids for direct uptake.
4. Protection of plants from pathogenic insects and diseases as biocontrol agents.
5. Stimulation of plant growth due to their plant growth promoting attributes.
6. Improved bioavailability of phosphorus.
7. Improved physicochemical conditions of soils.

In addition many cyanobacterial genera (e.g., *Anabaena* sp., *Anabaenopsis* sp., *Calothrix* sp., *Glactothece* sp., *Nostoc* sp., *Plactonema* sp., *Synechocystis* sp., *Cylindromum* sp., *Chlorogloeopsis* sp., *Calothrix* sp. etc.) produce plant growth promoting chemicals such as auxins, gibberelins and cytokinnins (Reddy et al. 2002). In spite of so many potential uses, lack of awareness and improper extension machinery contribute to the limited use of these important biofertilizers.

#### 15.5.1.4 Seaweeds

Seaweeds are a rich source of various macro- and micronutrients and have a great potential for agricultural applications. They are being used in agriculture since antiquity, but recent demands of organic farming and organic food, the application of organic treatments like seaweed extracts in agriculture has increased (Nabti et al. 2017). They are being used because of the presence of high amounts of macronutrients, micronutrients, vitamins, amino acids, and growth regulators. Among the seaweeds brown algae members are most commonly used as biofertilizers. Some of the examples are *Fucus* sp., *Laminaria* sp., *Macrocystis pyrifera*, *Sargassum* sp., *Ascophyllum nodusum*, *Dictyoteris australis*, *Durvillea potatorum*, *Ecklonia maxima*, and *Turbinaria*. Seaweed biofertilizers are considered superior than the chemical fertilizers and farmland manures because the benefits of seaweeds application in agricultural field are numerous and diverse. Some of them are used for stimulation of seed germination, enhancement of health and growth of plants namely shoot and root elongation, improved water and nutrient uptake, frost and saline resistance. Seaweeds can also be used as biocontrol agents and for resistance toward phytopathogenic organisms, remediation of pollutants from contaminated soil (Nabti et al. 2017; Hernández-Herrera et al. 2014).

#### 15.5.1.5 Mycorrhizal Fungi

In today's era, when people are looking for *sustainable agriculture*, mycorrhizae are effective candidate for biofertilizer. Mycorrhizae are symbiotic associations between fungi and roots of plants. The fungus receives carbohydrates from green plants while the roots obtain mineral nutrients that the fungus has absorbed from the soil. Mycorrhizae show one of the most ancient associations between plant root and filamentous fungi. The fossil record suggests that mycorrhizal association would have been established more than 450 million years ago (Delaux et al. 2015). They are ubiquitous soil fungi distributed in wide range of habitats. These mycorrhizal associations are widespread throughout the vascular plant kingdom and they colonize more than 80% of land plants (Delaux et al. 2015) including virtually all plant species of economic importance ranging from bryophytes, pteridophytes, gymnosperms, and up to angiosperms. Mycorrhizae can be categorized on the basis of symbiosis type and partner taxa such as Arbuscular Mycorrhiza, Fine Root Endophytes, Ectomycorrhiza, Ectendomycorrhiza, Arbutoid Mycorrhiza, Monotropoid Mycorrhiza, Orchid mycorrhiza, Ericoid Mycorrhiza, Sheathed Ericoid Mycorrhiza, Sebacinalean Endophytes, Dark Septate Endophytes, Fire-Associated Mutualism, and Feremycorrhiza (Kariman et al. 2018).

There are a number of features which make mycorrhizae a good source for organic farming:

1. It is ubiquitous in nature and can be easily isolated.
2. Its carrier-based inoculum (dried powder form) can be made with ease and have long shelf life in comparison to other sources of organic farming.
3. It is known to increase surface area of plant root (100–1000 times) which leads to enhanced water and nutrient uptake by the plant.
4. It is well known for phosphate solubilization which is an important macronutrient.
5. It improves uptake of some micronutrients needed by the plant like iron, zinc, copper etc.
6. Mycorrhiza also helps the plants in combating various abiotic stresses, e.g., heavy metal stress, salt stress, drought stress etc. It also helps plant to fight against biotic stress (pathogens) by secreting certain compounds and in bioremediation.
7. It also known to boost uptake of nitrogen; however, AMF themselves do not fix nitrogen.
8. Mycorrhiza also increase physical, chemical, and biological properties of soil.
9. It promotes augmentation of Plant Growth Promoting Rhizobacteria (PGPR).
10. AMFs have shown vast potential in revival and reclamation of wastelands produced due to various natural and anthropogenic activities.

#### 15.5.1.6 Composting

Nowadays, not only chemical pesticides and fertilizers, but also generation of huge amount of solid waste around the globe is a major ecological and technical problem, and one of such biggest challenges is crop residues, which are leftover material generated as part of any agricultural and allied activities. When crop residues are applied directly to the soil plant system for crop production, it can create hazards related to nutrient management. Composting technology has been recognized as the most eco-friendly and cost-effective alternative to convert these leftovers into the products that conditions soil and nourishes plants. Composting enhances soil fertility and soil health resulting in increased agricultural productivity, improved soil biodiversity, reduced ecological risks, and better environment. Composting can be considered as a highest form of recycling process that directly or indirectly benefits mankind. Composting has evolved during the twentieth century from an art to a science (Howard 1943).

Composting as a process involves the biological decomposition of organic matter. On the basis of organism used, process or space used, composting can be further subdivided into various types such as vermicomposting, microbial composting, aerobic composting, anaerobic composting, fungal composting, onsite composting, offsite composting, vessel composting, tank composting, etc. Pit methods, Perforated Tank method, and Windrow method are some of the common microbial composting methods which can be employed to prepare compost using compost inoculants.

Vermicomposting has manifold relevance to today's rural and urban environment which may be a very efficient option to handle solid waste in an environmentally friendly way. It is extremely useful for organic farming and is also an important technology for solid waste management. Vermicomposting involves bio-oxidation and stabilization of the waste as a result of the interactions between some species of earthworms and microorganisms (Domínguez et al. 2010). The species of earthworm used most often to produce vermicompost are *Eisenia fetida* (Red wiggler), *Lumbricus rubellus*, and *Eisenia hortensis* (European night crawler). The earthworms perform different roles in the soil such as that of a turner, mixer, aerator, screener, and accelerator that result in enhanced soil fertility, improved plant growth in an economic and eco-friendly way.

### *15.5.2 Biocontrol Agents for Reducing Pesticide Consumption*

Biological control is an approach to reduce populations of harmful organisms with natural enemies. It is defined many times as "the use of natural or modified organisms, genes, or gene products to reduce the effects of undesirable organisms (pests) and to favor desirable organisms such as crops, trees, animals, and beneficial insects and microorganisms." This definition provides a scientific framework that covers all the strategies toward biological control for all categories of pests. Biological control agents have poor storage, are highly target specific, have slow speed to kill, have long persistence through secondary cycling and thus have lower frequency of application, are environmentally friendly and low hazard for humans and livestock (Doelle et al. 2009). Parasitism, competition, pathogenesis, production of allelochemicals and making host plant resistant to pest are the most common actions of agents of biological control. Biological control agents include invertebrates predators (spiders and insect herbivores, mites, flies, ants, beetles, dragonflies, and water bugs); vertebrates predators (amphibians and fishes, reptilians, birds, rodents, and mammals); parasites (parasitoids); pathogens (entomopathogenic and nematopathogenic protozoans, nematodes, earthworms, fungi, bacteria, virus, and virus-like particles); pheromones and botanicals (plant products/extracts such as nicotine, azadirachtin, sesamin and seasamolin etc.) (Kwenti 2016). Apart from these biological control agents humans are practicing since long time to use mechanical methods to control pest. Some of the popular mechanical methods used for pest control are installation of bird shelter, using light trap, pheromone traps and colored plates. Biocontrol agents kill or suppress pathogens and pests by parasitizing the pest or pathogen, competing for space or nutrients, producing toxins to kill and/or inducing a change (physiological or biochemical) in the host plant making it less susceptible to tolerance to pest and pathogen attack.

There are three broad and somewhat overlapping types of biological control methods: conservation, classical biological control (introduction of natural enemies to a new locale), and augmentation (Gurr and Wratten 2012). Conservation process involves manipulation or creation of the environment to favor natural enemies, either by removing or mitigating adverse factors. These modifications include construction of artificial structures, provision of supplementary food and alternative hosts, improvement of pest–natural enemy synchronization, control of honeydew-feeding ants, and modification of adverse agricultural practices (Barbosa 1998). Classical biological control is an approach to search useful natural enemies, introduce them into the area of the target pest, and permanently establish them so that they can provide continuous pest management without any human intervention or with very less involvement. There are many examples of successful classical biological control programs, e.g., control of the cottony cushion scale, *Icerya purchasi* in California with the predatory coccinellid, *Rodolia cardinalis* imported from Australia in 1988; control of prickly pear, *Opuntia* spp., in Australia with the pyralid *Cactoblastis cactorum* from Argentina (1920s); the control of the coconut moth, *Levuana iridescens*, in Fiji Islands with the tachinids by *Besa remota* imported from Malaya in 1925 (Caltagirone 1981). Management of papaya mealy bug (*Paracoccus marginatus*) in Tamil Nadu, Maharashtra, and Karnataka by introduction of *Acerophagus papaya* is the recent example of classical biological control (Mani et al. 2012). Augmentation involves various techniques, including periodic releases and environmental manipulation to increase populations or beneficial effects of natural enemies (Hoy 2008). The extent to which the biological control agents can be utilized varies from crop to crop and from location to location (Kerkut 1985).

Ladybird beetles, lacewings, or parasitoids such as *Trichogramma* are frequently released in large numbers (inundative release). Introduction of *Trichoderma* spp. as an effective biological control agent of the plant pathogens and as plant growth enhancer has come as a boon to agriculture. As most of the crops are infected by the seed-or soilborne plant pathogens that primarily attack the vulnerable seeds or seedlings, the antagonists can be applied directly to the targeted area, i.e., to seeds or seedlings, and a single application (seed treatment, biopriming, furrow treatment) can significantly reduce crop losses (Maheshwari and Dubey 2008). A number of successful commercialized products based on different species of *Trichoderma* have been commercialized in India and elsewhere. Naturally occurring chemicals extracted from plants have long been proclaimed as an alternative to chemical pesticides. Approximately 2000 plant species have been documented to possess pest management properties, out of which 1005 species of plants exhibit insecticide properties, 384 with antifeedant properties, 297 with repellent properties, 27 with attractant properties, and 31 with growth inhabiting properties (Sharma and Gaur 2017). Many plants such as *Azadirachta*, *Cymbopogon* etc. have already been exploited for the commercial production of biopesticides.

### *15.5.3 Other Approaches*

There are many other approaches apart from those discussed above and one of the methods is intercropping. It is the agricultural practice of cultivating two or more crops in the same space at the same time. This method is an old and commonly used cropping practice which is practiced around the globe and it aims to match crop demands efficiently according to the available growth resources and labor (Lithourgidis et al. 2011). There are many advantages of intercropping such as better yield, improved soil fertility because of biological nitrogen fixation with the use of legumes, increased soil conservation, reduced pest incidence, suppression of weeds, reduced fertilizer and pesticide requirements, thus minimizing environmental impacts of agriculture. However, there are some disadvantages of intercropping as well, for example the selection of the appropriate crop species and their combination is very important, appropriate sowing densities and extra work during crop management practices, including harvesting. Wheat and maize are the two most widely used crops for intercropping. These are alternated with various other crops such as wheat/maize, wheat/chickpea, maize/chickpea, wheat/soybean, peanut/maize, pea/barley, etc. (Banik et al. 2006; Zhang and Li 2003; Hauggaard-Nielsen et al. 2001; Li et al. 2001, 2004).

## 15.6 Conclusions

Ever since the farming practices have been initiated there is depletion of soil health and environment. Human beings are using various chemicals to enrich the soil for better crop production. Although fertilizers and pesticides are being added to meet the food demand, they have turned out to be a point of concern. They are influencing biological, chemical, physical properties of soil, deteriorating soil health, contaminating the water, causing environmental pollution etc. Thus, a potential alternative is required which can minimize the use of fertilizers and pesticides and one of the options is sustainable agriculture. There have been examples from around the world where it has become possible to increase the crop yield in a sustainable way. Farmers around the globe must be trained with such methods and techniques and the use of inorganic fertilizers and pesticides must be stopped.

**Acknowledgments** The authors are thankful to Principal Maitreyi College, Zakir Husain Delhi College and PGDAV College University of Delhi, to provide all the necessary facilities to compile this manuscript.

## References

Aktar MW, Sengupta D, Chowdhury A (2009) Impact of pesticides use in agriculture: their benefits and hazards. Interdisc Toxicol 2(1):1–12. https://doi.org/10.2478/v10102-009-0001-7

Banik P, Midya A, Sarkar BK, Ghose SS (2006) Wheat and chickpea intercropping systems in an additive series experiment: advantages and weed smothering. Eur J Agron 24(4):325–332. https://doi.org/10.1016/j.eja.2005.10.010

Barbosa PA (1998) Conservation biological control. Elsevier, Amsterdam, p 396. ISBN: 0080529801, 9780080529806

Bhardwaj D, Ansari MW, Sahoo RK (2014) Biofertilizers function as key player in sustainable agriculture by improving soil fertility, plant tolerance and crop productivity. Microb Cell Factories 13:66. http://www.microbialcellfactories.com/content/13/1/66

Bhowmik SN, Das A (2018) Biofertilizers: a sustainable approach for pulse production. In: Meena RS et al (eds) Legumes for soil health and sustainable management. Springer, Singapore. https://doi.org/10.1007/978-981-13-0253-4_14

Caltagirone LE (1981) Landmark examples in classical biological control. Annu Rev Entomol 26:213–232

Chen JH (2006) The combined use of chemical and organic fertilizers and/or biofertilizer for crop growth and soil fertility. International workshop on sustained management of the soil-rhizosphere system for efficient crop production and fertilizer use, 16–20 October 2006. Land Development Department, Bangkok 10900, Thailand, pp 1–11

Delaux PM, Radhakrishnan GV, Jayaraman D, Cheema J, Malbreil M, Volkening JD, Sekimoto H, Nishiyama T, Melkonian M, Pokorny L, Rothfels CJ, Sederoff HW, Stevenson DW, Surek B, Zhang Y, Sussman MR, Dunand C, Morris RJ, Roux C, Wong GK, Oldroyd GE, Ané JM (2015) Algal ancestor of land plants was preadapted for symbiosis. Proc Natl Acad Sci USA 112 (43):13390–13395. https://doi.org/10.1073/pnas.1515426112

Doelle HW, Rokem JS, Berovic M (2009) Biotechnology-volume VI: fundamentals in biotechnology. EOLSS Publications, Oxford, p 382. ISBN: 1848262604, 9781848262607

Domínguez J, Aira M, Gómez-Brandón M (2010) Vermicomposting: earthworms enhance the work of microbes. In: Insam H, Franke-Whittle I, Goberna M (eds) Microbes at work. Springer, Berlin, pp 93–114. https://doi.org/10.1007/978-3-642-04043-6_5

Epule TE (2019) Contribution of organic farming towords global food security: an overview. In: Chandran S, Unni MR, Thomus S (eds) Organic farming: global perspectives and methods. Publishing series in food science, technology and nutrition. Woodhead Publishing, Cambridge, pp 1–16. ISBN: 9780128132739

Gurr G, Wratten SD (2012) Biological control: measures of success. Springer, Dordrecht, p 429. ISBN: 9401140146, 9789401140140

Hauggaard-Nielsen H, Ambus P, Jensen ES (2001) Interspecific competition, N use and interference with weeds in pea–barley intercropping. Field Crop Res 70(2):101–109. https://doi.org/10.1016/S0378-4290(01)00126-5

Hernández-Herrera RM, Santacruz-Ruvalcaba F, Ruiz-López MA, Norrie J, Hernández-Carmona G (2014) Effect of liquid seaweed extracts on growth of tomato seedlings (*Solanum lycopersicum* L.). J Appl Phycol 26(1):619–628. https://doi.org/10.1007/s10811-013-0078-4

Howard A (1943) An agricultural testament. Oxford University Press, London

Hoy MA (2008) Augmentative biological control. In: Capinera JL (ed) Encyclopedia of entomology. Springer, Dordrecht. ISBN: 978-1-4020-6359-6

Kariman K, Barker SJ, Tibbett M (2018) Structural plasticity in root-fungal symbioses: diverse interactions lead to improved plant fitness. Peer J 6:e6030. pp 38. https://doi.org/10.7717/peerj.6030

Kerkut GA (1985) Insect control, volume 12: comprehensive insect physiology, biochemistry & pharmacology. Elsevier, Amsterdam, p 864. ISBN: 148328624X, 9781483286242

Khosla R (2017) Biofertilizers and biocontrol agents for organic farming. Kojo Press, New Delhi, p 137. ISBN: 978-81-927567-90

Kwenti TE (2016) Biological control of parasites. In: Khater H (ed) Natural remedies in the fight against parasites. Intechopen, Rijeka, pp 23–58. https://doi.org/10.5772/68012

Li L, Sun J, Zhang F, Li X, Yang S, Rengel Z (2001) Wheat/maize or wheat/soybean strip intercropping: I. Yield advantage and interspecific interactions on nutrients. Field Crop Res 71(2):123–137. https://doi.org/10.1016/S0378-4290(01)00156-3

Li SM, Li L, Zhang FS, Tang C (2004) Acid phosphatase role in chickpea/maize intercropping. Ann Bot 94(2):297–303. https://doi.org/10.1093/aob/mch140

Lithourgidis AS, Dordas CA, Damalas CA, Vlachostergios DN (2011) Annual intercrops: an alternative pathway for sustainable agriculture. Aust J Crop Sci 5(4):396–410. ISSN: 1835-2693

Maheshwari DK, Dubey RC (2008) Potential microorganisms for sustainable agriculture: a techno-commercial perspective. I. K. International, New Delhi, p 482. ISBN: 8190746200, 9788190746205

Mani M, Shivaraju C, Shylesha AN (2012) *Paracoccus marginatus*, an invasive mealybug of papaya and its biological control – an overview. J Biol Control 26(3):201–216

Mendes R, Garbeva P, Raaijmakers JM (2013) The rhizosphere microbiome: significance of plant beneficial, plant pathogenic, and human pathogenic microorganisms. FEMS Microbiol Rev 37 (5):634–663. https://doi.org/10.1111/1574-6976.12028

Nabti E, Jha B, Hartmann A (2017) Impact of seaweeds on agricultural crop production as biofertilizer. Int J Environ Sci Technol 14(5):1119–1134. https://doi.org/10.1007/s13762-016-1202-1

Prashar P, Shah S (2016) Impact of fertilizers and pesticides on soil microflora in agriculture. In: Lichtfouse E (ed) Sustainable agriculture reviews, vol 19. Springer, Cham, pp 331–361. https://doi.org/10.1007/978-3-319-26777-7_8

Reddy PM, James EK, Ladha JK (2002) Nitrogen fixation in rice. In: Leigh GJ (ed) Nitrogen fixation at the millennium. Elsevier, Amsterdam, pp 421–445. ISBN: 978-0-444-50965-9

Sharma P, Gaur N (2017) Relative effectiveness of plant products against tobacco caterpillar (*Spodoptera litura*) and Bihar hairy caterpillar (*Spilarctia obliqua*). J Entomol Zool Stud 5 (6):1150–1156

Singh JS, Kumar A, Rai AN, Singh DP (2016) Cyanobacteria: a precious bio-resource in agriculture, ecosystem, and environmental Sustainability. Front Microbiol 7:529. https://doi.org/10.3389/fmicb.2016.00529

Talukdar NC, Thakuria D, Goswami C (2003) Organic farming and quality of organic food. Bioprospecting of commercially important plants. In: Borah RC, Talukdar A, Kataky JCS, Unni BG, Modi MK, Deka PC (eds) Proceedings of the national symposium on "Biochemical approaches for utilization and exploitation of commercially important plants", Jorhat, India, 12–14 Nov 2003, pp 61–72

Tomkins P, Bird C (2002) Chemicals, plants and man: the organic farming residue. In: Tomkins P, Bird C (eds) Secret life of plants, pp 240–258

World Commission on Environment and Development (1987) Our common future. Oxford University Press, Oxford, p 27. ISBN: 019282080X

Wu SC, Cao ZH, Li ZG, Cheung KC, Wong MH (2005) Effects of biofertilizer containing N-fixer, P and K solubilizers and AM fungi on maize growth: a greenhouse trial. Geoderma 125 (1–2):155–166. https://doi.org/10.1016/j.geoderma.2004.07.003

Yadav IC, Devi NL (2017) Pesticides classification and its impact on human and environment. Environ Sci Eng Toxicol 6:140–158

Zhang F, Li L (2003) Using competitive and facilitative interactions in intercropping systems enhances crop productivity and nutrient-use efficiency. Plant Soil 248(1–2):305–312. https://doi.org/10.1023/A:1022352229863

# Chapter 16
# Portraying Microbial Beneficence for Ameliorating Soil Health and Plant Growth

**Jay Kishor Prasad, Riddha Dey, Sandeep Kumar Gupta, and Richa Raghuwanshi**

**Abstract** Soil microflora is a vital component in improving plant growth as it performs many crucial and primary soil functions such as soil fertility, nutrient cycling, increased availability of limited nutrients, and decomposition of inorganic as well as organic matter. Soil microorganisms also positively influence physical properties of soil like its structure, porosity; aeration, and water infiltration which are favorably affected by their soil aggregate forming capabilities. Further the soil microflora plays an important role in practicing ecofriendly approaches like detoxification (bioremediation) of soils contaminated with toxins and undesirable components added due to anthropogenic activities. The global concern about chemical residues affecting soil health and environment has stimulated interest in the dynamic role of soil microbes in soil protection. Microbial interactions with the plants evoke various kinds of local and systemic response that not only improve plant's metabolic capability to resist abiotic stress but also indirectly affect the soil health as plant growth, microbial activity, and soil health are closely interlinked. The present review is focused on the central role of soil microbes in ameliorating the harmful effects of chemicals on soil health and plant growth.

**Keywords** Abiotic stress · PGPR · Plant growth · Soil health

J. K. Prasad · R. Dey · S. K. Gupta
Department of Botany, Institute of Science, Banaras Hindu University, Varanasi, Uttar Pradesh, India

R. Raghuwanshi (✉)
Department of Botany, Mahila Mahavidyalaya, Banaras Hindu University, Varanasi, Uttar Pradesh, India



B. Giri, A. Varma (eds.), *Soil Health*, Soil Biology 59,
https://doi.org/10.1007/978-3-030-44364-1_16

## 16.1 Introduction

Soil health may be defined as the continuous capability of soil to act as a vital living system, within ecosystem and land-use boundaries, maintaining biological productivity (Doran et al. 1996). Abiotic stresses and the fluctuating climate are the major factors deteriorating soil health by imbalancing the nutrient cycling and degrading the physical and biochemical properties of soil thereby reducing agricultural productivity. Besides natural factor even anthropogenic activities have led to an approximate loss of 24 billion tons of fertile soil from the world's crop lands (FAO 2011). The intensity of degradation caused by anthropogenic activities extends from highly degraded (25%) to moderately degraded (36%) lands of which only 10% of land is reported to be improving. The increasing population with changing consumption patterns has further intensified the problem by increasing the demand for energy by 45%, food by 50%, and water by 30% (IFPRI 2012). Although the use of agricultural chemicals increased the productivity of the available lands, high energy demands and environmental costs associated resulted in the search for alternative methods to improve soil fertility and productivity in degraded lands. No doubt that the traditional agricultural practices like composts, green and farm yard manures, and crop management practices like natural fallow, intercropping, relay cropping and crop rotations improve soil health but still the adaptive capability of plants is unable to keep pace with the rapid global environmental changes taking place due to anthropogenic activities. Under such a scenario, microorganisms which are critical in maintaining soil quality and health can be exploited as a tool to promote plant growth and maintain environmental sustainability (Jeffries et al. 2003). Plant growth promoting microbes have risen as an alternative tool among the sustainable agricultural practices to minimize the application of chemicals. Bacteria that promote plant growth either directly (nitrogen fixation, phosphate solubilization, siderophore production, and phytohormone production) or indirectly (antibiotic production, suppression of plant pathogens, induction of resistance in host against biotic and abiotic stresses, modifying plant–microbe interactions like mycorrhizal and *Rhizobium* symbiosis) are referred as plant growth promoting rhizobacteria (PGPR). Besides PGPR, the arbuscular mycorrhizae (AM), which is a wide spread mutualistic symbiosis between land plants and fungi benefit soil by their soil aggregation capabilities and plants by enhancing nutrient uptake and resistance against various stresses. Thus plant–microbe interactions not only provide a fundamental support to the plant in acquiring nutrients and imparting resistance but also affect soil health (Nguyen et al. 2016; Turner et al. 2013) as these three factors are dependent on each other. Any ecosystem functioning is largely governed by soil microbial number and activities which get reduced on addition of chemical and this reduction becomes critical for the reestablishment of the vegetation cover in a degraded soil. The present review discusses the role of microbes in terms of replacing chemicals, degrading chemicals, and facilitating plant growth in degraded soils.

## 16.2 Soil Structure and Aggregation

Soil structure denotes the three dimensional arrangement of organic or mineral complexes (aggregates) and the pore size (Rillig and Mummey 2006). Soil aggregates are clusters of soil particles that are clumped together by moist clay, organic matter (like roots), gums (from bacteria and fungi), and by fungal hyphae. The aggregates are made up of particles of different sizes, varying from 2 μm about 200 μm diameter in size. Soil pores are the spaces between soil aggregates and these are necessary for storing air, water, nutrients, microbes, and organic matter. Soil aggregation is an integral process in the development of soil structure and has a strong impact on the biological, hydrological, and mechanical properties of soils. Plant growth can be severely disturbed if the soil structure is not favorable for seedling emergence or root development (Marshall and Holmes 1988). The binding of soil particles into stable aggregates is essential for the production of optimum soil tilth. The soil profile depicted in Fig. 16.1 shows the different layers having different make up, texture, age, and characteristic. The O horizon, primarily composed of decaying organic matter, hosts most of the roots and soil microbes which contribute immensely in maintaining soil quality. Aggregate formation around roots involves the adherence of fine soil particles to living root hairs, bacteria, and fungal hyphae (Mishustin 1945; Hubbell and Chapman 1946). Soil biodiversity is a property that is vital for the continued capacity of the soil to produce crops. Thus, the ecological attributes associated with the soil biota, i.e., its diversity, food web structure, activity, and the range of functions it performs are important in maintaining soil health.

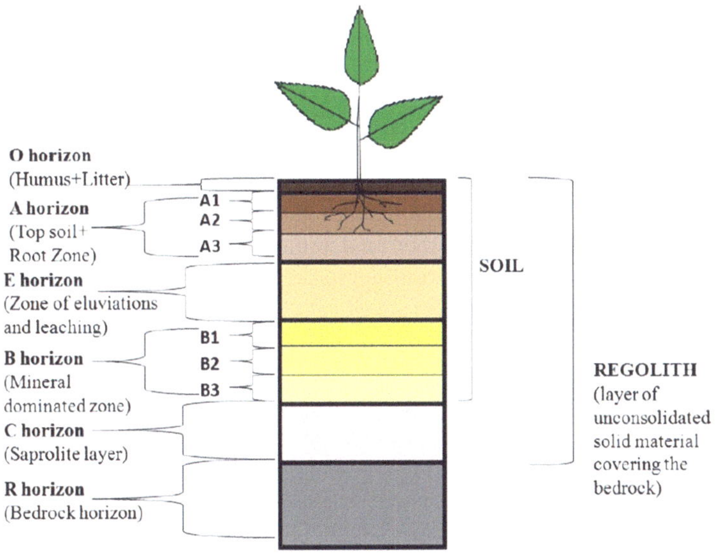


**Fig. 16.1** Soil profile diagram

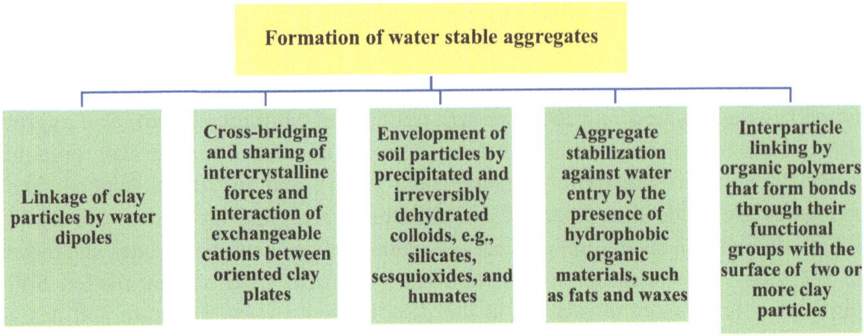


**Fig. 16.2** Mechanisms in formation of water-stable soil aggregates

The two ways proposed for the formation of soil aggregates by bacteria are:

1. Bacteria produce polysaccharides which makes many points of contact with soil particles producing aggregates. Bacterial polysaccharides resist decomposition (Harris et al. 1964).
2. Bacteria develop a small electrostatic charge which attracts to the electrostatic charge on clay surfaces, bringing together small aggregates of soil (Naseem and Bano 2014).

The mechanisms involved in the development of water-stable soil aggregates (Harris et al. 1964) is depicted in Fig. 16.2.

Soil aggregation is an important index for evaluation of soil physical properties as it sustains soil health and fertility, reduces soil erosion and mediates the air permeability, water infiltration and nutrient cycling (Spohn and Giani 2011; Zhang et al. 2012). Soil aggregates are highly stable in soil rich in organic carbon. Several factors that interact like microbial activities, nutrient reserves, moisture availability, exchangeable ions, environment, pedogenic processes results in complex dynamics of soil aggregation (Kay 1990). The abundant water stable aggregate of size 0.25–0.1 mm resides at the upper soil surface layer (0–15 cm), determines the potential for sheet erosion and crust formation (Ahemad et al. 2012). Soil aggregates help in retaining soil organic carbon and protect against the decomposition of organic matter (Six et al. 2000). Soil aggregate stability is a good index of soil erodibility (Diaz-Zorita et al. 2002) as well.

However, under water stress conditions soil aggregation can be challenging and hence rhizobacteria play a dynamic role in overcoming this barrier. PGPR have different kinds of mechanisms that directly help to increase the soil organic carbon and maintain soil fertility. PGPR produces exopolysaccharides (EPS) which are most important part of extracellular matrix that often represent 40–95% of bacterial weight (Flemming and Wingender 2001). Bacteria produce EPS in two forms: (1) slime EPS and (2) capsular EPS (Vanhooren and Vandamme 1998). EPS are found in diverse complex structures (Kumon et al. 1994). The important roles exhibited by EPS are (1) protection, (2) surface attachment, (3) biofilm formation,

(4) microbial aggregation, (5) plant–microbe interaction, and (6) bioremediation (Manca de Nadra et al. 1985). Exopolysaccharides are the active constituents of soil organic matter which help in improving soil health and fertility (Gouzou et al. 1993). Histochemical investigations confirm that polysaccharides are the most important component of root mucilage for binding soil particles. The mucilage from the rhizosheath bacterium *Cytophaga* sp. has been shown to bound to soil by component protein (Watt et al. 1993). Earlier, Reid and Goss (1981, 1982) showed that polysaccharides from growing roots played an important role in stability of microaggregates in a silt loam and a sandy loam soil. Bacterial and fungal hyphae together with fine roots of plants are involved in the formation of microaggregates and their turnover. Residues of root and shoot form the cavitation sites for the growth of fungi and bacteria. Microbial activity is activated by the biological macroaggregates forming around growing roots in soil. Soil fauna especially earthworms, termites, and ants play an important role in formation of biological macroaggregates in soils in diverse climates (Six et al. 2002).

Role of glomalin protein secreted by arbuscular mycorrhizal fungi (AMF) in soil aggregation is well documented (Wright et al. 1996) as these symbionts produce the protein 2–3 folds higher than other fungi (Wright and Upadhyaya 1996). Among all soil factors Jastrow et al. (1998) demonstrated that AMF hyphae have a direct effect on soil aggregation through production of glomalin protein which helps in soil binding. Role of AMF in soil engineering has mostly been studied with respect to soil aggregation, as these aggregates physically protect the carbon inside and contribute to the soil carbon pool (Jastrow 1996; Six et al. 2000). Based on C dating Rillig et al. (2001) concluded that glomalin had a residence time in soil ranging from 6 to 42 yrs., as it is not readily metabolized by soil organisms. Rillig et al. (2001) demonstrated that glomalin actually accumulates in soil to the extent that it represents 4 to 5% of the soil carbon. Mycorrhizal helper bacteria like *Pseudomonas* species stimulating the formation of arbuscular mycorrhizas, are also known to affect the stability of macroaggregates (Krishna et al. 1982). *Lolium perenne* grasses and most of the C4 plants support more external hyphae and stabilize soil more effectively than C3 grasses like *Trifolium repens* (Tisdall and Oades 1979; Miller and Jastrow 1990).

## 16.3 Soil Degradation

Decline in soil quality in terms of its physical, chemical, or biological state is referred to as soil degradation. While soil degradation may be a natural process, it is highly exuberated by human activity. Physical factors like rainfall, tillage, floods, surface runoff, wind erosion, and mass movements resulting in the loss of top fertile soil decrease the soil quality. Alkalinity or acidity resulting due to mining and excess use of chemicals also leads to soil deterioration in the long term. Extensive cultivation and improper agricultural practices like mono-cropping, poor manuring, excess use of chemical fertilizers, excessive irrigation, overgrazing, and deforestation also

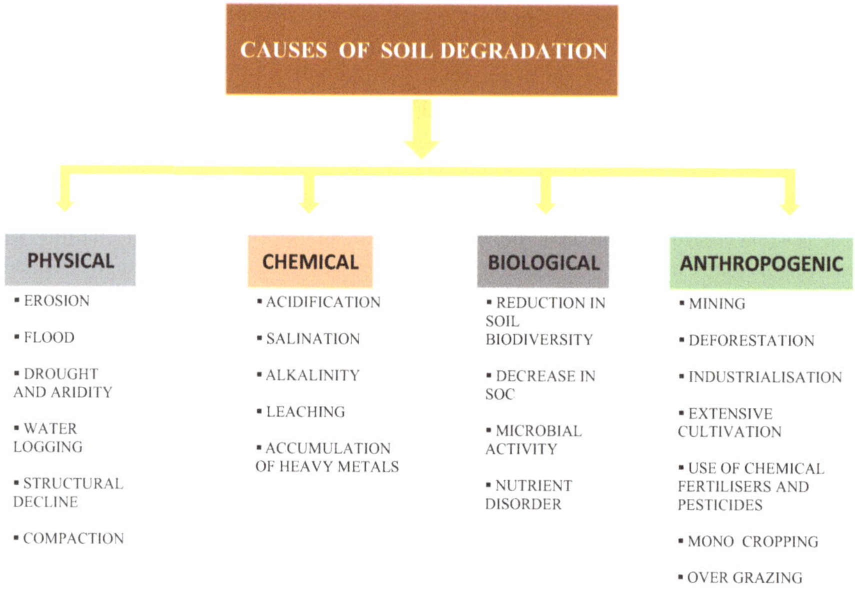


**Fig. 16.3** Various causes of soil degradation

accelerate the process of soil degradation. Soil degradation leads to adverse changes in soil structural attributes, organic matter and nutrient status thereby affecting crop productivity. The various causes of soil degradation are depicted in Fig. 16.3.

Any factor adversely affecting the soil microfloral diversity and count also reduces the biological or microbial activity of soil. Mono-cropping often leads to increasing attack of pests and diseases, thereby affecting the soil microbial equilibrium and in turn its activity. Similarly organic matter is rapidly decomposed and leached or eroded by heavy rains or floods (Bennett 1993). Salinity also lowers productivity through its adverse effects on the availability of nutrients. The adverse effect of alkalinity on availability of nutrients is due to deflocculating effect of sodium ions. Dumping of unprocessed industrial waste in open grounds often leads to the accumulation of heavy metals such as Lead (Pb), Cadmium (Cd), and Arsenic (As) in the soil. Excessive and unrestricted application of chemical fertilizers and pesticides creates not only negative impacts on soil fertility and soil health but also a major part of it is inaccessible to plants (Bhandari 2014), which further leads to decrease in soil organic carbon, loss of beneficial microbial flora, increased acidity/alkalinity, and hardening of soil (Nicholson and Hirsch 1998; Yang et al. 2000; Bohme et al. 2005).

Soil erosion is another major cause of soil degradation. Soil erosion mostly disturbs the upper fertile layer of soil, making it deficient in essential minerals and resulting in productivity loss. Destruction of forests accompanied by reduced rainfall frequency is also a factor for soil erosion causing damage to agriculture property.

Soil erosion not only makes an area vulnerable to nutrient loss but also destroys microbial dynamics affecting the ecosystem at large (Lal 1987).

## 16.4 Microbial Assistance in Plant Growth in Degraded Soils

Plant's response to any microbial treatment is the result of the complex interactions with the rest of the soil microbial inhabitants and soil condition. Soil health in turn is influenced by the plant roots as the root exudates are not only rich in nutrients (amino acids, nucleotides, fatty acids, phenols, plant growth regulators, organic acids, putrescine, sterols, sugars and vitamins) compared to bulk soil but also provide a surface for microbial attachment and proliferation. This leads to enrichment of microorganisms (multiple folds than the bulk soil) such as bacteria, fungus, cyanobacteria, algae and protozoa, among which growth promoting rhizobacteria is most significant (Uren 2007). These microbes can be used as a replacement to chemicals used in agriculture or may help in overcoming the adverse effects of chemicals as discussed below.

### *16.4.1 Fertilizer-Affected Soils*

Non-sustainable intensive agricultural practices like cultivating fertilizer-responsive crops or short duration crops with high yield often make the soils scarce in organic matter and nutrients and accelerate the loss of plant nutrients to a much greater extent compared to what is supplemented through fertilizers. Long-term fertilization alters the soil organic carbon, nitrogen content, pH, soil moisture and thus shifts the microbial population (Bohme et al. 2005; Bunemann and McNeill 2004; Wu et al. 2012). A survey of long-term monitoring of 10 field sites in China revealed that soil pH decreased substantially by 0.45–2.20 units over a 8 to 25 years period in plots fertilized with mineral N, P, and K fertilizers (Guo et al. 2010). Use of large amount of potassium fertilizers causes Ca, Fe, and Zn imbalance which disrupts the nutrient balance and uptake by plants. Usage of fertilizers more than the recommended amount leads to soil compaction in the long term. High compaction decreases porosity and aeration while increasing bulk density and soil penetration resistance. Restricted root development further limits the water and nutrient uptake adversely affecting the yields (Massah and Azadegan 2016). Moreover overuse of fertilizers shows negative effects on organisms like worms and soil mites which are responsible for maintaining soil health. Chemical fertilization creates nutrient channels or patches which results in nutrient gradients in the soil affecting the microbial populations and diversity (Li et al. 2013).

The use of microbes as biofertilizers is a potential method to decrease the negative impact in the environment caused by the continuous overuse of chemical fertilizers. PGPRs mostly influence plant growth either by facilitating uptake of nutrients from the soil or by synthesizing plant hormones (Glick et al. 2007) like IAA, which promote root growth enhancing the nutrient absorption. Diazotrophs convert atmospheric nitrogen into usable organic forms in rhizospheric soil. *Rhizobium* inoculation helps to enhance nodulation, plant growth, and produces enhanced grain yield by 10–15 fold (Nutmen and Roberts 1969). *Azotobacter* plays an imperative role in nitrogen fixation as well as releases considerable amounts of biologically active substances like vitamins such as thiamine and riboflavin (Revillas et al. 2000). It also helps in secretion of ammonia in the rhizosphere along with the root exudates, which results in modification of nutrient uptake by the plants (Narula and Gupta 1986). *Azospirillum* is reported to fix atmospheric nitrogen, produce plant growth promoting substances IAA and IBA, and increase the rate of mineral uptake by plant root enhancing the plant yield (Gadagi et al. 2004). Blue-green algae, a diverse group of prokaryotes often form complex associations with bacteria and green algae forming cyanobacterial mats used as biofertilizers in modern agriculture to supplement nitrogen. Besides these well-known microbes, a number of microbes (*Serratia, Enterobacter cloacae, Burkholderia ambifera, Pantoea ananatis*) are being studied for their diazotrophic role.

Many PGPR help in solubilizing insoluble phosphate either by producing organic acids that lowers the soil pH, by chelation of cations or by competing with phosphate for adsorption sites in the soil (Nahas 1996). *Pseudomonas* degrade organic nitrogenous compounds and also enhance circulation of nitrogen and phosphorous in soil (Hayat et al. 2010). It is a visual observation that the plant growth regulators result in vigorous development of greener and larger leaves, despite the unfavorable conditions (Nagy and Pinter 2015). Plant growth promoting microbes like *Pseudomonas*, *Bacillus*, and *Trichoderma* show beneficial effect upon plant growth and can be used as biofertilizers for agriculture (Broughton et al. 2003; Saxena et al. 2016a, b).

Arbuscular mycorrhizal fungi are primarily responsible for increased nutrient uptake by plants. Extraradical hyphae from the mycorrhizal plant extend up to several centimeters and increase the effectiveness of absorption of immobile elements by 60-folds (Bieleski 1973). Though AM fungi are able to absorb and transfer all the 15 macro- and micronutrients necessary for plant growth, this behavior is particularly evident with soil nutrients that are more immobile such as P, Zn, and Cu (Khan et al. 2000). When a mixture of PGPR strains *Bacillus amyloliquefaciens, Bacillus pumilus* T4 and AMF (*Glomus intraradices)* were inoculated in tomato, it supplemented 75% of the recommended fertilizers, increasing yield (Adesemoye et al. 2009). Similarly, inoculation of *Glomus coronatum* with *Rhizobium sp.* NR9 showed increased plant growth, nutrient uptake, and mycorrhizal root colonization (Barea et al. 1997). Soil microbial components and their interactions can be used for restoring the soil's natural fertility and the microbes being self-replicating minimizes the need for repeated application as is with chemical fertilizers.

### 16.4.2 Pesticide-Affected Soils

Pesticides are bioactive, toxic, and xenobiotic substances which directly or indirectly affect agroecosystem quality and soil productivity (Imfeld and Vuilleumier 2012). Pesticides and fertilizers have become an integrated part of the modern agricultural practices that has been able to manage the rising food demand but the soil health has been severely compromised due to its adverse effects. An actively growing living system shows rapid metabolism of the chemical pesticides which dilutes its adverse effects compared to a relatively static soil system where it persists for prolonged period (Edwards 1975). Furthermore, the breakdown of pesticide molecules results in release of many toxic molecules in the soil. Some essential biochemical reactions that occur in soil like nitrogen fixation, mineralization, nitrification, denitrification, and ammonification are affected by the pesticides through alteration of specific soil microorganisms or by activating/deactivating the enzymes involved (Kinney et al. 2005; Menon et al. 2005; Niewiadomska 2004). The excess uses of chemical pesticides like unden and lindane (156 and 125 g $ha^{-1}$) inhibit the growth of microbes and alter the physicochemical structure of the soil decreasing the crop yields in future (Glover-Amengor and Tetteh 2008). Similar reports on inhibition of nitrification by higher dose of hexazinone pesticide have been reported in a soddy podzolic soil (Bliev et al. 1985). Pesticides adversely affect the nodulation by affecting virulence of attacking nodular bacteria. Niewiadomska (2004) and Niewiadomska and Klama (2005) reported the adverse effects of carbendazime, thiram (fungicides), and imazetapir (herbicide) on nitrogenase activity of *Rhizobium leguminosarum*, *Sinorhizobium meliloti*, and *Bradyrhizobium* sp., in clover, lucerne, and serradella plants. Herbicides have been reported to affect *Bradyrhizobium japonicum* growth in vitro and reduce the nodulation of soybeans under greenhouse conditions (Malik and Tesfai 1985, 1993). Application of pesticides affects the efficiency and activity of nitrogenase enzyme. Pesticides also show inhibitory effects on rates of decomposition of organic matter and mineralization in agricultural and grassland ecosystems (Perfect et al. 1981; Pimentel and Warneke 1989), forest areas (Weary and Merriam 1978), and desert ecosystem (Santos and Whitford 1981). Metalaxyl (fungicide) significantly decreased total C and N contents in soil during 0–30 days of incubation (Sukul 2006). Similar results are described by Das and Mukherjee (1994, 1998a, b, 2000a, b) in laterite, alluvial, and in rhizospheric soil of rice. Nitrification was repressed and phosphate solubilization was decreased in the soil for 30 days by application of pesticides (lindane, captan, and malathion) (Ogunseitan and Odeyemi 1985). Pesticides falling on soil disturbs the microbial metabolism or enzymatic activities. The activity of soil enzymes like dehydrogenase, hydrolases, and oxidoreductases are negatively affected by pesticides (Engelen et al. 1998; Liu et al. 2008; Topp et al. 1997). On application of chlorimuron ethyl and furadan the urease activity is enhanced up to 14–18% and 13–21%, respectively. Mefenoxam, metalaxyl inhibit alkaline phosphatase and stimulates acid phosphatase (Monkiedje and Spiteller 2002). It is difficult to quantify the net impact of pesticides on biochemical reactions in soil due to greater soil resilience, nature and concentration of pesticide, its activity and metabolism in soil. But in most cases, application of pesticides disturbs the microbial biochemical equilibrium and cycling of biological elements.

Some rhizospheric bacteria and fungi degrade the toxic pesticides converting them into their nontoxic or natural compound (Vargas 1975) which may further be used as energy or nutrient source (Ishaq et al. 1994; Megadi et al. 2010; Mohamed et al. 2011; Tancho et al. 1992). Simultaneously due to co-metabolism the pesticides which do not serve as the source of energy or nutrient for soil microflora are also degraded by the microorganisms (Bollag and Liu 1990). A wide-ranging soil bacteria when constantly exposed to high amounts of toxic and persistent pesticides in agricultural soils develop a capacity not only to survive the presence of these highly toxic substances but may also use them as energy and nutrient source. This leads to complete or partial transformation/mineralization of these pesticides to a level in soil that are either no more toxic or considerably less toxic than the parent molecule resulting in bioremediation of such contaminated sites. Some PGPRs well studied for their mechanism of growth promotion with pesticides degrading property are summarized in Table 16.1.

Biocontrol strategy for disease management elucidated long back by Lenné and Parbery (1976) controls the plant pathogens, restoring the lost homeostasis of the environment. The potential of biocontrol agents (e.g., *Trichoderma*) for controlling the phytopathogens like *Alternaria*, *Colletotrichum*, *Phytophthora*, *Pythium*, *Rhizoctonia*, *Sclerotinia*, *Verticillium* (Begum et al. 2008; Imtiaj and Lee 2008; Jain et al. 2012; Saxena et al. 2016a, b) is well established. The mechanisms involved have been attributed to be antibiosis, mycoparasitim, competition for space and nutrients, enhancing the mechanical strength along with its ability to induce systemic resistance in the plants against the pathogens (Harman 2006; Shoresh et al. 2010; Hermosa et al. 2012; Saxena et al. 2015).

Role of mycorrhizal fungi in imparting resistance against biotic stress has mostly been discussed with its effect in inducing physiological changes in hosts, increased lignification, production of antifungal chitinase, arginine accumulation, and changed root exudation (Dehne and Schonbeck 1979; Harrier and Watson 2004; Ramaiah and Bagyaraj 1989). Mycorrhizal associations lower the competitive ability of weeds relative to crop (Van der Heijden et al. 2008).

### *16.4.3 Flood-Affected Soils*

Soil inundation sets in motion a variety of chemical, physical, and biological processes that alter the capacity of soils to support plant growth. Flooding with moving water often removes soil by scouring or adds soil by transport and silting (Brinson et al. 1981). Changes in soil structure following flooding typically include breakdown of aggregates, deflocculation of clays, and destruction of cementing agents (Ponnamperuma 1984). Major chemical changes include decrease or disappearance of $O_2$, accumulation of $CO_2$, increased solubility of mineral substances,

**Table 16.1** Reports on pesticide degrading and growth promoting PGPRs

| S. No. | PGPR | Mechanism of plant growth promotion | Pesticide degraded | Nature of pesticide | References |
|---|---|---|---|---|---|
| 1 | *Azotobacter chroococcum, Azotobacter vinelandii* | Phosphate solubilization, siderophore, IAA, GA, HCN and ammonia production | Endosulfan, Chlorpyrifos | Insecticide | Gurikar et al. (2015) |
| 2 | *Achromobacter xylosoxidans* | Phosphate solubilization, IAA, HCN, ammonia and catalase production | Endosulfan | Insecticide | Singh and Singh (2011), Saranya et al. (2015) |
| 3 | *Pseudomonas aeruginosa* | Phosphate solubilization, siderophore, IAA, HCN and ammonia production, antagonistic activity | Organophosphate profenofos | Herbicide, Insecticide | Malghani et al. (2009), Kumari et al. (2018) |
| 4 | *Pseudomonas putida* | Phosphate solubilization, siderophore and HCN production | Organophosphate cadusafos | Herbicides, Nematicide | Abo-Amer (2012) |
| 5 | *Providencia stuartii* | IAA production, ACC deaminase and protease enzymes activities | Chlorpyrifos | Insecticide | Rani et al. (2008) |
| 6 | *Acinetobacter johnsonii MA19* | Siderophore and IAA production | Malathion | Insecticide | Xie et al. (2009) |
| 7 | *Acinetobacter sp., Flavobacterium sp., Serratia marcescens, Stenothrophomonas maltophilia, Penicillium sp.* | Phosphate solubilization, siderophore, IAA, HCN and ammonia production | 2,4dichlorophenoxyacetic acid (2,4-D) | Herbicide | Silva et al. (2007) |
| 8 | *Bacillus subtilis GB03, Bacillus subtilis FZB24, Bacillus amyloliquefaciens, Bacillus pumilus* | Phosphate solubilization, siderophore and IAA production | Acibenzolar-S-methyl, Metribuzin, Propamocarb, Napropamide | Fungicide, Herbicide, Insecticide | Charalampos et al. (2012) |
| 9 | *Enterobacter cloacae K7* | Phosphate solubilization, nitrogen fixation, phytohormone production. | Glyphosate | Herbicides | Kryuchkova et al. (2014) |
| 10 | *Rhizobium strain MRP1* | Siderophore, IAA, HCN, ammonia and EPS production | Metribuzin, glyphosate, Kitazin | Herbicide, Fungicide | Ahemad and Khan (2011) |

(continued)

**Table 16.1** (continued)

| S. No. | PGPR | Mechanism of plant growth promotion | Pesticide degraded | Nature of pesticide | References |
|---|---|---|---|---|---|
| 11 | *Bacillus subtilis GB03, Bacillus subtilis FZB24, Bacillus amyloliquefaciens IN937a Bacilluspumilus SE34* | Phosphate solubilization, siderophore, IAA and HCN production, ACC deaminase activity | Acibenzolar-S-methyl, Propamocarb hydrochloride | Fungicide | Charalampos et al. (2012) |
| 12 | *Stenotrophomonas maltophilia* | Phosphate- solubilization, siderophore and IAA production, ACC deaminase activity | Methomyl | Insecticide | Saptanmasi et al. (2008); Mohamed et al. (2011) |
| 13 | *Azotobacter salinestris* | Phosphate solubilization, IAA and GA production | Pendimethalin, Glyphosate, Phorate, Chloropyrifos | Herbicides | Chennappa et al. (2014) |
| 14 | *Azotobacter vinelandii* | Phosphate solubilization, siderophore, HCN and ammonia production | Glyphosate and atrazine | Herbicides | Shahid et al. (2019) |
| 15 | *Bacillus pumilus SE34* | Siderophore and IAA production | Acibenzolar-S methyl | Fungicide | Myresiotis et al. (2012), Kaushal et al. (2019) |

reduction of Fe and Mn, anaerobic decomposition of organic matter, and formation of toxic compounds (Gambrell et al. 1991; Janiesch 1991). Flooding eliminates soil $O_2$ because water occupies previously gas-filled pores. The $O_2$ concentration remains high in only the few millimeters of surface soil that are in contact with oxygenated water. Well-drained soils are characterized by redox potentials of +300 mV or higher, whereas flooded soils have redox potentials of -300 mV or lower (Pezeshki and Chambers 1985a, 1985b). The aerobic organisms typical of well-drained soils are replaced in flooded soils by anaerobes, primarily bacteria, which cause denitrification and reduction of Mn, Fe, and S. Many potentially toxic compounds accumulate in flooded soils. Soil flooding or waterlogging causes major changes in the normal functioning of plant roots (Jackson and Drew 1984) as the gas diffusion rates get reduced in flooded soil (Jackson 1985) and at the same time, respiration by microorganisms and plant roots leads to a rapid buildup of anaerobic conditions in the soil. Methane, ethane, propylene, fatty acids, hydroxy and decarboxylic acids, unsaturated acids, aldehydes, ketones, diamines, mercaptans, and heterocyclic compounds are products of anaerobic metabolism of microbes. Similarly ethanol, acetaldehyde, and cyanogenic compounds are produced by roots (Fulton and Erickson 1964; Rowe and Catlin 1971; Ponnamperuma 1984). Ethylene, a gaseous stress hormone is produced in abundance by flooded plants (Kozlowski and Pallardy 1984) and by microbial metabolism (Lynch 1975; Lindberg et al. 1979). Overall the toxic conditions of soil affects some of the vital processes like ion uptake in root (Jackson and Drew 1984) affecting plant growth.

Most of the studies done on role of microbes in plant growth promotion in flooded soil are with respect to the activity of ACC deaminase (ACCD) enzyme as this reduces the ethylene induced negative response to a great extent in plants (Glick 2005). ACC deaminase producing PGPR (Barnawal et al. 2012; Grichko and Glick 2001a; Li et al. 2013) mitigate the stress by lowering the ethylene level in plants (Saleem et al. 2007) and thereby ameliorating the stress effects. Studies have shown that flooded plants inhabiting ACCD producing microbes are able to overcome the flood response partially. Even plants are genetically engineered to express this enzyme in root-specific manner resulting in less accumulation of ethylene in the roots and thereby minimizing the adverse effects of flooding (Grichko and Glick 2001a, b). Most of the phyllospheric, rhizospheric, and endophytic microbes with ACCD producing enzymes are non-specific toward their host which make them further beneficial. They exhibit a wide range (>100 fold) in ACCD activity and act as a sink for the ACC produced in response to stress. Grichko and Glick (2001) studied the effect of inoculation with ACCD PGPR on tomato subjected to flooding. Seeds of wild type tomato plants when inoculated with *Enterobacter cloacae* CAL2, *Pseudomonas putida* UW4 and *P. putida* (ATCC17399/pRKACC), respectively, showed substantial tolerance to flooding stress implying that bacterial ACCD lowered the effects of stress induced ethylene. Different ACCD producing bacteria have been reviewed by different authors (Glick et al. 2007; Glick 2014; Raghuwanshi and Prasad 2018) for their efficacy in plant protection against yield loss caused by flooding.

### 16.4.4 Salt-Affected Soils

High levels of salinity in soils are due to inadequate irrigation management and excess fertilization leading to secondary salinization which has almost affected 20% of globally irrigated land (Glick et al. 2007). An arid and semi-arid condition also leads to salinization of land and water resources. Salts in the soil occurring as ions are released from breaking down of minerals, i.e., weathering. They may also increase through irrigation water, fertilizers, or sometimes move upward in the soil from shallow groundwater. When precipitation is inadequate to leach ions from the soil profile, salts accumulate in the soil which leads to soil salinity (Blaylock 1994). Water soluble salts are present in all types of soils. Plants uptake essential nutrients in the ionic form of salts, but excessive accumulation strongly decrease the plant growth. During the last century global natural resources have faced serious physical, chemical, and biological land degradation processes resulting in inorganic/organic contamination, compaction and diminished microbial activity/diversity. Around 800 million hectares of land is estimated to be affected by salinity throughout the world (FAO 2008). Approximately seven million hectares of land in India comes under saline soil (Patel et al. 2011) and it continues to expand by the introduction of irrigation in new areas. Salinization can be checked by leaching of salt from root zone, altered farm management practices and use of salt tolerant plants. Salinity stress creates an oxidative burst in cells resulting in an increased accumulation of reactive oxygen species which affects the plasma membrane, cell metabolism, and homeostasis. Studies on interaction of PGPR with other microbes and their effects on the physiological response of crop plants under different soil salinity regimes are still in initial stage.

Association of plants with PGPR equipped with ACCD activity can have a tremendous effect on mitigating plant growth inhibition resulting from stress ethylene formed during salt stress. Under salt stress condition, much of the ACC exudes out from plant roots where ACCD bacteria sequester and degrade ACC to α-ketobutyrate and ammonia, thus, decreasing the building up of stress ethylene in plants. Many reports have shown that high ACC deaminase activity could be one of the primary mechanisms by which bacteria support plant growth under salt stress (Mayak et al. 2004; Saleem et al. 2007). Studies done on maize plant growing in saline–sodic soil when treated with fertilizer along with ACCD producing *Pseudomonas* strains showed 198% augmented plant dry weight (Zafar-ul-Hye et al. 2014). Improvement in the growth of groundnut and red pepper plants by ACCD containing *Pseudomonas fluorescens* TDK1 and *Bacillus* sp. under salt stress have been reported in earlier studies (Saravanakumar and Samiyappan 2007; Siddikee et al. 2011). Similarly, Nadeem et al. (2009) also reported a protective effect of ACCD containing bacteria *Pseudomonas syringae*, *Enterobacter aerogenes* and *Pseudomonas fluorescens* on the growth of maize under salt stress conditions. Another experimental report of Gamalero et al. (2008) also suggested a plant growth stimulatory effect of *Gigaspora rosea* BEG9 and *Pseudomonas putida* UW4 on the growth of cucumber under salt stress condition. ACCD activity has also been observed in *Achromobacter, Azospirillum, Agrobacterium, Burkholderia, Enterobacter, Pseudomonas*, and *Ralstonia* (Blaha et al. 2006). PGPR isolated from saline habitats are well adapted to tolerate salt and hence can be helpful in increasing plant resistance to salt stress. Reviews of Mosse (1986) and Sylvia and Williams (1992) point that most of the effect of AM fungi in protecting plants from salt stress is based on the compensation of plant damage by improved nutrient acquisition. When salt concentrations in soils are high then sodium ions compete with potassium ions for binding sites. A high $Na^{+}$: $K^{+}$ ratio generated due to salinity disrupts the ionic balance in cytoplasm further disrupting the plant metabolic processes (Giri et al. 2007). Mycorrhizal colonization not only enhances $K^{+}$ absorption but also prevents $Na^{+}$ translocation to shoot tissues (Giri et al. 2007; Sharifi et al. 2007). However, contrasting studies have been reported in this regard (Allen and Cunningham 1983).

### 16.4.5 Metal-Affected Soils

Rhizoremediation has been defined as the removal of soil contaminants by the help of microbes found in rhizosphere (Segura and Ramos 2013). Rhizoremediation technique uses combined degradative potential of plants and their associative rhizospheric microorganisms (Zhuang et al. 2007). Microbes produce a wide range of hydrolytic enzymes that accelerate the degradation process and helps in eco-restoration of polluted site (Daane et al. 2001). Microbial remediation of metal has gained considerable interest with the understanding of the microbe–plant interaction. The dynamic microenvironment of rhizosphere harboring microbes not only has considerable potential for plant growth promotion but also detoxification of harmful chemical compounds (De Souza et al. 1999).PGPR that solubilize inorganic phosphate and synthesize growth promoting substances can be utilized in the plant-assisted bioremediation of metal affected soil (Khan et al. 2009). The heavy metal resistant microbes in rhizosphere affect trace metal mobility and availability to the plants through the secretion of chelating agents, acidification, phosphate solubilization, and redox changes (Abou-Shanab et al. 2003; Idris et al. 2004). *Bacillus megaterium* has been reported for microbial assisted phytoremediation of lead contaminated soils (Gullap et al. 2014).

*Alcalegenes sp.* RZS2, *Pseudomonas aeruginosa* RZS3 has been reported for its bioremediation and growth promotion activities against Zn-induced oxidative stress in wheat, as it improves the availability of necessary nutrient, elicits the antioxidant defense system of the plant, and lowers the Zn metal uptake (Islam et al. 2014).

Copper being a trace element at elevated levels in soil affects the plants and microbial processes and hence its removal from the polluted soil becomes essential. A study revealed that inoculation of *Vicia faba* with copper accumulating PGPRs *Enterobacter cloacae* and *Pseudomonas sp* elevated copper tolerance in *Vicia faba* by altering antioxidant enzyme system (Fatnassi et al. 2015). Few soil microbes like *Kocuria sp.* CRB15, *Enterobacter clocae* and *Pseudomonas sp.* have been reported for their strong metal binding capacity which prevents their bioaugmentation in plants (Hansda and Anshumali 2017; Fatnassi et al. 2015).

Cadmium, a toxic heavy metal highly mobile in soil gets accumulated in crops which when consumed by animals and humans affect their survival (Sanita di Toppi and Gabrielli 1999). Cadmium contamination also alters the microbial diversity of soil (Roberts 2003; Lenntech 2009; Shukla et al. 2010). Interest has therefore increased in developing a system that could remove or neutralize the toxic effects of metals found in soils, sediments, and wastewaters. Besides phytoremediation some microorganisms, like *Bacillus subtilis*, *Citrobacter spp*., and *Pseudomonas spp*. have been found effective in reducing/remediating the toxicity of metals like cadmium (Sheng-wang et al. 2008; Chunxiao et al. 2009). Losses due to the toxic effects of cadmium on plant physiology are well documented (Vassilev et al. 2004). Glutathione having several roles in the cell metabolism such as scavenging ROS, redox state regulation, transport of amino acids, and sulfur storage (Meister 1995; Noctor and Foyer 1998) gets augmented in plants grown in Cd-contaminated soil when inoculated with Cd-tolerant strains of *Rhizobium* (Figueira et al. 2005). It has been reported that Cd content in the soil can be decreased by the help of certain *Pseudomonas* strains (MKRh1, MKRh3, and MKRh4) (Ganesan 2008) and *Rhizobium leguminosarum* (Sofia et al. 2012) through enzymatic solubilization.

Chromium contamination in soil occurs due to the disposal of chromium contaminated industrial wastewater or sludge that excess the quality standard. Chromium concentration in soil ranges between 1 and 300 mg/kg while the maximum health standard is 2.5 mg/kg. For use of microbes as bioinoculant for microbial-assisted phytoremediation, the microorganisms must be able to promote plant growth even under heavy metal stress. Many previous studies have put forward the role of *Pseudomonas aeruginosa, Enterobacter, Bacillus,* and *Shewanella* in alleviating Cr (VI) from the contamination site (Wei-hua et al. 2009, Samuel et al. 2013; Ahemad 2015). Addition of *Bacillus subtilis* in chromium-contaminated soil could reduce up to 5–11% of the Cr content (Purwanti et al. 2017). Upadhyay et al. (2017) reported that *Bacillus* sp. MNU16 isolated from the coal mine areas reduces chromium (VI) toxicity of the contaminated soil.

Nickel mobilizing rhizobacteria *Pseudomonas* sp. SRI2, *Psychrobacter* sp. SRS8 and *Bacillus* sp. SN9 help plants in defending growth inhibition caused by Ni. *Bacillus* sp.SN9 while significantly increased the Ni concentration in the root and shoot tissues of *Brassica juncea* and *B. oxyrrhina*, at the same time promoted plant growth against the toxic effects of Ni in soils by producing IAA, siderophore and solubilization of phosphate (Ma et al. 2009). Most of the findings reveal that inoculation of Ni mobilizing PGPR increases the efficiency of phytoextraction in plants protecting them from heavy metal. *Bacillus subtilis* SJ-101 is reported to decrease Ni content through microbial assisted phytoremediation (Zaidi et al. 2006). Burd et al. (2000) showed that *Kluyvera ascorbita* SUD165 enables plant growth even in Ni-contaminated soil.

Mercury is a heavy metal and toxic pollutant entering the ecosystem through industries associated with production of paints, disinfectants, fungicidal bactericidal agents, and through mercury mining (Moreno et al. 2008). Mercury tolerance is a widespread character in PGPRs isolated from mercury-contaminated environments (Abou-Shanab et al. 2007; De Andrés et al. 2007; Ruiz-Díez et al. 2012). *Enterobacter cloacae, Enterobacter ludwigii,* and *Klebsiella pneumonia* having good PGPR traits have been known to impart mercury tolerance in the wheat crop (Mishra et al. 2016).

Arsenic removal by detoxification has been reported by several bacteria possessing *ars* gene that encode protein for detoxification like *Escherichia coli, Staphylococcus aureus*, and *Staphylococcus xylosis* (Tamaki and Frankenberger 1992; Cervantes et al. 1994). *Pseudomonas* species strain As-1, isolated from an electroplating industry showed great tolerance toward high concentrations of As when compared to reference organisms *P. aeruginosa, P. putida, B. subtilis, A. eutrophus, E. coli, S. aureus, B. subtilis,* and *S. cerevisiae* (Carlin et al. 1995; Wysocki et al. 1997; Canovas et al. 2003; Podolskaia et al. 2002).

## 16.5 Conclusion and Future Prospects

Deterioration of soil health quality due to natural and anthropogenic interventions has pressing ecological concerns. Plant growth promoting microbes functioning as a vital living system within an ecosystem hold immense possibilities to improve soil health through their multifaceted roles. Maintaining soil health by microbial interventions is a cost-effective, renewable, and nonintrusive practice. In the past 25 years, plant researchers have developed efficacious use of PGPRs in areas of plant growth promotion and soil health but a greater need is to put the endophytic and rhizospheric microbes together with mycorrhizal fungi to understand their mechanism of interactions in promoting plant growth and soil health. Understanding the microbial functioning through metabolomic approaches may help in overcoming the barriers in its effective application.

## References


Abo-Amer AA (2012) Characterization of a strain of *Pseudomonas putida* isolated from agricultural soil that degrades cadusafos (an organophosphorus pesticide). World J Microbiol Biotechnol 28:805–814

Abou-Shanab RA, Angle JS, Delorme TA, Chaney RL, van Berkum P, Moawad H, Ghanem K, Ghozlan HA (2003) Rhizobacterial effects on nickel extraction from soil and uptake by *Alyssum murale*. New Phytol 158:219–224

Abou-Shanab RAI, Van Berkum P, Angle JS (2007) Heavy metal resistance and genotypic analysis of metal resistance genes in gram-positive and gram-negative bacteria present in Ni-rich serpentine soil and in the rhizosphere of *Alyssum murale*. Chemosphere 68:360–367

Adesemoye AO, Torbert HA, Kloepper JW (2009) Plant growth-promoting rhizobacteria allow reduced application rates of chemical fertilizers. Microb Ecol 58:921–929

Ahemad M (2015) Enhancing phytoremediation of chromium-stressed soils through plant-growth-promoting bacteria. J Genet Eng Biotechnol 13:51–58

Ahemad M, Khan MS (2011) Ecotoxicological assessment of pesticides towards the plant growth promoting activities of Lentil (*Lens esculentus*)specific *Rhizobium* sp. strain MRL3. Ecotoxicology 20:661–669

Ahemad S, Chaudhari SK, Dagar JC, Basak N (2012) Soil aggregates as indicator of soil health in waterlogged sodic soil. J Soil Salinity Water Quality 4(2):92–96

Allen EB, Cunningham GL (1983) Effects of vesicular-arbuscular mycorrhizae on *Distichlis spicata* under three salinity levels. New Phytol 93:227–236

Barea JM, Aguilar CA, Azcon R (1997) Inter-actions between mycorrhizal fungi and rhizosphere microorganisms within the context of sustainable. In: Gange AC, Brown VK (eds) Multitrophic interactions in terrestrial systems. Blackwell Science, Oxford, UK, pp 65–77

Barnawal D, Bharti N, Maji D, Chanotiya CS, Kaira A (2012) 1-Aminocyclopropane-1-carboxylic acid (ACC) deaminase-containing rhizobacteria protect *Ocimum sanctum* plants during waterlogging stress via reduced ethylene generation. Plant Physiol Biochem 58:227–235

Begum MM, Sariah M, Abidin ZMA, Puteh BA, Rahman AM (2008) Antagonistic potential of selected fungal and bacterial biocontrol agents against *Colletotrichum truncatum* of soybean seeds. Pertanica J Trop Agric Sci 31:45–53

Bennett WF (1993) Nutrient deficiencies and toxicities in crop plants. APS Press, St Paul, MN, p 202

Bhandari G (2014) An overview of agrochemicals and their effects on environment in Nepal. Appl Ecol Environ Sci 2(2):66–73

Bieleski RL (1973) Phosphate pools, phosphate transport, and phosphate availability. Annu Rev Plant Physiol 24:225–252

Blaha D, Prigent-Combaret C, Mirza MS, Moënne-Loccoz Y (2006) Phylogeny of the 1-aminocyclopropane-1-carboxylic acid deaminase-encoding gene acdS in phytobeneficial and pathogenic Proteobacteria and relation with strain biogeography. FEMS Microbiol Ecol 56:455–470

Blaylock AD (1994) Soil salinity, salt tolerance and growth potential of horticultural and landscape plants. Co-operative Extension Service, Department of Plant Soil and Insect Sciences College of Agriculture, University of Wyoming, Laramie Wyoming, pp 2–94

Bliev UK, Martynov AN, Zarkov AV, Maximova LI (1985) Effect of verpa preparation on the fertility of soddy podzolic soil. Agrochimica 2:97–100

Bohme L, Langer U, Bohme F (2005) Microbial biomass, enzyme activities and microbial community structure in two European long term field experiments. Agric Ecosys Environ 109:141–152

Bollag JM, Liu SY (1990) A biological transformation processes of pesticides. In: Cheng HH (ed) Pesticide in the environment. Soil Science Society of America, Inc., Madison, pp 169–211

Brinson MM, Swift BL, Plantico RC, Barclay JS (1981) Riparian ecosystems: their ecology and status, OBS, vol 81. USA Fish Wildlife Service, Washington, DC, p 17

Broughton WJ, Zhang F, Perret X, Staehelin C (2003) Signals exchanged between legumes and *Rhizobium*: agricultural uses and perspectives. Plant Soil 252:129–137

Bunemann EK, McNeill A (2004) Impact of fertilizers on soil biota. In: Lines R (ed) Proceedings current research into soil biology in agriculture. Kelly Tamworth. pp 64–71

Burd GI, Dixon DG, Glick BR (2000) Plant growth-promoting bacteria that decrease heavy metal toxicity in plants. Can J Microbiol 46:237–245

Canovas D, Cases I, de Lorenzo V (2003) Heavy metal tolerance and metal homeostasis in *Pseudomonas putida* as revealed by complete genome analysis. Environ Microbiol 5:1242–1256

Carlin A, Shi W, Dey S, Rosen BP (1995) The ars operon of *Escherichia coli* confers arsenical and antimonial resistance. J Bacteriol 177:981–986

Cervantes C, Ji G, Ramirez JL, Silver S (1994) Resistance to arsenic compounds in microorganisms. FEMS Microbiol Rev 15:355–367

Charalampos K, Myresiotis VZ, Papadopoulou-Mourkidou E (2012) Biodegradation of soil-applied pesticides by selected strains of plant growth promoting rhizobacteria (PGPR) and their effects on bacterial growth. Biodegradation 23:297–310

Chennappa G, Adkar-Purushothama CR, Naik MK, Suraj U, Sreenivasa MY (2014) Impact of pesticides on PGPR activity of *Azotobacter sp.* isolated from pesticide flooded paddy soils. Greener J Agric Sci 4(4):117–129

Chunxiao J, Hongwen S, Tieheng S, Qingmin Z, Yanfeng Z (2009) Immobilization of cadmium in soil by UV affected *Baccillus subtilis* 38 bioaugmentation and NovoGro amendment. J Hazard Mater 167:170–177

Daane L, Harjono I, Zylstra G, Häggblom M (2001) Isolation and characterization of polycyclic aromatic hydrocarbon-degrading bacteria associated with the rhizosphere of salt marsh plants. Appl Environ Microbiol 67:2683–2691

Das AC, Mukherjee D (1994) Effect of insecticides on the availability of nutrients, nitrogen fixation, and phosphate solubility in the rhizosphere soil of rice. Biol Fertil Soils 18:37–41

Das AC, Mukherjee D (1998a) Insecticidal effects on soil microorganisms and their biochemical processes related to soil fertility. World J Microbiol Biotechnol 14:903–909

Das AC, Mukherjee D (1998b) Persistence of phorate and carbofuran in relation to their effect on the mineralization of C, N, and P in alluvial soil. Bull Environ Contam Toxicol 61:709–715

Das AC, Mukherjee D (2000a) Soil application of insecticides influences microorganisms and plant nutrients. Appl Soil Ecol 14:55–62

Das AC, Mukherjee D (2000b) Influence of insecticides on microbial transformation of nitrogen and phosphorus in Typic Orchragualf soil. J Agric Food Chem 48:3728–3732

De Andrés F, Walter I, Tenorio JL (2007) Revegetation of abandoned agricultural land amended with biosolids. Sci Total Environ 378:81–83

De Souza MP, Chu D, Zhao M, Zayed AM, Ruzin SE, Schichnes D, Terry N (1999) Rhizosphere bacteria enhance selenium accumulation and volatilization by Indian mustard. Plant Physiol 119 (2):565–573

Dehne HW, Schonbeck F (1979) Investigations on the influences of endotropic mycorrhizal on plant diseases, II phenol metabolism and lignifications. Phytopathology 95:210–216

Diaz-Zorita M, Perfect E, Grove JH (2002) Disruptive methods for assessing soil structure. Soil Tillage Res 64:3–22

Doran JW, Sarrantonio M, Liebig MA (1996) Soil health and sustainability. Adv Agron 56:1–54

Edwards CA (1975) Factors that affect the persistence of pesticides in plants and soils. Pure Appl Chem 42:39–56

Engelen B, Meinken K, Von Wintzingerode F, Heuer H, Malkomes HP, Backhaus H (1998) Monitoring impacts of a pesticide treatment on bacterial soil communities by metabolic and genetic fingerprinting in addition to conventional testing procedures. Appl Environ Microbiol 64:2814–2821

FAO (2008) FAO land and plant nutrition management service. http://www.fao.org/ag/agl/agll/spush

FAO (2011) FAOSTAT Statistical databases and data-sets of the Food and Agriculture Organization of the United Nations. http://faostat.fao.org/default.aspx

Fatnassi IC, Chiboub M, Saadani O, Jebara M, Jebara SH (2015) Impact of dual inoculation with *Rhizobium* and PGPR on growth and antioxidant status of *Vicia faba* L. under copper stress. C R Biol 338:241–254

Figueira EMAP, Lima AIG, Pereira SAI (2005) Cadmium tolerance plasticity in *Rhizobium leguminosarum* bv. viciae: glutathione as a detoxifying agent. Can J Microbiol 5:1–6

Flemming HC, Wingender J (2001) Relevance of microbial extracellular polymeric substances (EPSs)-parts I: structural and ecological aspects. Water Sci Technol 43:1–8

Fulton JM, Erickson AE (1964) Relation between soil aeration and ethyl alcohol accumulation in xylem exudate of tomatoes. Proc Soil Sci Soc Am 28:610–614

Gadagi RS, Krishnaraj P, Kulkarni J, Sa T (2004) The effect of combined *Azospirillum* inoculation and nitrogen fertilizer on plant growth promotion and yield response of the blanket flower *Gaillardia pulchella*. Acta Sci Pol Hortorum Cultus 100:323–332

Gamalero E, Berta G, Massa N, Glick BR, Lingua G (2008) Synergistic interactions between the ACC deaminase-producing bacterium *Pseudomonas putida* UW4 and the AM fungus *Gigaspora rosea* positively affect cucumber plant growth. FEMS Microbiol Ecol 64:459–467

Gambrell RP, De Laune RD, Patrick JRWH (1991) Redox processes in soils following oxygen depletion. In: Jackson MB, Davies DD, Lambers H (eds) Plant life under oxygen stress. Academic, The Hague, pp 101–117

Ganesan V (2008) Rhizoremediation of cadmium soil using a cadmium resistant plant growth promoting rhizopseudomonad. Curr Microbiol 56:403–407

Giri B, Kapoor R, Mukerji KG (2007) Improved tolerance of *Acacia nilotica* to salt stress by arbuscular mycorrhizal, *Glomus fasciculatum*, may be partly related to elevated $K^+/Na^+$ ratios in root and shoot tissues. Microbiol Ecol 54:753–760

Glick BR (2005) Modulation of plant ethylene levels by the bacterial enzyme ACC deaminase. FEMS Microbiol Lett 251:1–7

Glick BR (2014) Bacteria with ACC deaminase can promote plant growth and help to feed the world. Microbiol Res 169:30–39

Glick BR, Cheng Z, Czarny J, Duan J (2007) Promotion of plant growth ACC deaminase containing soil bacteria. Eur J Plant Pathol 119:329–339

Glover-Amengor M, Tetteh M (2008) Effect of pesticide application rate on yield of vegetables and soil microbial communities. West Afr J App Ecol 12

Gouzou L, Burtin G, Philippy R, Bartoli F, Heulin T (1993) Effect of inoculation with *Bacillus polymyxa* on soil aggregation in the wheat rhizosphere: preliminary examination. Geoderma 56:479–490

Grichko VP, Glick BR (2001a) Amelioration of flooding stress by ACC deaminase-containing plant growth-promoting bacteria. Plant Physiol Biochem 39:11–17

Grichko VP, Glick BR (2001b) Flooding tolerance of transgenic tomato plants expressing the bacterial enzyme ACC deaminase controlled by the 35S, rolD or PRB-1b promoter. Plant Physiol Biochem 39:19–25

Gullap MK, Dasci M, Erkovan HI, Koc A, Turan M (2014) Plant growth-promoting rhizobacteria (PGPR) and phosphorus fertilizer-assisted phytoextraction of toxic heavy metals from contaminated soils. Commun Soil Sci Plant Anal 45(19):2593–2606

Guo JH, Liu XJ, Zhang Y, Shen JL, Han WX, Zhang WF, Christie P, Goulding KWT, Vitousek PM, Zhang FS (2010) Significant acidification in major Chinese croplands. Science 327:1008

Gurikar C, Naik MK, Sreenivasa YM (2015) *Azotobacter*—PGPR activities with special reference to effect of pesticides and biodegradation. In: Singh DP, Singh HB, Prabha R (eds) Microbial inoculants in sustainable agricultural productivity. Springer Nature, Switzerland, pp 229–244

Hansda A, Anshumali VK (2017) Cu-resistant *Kocuria* sp. CRB15: a potential PGPR isolated from the dry tailing of Rakha copper mine. Biotech 7:132

Harman GE (2006) Overview of mechanisms and uses of *Trichoderma* spp. Phytopathology 96:190–194

Harrier LA, Watson CA (2004) The potential role of arbuscular mycorrhizal (AM) fungi in the bio-protection of plants against soil borne pathogens in organic and/or other sustainable farming systems. Pest Manag Sci 60:149–157

Harris RF, Chesters G, Allen ON, Attoe OJ (1964) Mechanisms involved in soil aggregate stabilization by fungi and bacteria. Soil Sci Soc Am Proc 28:529–532

Hayat R, Ali S, Amara U, Khalid R, Ahmed I (2010) Soil beneficial bacteria and their role in growth promotion: a review. Annu Rev Microbiol 60:579–598

Hermosa R, Viterbo A, Chet I, Monte E (2012) Plant beneficial effects of *Trichoderma* and of its genes. Microbiology 158:17–25

Hubbell DS, Chapman JE (1946) The genesis of structure of two in two calcareous soil. Soil Sci 62:271–281

Idris R, Trifonova R, Puschenreiter M, Wenzel W, Sessitsch A (2004) Bacterial communities associated with flowering plants of the Ni hyperaccumulator Thaspigoesingense. Appl Environ Microbiol 70:2667–2677

Imfeld G, Vuilleumier S (2012) Measuring the effects of synthetic pesticides on bacterial communities in soil: a review. Eur J Soil Biol 49:1–4

Imtiaj A, Lee ST (2008) Antagonistic effect of three *Trichoderma* species on the *Alternaria porri* pathogen of onion blotch. World J Agric Sci 4:13–17

International food policy research institute (IFPRI) (2012) Global food policy report. www.ifpri.org › publication › 2012-global-food-policy-report

Ishaq A, Khan JA, Ahmed N (1994) Biodegradation of a pesticide alpha-cyano, 3 phenoxybenzyl-2,2-dimethyl-3 (2,2-dichlorophenyl) by *Pseudomonas aeuroginosa*. Pak J Agric Res 15:242–250

Islam F, Yasmeen T, Ali Q, Ali S, Arif MS, Hussain S, Rizvi H (2014) Influence of *Pseudomonas aeruginosa* as PGPR on oxidative stress tolerance in wheat under Zn stress. Ecotoxicol Environ Saf 104:285–293

Jackson MB (1985) Ethylene and responses of plants to soil water logging and submergence. Annu Rev Plant Physiol 36:145–174

Jackson MB, Drew MC (1984) Effects of flooding on growth and metabolism of herbaceous plants. In: Kozlowski TT (ed) Flooding and plant growth. Academic, New York, pp 47–128

Jain A, Singh S, Sarma BK, Singh HB (2012) Microbial consortium mediated reprogramming of defence network in pea to enhance tolerance against *Sclerotinia sclerotiorum*. J Appl Microbiol 112:537–550

Janiesch P (1991) Ecophysiological adaptations of higher plants in natural communities to waterlogging. In: Rozema J, Verkleij JAC (eds) Ecological responses to environmental stresses. Kluwer Academic Publishers, The Netherlands, pp 453–477

Jastrow JD (1996) Soil aggregate formation and the accrual of particulate and mineral-associated organic matter. Soil Biol Biochem 28:665–676

Jastrow JD, Miller RM, Lussenhop J (1998) Contributions of interacting biological mechanisms to soil aggregate stabilization in restored prairie. Soil Biol Biochem 30:905–916

Jeffries P, Gianinazzi S, Perotto S, Turnau K, Barea JM (2003) The contribution of arbuscular mycorrhizal fungi in sustainable maintenance of plant health and soil fertility. Biol Fertil Soils 37:1–16

Kaushal M, Mandyal P, Kaushal R (2019) Field based assessment of *Capsicum annuum* performance with inoculation of rhizobacterial consortia. Microorganisms 7:89

Kay BD (1990) Rates of change of soil structure under different cropping system. Adv Soil Sci 12:1–52

Khan AG, Kuek C, Chaudhary TM, Khoo CS, Hayes WJ (2000) Plants, mycorrhizae and phytochelators in heavy metal contaminated land remediation. Chemosphere 41:197–207

Khan MS, Zaidi A, Wani PA, Oves M (2009) Role of plant growth promoting rhizobacteria in the remediation of metal contaminated soils. Environ Chem Lett 7:1–19

Kinney CA, Mandernack KW, Moiser AR (2005) Laboratory investigations into the effects of the pesticides mancozeb chlorothalonil and prosulfuron on nitrous oxide and nitric oxide production in fertilized soil. Soil Biol Biochem 37:837–850

Kozlowski TT, Pallardy SG (1984) Effect of flooding on water, carbohydrate, and mineral relations. In: Kozlowski TT (ed) Flooding and plant growth. Academic, New York, pp 165–193

Krishna KR, Balakrishna AN, Bagyaraj DJ (1982) Interaction between a vesicul arbuscular mycorrhizal fungus and *Streptomyces cinnamomeous* and their effects in finger millet. New Phytol 92:401–405

Kryuchkova YV, Burygin GL, Gogoleva NE, Gogolev YV (2014) Isolation and characterization of a glyphosate degrading rhizosphere strain, *Enterobacter cloacae* K7. Microbiol Res 169:99–105

Kumari P, Meena M, Upadhyay RS (2018) Characterization of plant growth promoting rhizobacteria (PGPR) isolated from the rhizosphere of *Vigna radiata* (mung bean). Biocatal Agric Biotechnol 16:155–162
Kumon H, Tomoshika K, Matunaga T, Ogawa M, Ohmori HA (1994) Sandwich cup method for the *Pseudomonas* exopolysaccharides. The IR spectrum of the polymer proved the presence. Microbiol Immunol 38:615–619
Lal R (1987) Managing the soils of Sub-Sahara Africa. Science 236:1069–1086
Lenné JM, Parbery DG (1976) Phyllosphere antagonists and appressoria formation in *Colletotrichum gloeosporioides*. Trans Br Mycol Soc 66:334–336
Lenntech WT (2009) Chemical properties, health and environmental effects of copper. Lenntech Water Treatment and Purification Holding BV, Delft, The Netherlands, p 2009
Li F, Liu M, Li Z, Jiang C, Han F, Che Y (2013) Changes in soil microbial biomass and functional diversity with a nitrogen gradient in soil columns. Appl Soil Ecol 64:1–6
Lindberg T, Granhall U, Berg B (1979) Ethylene formation in some coniferous forest soils. Soil Biol Biochem 11(6):637–643
Liu J, Xie J, Chu Y, Sun C, Chen C, Wang Q (2008) Combined effect of cypermethrin and copper on catalase activity in soil. J Soil Sediment 8:327–332
Lynch JM (1975) Ethylene in the soil. Nature 256:576–577
Ma Y, Rajkumar M, Freitas H (2009) Isolation and characterization of Ni mobilizing PGPB from serpentine soils and their potential in promoting plant growth and Ni accumulation by *Brassica* spp. Chemosphere:1–56
Malghani S, Chatterjee N, Hu X, Zejiao L (2009) Isolation and characterization of a profenofos degrading bacterium. J Environ Sci 21:1591–1597
Malik MAB, Tesfai K (1985) Pesticidal effect on soybean-rhizobia symbiosis. Plant Soil 85:33–41
Malik MAB, Tesfai K (1993) Compatibility of *Rhizobium japonicum* with commercial pesticides in vitro. Bull Environ Contam Toxicol 31:432–437
Manca de Nadra MC, Strasser AM, de Saad AA, de Ruiz HP, Oliver G (1985) Extracellular polysaccharide production by *Lactobacillus bulgaricus* CRL 420. Milchwissenschaft 40:409–411
Marshall TJ, Holmes JW (1988) Soil physics, 2nd edn. Cambridge University Press, Cambridge
Massah J, Azadegan B (2016) Effect of chemical fertilizers on soil compaction and degradation. Agr Mech Asia Afr Latin Am 47(1):44–50
Mayak S, Tirosh T, Glick BR (2004) Plant growth-promoting bacteria that confer resistance in tomato to salt stress. Plant Physiol Biochem 42:565–572
Megadi VB, Tallur PN, Hoskeri RS, Mulla SI, Ninnekar HZ (2010) Biodegradation of pendimethalin by *Bacillus circulans*. Indian J Biotechnol 9:173–177
Meister A (1995) Glutathione metabolism. Methods Enzymol 251:3–13
Menon P, Gopal M, Parsad R (2005) Effects of chlorpyrifos and quinalphos on dehydrogenase activities and reduction of Fe3þ in the soils of two semi-arid fields of tropical. Indian Agric Ecosyst Environ 108:73–83
Miller RM, Jastrow JD (1990) Hierarchy of root and mycorrhizal fungal interactions with soil aggregation. Soil Biol Biochem 22:579–584
Mishra G, Sapre I, Sharma S, Tiwari A (2016) Alleviation of mercury toxicity in wheat by the interaction of mercury tolerant plant growth promoting rhizobacteria. J Plant Growth Regul 35 (4):1000–1012
Mishustin EN (1945) The unstable part of the soil macrostructure. Pochvovedenie 2
Mohamed AT, El Hussein AA, El Siddig MA, Osman AG (2011) Degradation of oxyfluorfen herbicide by soil microorganisms biodegradation of herbicides. Biotechnology 109:274–279
Monkiedje A, Spiteller M (2002) Effects of the phenylamide fungicides, mefenoxam and metalaxyl, on the biological properties of sandy loam and sandy clay soils. Biol Fertil Soils 35:393–398
Moreno FN, Anderson CWN, Stewart RB, Robinson BH (2008) Phytofiltration of mercury contaminated water volatization and plant accumulation aspects. Environ Exp Bot 62(1):78–85
Mosse B (1986) Mycorrhiza in sustainable agriculture. Biol Agric Hortic 3:191–209

Myresiotis CK, Karaoglanidis GS, Vryzas V, Mourkidou P (2012) Evaluation of plant-growth-promoting rhizobacteria, acibenzolar-*S*-methyl and hymexazol for integrated control of *Fusarium* crown and root rot on tomato. Pest Manag Sci 68(3):404–411
Nadeem SM, Zahir ZA, Naveed M, Arshad M (2009) Rhizobacteria containing ACC-deaminase confer salt tolerance in maize grown on salt-affected fields. Can J Microbiol 55:1302–1309
Nagy PT, Pinter T (2015) Effects of foliar biofertilizer sprays on nutrient uptake, yield, and quality parameters of *Blaufrankish* (*Vitis vinifera L.*) grapes. Commun Soil Sci Plant Anal 46 (51):219–227
Nahas E (1996) Factors determining rock phosphate solubilization by microorganisms isolated from soil. World J Microbiol Biotechnol 12:567–572
Narula N, Gupta KG (1986) Ammonia excretion by *Azotobacter chroococcum* in liquid culture and soil in the presence of manganese and clay minerals. Plant Soil 93:205–209
Naseem H, Bano A (2014) Role of plant growth-promoting rhizobacteria and their exopolysaccharide in drought tolerance of maize. J Plant Interact 9:689–701
Nguyen D, Rieu I, Mariani C, van Dam NM (2016) How plants handle multiple stresses: hormonal interactions underlying responses to abiotic stress and insect herbivory. Plant Mol Biol 91:727–740
Nicholson PS, Hirsch PR (1998) The effects of pesticides on the diversity of culturable soil bacteria. J Appl Microbiol 84:551–558
Niewiadomska A (2004) Effect of carbendazim, imazetapir and thiram on nitrogenase activity, the number of microorganisms in soil and yield of red clover (*Trifolium pretense* L). Pol J Environ Stud 13:403–410
Niewiadomska A, Klama J (2005) Pesticide side effect on the symbiotic efficiency and nitrogenase activity of Rhizobiaceae bacteria family. Pol J Microbiol 54:43–48
Noctor G, Foyer C (1998) Ascorbate and glutathione: keeping active oxygen under control. Annu Rev Plant Physiol Plant Mol Biol 49:249–279
Nutmen FJ, Roberts FM (1969) The stimulating effect of some fungicides on *Glomerella cingulata*in relation to the control of coffee berry disease. Ann Appl Biol 64(2):335–344
Ogunseitan OA, Odeyemi O (1985) Effect of lindane, captan and malathion on nitrification, sulphur oxidation, phosphate aerobic heterotrophic soil bacteria to the generic level by solubilization and respiration in tropical soil. Environ Pollut 37:343–354
Patel BB, Bharat B, Patel RS (2011) Studies on infiltration of saline–alkali soils of several parts of Mehsana and Patan districts of North Gujarat. J Appl Technol Environ Sanit 1(1):87–92
Perfect TJ, Cook AG, Critchley BR, Smith AR (1981) The effect of crop protection with DDT on the microarthropod population of a cultivated forest soil in the sub-humid tropics. Pedobiologia 21:7–18
Pezeshki SR, Chambers JL (1985a) Stomatal and photosynthetic response of sweet gum (*Liquidambar styraciflua*) to flooding. Can J For Res 15:371–375
Pezeshki SR, Chambers JL (1985b) Responses of cherrybark oak (*Quercus falcata* var. *pagodaefolia*) seedlings to short-term flooding. For Sci 31:760–771
Pimentel D, Warneke A (1989) Ecological effects of manure, sewage sludge, and other organic wastes on arthropod populations. Agric Zool Rev 3:1–30
Podolskaia VI, Gruzina TG, Ulberg ZP, Sokolovskaia AS, Grishchenko NI (2002) Effect of arsenic on bacterial growth and plasma membrane ATPase activity. Prikl Biokhim Mikrobiol 38:57–62
Ponnamperuma FN (1984) Effects of flooding on soils. In: Kozlowski TT (ed) Flooding and plant growth. Academic, Orlando FL, pp 9–45
Purwanti IF, Kurniawan SB, Tangahu BV, Rahayu NM (2017) Bioremediation of trivalent chromium in soil using bacteria. Int J Appl Eng Res 12(20):9346–9350
Raghuwanshi R, Prasad JK (2018) Perspectives of rhizobacteria with ACC Deaminase activity in plant growth under abiotic stress. In: Giri B, Prasad R, Varma A (eds) Root biology, Soil biology, vol 52. Springer Nature, Cham, pp 303–321
Ramaiah, Bagyaraj DJ (1989) Root diseases and mycorrhizae, a review. J Phytol Res 2:1–6

Rani MS, Lakshmi KV, Devi PS, Madhuri RJ, Aruna S, Jyothi K, Narasimha G, Venkateswarlu K (2008) Isolation and characterization of a chlorpyrifos degrading bacterium from agricultural soil and its growth response. Afr J Microbiol Res 2:26–31

Reid JB, Goss MJ (1981) Effect of living roots of different plant species on the aggregate stability of two arable soils. J Soil Sci 32:521–541

Reid JB, Goss MJ (1982) Suppression of decomposition of $^{14}C$ labelled plant roots in the presence of living roots of maize and perennial ryegrass. J Soil Sci 33:387–398

Revillas JJ, Rodelas B, Pozo C, Martínez-Toledo MV, González-López J (2000) Production of B-group vitamins by two *Azotobacter* strains with phenolic compounds as sole carbon source under diazotrophic and a diazotrophic conditions. J Appl Microbiol 89(3):486–493

Rillig MC, Mummey DL (2006) Mycorrhizas and soil structure. New Phytol 171:41–53

Rillig MC, Wright SF, Nichols KA, Schmidt WF, Torn MS (2001) Large contribution of arbuscular mycorrhizal fungi to soil carbon pools in tropical forest soils. Plant Soil 233:167–177

Roberts M (2003) Review of risks from metals in the UK. Chemical Stakeholder Forum, Fourteenth meeting 20

Rowe RN, Catlin PB (1971) Differential sensitivity to water logging and cyanogenesis by peach, apricot, and plum roots. J Am Soc Hortic Sci 96:305–308

Ruiz-Díez B, Quiñones MA, Fajardo S, López MA, Higueras P, Fernández-Pascual M (2012) Mercury-resistant rhizobial bacteria isolated from nodules of leguminous plants growing in high Hg-contaminated soils. Appl Microbiol Biotechnol 96:543–554

Saleem M, Arshad M, Hussain S, Bhatti AS (2007) Perspective of plant growth promoting rhizobacteria (PGPR) containing ACC deaminase in stress agriculture. J Ind Microbiol Biotechnol 34:635–648

Samuel J, Paul ML, Ravishankar H, Mathur A, Saha DP, Natarajan C (2013) The differential stress response of adapted chromite mine isolates *Bacillus subtilis* and *Escherichia coli* and its impact on bioremediation potential. Biodegradation 24:829–842

Sanita di Toppi L, Gabrielli R (1999) Response to Cadmium in higher plants. Environ Exp Bot 41:105–130

Santos PF, Whitford WG (1981) The effects of microarthropods on litter decomposition in a Chihuahuan desert ecosystem. Ecology 62:654–663

Saptanmasi B, Karayilanoglu T, Kenar L, Serdar M, Kose S, Aydin A (2008) Bacterial biodegradation of aldicarb and determination of bacterium which has the most biodegradative effect. Turk J Biochem 33:209–214

Saranya A, Krishnan DU, Ramasamy S (2015) Optimization of lipase production by Myroides odoratimus SKS05-GRD and bioremediation of diesel hydrocarbons. Res J Biotechnol 10 (7):1–10

Saravanakumar D, Samiyappan R (2007) ACC deaminase from *Pseudomonas fluorescens* mediated saline resistance in groundnut (*Arachis hypogea*) plants. J Appl Microbiol 102(5):1283–1292

Saxena A, Raghuwanshi R, Singh HB (2015) *Trichoderma* species mediated differential tolerance against biotic stress of phytopathogens in *Cicer arietinum* L. J Basic Microbiol 55:195–206

Saxena A, Raghuwanshi R, Gupta VK, Singh HB (2016a) Chilli anthracnose: the epidemiology and management. Front Microbiol 7:1527

Saxena A, Raghuwanshi R, Singh HB (2016b) Elevation of defense network in chilli against *Colletotrichum capsici* by phyllospheric *Trichoderma* strain. J Plant Growth Regul 35:377–389

Segura A, Ramos JL (2013) Plant-bacteria interactions in the removal of pollutants. Curr Opin Biotechnol 24:467–473

Shahid M, Zaidi A, Ehtram A, Khan SM (2019) In vitro investigation to explore the toxicity of different groups of pesticides for an agronomically important rhizosphere isolate *Azotobacter vinelandii*. Pest Biochem Physiol 157:33–34

Sharifi M, Ghorbanli M, Ebrahimzadeh H (2007) Improved growth of salinity stressed soybean after inoculation with pre-treated mycorrhizal fungi. J Plant Physiol 164:1144–1151

Sheng-wang PAN, Shi-qiang WEI, Xin YUAN, Sheng-xian CAO (2008) The removal and remediation of phenanthrene and pyrene in soil by mixed cropping of Alfalfa and Rape. Agric Sci China 7:1355–1364

Shoresh M, Harman GE, Mastouri F (2010) Induced systemic resistance and plant responses to fungal biocontrol agents. Annu Rev Phytopathol 48:21–43

Shukla KP, Nand Kumar S, Shivesh S (2010) Bioremediation: developments, current practices and perspectives. Gen Eng Biotech J 3:1–20

Siddikee MA, Glick BR, Chauhan PS, Yim WJ, Sa T (2011) Enhancement of growth and salt tolerance of red pepper seedlings (*Capsicum annuum* L.) by regulating stress ethylene synthesis with halotolerant bacteria containing 1-aminocyclopropane-1-carboxylic acid deaminase activity. Plant Physiol Biochem 49:427–434

Silva TM, Stets MI, Mazzetto AM, Andrade FD, Pileggi SAV, Favero PR, Cantu MD, Carrilho E, Carneiro PIB, Pileggi M (2007) Degradation of 2,4-D herbicide by microorganisms isolated from Brazilian contaminated soil. Braz J Microbiol 38:522–525

Singh NS, Singh DK (2011) Biodegradation of endosulfan and endosulfan sulfate by *Achromobacter xylosoxidans* strain C8B in broth medium. Biodegradation 22:845–857

Six J, Elliott ET, Paustian K (2000) Soil macroaggregate turnover and microaggregate formation: a mechanism for C sequestration under no-tillage agriculture. Soil Biol Biochem 32:2099–2103

Six J, Feller C, Denef K, Ogle SM, de Moraes JC, Albrecht A (2002) Soil organic matter, biota and aggregation in temperate and tropical soils effects of no-tillage. Agronomie 22:755–775

Sofia C, Sofia P, Ana L, Etelvina F (2012) The influence of Glutathione on the tolerance of *Rhizobium leguminosarum* to Cadmium. In: Zaidi A, Wani P, Khan M (eds) Toxicity of heavy metals to legumes and bioremediation. Springer, Vienna, pp 89–100

Spohn M, Giani L (2011) Impact of land use change on soil aggregation and aggregate stabilizing compound as dependent on time. Soil Biol Biochem 43:1081–1088

Sukul P (2006) Enzymes activities and microbial biomass in soil as influenced by metalaxyl residues. Soil Biol Biochem 38:320–326

Sylvia DM, William SE (1992) Vesicular arbuscular mycorrhizae and environmental stresses. In: Bethlenfalvey GJ, Linderman RG (eds) Mycorrhizae in sustainable agriculture. ASA Spec Publ, Madison WI, pp 101–124

Tamaki S, Frankenberger WT (1992) Environmental biochemistry of arsenic. Rev Environ Contam Toxicol 124:79–110

Tancho A, Merckx R, Van Look K, Vlassak K (1992) The effect of carbofuran and monocrotophos on heat output, carbon and nitrogen mineralization of northern Thailand soils. Sci Total Environ 123:241–248

Tisdall JM, Oades JM (1979) Stabilization of soil aggregates by the root system of rye grass. Aust J Soil Res 17:429–441

Topp E, Vallaeys T, Soulas G (1997) Pesticides: microbial degradation and effects on microorganisms. In: Van Elsas JD, Trevors JT, EMH W (eds) Modern soil microbiology. Mercel Dekker, New York, pp 547–575

Turner TR, James EK, Poole PS (2013) The plant microbiome. Genome Biol 14:209

Upadhyay N, Vishwakarma K, Singh J, Mishra M, Kumar V, Rani R, Mishra RK, Chauhan DK, Tripathi DK, Sharma S (2017) Tolerance and reduction of chromium (VI) by *Bacillus* sp. MNU16 isolated from contaminated coal mining soil. Front Plant Sci 8:778

Uren NC (2007) Types, amounts, and possible functions of compounds released into the rhizosphere by soil-grown plants. In: Pinton R, Varanini Z, Nannipieri P (eds) The rhizosphere: biochemistry and organic substances at the soil–plant interface. CRC, Boca Raton, FL, pp 1–22

Van der Heijden MGA, Rinaudo V, Verbruggen E, Scherrer C, Barberi P, Giovannetti M (2008) The significance of mycorrhizal fungi for crop productivity and eco system sustainability in organic farming systems. Proceedings of the 16th IFOAM Organic World Congress June 16–20 Modena Italy 1–4

Vanhooren P, Vandamme EJ (1998) Biosynthesis, physiological role, use and fermentation process characteristics of bacterial exopolysaccharides. Rec Res Develop Ferment Bioeng 1:253–300

Vargas JM (1975) Pesticide degradation. J Arboric 1(12):232–233

Vassilev A, Schwitzguebel JP, Thewys T, Van dar Leloe D, Vangronsveld J (2004) The use of plant for metal contamination soil. Sci World J 4:9–34

Watt M, Mc Cully ME, Jeffree CE (1993) Plant and bacterial mucilages of the maize rhizosphere: comparison of their soil binding properties and histochemistry in a model system. Plant Soil 151:151–165

Weary GC, Merriam HG (1978) Litter decomposition in a red maple woodlot under natural conditions and under insecticide treatment. Ecology 59:180–184

Wei-hua X, Yun-guo L, Guang-ming XL, Hua-xiao S, Qing-qing P (2009) Characterization of Cr (VI) resistance and reduction by *Pseudomonas aeruginosa*. Trans Nonferrous Met Soc China 19:1336–1339

Wright SF, Upadhyaya A (1996) Extraction of an abundant and unusual protein from soil and comparison with hyphal protein from arbuscular mycorrhizal fungi. Soil Sci 161:575–586

Wright SF, Franke Snyder M, Morton JB, Upadhyaya A (1996) Time-course study and partial characterization of a protein on hyphae of arbuscular mycorrhizal fungi during active colonization of roots. Plant Soil 181:193–203

Wu F, Gai Y, Jiao Z, Liu Y, Ma X, An L, Wang W, Feng H (2012) The community structure of microbial in arable soil under different long-term fertilization regimes in the Loess Plateau of China. Afr J Microbiol Res 6:6152–6164

Wysocki R, Bobrowicz P, Ulaszewski S (1997) The *Saccharomyces cerevisiae* ACR3 gene encodes a putative membrane protein involved in arsenite transport. J Biol Chem 272:30061–30066

Xie S, Liu J, Li L, Qiao C (2009) Biodegradation of malathion by *Acinetobacter johnsonii* MA19 and optimization of co-metabolism substrates. J Environ Sci (China) 21:76–82

Yang YH, Yao J, Hu S, Qi Y (2000) Effects of agricultural chemicals on DNA sequence diversity of soil microbial community: a study with RAPD marker. Microb Ecol 39:72–79

Zafar-ul-Hye M, Farooq HM, Zahir ZA (2014) Combined application of ACC-deaminase biotechnology and fertilizers improves maize productivity subjected to drought stress in salt affected soils. Int J Agric Biol 16:591–596

Zaidi S, Usmani S, Singh BR, Musarrat J (2006) Significance of *Bacillus subtilis* strain SJ-101 as a bioinoculant for concurrent plant growth promotion and nickel accumulation in *Brassica juncea*. Chemosphere 64:991–997

Zhang S, Li Q, Zhang X, Wei K, Chen L, Liang W (2012) Effect of conservation tillage on soil aggregation and aggregate binding agents in black soil of Northeast China. Soil Tillage Res 124:196–202

Zhuang X, Chen J, Shim H, Bai Z (2007) New advances in plant growth-promoting rhizobacteria for bioremediation. Environ Int 33(3):406–413

# Chapter 17
# Role of Soil Organisms in Maintaining Soil Health, Ecosystem Functioning, and Sustaining Agricultural Production

**Neemisha**

**Abstract** Soil is the fundamental natural resource on which all living and nonliving things are evolved and sustained for carrying out vital life processes. Sustainable agriculture directly relies on soil health which is defined as the capacity of the soil to ascertain environmental quality, sustain biological productivity, and maintain health of all living beings. The biotic and abiotic components of an ecosystem interact with each other for proper functioning of all processes. Soil organisms consist of different living forms dwelling in the soil which spend at least one part of their life cycle in soil. These include soil megafauna, macrofauna, mesofauna, microfauna, and microflora. The life within the soil is hidden and so often suffers from being out of sight and out of mind, but their role in maintaining soil health and ecosystem productivity is indispensable. This chapter is an attempt to highlight the contribution of soil organisms in sustaining soil health, fertility, and agricultural productivity.

**Keywords** Biogeochemical cycles · Decomposition of organic matter · Soil macrofauna · Soil mesofauna · Soil organism · Microbial aggregates

## 17.1 The Living Soil

Optimal soil health represents a balance between productivity, environmental quality, and health of living beings residing on it (Doran 2002). Nowadays, the term soil health is directly used in connection to sustainable agriculture (Kibblewhite et al. 2008; Cardoso et al. 2013; Frac et al. 2018) and is defined as the capacity of the soil to maintain environmental quality, sustain biological productivity, and promote animal, human, and plant health (Doran and Parkin 1994; Frac et al. 2018; Doran and Jones 1996). The four spheres of an ecosystem, i.e., biosphere, hydrosphere, lithosphere, and atmosphere interact reciprocally using biological, chemical,

Neemisha (✉)
Department of Soil Science, Punjab Agricultural University, Ludhiana, Punjab, India
e-mail: neemisha14@pau.edu



B. Giri, A. Varma (eds.), *Soil Health*, Soil Biology 59,
https://doi.org/10.1007/978-3-030-44364-1_17

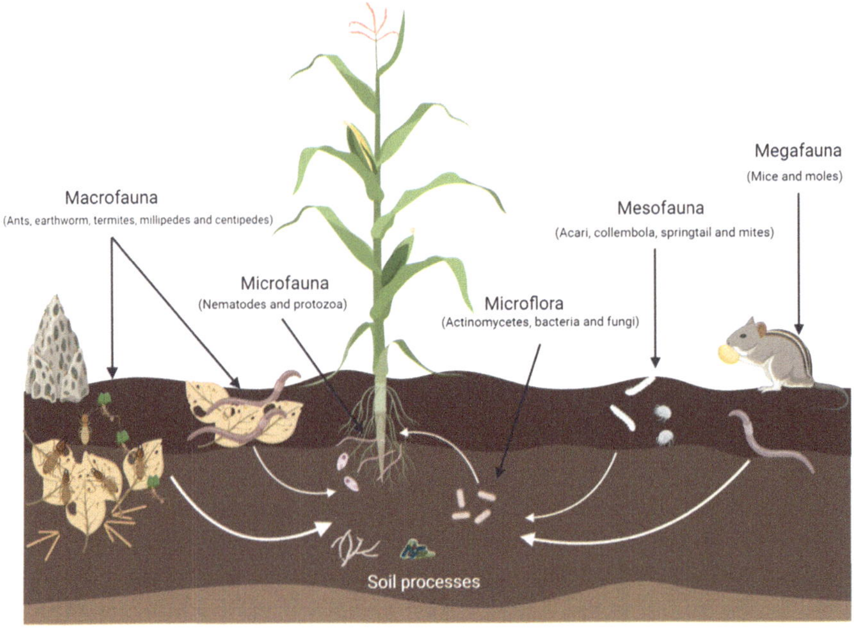


**Fig. 17.1** Diagrammatic representation of soil organisms and their functions in soil

biochemical, and physical transformations and translocations (Biological and physical). The soil ecosystem consists of both abiotic and biotic components, and these components interact with each other for the proper functioning of the ecosystem. The biotic components include soil macroorganisms (moles, rats, earthworms), microorganisms (bacteria, fungi, actinomycetes), plant roots, and algae. The abiotic components include mineral particles, nutrients, gases, water, and nonliving organic matter. The term soil biology represents diverse groups of organisms dwelling in the soil which spend at least one part of their life cycle in soil (Fig. 17.1). Although soil microorganisms are large in number as compared to other organisms, macrofauna has received more attention as compared to microorganisms. A detailed classification of soil organisms (Swift et al. 1979; Barrios 2007) their types and functions performed in the ecosystem is represented in Table 17.1.

## 17.2 Soil Macrofauna

Soil macrofauna includes the organisms which range in size from 500 μm to 50 mm (Swift et al. 1979). Soil macrofauna represents an invertebrate group that is found within terrestrial soil samples which has the majority of its individuals visible to the naked eye (IBOY 2000). Soil macrofauna groups include organisms like ants, centipedes, Coleoptera (adults and larvae), Diptera larvae, Dermaptera,

**Table 17.1** Classification of soil organisms

| Classes | Organisms | Types/habit | Functions |
|---|---|---|---|
| Megafauna | Soil dwelling animals | Mice, moles, groundhogs | • Soil turnover and distribution<br>• Loosens soil structure<br>• Improves aeration and drainage<br>• Distributes soil microorganisms |
| Macrofauna (500 μm–50 mm) | Earthworms | Epigeic<br>Endogeic<br>Anecic | • Improves soil structure, aeration, and water infiltration<br>• Casts are rich in organic matter, microbes, and nutrients which improves nutrient cycling<br>• Modify food quality through its passage in the gut<br>• Mineralization of organic matter and release of nutrients<br>• Influences soil chemical composition through the deposition of excrement<br>• Disseminates bacteria and spores in soils |
| | Ants | Microarthropds<br>Decaying organic debris<br>Seeds<br>Plant secretions<br>Aphid secretions | • Important in mixing soil from lower depths with surface soil<br>• Redistribute nutrients around the ecosystem<br>• Turn and aerate the soil<br>• Consume a variety of foods |
| | Termites | Wood feeders<br>Plant debris feeders<br>Soil humus feeders<br>Fungus growers | • Formation of biogenic structure<br>• Burrowing activities<br>• Eat soil organisms, organic matter and maintain equilibrium |
| | Mollusks | Soft-bodied animals enclosed within a calcium carbonate shell<br>Snails, slugs | • Significant role in the food web and ecosystem processes<br>• Nutrient cycling<br>• Decomposition of SOM (cellulases)<br>• Affects microbial and microfauna diversity |
| | Solitary wasps | Construct nests in soil | • Prey on insects and spiders to feed developing eggs |
| | Millipedes | Calcareous skeleton | • Saprophagus feeders<br>• Consume dead and decaying organic debris<br>• Important in calcium cycling<br>• Feces acts as food for microbes<br>• Their droppings contribute to humus and soil formation |
| | Centipedes | Multilegged creatures | • Active predators of various microarthropods in soil and surface litter |

(continued)

**Table 17.1** (continued)

| Classes | Organisms | Types/habit | Functions |
|---|---|---|---|
| | Scorpions | Predatory arachnids with burrowing habits | • Mobile predators of other arthropods, lizards, mice, and birds |
| | Spiders | Air breathing arthropods | • Predators of soil insects<br>• Considered as indicators of the sustainability and adequate management of agricultural landscapes |
| | Bettles | Predacious<br>Leaf feeding<br>saprophagous | • Feed on other insects, animal feces, and plant roots |
| | Isopods | Cryptozoans surface soil dwellers under stones, bark, and litter layers | • Feed on roots, vegetation, and decaying plant litter<br>• Appear at night and otherwise roll back to balls to avoid desiccation |
| Mesofauna (80 µm–2 mm) | Collembola<br>Acari<br>Springtails | Opportunistic microarthropods | • Feeds on bacteria, fungi, mineral soil particles, organic matter, protozoa, and nematodes<br>• Important biological control agents for crop and feeds on pathogenic fungi<br>• Important food sources for predacious mites, beetles and ants |
| | Tardigrades (Water bears) | Found in the top few centimeters of soil | • Feed on algae, microorganisms, and organic matter<br>• Undergo dormancy induced environmental stress<br>• Similarities of both nematodes and microarthropods |
| | Rotifers | Rotifers are commonly found on mosses, lichens, and tree trunks | • Require moisture for their survival<br>• Serves as a major food source<br>• Decomposition of organic matter |
| | Mites | Oribatids<br>Prostigmatics<br>Mesostigmatics<br>Astigmatics | • Oribatids feed on detritus and fungi<br>• Prostigmatics feeds on fungi, algae, and other organisms<br>• Mesostigmatics acts as predators of nematodes and other microarthropods<br>• Astigmatics are mainly found in moist soil with high organic matter content |

(continued)

**Table 17.1** (continued)

| Classes | Organisms | Types/habit | Functions |
|---|---|---|---|
| | Protura | Wingless insects lacking antennae and eyes | • Lesser importance in the soil community<br>• Little influence on soil processes |
| | Diplurans | Delicate insects with long antennae and two small cerci | • Feed on decaying vegetation<br>• Predacious on nematodes, collembola, and enchytraeids |
| | Pseudoscorpions | Small scorpion lacking tails and stingers | • Feeds on nematodes, micro arthropds, and enchytraeids |
| | Symphylids | Insects resembling centipedes | • Feeds on vegetation and soft soil animals |
| | Pauropoda | Colorless many legged insects with branched antennae | • Feeds on fungi and other soil organisms |
| | Enchytraeids | Potworms | • Feeds on fungi, algae, bacteria, and other soil organisms<br>• Usually smaller than earthworms |
| Microfauna (5–120 μm) | Protozoa | Mostly feeds on bacteria | • Nutrient turnover in the rhizosphere<br>• Production of plant growth-promoting substances<br>• Phagotrophic with bacteria, fungi, algae, and other fine particulate organic matter |
| | Nematodes | Pathogenic Herbivorous<br>Microbivorus omnivorous<br>Carnivorous | • Nutrient turnover<br>• Soil organic matter decomposition |
| | Algae<br>Lichens | Different types | • Soil formation from rocks<br>• Soil aggregate stability<br>• fix atmospheric nitrogen<br>• Add organic matter to the soil<br>• Liberate oxygen to the soil environment |
| Microflora (1–100μm) | Fungi | Yeasts<br>Molds<br>Mushrooms | • Decomposition of organic matter<br>• Promotion of soil aggregation, nutrient cycling, and biocontrol of plant pathogens<br>• Plant diseases<br>• Degradation of wastes and harmful chemicals |

(continued)

**Table 17.1** (continued)

| Classes | Organisms | Types/habit | Functions |
|---|---|---|---|
| | Bacteria | Photoautotrophs<br>Photoheterotrophs<br>Chemoautotrophs<br>Chemoheterotrophs | • Nutrient cycling<br>• Organic matter decomposition<br>• Production of industrially important secondary metabolites<br>• Plant diseases<br>• Degradation of wastes and harmful chemicals |
| | Actinomycetes | Terrestrial<br>Aquatic<br>Marine | • Decomposition of soil organic matter<br>• Release of nutrients<br>• Mostly develop during later stages of decomposition<br>• Very important in curing of compost<br>• Inhibit the growth of several plant pathogens in the rhizo-sphere<br>• Biological buffering of soils |
| | Viruses<br>Viroids<br>Mycoplasmas<br>Prions | Variable | • Influence ecology of soil bio-logical communities<br>• Turnover of nutrients and gases |

earthworms, Isopoda, Lepidoptera larvae, millipedes, slugs, snails, spiders, termites and majority of insects (Ruiz and Lavelle 2008; Menta 2012). Among these organisms, earthworms, termites, and ants are most abundant in soils. Earthworms act as decomposers by fragmentation of dead organic matter and obtain food by growing on it. Earthworms are classed into three main ecological types (Lee 1985): endogeic (live and feed in soil), anecic (feed on surface litter and mix it with soil), and epigeic (live within the litter layers where they are subject to adverse environmental conditions viz occasional drought, extreme temperatures, and high predator densities). They also play a major role in nutrient recycling, improve infiltration rate, water holding capacity, and reduce surface water erosion. Earthworms also contribute in mixing and soil aggregation as they excrete organic matter in form of feces in the soil. While making bio-structures in soil, earthworms fragment litter, mix organic matter with soil, and release their gut contents to the soil which are rich in organic matter, nutrients, and water (Bhadauria and Saxena 2010). The earthworms eat organic matter and excrete partially digested material which is rich in NPK, micronutrients, and beneficial soil microorganisms. Due to all these enrichments they can foster high populations of microorganisms (Nechitaylo et al. 2010; Mora et al. 2005; Haynes and Fraser 1998). The existence of microbial diversity and species richness depends upon the species, microhabitat (gut or casts), and surrounding environment of earthworms (Medina-Sauza et al. 2019). These effects can be positive (Hoeffner et al. 2018), neutral (de Menezes et al. 2018), or negative (Gopal et al. 2017). Earthworms generally exert beneficial effects on plant growth and soil health. Ayuke et al. (2011) investigated the effects of organic and inorganic inputs on soil abundance, biomass and taxonomic diversity of macrofauna as well as aggregate formation in Central Kenya. Earthworms play a pivotal role in improving soil fertility by depositing nutrients (brought from deep soil) on the soil surface as castings, checks nutrient leaching, improves soil structure, and incorporates organic matter into the soil.

## 17.3 Soil Mesofauna

Mesofauna includes organisms that range in size from 80 μm to 2 mm. Mesofauna is mainly grouped as Acari, Collembola, Insecta, Isopoda and Myriapoda, and among these, Acari and Collembola are the most abundant and diverse (Culliney 2013). Other organisms include Diplurans, Enchytraeids, Mites, Pseudoscorpions, Pauropoda, Protura, Rotifers, Symphylids, Tardigrades etc (Franzluebbers 2009, Table 17.1). These organisms directly contribute to several ecosystem processes such as organic matter decomposition and nutrient cycling. Soil ecological conditions control the population density of soil invertebrates (Neher and Barbercheck 1998). Dominant mesofauna species are morphologically and physiologically adapted for living near the soil surface (Ubugunova et al. 2007). Unlike macrofauna, mesofauna cannot modify soil structure and thus use existing channels, pore space, and cavities for movement in the soil (Neher and Barbercheck 1998). Mesofauna which feeds on microbes regulates decomposition rate by affecting microbial growth, metabolic activities and alter microbial community (Neher and Barbercheck 1998). In a forest ecosystem, soil microarthropods are important in nutrient cycling as well as decomposition of the organic matter; however, a little variation in the microarthropod population can significantly affect mobilization of nutrients (Heneghan and Bolger 1998). The major function of Collembola is plant litter decomposition and soil microstructure formation (Heneghan et al. 1998; Rusek 1998). Moisture plays an important role in determining the survival, abundance, and distribution of Collembola in soil, e.g., Collembola and Acari population remain low during drought conditions (Wiwatwitaya and Takeda 2005). However, certain factors such as soil acidification (Collembola), soil-water availability (Acari), nitrogen supply (Oribated mites), and intensive farming (Collembola) have a negative impact on soil mesofaunal diversity (O'Lear and Blair 1999; Deleporte and Tillier 1999; Rusek 1998; Bedano and Domínguez 2016).

## 17.4 Soil Microfauna

Soil organisms less than 1 mm constitutes microfauna. They include effective predators such as protozoa and nematodes which feed on soil bacteria, fungi, and other microfauna (Table 17.1). Protozoa play a major role in nutrient mineralization by delivering extra nitrogen in the form of ammonium which is easily utilized (by bacteria and plants) in the soil near the root system of the plant. Another important function of protozoa is to maintain the population of bacteria by eating bacteria, which further enhance the bacterial growth, soil aggregation, and decomposition rates. Protozoa are also involved in the control of diseases by eating and competing with pathogens and acts as a food source for pathogens. Like protozoa, nematodes are also involved in nutrient mineralization and release of usable forms of ammonium in soil. Quality of soil is directly linked with the occurrence of a variety of nematodes performing different functions in the soil food web. Nematodes also act as biocontrol agents as they control disease development by blocking access to roots. They also serve as food source for soil insects, microarthropods, predatory nematodes, and maintain equilibrium between bacteria and fungi (Tugel and Lewandowski 2010).

## 17.5 Soil Microflora

Soil contains diverse organisms like bacteria, archaea, fungi, algae, insects, annelids, and other invertebrates which show an intimate relationship to each other and with plants (Glick 2010). Soil microorganisms play a very important role in developing a healthy structure of the soil. Soil microbes secrete sticky substances such as mucilage and polysaccharides which help in cementing the soil aggregates. This cementing action of gummy substances prevents crumbling in aggregates on exposure to water. Fungi exhibit threadlike structures which add stability to soil structure because these filaments have a large surface area which spread throughout the soil (Johns 2017). The contribution of microbes in changing the physical structure of the soil is less as compared to macroorganisms. Soil microorganisms have been classified into three types depending upon their functions in soil as ecosystem engineers, chemical engineers, and biological regulators (Bagyaraj and Ashwin 2017).

(a) Ecosystem engineers: Ecological engineers which create or modify habitats for other organisms by constructing resistant pores and soil aggregates. These habitats act as hot spot for the proliferation of mesofauna, microfauna, microbes etc. and also act as a regulator of resources for other organisms.
(b) Chemical engineers: Chemical engineers bring about decomposition of organic matter and provide readily available nutrients (Gardi and Jeffery 2009).
(c) Biological regulators: Biological regulators act as predators of plants, invertebrates and microorganism and regulate their dynamics in space and time. These include mites, nematodes, pot worms, and springtails. This group of organisms is relatively less explored as compared to rest of the two groups.

Due to their capability to adapt to adverse or extreme environmental conditions and high plasticity, fungi are considered as the most successful inhabitants of soil (Sun et al. 2005; Frac et al. 2018). Based on their functions in soil (Swift 2005; Gardi and Jeffery 2009; Frac et al. 2018), fungi can be grouped as: Ecosystem regulators, Biological controllers, Decomposers and compound transformers. Actinomycetes are aerobic, spore-forming Gram-positive bacteria that bear the talent of degrading recalcitrant substances which are very hard to decompose. Moreover, they produce the a characteristic earthy smell of soil and one of the most exploited organisms for production of antibiotics (Tugel and Lewandowski 2010). These microorganisms (bacteria, fungi and actinomycetes) play an important role in cycling of organic matter, control of plant pathogens (Jose and Jha 2016), production of extracellular enzymes and metabolites (Gomez-Escribano et al. 2016; Katz and Baltz 2016; Charousova et al. 2017), nitrogen fixation, P solubilization, siderophore production, improving availability of minerals and nutrients, formation and stabilization of compost piles (Bhatti et al. 2017) and bioremediation (Radwan et al. 1998).

## 17.6 Role of Soil Biology in Maintaining the Sustainability of Agriculture

### *17.6.1 Microbes for Conservation of Soil Structure*

Sustainable agriculture mainly relies on the adoption of management practices that minimize soil erosion and conserve soil fertility. The success of these practices is judged by their ability to create soil aggregates which act as a sink of nutrients and prevent soil erosion (Elliott and Coleman 1988; Pankhurst and Lynch 1995). The soil aggregates are composed of soil mineral (sand, silt, and clay) fragments, gases, water and solutes, decaying, and living organisms bound together. According to Forster (1990), in sand dunes, three types of aggregates are found in the soil which include: (1) Microbial aggregates (1–12 mm in diameter): These aggregates are usually formed by the interaction of soil particle with bacteria, actinomycetes, fungi, cyanobacteria, and algae; (2) Root-microbial aggregates: Root microbial aggregates are formed by the combined action of microbes and roots (structural components and exudates); (3) Debris-microbial aggregates (up to 6 mm in diameter): These are formed by the decomposition of macroscopic plant debris.

#### 17.6.1.1 Mechanism of Aggregate Formation

Soil aggregates act as microhabitats for the proliferation of a variety of microorgan isms. The process of aggregate formation occurs via physical attachment of microbes or their products with mineral and organic debris, and this results in the formation of microaggregates. These microaggregates are subsequently trapped by extraradical hyphae and plant roots to form macroaggregates (Oades and Waters 1991). The macroaggregate are further stabilized by cementing with polysaccharides or other organic substances to result in the formation of inter-microaggregate and inter-macroaggregate. Several microbes have been reported to contribute to soil aggregate formation by the production of specific exudates (Table 17.2). Some of these

**Table 17.2** The microbial products which help in soil aggregate formation

| Microbes | Group | Exudates | Mechanism |
|---|---|---|---|
| Fungi | Mycorrhizal fungi | Glomalin | Glomalin binds to the hyphal wall and transports water nutrients and for Arbuscular mycorrhizae |
| | | Hydrophobic compounds | Transports water and nutrients for fungi |
| | | Hydrophobins | Increase hydrophobic soil organic matter and add stability to aggregates |
| | Saprophytic fungi (Basidiomycetes and Trichocomaceae) | Extracellular exudates | Bing to soil particles and increase stability |
| Bacteria | *Azotobacter, Bacillus, Chlamydomonas, Rhizobium* | Amino acids, mucigels, Polysaccharides, polyuronic acids | Act as glue/gum |

organisms along with their contributing properties are explained in the subheads below:

#### 17.6.1.2 Bacteria in Soil Aggregation

Bacteria exist in the soil as individual cells, as colonies or biofilms inside the pores of microaggregates (Degens 1997; Rashid et al. 2016). Bacteria produce polysaccharides, polyuronic, and amino acids that are negatively charged and help in attaching to clay particles to form aggregates. Some microbial species (e.g. *Azotobacter vinelandii, Bacillus megaterium, Chlamydomonas sajao* and *Rhizobium* sp) have been reported to increase aggregate stability through production of extracellular compounds or polysaccharides (Metting 1986; Alami et al. 2000; Ortiz et al. 2015; Mengual et al. 2014).

#### 17.6.1.3 Fungi in Soil Aggregation

Arbuscular Mycorrhizal (AM) fungi bear extra-radical hyphae to modify the morphological structure, biochemical nature, and rhizosphere microbes in its vicinity (Borie et al. 2008; Peng et al. 2013; Rashid et al. 2016) and adds stability to soil aggregates (Daynes et al. 2012). Saprophytic, ectomycorrhizal, AM, and other fungi form soil aggregates by complicated direct and indirect mechanisms. The direct mechanism includes the formation of hyphae and mycelium which entangle soil particles and bring them together to form aggregates (Rashid et al. 2016). Indirect mechanisms include the exudation of polysaccharides glomalin, mucilages, hydrophobins, and extracellular compounds from hyphae into the soil (Caesar-TonThat 2002). Glomalin, a hydrophobic protein, existing in the hyphal wall is responsible for the transport of water and nutrients for AM Fungi (Rashid et al. 2016). Glomalin protein combines with organic matter and minerals to form clumps of hydrophobic soil aggregates. The adhesive properties of glomalin imparts initiation and stabilization to aggregates (Miller and Jastrow 2000; Leifheit et al. 2015) and also helps in binding of soil particles (Wright and Upadhyaya 1996). This protein protects fungal hyphae and their spores from adverse environmental conditions and attack by microbes. Saprophytic fungi produce extracellular exudates, which bind to soil particles and increase their stability (Agerer 2001). Certain small proteins known as hydrophobins are known to increase hydrophobic soil organic matter. Hydrophobins are known to affect the soil wettability and water repellency and hence reduce the breakage of dry soil aggregates (Six et al. 2004; Diehl 2013) (Table 17.2).

### *17.6.2 Role of Soil Microbes in Biogeochemical Cycles*

All living forms of life are dependent on a stockpile of essential elements from the earth that are necessary for life (Gougoulias et al. 2014). Earth is considered as a closed system where recycling of mineral elements is a continuous process which avoids their exhaustion, thus ensuring a finite supply of essential elements such as carbon (C), hydrogen (H), nitrogen (N), oxygen (O), phosphorus (P), and sulfur (S) (Gougoulias et al. 2014). The quantity of all the compounds found on the surface of the earth represents a net balance between their rate of formation and utilization in the biological and geological processes. The key steps of the biogeochemical cycles are mentored by soil microorganisms and enzymes produced by them (Modi 2013; Schaller 2009). Relatively small size and large surface–volume ratio of microbes allows a rapid exchange of substrates and waste products among microbes and the environment. Transformations of different elements such as N, P, S, and C with microorganisms involved and their functions are tabulated in Table 17.3.

Nitrogen is an essential component of biological systems and the main nutrient limiting life on terrestrial systems (Ward and Jensen 2014). The acquisition of nitrogen and its cycling play a pivotal role in shaping the microbial communities and directly control ecosystem productivity (Kuypers et al. 2018). Earth's atmosphere contains abundant nitrogen gas ($N_2$); however, due to extreme stability, its conversion to other forms requires high energy. The utilization of this gaseous nitrogen consists of four major steps: nitrogen fixation, ammonification, nitrification, and denitrification (Modi 2013). Different groups of microorganisms are involved in these transformations and in this way net $N_2$ concentration in the ecosystem is regulated. Microorganisms are integral to the soil phosphorus (P) cycle and play a crucial role in mediating the P availability to plants (Richardson and Simpson 2011). Several microbes (bacteria, fungi, actinomycetes, and algae) possess P solubilization and mineralization ability (Alori et al. 2017). The only possible way of increasing plant available phosphorus is by microbial P solubilization and mineralization (Bhattacharyya and Jha 2012). Sulfur is an element whose transformation

**Table 17.3** Role of microorganisms in different biogeochemical cycles

| Nutrient cycle | Processes involved | Soil microorganisms | Functions |
|---|---|---|---|
| Nitrogen cycle | Symbiotic nitrogen fixation | Symbiotic $N_2$ fixers *Rhizobium, Bradyrhizobium, Mesorhizobium* | Symbiotic fixation of atmospheric nitrogen by microorganisms in roots nodules of leguminous plants |
| | Nonsymbiotic nitrogen fixation | Aerobic heterotrophic *Azomonas, Azotobacter, Beijerinckia, Derxia, Pseudomonas,* Aerobic autotrophic Blue-green bacteria, *Anabaena, Calothrix, Nostoc, Oscillatoria* Anerobic heterotrophic *Aerobacter, Clostridium* Anerobic autotrophic *Chlorobium, methanobacterium Rhodospirillium* | Nonsymbiotic fixation of atmospheric nitrogen in cellular form in bulk and rhizospheric soil |
| | Ammonification | Bacteria *Bacillus, Clostridium, Proteus, Pseudomonas,* Actinomycetes *Streptomyces* Fungi *Alternaria, Aspergillus, Mucor, Penicillium* | Ammonia released with degradation of organic matter of dead plants and animals through proteolysis and amino acid degradation. |
| | Nitrification | Nitrifying microorganism *Nitrosomonas, Nocardia, Nitrosospira, Nitrosolobus, Nitrosococcus, Nitrobacter* | Facilitate the uptake of nutrients by converting the nutrients into plant consumable form such as conversion or oxidation of ammonia into nitrate |
| | Denitrification | Denitrifying microorganism *Bacillus, Bacillus licheniformis Chlorobacterium, Micrococcus denitrificans, Paracoccus, Pseudomonas denitrificans, Serratia, Thiobacillus denitrificans* | Degrade the availability of nutrients in soil such as reduction of plant available nitrate form into nitrogen and nitrous oxide gases. |
| Phosphorus cycle | Mineralization (phosphatases) and Solubilization | Bacteria *Azotobacter, Bacillus, Burkholderia, Enterobacter, Erwinia, Flavobacterium, Pseudomonas* | Mineralization of organic nitrogen compounds with the degradation of organic phosphorus into insoluble inorganic phosphates. |

(continued)

**Table 17.3** (continued)

| Nutrient cycle | Processes involved | Soil microorganisms | Functions |
|---|---|---|---|
| | | Fungi<br>*Alternaria, Aspergillus, Cephalosporium, Cladosporium, Fusarium, Glomus, Helminthosporium, Micromonospora, Penicillium, Pichia fermentans, Pythium, Rhizoctonia, Rhizopus, Saccharomyces, Sclerotium, Torula, Trichoderma*<br>*Actinomycetes*<br>*Micromonospora, Streptomyces* | Organic and inorganic forms of phosphates are converted into plant available form, i.e., soluble inorganic phosphates and assimilation into organic phosphates. |
| Sulfur cycle | Liberation of $H_2S$ with degradation of organic compounds | *Beggiatoa, Chlorabiaceae, Chromatiaceae, Desulfococcus, Desulphurase, Desulfotomaculum, Desulfosarcina, Desulfonema, Thermoproteus, Thermococcus* | Sulfur-containing amino acids are released with degradation of proteins. |
| | Oxidation of $H_2S$ to elemental sulfur | Photosynthetic sulfur bacteria<br>*Chlorobium, Chromatium, Chloroflexus, cyanobacteria*<br>Non-sulfur purple bacteria<br>*Rhodospirillium, Rhodopsudomonas, Rhodomicrobium* | Hydrogen sulfide degraded to elemental sulfur. |
| | Oxidation of elemental sulfur to sulfates | *Achromatium, Macromonas, Paracoccus, Pseudomonas, Sulfolobus, Thermothrix, Thiomicrospira, Thiobacillus thioxidans, Thiobacterium, Thiothrix* | Oxidation of elemental sulfur and organic sulfur or sulfates into plant consumable form with the help of chemolithotrophic bacteria. |
| | Reduction of sulfates | Sulfur-reducing bacteria<br>*Desufovibrio, Desulphotomaculum* | Reduction of sulfate sulfur into hydrogen sulfide gas. |
| Carbon cycle | Carbon dioxide fixation | Plants, phytoplankton autotrophic and heterotrophic bacteria | Oxidized form of carbon is reduced into organic carbon compounds through the process of photosynthesis. |
| | Restoration of oxidized form | Decomposers cellulolytic bacteria and fungi<br>*Cellulomonas* species | With the mineralization of organic matter, $CO_2$ released in the atmosphere for restoration.<br>Process of respiration<br>Accidental forest fire<br>Fuel burning |

(continued)

**Table 17.3** (continued)

| Nutrient cycle | Processes involved | Soil microorganisms | Functions |
|---|---|---|---|
| | Organic matter decomposition | Fungi<br>*Trichoderma, Aspergillus, Penicillium, Fusarium, Rhizocotonia, Pleurotus* etc.<br>Bacteria<br>*Clostridium, Streptomyces, Bacillus, Nocardia, Xanthomonas* etc. | Decomposition of cellulose, hemicelluloses and lignin |

and fate in the environment is critically dependent upon microbial activities. All living beings contain sulfur in the form of amino acids, coenzymes, vitamins and can metabolize sulfur compounds during various catabolic, anabolic, or excretory processes (Troper 1984). Sulfur cycle involves a series of oxidation–reduction and assimilation–dissimilation steps (Bremner and Steele 1978). The fate and transformation of sulfur in nature are critically dependent on microorganisms (Klotz et al. 2011). The carbon cycle is one of the most important cycles in the ecosystem. The terrestrial carbon cycle is maintained by the process of conversion of oxidized form of carbon to reduced form by the process of photosynthesis (Gougoulias et al. 2014; Modi 2013) by plants, photo autotrophic and chemoautotrophic bacteria that convert carbon dioxide to organic material (Trumbore 2006). The reduced form is then oxidized by respiration (autotrophic and heterotrophic organisms), fires and decomposition by microorganisms (Liang and Balser 2011). Thus in nutshell nutrient cycles regulates the flow of nutrients, helps in the storage of elements, establishes equilibrium in nature, facilitates the transfer of elements, and provides available nutrients to all living organisms.

## 17.7 Decomposition of Organic Matter

Organic matter is the non-mineral, solid portion of the soil originating from plant and animal residues (Aust and Lea 1991). Organic matter play several important roles in soil which includes stabilization of soil aggregates, easy cultivation, improving soil buffering and water-holding capacities, release of nutrients, and adsorption of heavy metals (Carter and Stewart 1996). The major portion of organic matter is contributed by plant residues and their main constituents are carbohydrates (which range from simple sugars to starch to cellulose), lignins, polyphenols, and proteins. The process of decomposition of organic matter by soil organisms (Fig. 17.2) is divided into different steps (Brady and Weil 2015).

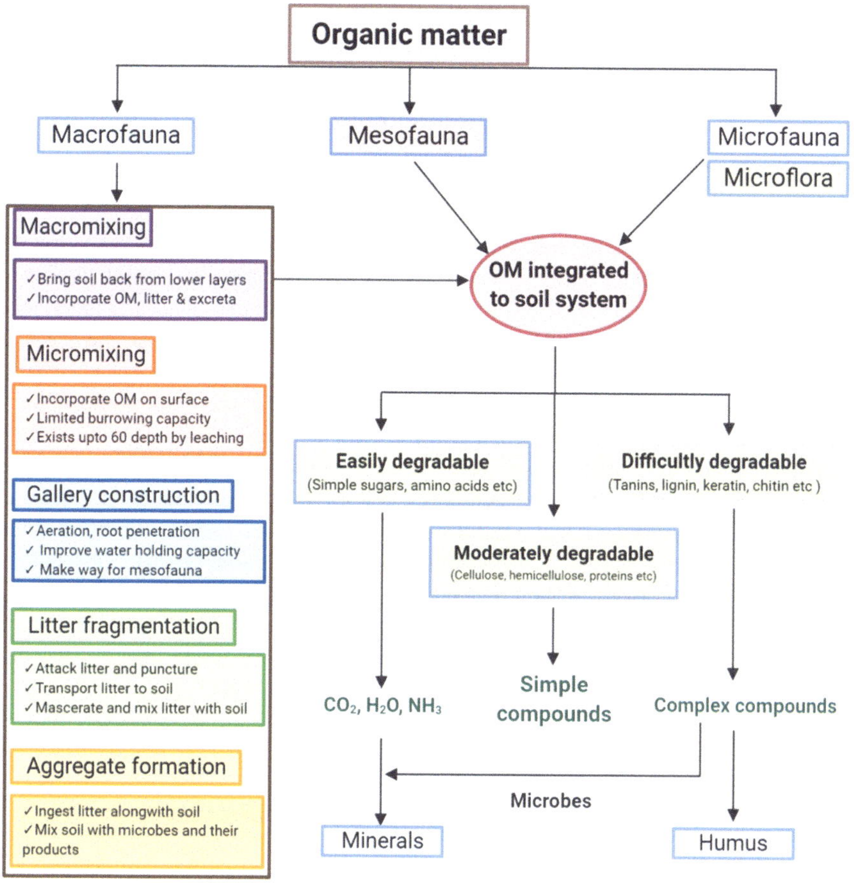


**Fig. 17.2** Schematic diagram showing role of soil organisms in decomposition of organic matter

### *17.7.1 Reduction in Surface Area by Macrofauna*

Macrofauna attack the litter and puncture holes in leaf epidermis via a process known as fenestration. The physical fragmentation results in destruction of leaf protective layer which expose cell contents. This improves aeration, water-holding capacity, and movement of particulate and soluble substances (Culliney 2013). Saprophagous animals result in comminution of plant litter by ingestion during which certain catabolic changes occur while passing through the digestive system. Thus, the excreta produced by the organism is different in chemical composition and smaller in size than the ingested food (Swift et al. 1979; Culliney 2013). The feeding activity of soil organisms significantly contribute to total annual litter turnover such as Collembola and oribatid mites (20%), Isopoda and Diplopoda (3–10%), termites (60%), and Symphyla (2%) (Berthet 1967; Edwards 1974; Petersen 1994; Collins 1981). Macrofauna helps to macerate, pulverize, transport, and mix the litter with the soil. If macrofauna is removed from the soil the process of decomposition slows down.

### 17.7.2 Enzymatic Degradation of Carbonaceous Materials

In this step, different carbonaceous materials undergo enzymatic degradation to produce carbon dioxide, water, energy, and decomposer biomass. Plant litter is a mixture of easily degradable (simple sugars, starch, amino acids), moderately degradable (cellulose, hemicelluloses, proteins) and recalcitrant forms that are difficult to degrade (lignins, tannins, chitin, subarin, keratin) (Coleman et al. 2004). Microbes cannot transport these carbonaceous materials into their cytoplasm so they depend on the activity of extracellular enzymes that are secreted into their immediate environment (Burns 2010). Different enzymes involved in the degradation include $\beta$-glucosidase (convert cellulose to glucose), cellobiohydrolase (convert cellulose to disaccharides), xylosidase (hemi-cellulose degraders), $\alpha$-glucosidase (convert soluble saccharides to glucose), N-acetyl-glucosaminidase (chitin degraders), leucine-amino-peptidase (convert protein to amino acids), phosphatase (convert phosphate group to phosphate ions) etc. Certain strains of bacteria, actinomycetes and fungi have been reported which can degrade carbonaceous materials using cellulase, hemicellulase, and laccase enzymes.

### 17.7.3 Essential Elements Release or Immobilization

A series of cyclic transformations take place in nature in the form of nutrient cycling where several organisms are involved for the release or immobilization of nutrients. The mechanism of action of each transformation is specific and unique. Soil macro- and mesofauna constitute an important nutrient pool in the soil including $K^+$, $PO_4^{3-}$, N, $Na^+$, and $Ca^{2+}$ (Hassall 1977; Cornaby et al. 1975; Teuben and Verhoef 1992; Seastedt and Tate 1981). Soil macrofauna immobilizes ions temporarily and thus prevents their leaching. The feces of soil organisms are rich in $NO_3^-$ nitrogen (Collembola, Teuben and Verhoef 1992), total nitrogen (*Glomeris marginata*, Bocock 1963; Marcuzzi 1970), sulfur (*Oniscus asellus*, Morgan and Mitchell 1987; Morgan 1983) and carbon and nitrogen (termites, Zaady et al. 2003).

### 17.7.4 Formation of Resistant Compounds

In this step, different compounds are formed either through microbial synthesis or through modification of compounds and result in the formation of a complex which is very resistant to microbial degradation.

## 17.8 Use of Microbes or Microbial Amendments

The use of microbes, either directly or indirectly as inoculants or amendments, can promote plant growth, yield, and soil health. The direct mechanisms include nitrogen fixation, P solubilization, phytohormone production, and siderophore production by PGPRs and AM fungi (Adesemoye et al. 2008). The indirect mechanisms include the use of biocontrol bacteria (Bashan et al. 1993) that promote plant growth by controlling plant pathogens through phytotoxic substances (antibiotics, cellulase, chitinase, ethylene, hydrogen cyanide, pectinase). Another group consists of Stress Homeoregulating Bacteria (Sgroy et al. 2009) which help the plant to overcome biotic and abiotic stresses either directly by releasing stress-related phytohormones (Abscisic acid, jasminoic acid, salicyclic acid). Indirect mechanisms include plant growth regulators and stress signaling molecules (Sharma et al. 2017). Microbial amendments can be classified as either specific or general strategy. Biostimulants are microbial amendments that improve soil health and fertility by supplying growth factors and nutrients. They stimulate the soil and plant metabolic activities for improving crop growth, productivity, nutrient mineralization, decomposition of organic matter and also help in the proliferation of microbial activity (Chen et al. 2002; Lehman et al. 2015; Mishra et al. 2016; Subler et al. 1998). The restoration of degraded soil could be achieved by a thorough understanding of mechanisms of action of microbial amendments.

## 17.9 Breakdown of Toxic Compounds

Many toxic compounds find their way into the soil. These toxic substances can be produced by microorganisms as metabolic by-products, some are applied by humans as agrichemicals to kill pests and some are deposited in soil because of unintentional human activities or as effluents from factories. If these chemicals remain in the soil system they can be of potential harm for the ecosystem and its diversity (Brady and Weil 2015). Fortunately, most of the biologically produced toxins are utilized as food by some groups of microbes. However, xenobiotics are resistant to microbial attack by commonly occurring microorganisms. These toxic compounds are degraded by a specific group of fungi and bacteria which are mostly localized on the surface soil. Several bacterial genera have been isolated and characterized for degradation for xenobiotics from the soil as *Arthrobacter, Flavobacterium, Pseudomonas, Sphingobium, Xanthobacter* etc. (Bhatt et al. 2017). Wittich et al. (2007) isolated a bacterium *Sphingobium aromaticiconvertens sp nov* which could degrade monochlorinated dibenzofurans from aerobic river Elbe sediments. *Sphingobium yanoikuyae* XLDN2-5 is an efficient carbazole-degrading strain. Anan'ina et al. (2011) isolated moderately halotolerant *Rhodococcus* naphthalene-degrading bacteria from soil samples and slime pit bottom sediment of the Verkhnekamsk salt mining region of Russia. A halotolerant bacterial strain of *Ochrobactrum* sp. VA1 capable of degrading polycyclic aromatic hydrocarbons was isolated from marine water samples (Arulazhagan and Vasudevan 2011). This culture was able to degrade anthracene (88%), benzo (k) fluoranthene (57%), benzo(e)pyrene (50%), fluorene (97%), naphthalene (90%), phenanthrene (98%), pyrene (84%) and at a 30 g/L NaCl concentration. Chien et al. (2014) determined the ability of *Pseudomonas putida* strain TP1 and *Pseudomonas aeruginosa* strain TP6 for degradation of TNT and found that after 22-days incubation, more than 90% of the TNT was degraded. *Methylobacillus* sp. utilizing dibutyl phthalate (DBP) as sole carbon and energy source was isolated and its chemical pathway for DBP degradation was proposed by Kumar and Maitra (2016).

## 17.10 Conclusion

Soil biology represents living organisms dwelling in the soil which vary from small mammals, large invertebrates, arthropods, to microorganisms. Megafauna and macrofauna contribute in comminution of soil organic matter, improve aeration, water infiltration and hence improve soil structure. Mesofauna forms an important component of food chain and help to maintain biological equilibrium in soil. Microfauna are important predators of bacteria and algae and microflora is an important component for the cycling of nutrients in the ecosystem and critical players for decomposition of organic matter. In soil, all these organisms work alone or together for the proper functioning of biogeochemical cycles, organic matter decomposition, promoting plant growth, and maintaining soil health. Thus, the soil biota plays a critical role in sustaining agricultural productivity by making soil healthy and productive.

**Acknowledgment** I am highly thankful to Dr Arun Kumar, Senior Scientist, CSIR-IHBT, Palampur for helping me in preparing figures.

## References

Adesemoye A, Torbert H, Kloepper J (2008) Enhanced plant nutrient use efficiency with PGPR and AMF in an integrated nutrient management system. Can J Microbiol 54:876–886

Agerer R (2001) Exploration types of ectomycorrhizae. Mycorrhiza 11:107–114

Alami Y, Achouak W, Marol C, Heulin T (2000) Rhizosphere soil aggregation and plant growth promotion of sunflowers by an exopolysaccharide-producing *Rhizobium* sp. strain isolated from sunflower roots. Appl Environ Microbiol 66:3393–3398

Alori ET, Glick BR, Babalola OO (2017) Microbial phosphorus solubilization and its potential for use in sustainable agriculture. Front Microbiol 8:971

Anan'ina LN, Yastrebova OV, Demakov VA, Plotnikova EG (2011) Naphthalene degrading bacteria of the genus Rhodococcus from the Verkhnekamsk salt mining region of Russia. Antonie van Leeuwen 100:309–316

Arulazhagan P, Vasudevan N (2011) Biodegradation of polycyclic aromatic hydrocarbons by a halotolerant bacterial strain *Ochrobactrum* sp VA1. Mar Pollut Bull 62:388–394
Aust WM, Lea R (1991) Soil temperature and organic matter in a disturbed forested wetland. Soil Sci Soc Am J 55:1741–1746
Ayuke FO, Brussard L, Vanlauwe B, Lelei DK, Kibunja C, Pulleman MM (2011) Soil fertility management: impacts on soil macrofauna, soil aggregation and soil organic matter allocation. Appl Soil Ecol 48(1):53–62
Bagyaraj DJ, Ashwin R (2017) Soil biodiversity: role in sustainable horticulture. Biodiversity Hortic Crops 5:1–18
Barrios E (2007) Soil biota, ecosystem services and land productivity. Ecol Econ 64:269–285
Bashan Y, Holguin G, Lifshitz R (1993) Isolation and characterization of plant growth-promoting rhizobacteria. In: Glick BR, Thompson JE (eds) Methods in plant molecular biology and biotechnology. CRC, Boca Raton, FL, pp 331–345
Bedano JC, Domínguez A (2016) A large-scale agricultural management and soil meso- and macrofauna conservation in the argentine pampas. Sustainability 8:653
Berthet P (1967) The metabolic activity of Oribatid mites (Acarina) in different forest floors. In: Petrusewicz K (ed) Secondary productivity of terrestrial ecosystems (principles and methods), vol II. Państwowe Wydawnictwo Naukowe, Warsaw, Poland, pp 709–725
Bhadauria T, Saxena KG (2010) Role of earthworms in soil fertility maintenance through the production of biogenic structures. Appl Environ Soil Sci 2010:816073
Bhatt P, Tiwari S, Gangola S, Khati P, Kumar G, Sharma A (2017) Removal of xenobiotics from environment using microbial metabolism. Scientific India 5:33–34
Bhattacharyya PN, Jha DK (2012) Plant growth-promoting rhizobacteria (PGPR): emergence in agriculture. World J Microbiol Biotechnol 28:1327–1350
Bhatti AA, Haq S, Bhat RA (2017) Actinomycetes benefaction role in soil and plant health. Microb Pathog 111:458–467
Bocock KL (1963) The digestion and assimilation of food by *Glomeris*. In: Doeksen J, van der Drift J (eds) Soil organisms. North-Holland Publishing Company, Amsterdam, The Netherlands, pp 85–91
Borie F, Rubio R, Morales A (2008) Arbuscular mycorrhizal fungi and soil aggregation. J Soil Sci Plant Nutr 8:9–18
Brady NC, Weil RR (2015) The nature and properties of soil, 15th edn. Pearson, Upper Saddle River, NJ
Bremner JM, Steele CG (1978) Role of microorganisms in the atmospheric sulfur cycle. Adv Microbiol Ecol 2:155–201
Burns RG (2010) How do microbial extracellular enzymes locate and degrade natural and synthetic polymers in soil. In: Xu J, Huang PM (eds) Molecular environmental soil science at the interfaces in the Earth's critical zone. Springer, Berlin, Heidelberg
Caesar-TonThat AJ (2002) Soil binding properties of mucilage produced by a basidiomycete fungus in a model system. Mycol Res 106:930–937
Cardoso EJBN, Vasconcellos RLF, Bini D, MYH M, dos Santos CA, PRL A et al (2013) Soil health: looking for suitable indicators. What should be considered to assess the effects of use and management on soil health? Sci Agric 70:274–289
Carter MR, Stewart BA (1996) Structure and organic matter storage in agriculture soils. CRC, Boca Raton, FL
Charousova I, Medo J, Halenarova E, Javorekova S (2017) Antimicrobial and enzymatic activity of actinomycetes isolated from soils of coastal islands. J Adv Pharm Technol Res 8:46–51
Chen S, Subler S, Edwards CA (2002) Effects of agricultural biostimulants on soil microbial activity and nitrogen dynamics. Appl Soil Ecol 19:249–259
Chien CC, Kao CM, Chen DY, Chen SC, Chen CC (2014) Biotransformation of trinitrotoluene (TNT) by Pseudomonas spp. isolated from a TNT contaminated environment. Environ Toxicol Chem 33:1059–1063
Coleman DC, Crossley DA Jr, Hendrix PF (2004) Fundamentals of soil ecology, 2nd edn. Elsevier, Burlington, MA, p 386

Collins NM (1981) The role of termites in the decomposition of wood and leaf litter in the Southern Guinea savanna of Nigeria. Oecologia 51:389–399
Cornaby BW, Gist CS, Crossley DA Jr (1975) Resource partitioning in leaf-litter faunas from hardwood and hardwood-converted-to-pine forests. In: Howell FG, Gentry JB, Smith MH (eds) Mineral cycling in southeastern ecosystems. Technical Information Center, Office of Public Affairs, US Energy Research and Development Administration, Washington, DC, pp 588–597
Culliney TW (2013) Role of arthropods in maintaining soil fertility. Agriculture 3:629–659
Daynes CN, Zhang N, Saleeba JA, McGee PA (2012) Soil aggregates formed in vitro by saprotrophic Trichocomaceae have transient water-stability. Soil Biol Biochem 48:151–161
de Menezes AB, Prendergast-Miller MT, Macdonald LM, Toscas P, Baker G, Farrell M et al (2018) Earthworm-induced shifts in microbial diversity in soils with rare versus established invasive earthworm populations. FEMS Microbiol Ecol 94:051
Degens BP (1997) Macro-aggregation of soils by biological bonding and binding mechanisms and the factors affecting these: a review. Soil Res 35:431–460
Deleporte S, Tillier P (1999) Long-term effects of mineral amendments on soil fauna and humus in an acid beech forest floor. For Ecol Manag 118:245–252
Diehl D (2013) Soil water repellency: dynamics of heterogeneous surfaces. Colloids Surf A Physicochem Eng Asp 432:8–18
Doran JW (2002) Soil health and global sustainability: translating science into practice. Agric Ecosyst Environ 88:119–127
Doran JW, Jones AJ (1996) Soil quality and health: indicators of sustainability. In: Doran JW, Jones AJ (eds) Methods for assessing soil quality, SSSA Special Publication Number 49. Soil Science Society of America, Madison, WI, pp XI–XIV
Doran JW, Parkin TB (1994) Defining and assessing soil quality. In: Doran JW, Coleman DC, Bezdicek DF, Stewart BA (eds) Defining soil quality for a sustainable environment. Soil Science Society of America, Madison, WI, pp 3–21
Edwards CA (1974) Macroarthropods. In: Dickinson CH, Pugh GJE (eds) Biology of plant litter decomposition, vol 2. Academic, London, UK, pp 533–554
Elliott ET, Coleman DC (1988) Let the soil work for us. Ecol Bull 39:23–32
Forster SM (1990) The role of microorganisms in aggregate formation and soil stabilization: types of aggregation. Arid Soil Res Rehabil 4:85–98
Frac M, Hannula SE, Bełka M, Je-dryczka M (2018) Fungal biodiversity and their role in soil health. Front Microbiol 9:707
Franzluebbers AJ (2009) Soil biology. Agric Sci 1(1):47
Gardi C, Jeffery S (2009) Soil biodiversity. Joint Research Center, European Commission Luxembourg, Brussels, p 27
Glick BR (2010) Using soil bacteria to facilitate phytoremediation. Biotechnol Adv 2:367–374
Gomez-Escribano JP, Alt S, Bibb MJ (2016) Next generation sequencing of actinobacteria for the discovery of novel natural products. Mar Drugs 14:E78
Gopal M, Bhute SS, Gupta A, Prabhu SR, Thomas GV, Whitman WB et al (2017) Changes in structure and function of bacterial communities during coconut leaf vermicomposting. Antonie Leeuwenhoek 110:1339–1355
Gougoulias C, Clark JM, Shaw LJ (2014) The role of soil microbes in the global carbon cycle: tracking the below-ground microbial processing of plant-derived carbon for manipulating carbon dynamics in agricultural systems. J Sci Food Agric 94:2362–2371
Hassall M (1977) Consumption of leaf litter by the terrestrial isopod *Philoscia muscorum* in relation to food availability in a dune grassland ecosystem. In: Lohm U, Persson T (eds) Soil organisms as components of ecosystems. Swedish Natural Science Research Council, Stockholm, Sweden, pp 550–553
Haynes RJ, Fraser PM (1998) A comparison of aggregate stability and biological activity in earthworm casts and uningested soil as affected by amendment with wheat or Lucerne straw. Eur J Soil Sci 49:629–636
Heneghan L, Bolger T (1998) Soil microarthropod contribution to forest ecosystem processes: the importance of observational scale. Plant Soil 205:113–124

Heneghan L, Coleman DC, Zou X, Crossley DA, Haines BL (1998) Soil microarthropod community structure and litter decomposition dynamics: a study of tropical and temperate sites. Appl Soil Ecol 9:33–38
Hoeffner K, Monard C, Santonja M, Cluzeau D (2018) Feeding behaviour of epi-anecic earthworm species and their impacts on soil microbial communities. Soil Biol Biochem 125:1–9
IBOY (2000) Soil macrofauna: an endangered resource in a changing world. Report of an international workshop held at IRD, Bondy (France) 19–23 June 2000
Johns C (2017) Living soils: the role of microorganisms in soil health. Future Directions International 1–7
Jose PA, Jha B (2016) New dimensions of research on Actinomycetes: quest for next generation antibiotics. Front Microbiol 7:1295
Katz L, Baltz RH (2016) Natural product discovery: past, present, and future. J Ind Microbiol Biotechnol 43:155–176
Kibblewhite MG, Ritz K, Swift MJ (2008) Soil health in agricultural systems. Philos Trans R Soc B Biol Sci 363:685–701
Klotz MG, Bryant DA, Hanson TE (2011) The microbial sulfur cycle. Front Microbiol 2:241
Kumar V, Maitra SS (2016) Biodegradation of endocrine disruptor dibutyl phthalate (DBP) by a newly isolated *Methylobacillus* sp. V29b and the DBP degradation pathway. 3 Biotech 6:200
Kuypers MMM, Marchant HK, Kartal B (2018) The microbial nitrogen-cycling network. Nat Rev Microbiol 16:263–276
Lee KE (1985) Earthworms: their ecology and relationships with soils and land use, vol 411. Academic, New York
Lehman RM, Cambardella CA, Stott DE, Acosta-Martinez V, Manter DK, Buyer JS, Maul JE, Smith JL, Collins HP, Halvorson JJ, Kremer RJ, Lundgren JG, Ducey TF, Jin VL, Karlen DL (2015) Understanding and enhancing soil biological health: the solution for reversing soil degradation. Sustainability 7:988–1027
Leifheit E, Verbruggen E, Rillig M (2015) Arbuscular mycorrhizal fungi reduce decomposition of woody plant litter while increasing soil aggregation. Soil Biol Biochem 81:323–328
Liang C, Balser TC (2011) Microbial production of recalcitrant organic matter in global soils: implications for productivity and climate policy. Nat Rev Microbiol 9:75
Marcuzzi G (1970) Experimental observations on the rôle of *Glomeris* spp (Myriapoda, Diplopoda) in the process of humification of litter. Pedobiologia 10:401–406
Mason MG, Ball AS, Reeder BJ, Silkstone G, Nicholls P, Wilson MT (2001) Extracellular heme peroxidases in actinomycetes: a case of mistaken identity. Appl Environ Microbiol 67:4512–4519
Medina-Sauza RM, Álvarez-Jiménez M, Delhal A, Reverchon F, Blouin M, Guerrero-Analco JA, Cerdán CR, Guevara R, Villain L, Barois I (2019) Earthworms building up soil microbiota, a review. Front Environ Sci 7:81
Mengual C, Roldán A, Caravaca F, Schoebitz M (2014) Advantages of inoculation with immobilized rhizobacteria versus amendment with olive-millwaste in the afforestation of a semiarid area with *Pinus halepensis* Mill. Ecol Eng 73:1–8
Menta C (2012) Soil fauna diversity - function, soil degradation, biological indices, soil restoration. In: Lameed GA (ed) Biodiversity conservation and utilization in a diverse world. IntechOpen, London. https://doi.org/10.5772/51091
Metting B (1986) Population dynamics of *Chlamydomonas sajao* and its influence on soil aggregate stabilization in the field. Appl Environ Microbiol 51:1161–1164
Miller R, Jastrow J (2000) Mycorrhizal fungi influence soil structure. In: Kapulnik Y, Douds DD (eds) Arbuscular mycorrhizas: physiology and function. Springer, Dordrecht, pp 3–18
Mishra J, Prakash J, Arora NK (2016) Role of beneficial soil microbes in sustainable agriculture and environmental management. Climate Change Environ Sustain 4:137–149
Modi HA (2013) Soil microbiology. Pointer Publishers, Jaipur
Mora P, Miambi E, Jimenez JJ, Decaens T, Rouland C (2005) Functional complement of biogenic structures produced by earthworms, termites and ants in the neotropical savannas. Soil Biol Biochem 37:1043–1048

Morgan CR (1983) Importance of organic sulfur constituents of forest soils and the role of the soil macrofauna in affecting sulfur flux and transformation. In: Lebrun P, André HM, de Medts A, Grégoire-Wibo C, Wauthy G (eds) New trends in soil biology. Imprimerie J Dieu-Brichart, Ottignies-Louvain-la-Neuve, Belgium, pp 75–85

Morgan CR, Mitchell MJ (1987) The effects of feeding by *Onisus asellus* on leaf litter sulfur constituents. Biol Fertil Soils 3:107–111

Nechitaylo TY, Yakimov MM, Godinho M, Timmis KN, Belogolova E, Byzov BA et al (2010) Effect of the earthworms *Lumbricus terrestris* and *Aporrectodea caliginosa* on bacterial diversity in soil. Microb Ecol 59:574–587

Neher DA, Barbercheck ME (1998) Diversity and function of soil mesofauna. In: W Collins & CO (eds) The Biodiversity in Agroecosystems. CRC Press, Boca Raton, FL pp 27–47

O'Lear HA, Blair JM (1999) Responses of soil microarthropods to changes in soil water availability in tallgrass prairie. Biol Fertility Soils 29:207–217

Oades JM, Waters AG (1991) Aggregate hierarchy in soils. Aust J Soil Res 29:815–828

Ortiz N, Armada E, Duque E, Roldán A, Azcón R (2015) Contribution of arbuscular mycorrhizal fungi and/or bacteria to enhancing plant drought tolerance under natural soil conditions: effectiveness of autochthonous or allochthonous strains. J Plant Physiol 174:87–96

Pankhurst CE, Lynch JM (1995) The role of soil microbiology in sustainable intensive agriculture. Adv Plant Pathol 11:229–247

Peng S, Guo T, Liu G (2013) The effects of arbuscular mycorrhizal hyphal networks on soil aggregations of purple soil in Southwest China. Soil Biol Biochem 57:411–417

Petersen H (1994) A review of collembolan ecology in ecosystem context. Acta Zool Fenn 195:111–118

Radwan SS, Al-Awadhi H, Sorkhoh NA, El-Nemr IM (1998) Rhizospheric hydrocarbon-utilizing microorganisms as potential contributors to phytoremediation for the oily Kuwaiti desert. Microbiol Res 153:247–251

Rashid MI, Mujawar LH, Shahzad T, Almeelbi T, IMI I, Oves M (2016) Bacteria and fungi can contribute to nutrients bioavailability and aggregate formation in degraded soils. Microbiol Res 183:26–41

Richardson AE, Simpson RJ (2011) Soil microorganisms mediating phosphorus availability. Plant Physiol 156:989–996

Ruiz N, Lavelle P (2008) Soil macrofauna field manual technical level. Food and Agriculture Organization of The United Nations

Rusek J (1998) Biodiversity of Collembola and their functional role in the ecosystem. Biodivers Conserv 7:1207–1219

Schaller K (2009) Soil enzymes: valuable indicators of soil fertility and environmental impacts. Bulletin UASVM Horticulture 66:2

Seastedt TR, Tate CM (1981) Decomposition rates and nutrient content of arthropods remains in forest litter. Ecology 62:13–19

Sgroy V, Cassán F, Masciarelli O, Del Papa MF, Lagares A, Luna V (2009) Isolation and characterization of endophytic plant growth-promoting (PGPB) or stress homeostasis-regulating (PSHB) bacteria associated to the halophyte *Prosopis strombulifera*. Appl Microbiol Biotechnol 85:371–381

Sharma IP, Chandra S, Kumar N, Chandra D (2017) PGPR: heart of soil and their role in soil fertility. In: Meena V, Mishra P, Bisht J, Pattanayak A (eds) Plant-soil-microbe nexus agriculturally important microbes for sustainable, vol I. Springer, Singapore, pp 57–61

Six J, Bossuyt H, Degryze S, Denef K (2004) A history of research on the link between (micro) aggregates, soil biota, and soil organic matter dynamics. Soil Tillage Res 79:7–31

Subler S, Dominguez J, Edwards CA (1998) Assessing biological activity of agricultural biostimulants: bioassays for plant growth regulators in three soil additives. Commun Soil Sci Plant Anal 29:859–866

Sun JM, Irzykowski WJ, Edryczka M, Han FX (2005) Analysis of the genetic structure of *Sclerotinia sclerotiorum* (lib.) de Bary populations from different regions and host plants by random amplified polymorphic DNA markers. J Integr Plant Biol 47:385–395

Swift MJ (2005) Human impacts on biodiversity and ecosystem services: an overview. In: Dighton J, White JF, Oudemans P (eds) The fungal community its organization and role in ecosystems. CRC, Boca Raton, FL, pp 627–641

Swift MJ, Heal OW, Anderson JM (1979) Decomposition in terrestrial ecosystems. University of California Press, Berkeley and Los Angeles, CA, p 372

Teuben A, Verhoef HA (1992) Direct contribution by soil arthropods to nutrient availability through body and faecal nutrient content. Biol Fert Soils 14:71–75

Troper HG (1984) Microorganisms and the sulphur cycle. Studies Inorganic Chemistry 5:251–265

Trumbore S (2006) Carbon respired by terrestrial ecosystems – recent progress and challenges. Glob Chang Biol 12:141–153

Tugel AJ, Lewandowski AM (2010) Soil biology primer. Available: www.statlab.iastate.edu/survey/SQI/SoilBiologyPrimer.htm Accessed November 4

Ubugunova VI, Lavrent'eva IN, Ubugunov LL, Nikheleeva TP (2007) Mesofauna in soils of the ivolga depression (Western Transbaikal region). Institute of General and Experimental Biology, Russian Academy of Sciences, Buryat Republic, Russia

Ward BB, Jensen MM (2014) The microbial nitrogen cycle. Front Microbiol 5:553

Wittich RM, Busse HJ, Kampfer P, Tiiola M (2007) Sphingobium aromaticiconvertens sp. nov a xenobiotic compound degrading bacterium from polluted river sediment. Int J Syst Evol Microbiol 57:306–310

Wiwatwitaya O, Takeda H (2005) Seasonal changes in soil arthropod abundance in the dry evergreen forest of Northeast Thailand, with special reference to collembolan communities. Ecol Res 20:59–70

Wright SF, Upadhyaya A (1996) Extraction of an abundant and unusual protein from soil and comparison with hyphal protein of arbuscular mycorrhizal fungi. Soil Sci 161:575–586

Zaady E, Groffman PM, Shachak M, Wilby A (2003) Consumption and release of nitrogen by the harvester termite *Anacanthotermes ubachi* Navas in the northern Negev desert, Israel. Soil Biol Biochem 35:1299–1303

# Chapter 18
# Bacterial Inoculants: How Can These Microbes Sustain Soil Health and Crop Productivity?

**Anu Kalia, Sat Pal Sharma, Sukhjinder Kaur, and Harleen Kaur**

**Abstract** An increasing trend for the use of microbial bioinoculants to accomplish sustainable agriculture has been witnessed across the globe. Bacterial inoculants, mostly composed of beneficial bacteria including the plant growth-promoting rhizobacteria (PGPRs), exhibit tremendous metabolic versatility for carrying out processes such as nitrogen fixation, phosphate, potassium, zinc, silica, and other substrate solubilization or mineralization, release of plant growth-promoting substances (PGPSs), antibiotic synthesis, and biodegradation of soil organic matter. These processes contribute toward maintenance of soil health. Appropriately screened and applied bacterial inoculants can be a prodigious tool for increasing crop productivity besides decreasing current intensive use of chemical or synthetic fertilizers. These inoculants can help in achieving the long-desired goal of sustainable productivity with a low eco-footprint such that environmental quality conducive for the health of humans, livestock, plants, and soil can be maintained. Because the soil microbial diversity, enumerated as microbial species richness and number, can be considered as an index of soil health and fertility, bacterial genera in the rhizosphere and endosphere of the plants will indicate the health of the soil and, therefore, have a regulating effect on crop productivity. This chapter discusses various mechanisms of action of these bacteria and the beneficial effects of plant growth-promoting bacterial inoculants to realize the concept of conservation of

A. Kalia (✉)
Electron Microscopy and Nanoscience Laboratory, Department of Soil Science, College of Agriculture, Punjab Agricultural University, Ludhiana, Punjab, India
e-mail: kaliaanu@pau.edu

S. P. Sharma
Department of Vegetable Science, College of Horticulture, Punjab Agricultural University, Ludhiana, Punjab, India

S. Kaur
Department of Soil Science, College of Agriculture, Punjab Agricultural University, Ludhiana, Punjab, India

H. Kaur
Department of Microbiology, College of Basic Sciences and Humanities, Punjab Agricultural University, Ludhiana, Punjab, India



B. Giri, A. Varma (eds.), *Soil Health*, Soil Biology 59,
https://doi.org/10.1007/978-3-030-44364-1_18

natural resources, particularly the soil as a natural resource for sustainable crop production.

**Keywords** Microbial inoculants · Endophytes · Plant growth-promoting rhizobacteria · Soil health · Sustainability

## 18.1 Introduction

The industrial cultivation of crops involves the application of chemicals, hormones, and antibiotics to achieve improved growth and yield compared to primitive agricultural practices (Alori and Fawole 2017; Gilchrist et al. 2007). Heavy doses of chemical fertilizers and pesticides are applied to achieve high yields and to maintain crop productivity levels. However, intensive application of agri-inputs can be detrimental to the soil microbiome, with repercussions involving altered soil fertility and quality standards (Alori et al. 2017). Further, the use of pesticides to curb and control the spread of unwanted pests and pathogenic microbes may also kill plant-friendly microbes that are helpful to establish homeostasis and improve the growth of plants (Kothe and Turnau 2018). Moreover, contamination and degradation of the soil is becoming a sporadic problem. It is, therefore, essential to look for a new technology that can reduce the health hazards posed to the atmosphere, human beings, and livestock or may help avert the negative impacts of industrial agriculture.

Microbe-based products can be successfully utilized for developing sustainable agricultural practices, because plant beneficial microbes may enhance plant growth through improvement in nutrient availability via numerous mechanisms (Zaidi et al. 2009) and thus may reduce reliance on chemical input (Alori and Babalola 2018). By versatile physiological and biochemical potentials and the intrinsic biological potential of rhizosphere processes, microbial agents allow supplementing or possibly replacing inorganic and other agri-chemicals for ensuring food security (Yang et al. 2017). In view of these aspects, this chapter summarizes the effect of bacterial inoculants on crop productivity and sustainable maintenance of plant health. The diversity and plant growth-promoting effects of bacterial inoculants, their influence on soil fertility, overall plant health, and the cycling of nutrients, are also discussed.

## 18.2 The Rhizobacteria Microbiome: Bacteria Associated with Plant Roots

The rhizosphere microbiome includes a functional gene pool (including viruses, prokaryotes, and eukaryotes) linked with diverse habitats of the host plant that can directly influence plant health and productivity. The rhizosphere is designated as that region of the soil which is being constantly affected by plant roots in the process of rhizodeposition (Moe 2013; Lakshmanan et al. 2014). Rhizodeposition, one of the main rhizosphere processes, includes the release of organic substances from low molecular weight (organic acids, amino acids, sugars, phenolic acids, flavonoids, etc.) to high molecular weight compounds (carbohydrates, enzymes, etc.). As the exudates released by plant roots are extremely variable, differences in plant species and ecological conditions can affect the type and amount of root exudates (Mimmo et al. 2011). Further, these exudates may alter the diversity of the root-associated microbiome. Thus, the plant roots can maintain a specific population of microorganisms in the soil surrounding the roots where there are more nutrients in the rhizosphere region as compared to the bulk soil, so the rhizosphere can show enormous biological and chemical activity.

Microorganisms are generally involved in interactions with numerous hosts, and these microbe–soil–plant interactions and the soil physicochemical parameters cumulatively affect plant growth and development (Igiehon and Babalola 2018). Exploring the microorganisms that reside near the plant (in the rhizosphere, on the rhizoplane, and in the endosphere) is a justified option to isolate and multiply microbes showing close proximity with the plant roots. The coexisting diverse rhizospheric macro- and microorganisms display varied interactions within themselves as well as with their host plant (Prashar et al. 2013). Mutually, the rhizosphere microorganisms can influence the plant by producing regulatory compounds. Thus, the rhizospheric microbiome acts as a highly evolved external functional milieu for plants (Spence and Bais 2013; Spence et al. 2014), acting as the plant's second genome (Berendsen et al. 2012).

The rhizosphere microbiomes of plants exhibit beneficial or even deleterious effects on plant development. The belowground and aboveground microbial diversity within the rhizosphere can stimulate numerous interactions, which can be effectively utilized in agriculture to improve both soil and plant health. Thus, consideration of rhizospheric and endophytic microbial interactions is imperative for current agricultural practices (Igiehon and Babalola 2018). The microorganisms associated with the rhizosphere are more vigorous and contribute to soil health and fertility phenomena. Plant growth-promoting rhizobacteria (PGPR) have a close association with the plant as these bacteria colonize the roots and may alter the plant's growth and development more explicitly than those occurring in the bulk soil. Further, rhizospheric microbial diversity can limit or completely restrain the growth of pathogenic microorganisms by producing antibiotics, antifungals, and insect-repelling or insecticidal compounds (Yang et al. 2017).

### *18.2.1 Rhizobacteria–Plant–Soil Interactions: Diversity of Predominant Bacterial Genera in the Rhizosphere Soil*

Plant-associated microorganisms are essential in agricultural, for ensuring food safety and contribute to environmental equilibrium (Montesinos 2003). The interactions between a plant and the rhizosphere microbiome occur through a coordinated process mediated by the host plant and the soil. The plant developmental stage and the type of genera or species of the host plant are essential in shaping the rhizosphere microbiome (Peiffer et al. 2013; Chaparro et al. 2013). Certain other factors such as prevailing climatic conditions, biotic or abiotic stress, and anthropogenic interventions determine the establishment of the rhizospheric and endophytic microbiome (Bulgarelli et al. 2012; Lundberg et al. 2012). Because different plant host species possess different surface receptors to perceive the secretory exudates of microbes, variations exist in the composition and number of active members of the rhizo-microbiome. Further, root exudates released by plants are a crucial driving force for many interactions occurring in the rhizosphere that govern both plants and the microbial community (Bais et al. 2006). Apart from utilizing the root exudates as substrates, plant-associated microorganisms also respond to these exudates so as to grow competitively in a complex interactive manner in the root environment (Lakshmanan et al. 2014). The plants provide numerous habitats (phyllosphere, rhizosphere, endosphere) for close interaction with microorganisms. These interactions are named neutralism, commensalism, synergism, mutualism, amensalism, competition, or parasitism depending upon the useful or harmful effects of the microorganisms on the host (Ho et al. 2017).

The rhizosphere thus consists of an enormously complicated microbial diversity (Avis et al. 2008), with populations ranging from $10^{10}$ to $10^{12}$ cells per gram of soil whereas the density is normally less than $10^8$ cells in the bulk soil (Prashar et al. 2013). The largest numbers of bacteria occur in the rhizosphere. Proteobacteria and Actinobacteria are the major groups of bacteria in the rhizosphere (Teixeira et al. 2010), of which the genus *Pseudomonas* is especially abundant. The predominance of gram-negative bacteria in the soil may result from the efficacy of these bacterial genera to use exudates released by the plant roots in greater amounts than gram-positive bacteria, which are rather inhibited (Steer and Harris 2000). The aerobic, spore-forming genus *Bacillus* constitutes the major gram-positive population in soil, followed by *Arthrobacter* and *Frankia* (Barriuso et al. 2008). A smaller number of aerobic bacteria exist in the rhizosphere due to decreased effective concentration of oxygen in the region (Garbeva et al. 2004). The most common rhizospheric bacterial genera include *Pseudomonas*, *Bacillus*, *Arthrobacter*, *Rhizobium*, *Agrobacterium*, *Alcaligenes*, *Azotobacter*, *Mycobacterium*, *Flavobacterium*, *Cellulomonas*, and *Micrococcus*. Recent reports have also demonstrated *Bacillus* to be dominant in the rhizosphere compared to other bacteria. *Bacillus*, *Pseudomonas*, *Enterobacter*, *Acinetobacter*, *Burkholderia*, *Arthrobacter*, and *Paenibacillus* are the PGPRs most frequently observed in the rhizosphere (Yang et al. 2017). Hence, it seems that the rhizosphere is among the richest soil environmental zones in terms of diversity.

### *18.2.2 Endophytic Bacteria: Communication Signals for Internalization in Plant Tissues*

Endophytes constitute a group of microbes that reside intercellularly or intracellularly in plant tissues without causing any detrimental effect on their host. These bacteria, recognized as the major source of bioactive natural products, reside inside their hosts, either completely or for a certain phase of their life cycle, without exhibiting any disease. Endophytic bacteria show multifarious interactions with hosts, involving antagonism, mutualism, and rarely parasitism (Kaur et al. 2019). Therefore, these bacteria are regarded as a subclass of rhizospheric bacteria, in fact, a specialized group that has acquired the ability to invade within the plant host, including aboveground and underground plant parts and even seeds. These bacteria share all the crucial traits consistent with those required for plant growth promotion of the host (Chebotar et al. 2015) and exhibit greater beneficial effects on plants than do the rhizospheric bacteria (Afzal et al. 2019). Endophytic bacteria have been isolated and characterized from diverse plant hosts including agronomic crops, plants growing in extreme environments, and wild and perennial plants (Nair and Padmavathy 2014; Yuan et al. 2014). Also, these bacteria have been isolated from various plant parts (roots, stems, leaves, seeds, fruits, tubers, ovules, nodules). However, roots exhibit the maximum diversity of bacterial endophytes compared to the aerial plant tissues (Afzal et al. 2019).

Various endophytic microorganisms have been categorized as plant growth-promoting bacteria (PGPB). These bacteria are currently being used to develop formulations of different bioproducts (e.g., biofertilizers and biofungicides) or to modify the phyto-microbiome by introduction of bacteria beneficial in agriculture (Zhang et al. 2019). Obligate endophytes rely on the metabolic activity of the plant for their existence and are usually seed transmitted. These endophytes move inside through vertical colonization or by the action of vectors in plants. In other words, these endophytes do not originate from the rhizosphere. In contrast, facultative endophytes are free in nature and may inhabit the plant tissues at certain stage of their life cycle through the rhizosphere. Root wounds are another possible entry mechanism for passive endophytes; thereafter, these bacteria adhere to the surface of roots as in the associative bacteria (Santos et al. 2018). Endophyte–plant interactions are controlled by expression of genes of both plant host and the microbiont with further modulation of the interactions by the third factor, the environment (Rosenblueth and Martínez-Romero 2006).

Endophytes are either transferred through seeds or can be recruited from the rhizospheric soil (Singh et al. 2017). Endophytes enter plant tissues primarily through the root zone; however, aerial portions of plants, such as flowers, stems, and cotyledons, may also be used for entry. Specifically, the bacteria enter tissues via germinating radicles, secondary roots, stomata, hydathodes, lenticels, nectaries, and nectar glands or as a result of foliar damage. The fate of the internalized microbe depends on its ability to travel through the plant tissues. Thus, internalized endophytes may either remain localized at the point of entry or spread throughout the plant body. These microorganisms can reside within cells, in the intercellular spaces, or in the vascular system (Zinniel et al. 2002). During invasion, degradation of the plant cell envelope is caused by microbe-derived enzymatic activity (endoglucanases, pectinases, cellulases), which helps in endorhizospheric colonization. The larger cellulose fibers are loosened by the action of endoglucanases, thereby assisting these bacteria to enter the plant cell. Additionally, the process of colonization is aided by exoglucanases (Singh et al. 2017). The endophytic bacteria colonize the plant passively through a bacterial motility mechanism. They navigate through soil, exhibiting chemotactic motility toward the plant roots, by the release of exudates by roots which act as a carbon source. After initial recognition between the plant and endophyte, colonization occurs, with reversible adsorption of microbial cells followed by an irreversible adhesion mediated by bacterial extracellular proteins. The survival of endophytic bacteria depends on various biotic and abiotic factors, after which the endophyte proliferates in internal tissues of the plant in search of more favorable niches (Santos et al. 2018).

## 18.3 Bacteria and Plant Growth Promotion: Mechanism of Action

The PGPRs constitute approximately 2–5% of the rhizosphere bacteria (Antoun and Prevost 2005). These bacteria are usually divided into two groups depending on the complex mechanisms of their modes of action. Foremost are those involved in nutrient cycling and plant growth stimulation, including nitrogen ($N_2$) fixation, P solubilization, Fe sequestration via excretion of siderophores and other chelating compounds, and the production of phytohormones. The PGPRs belonging to the other group are primarily involved in the biocontrol of plant pathogens mediated by mechanisms such as antibiosis, competition (for space and nutrients such as iron in the rhizosphere), and induced systemic resistance (mediated by ethylene) (Kang et al. 2013).

### *18.3.1 Direct Effects Imparted on the Inoculated Plant*

The growth regulators produced by plants are complex organic compounds that manage the proper growth of plants. A plant uses a substantial amount of energy and nutrients for the synthesis of these complex compounds. Besides synthesizing phytohormones, plants are capable of uptake of growth regulators when applied exogenously as extracted hormones or synthetic analogues (Gouda et al. 2018). The alternative sources of these compounds are those secreted by the bacteria having synthesis capacities up to 60 times more than are produced by the plant themselves (Camerini et al. 2008). The phytohormones exhibit a distinct positive effect on growth and development of the plant. Gibberellins, cytokinins, abscisic acid, ethylene, brassino-steroids, and auxins are among the general groups of phytohormones that alter the uptake of nutrients and water (Sureshbabu et al. 2016).

Auxin is a major regulator of plant growth, development, and stress response (Liu et al. 2014). It is well known for its role in development of lateral roots and is also a major player in nodule meristem elongation (Oldroyd et al. 2011). Many important plant–microbial interactions are centered on production of auxins, particularly indole-3-acetic acid (IAA), which is responsible for the division, expansion, and differentiation of plant cells and tissues and stimulates root elongation. IAA production is widespread among PGPRs in the soil and rhizosphere (Martinez-Viveros et al. 2010). Several IAA-producing bacterial species have been isolated previously, such as *Streptomyces* sp., *Bacillus*, *Pseudomonas syringae*, *Pseudomonas fluorescens*, *Agrobacterium tumefaciens*, *Alcaligenes faecalis*, *Azotobacter tumefaciens*, *Burkholderia*, *Enterobacter*, and *Rhizobium* (Yousefi 2018).

Besides auxins, several PGPRs can produce cytokinins and gibberellins, but the exact function and synthesis mechanism of these hormones in bacteria are not yet entirely understood (Kang et al. 2009). Enhanced shoot growth and root exudate production in plants from higher amounts of gibberellins have been reported by some of the PGPR strains (Jha and Saraf 2015; Becker et al. 2018). Cytokinin production has been documented in various PGPRs such as *Arthrobacter giacomelloi*, *Azospirillum brasilense*, *Bradyrhizobium japonicum*, *Bacillus licheniformis*, *Pseudomonas fluorescens*, and *Paenibacillus polymyxa*.

The process of nutrient cycling is primarily carried out by soil microorganisms. Soil organic matter includes 50% carbon and the rest is constituted by N, P, S, and other nutrients. Apart from the degradation of soil organic matter by microorganisms, these microbes also function for the availability of fixed nutrients such as phosphorus (P), zinc (Zn), potassium (K), and iron (Fe). The lowering of soil pH by organic acid production by microorganisms is the primary mechanism underlying solubilization of plant-unavailable nutrients (Ahmad et al. 2018). Nitrogen, a major plant nutrient, is generally available in lesser amounts because of heavy losses incurred through emission or leaching processes. Therefore, microorganisms possessing the ability to convert atmospheric N into available forms for plants have a vital function. Biological N fixation can be performed by several bacteria that may exhibit symbiotic or asymbiotic association with their host plants. Biological nitrogen fixation (BNF) is the key mechanism by which the majority of atmospheric N is fixed, and it takes place in well-defined structures called nodules. The process of BNF is mainly restricted to legume plant species–rhizobia interactions and certain trees and shrubs forming actinorrhizal roots by *Frankia* association. Among the symbiotic PGPR, *Rhizobium* sp., *Beijerinckia* sp., and *Klebsiella pneumoniae* are the predominant genera (Ahemad and Kibret 2014). Inoculation of biological $N_2$-fixing PGPR to crops and crop fields can revitalize growth-promoting activity, provide disease management, and may help maintain the nitrogen level in agricultural soil (Damam et al. 2016).

Phosphorus (P) is the second most important element crucial for growth and development of plants (Azziz et al. 2012). This element exists in both organic and inorganic forms in soil which are generally unavailable to plants. It is made available by various PGPR strains through the process of solubilization (through secretion of low molecular weight organic acids by soil bacteria) and mineralization (by bacteria capable of producing phytase enzymes for the mineralization of phytates) (Sharma et al. 2013). Phosphate-solubilizing PGPR including the genera *Arthrobacter*, *Agrobacterium*, *Bacillus*, *Beijerinckia*, *Burkholderia*, *Enterobacter*, *Microbacterium*, *Paenibacillus*, *Pseudomonas*, *Erwinia*, *Rhizobium*, *Rhodococcus*, *Mesorhizobium*, *Flavobacterium*, *Serratia*, and *Thiobacillus* upon soil inoculation can improve plant growth and yield (Oteino et al. 2015). *Bacillus*, *Burkholderia*, *Enterobacter*, *Pseudomonas*, *Serratia*, and *Staphylococcus* are among the P-mineralizing PGPR strains (Alori et al. 2017).

After phosphorus, potassium (K) is another crucial macronutrient; however, the amount of soluble K is generally very much less in soil because approximately 90% of K occurs in the form of insoluble rock and silicate minerals (Parmar and Sindhu 2013). The deficiency of potassium in crops has become a major constraint, without which the plants display poorly developed roots, low seed production, slow growth rate, and a lower yield (Kumar and Dubey 2012). *Acidithiobacillus* sp., *Bacillus edaphicus*, *Ferrooxidans* sp., *Bacillus mucilaginosus*, *Pseudomonas* sp., *Burkholderia* sp., and *Paenibacillus* sp. have been reported as K-solubilizing PGPR strains capable of releasing potassium in available form from potassium-bearing minerals (Liu et al. 2012). Hence, application of K-solubilizing PGPR as a bioinoculant may result in agriculture improvements (Setiawati and Mutmainnah 2016).

The micronutrients particularly iron (Fe) can be made available to plants by rhizomicrobes by secretion of siderophore compunds. These are low molecular weight peptide molecules containing side chains and functional groups that offer high-affinity ligands for binding of ferric ions (Goswami et al. 2016). Siderophores released by microorganisms through various mechanisms, such as chelation and release of Fe, direct uptake of Fe-complexed siderophores or can increase Fe uptake by plants by ligand-exchange reactions. The ferric–siderophore complex, a potent siderophore, has a significant role in plant iron uptake in the presence of other metals, such as nickel and cadmium (Beneduzi et al. 2012). Based on the iron-coordinating functional groups, structural features, and type of ligands, the siderophore-producing microorganisms can be divided into four major classes: carboxylate, hydroxamates, phenol catecholates, and pyoverdines (Crowley 2006). *Pseudomonas* strains capable of producing siderophores can significant increase germination and plant growth (Sharma and Johri 2003).

### *18.3.2 Indirect Benefits Imparted Through Various Interaction Mechanisms*

Indirect mechanisms enhance the host natural resistance by processes in which PGPRs can neutralize or reduce the harmful effects of phytopathogens on plants through production of inhibitory substances (Singh and Jha 2015). Under abiotic as well as biotic stress conditions, indirect mechanisms aid the plants to grow properly (Akhgar et al. 2014). In indirect mechanism, PGPRs cause the production of hydrolytic enzymes, the synthesis and secretion of antibiotics in response to plant pathogens, the production of siderophores, volatile organic compounds (VOCs), and exopolysaccharides (EPSs), and the initiation of systematic resistance against different pathogens and pests (Nivya 2015).

Microorganisms functioning as biocontrol agents can perform antibiosis, parasitism, or competition with pathogens for nutrients and space, production of antimicrobial compounds, and facilitation of plant defense by regulatory homeostatic mechanisms (Bhattacharyya and Jha 2012). Therefore, protection of plants from phytopathogenic attacks may result from one or more microbe–microbe or plant–microbe interactions (Vurukonda et al. 2018).

PGPRs contribute to the sustenance of intrinsic plant resistance to pathogenic organisms (Enebe and Babalola 2018) by an induced systemic resistance (ISR) mechanism. PGPRs stimulate the mechanism of this resistance mechanism at particular sites in plants where pathogenic attack has occurred (Kundan et al. 2015). Because the ISR is nonspecific against a particular pathogen, it keeps the plant from being infected by diverse types of phytopathogens. The ISR primarily employs two plant hormones, jasmonate and ethylene, to stimulate plant defense response to pathogens (Verhagen et al. 2004).

Plant growth-promoting rhizobacteria, apart from the aforementioned mechanisms, synthesize and release volatile organic compounds (VOCs) (Kai et al. 2009). These VOCs are lipophilic, low molecular weight compounds (less than 300 Da) that possess relatively lower boiling points. Because these volatiles occur in a wide range and over a long distance, these compounds act as ideal infochemicals (Wheatley 2002). Hence, these compounds exhibit a profound effect on the adjacent microorganisms, thereby altering the growth and development of microorganisms in a particular biological niche. These VOCs have been reported to be biologically valuable in various aspects, such as the localization of flowers to pollinators, attraction of herbivore predators (indirect defense), or by overpowering the pathogens directly or by complete growth inhibition. Accordingly, the VOC compounds may operate either inter- or intraspecifically (Piechulla and Pott 2003). Volatile compounds are chemically diverse, including alkanes, alkenes, alcohols, aldehydes, ammonia, esters, ketones, sulfides, and terpenoids-bearing compounds. Volatiles of diverse soil-inhabiting bacteria can alter pathogenic fungal growth (Fernando et al. 2005). Volatiles also are important in the inhibition of fungal sclerotial activity, limit ascospore production, and may help reduce disease severity. Other volatile compounds such as ammonia and hydrogen cyanide (HCN) produced by a number of rhizobacteria have been reported to be important in biocontrol.

Antibiosis has an active role in the biocontrol of plant diseases as it has been postulated to be important in disease suppression (Mallesh 2008). It often acts in association with competition and parasitism. HCN is a broad-spectrum antimicrobial compound involved in biological control of root diseases by plant-associated rhizobacteria (Ramette et al. 2003). Many PGPRs can produce HCN, which is a secondary metabolite that suppresses the growth and development of competing microorganisms by effective inhibition of many metal-containing enzymes, especially copper-containing cytochrome *c* oxidases (Hassanein et al. 2009). It also inhibits proper functioning of enzymes and natural receptors by reversible mechanism of inhibition.

## 18.4 How to Screen for the Potential Bacterial Cultures and Develop Inoculant Formulations?

Screening of bacteria for their in vitro potential for various plant growth-promoting traits may offer a consistent way for efficient PGPR selection in combination with growth-promoting potential under field conditions. Under field conditions, several other factors are involved that reduce PGPR ability to elicit beneficial effects on plant growth (Nelson 2004). The efficacy of the processes mediated by PGPRs are strongly varied by certain factors such as the competitive ability of the PGPR strain in the rhizosphere along with root colonization, metabolite production and release, and species and genotypes of plants residing in the rhizosphere (Shaikh et al. 2018). Root colonization is also considered to be a crucial step in the application of microorganisms for beneficial purposes such as biofertilization, phytostimulation, biocontrol, and phytoremediation (Lugtenberg et al. 2017). This complex process is influenced by various parameters such as bacterial traits, root exudates, and biotic and abiotic factors. The efficiency of PGPR depends on achieving specific cell densities, that is, establishing an effective population density of active cells in the plant rhizosphere. In general, PGPR cultures utilized for inoculation are prepared at cell densities of $10^8$–$10^9$ CFU $ml^{-1}$ for root dipping and soil inoculation (Martinez-Viveros et al. 2010). Inoculation efficacy depends on the rhizosphere competence of the bacteria for a particular host plant. The re-inoculation of microbial inoculants at regular intervals of crop growth is required to maintain effective microbial population, but the re-inoculation approach includes the higher cost of inoculum preparation along with its viable cells for longer periods.

Microbial inoculant formulations can exist as solid- or liquid-based formulations and may consist of a single culture or a consortium of cultures (Reddy and Saravanan 2013). The solid inoculant formulations can contain a range of carriers such as clay, coal, fly ash, peat, peat amended with chitin-containing materials, sawdust, wheat bran, or inorganic materials such as bentonites, kaolin, perlite, silicates, and vermiculite. The choice of carriers for preparation of inoculum is based on the type that provides the PGPR strain(s) a suitable microenvironment along with maintenance of their viability and sufficient shelf life. The dimensions of the granules or beads utilized for immobilization of a solid inoculum range from 75 to 250 mm (Malusá et al. 2012). Liquid microbial inoculants can be in the form of broth cultures, humic acid suspensions, or oil-in-water suspensions (mineral/organic oils). The powder- or liquid-based inoculants can be applied directly on the seeds, or in the form of foliar spray, or root dipping at the time of seedling transplantation, or can be used in seed beds directly (Reddy and Saravanan 2013). Thus, it is essential to develop effective strategies for better inoculation so that the inoculating bacteria gain the benefit for faster and effective colonization and exhibit higher competitiveness in the inoculated niche (Fig. 18.1). The plant growth-promoting potential of inoculants combined with compatibility and shelf life are the key factors required for efficient colonization and performance under field conditions (Lee et al. 2016).

## 18.5 Beneficial Bacteria and Effects of Their Inoculation in Different Crops

The role of microbial inoculation in agriculture is vital in improving crop productivity as well as maintaining soil management. Biofertilizers can promote plant growth and development, besides improving the soil structure and nutrient uptake by the plant, can enhance phytohormone activity, have ability to tolerate abiotic stresses, particularly drought tolerance, and can impart protection to crop against biotic stress agents (Mahdi et al. 2010). The growth-promoting effects of bacterial inoculants in different crops are summarized in Table 18.1.

## 18.6 Organic Agriculture, Soil Health, and Fertility Improvement

Soil is an important biological component for sustaining human existence as it is required for the cultivation of food crops. It is a natural habitat for diverse microbes, with cell numbers ranging from $10^8$ to $10^9$ bacteria per gram of the soil. Proliferation of microbial activity in the soil is vital for the maintenance of soil fertility (Kalia and Gupta 2005; Kalia and Gosal 2011). Deterioration in soil microbial diversity can cause a decline in soil fertility, which can be a major concern (Kalia and Gosal 2011). Organic farming aims at using alternative methods of weed, pest, and disease control, banning use of chemical fertilizers and emphasizing animal welfare (Srutek and Urbn 2008). Use of biofertilizers is considered a securing approach to achieve sustainability via organic farming as application of beneficial soil microbes as biofertilizers has a wide range of functions in controlling soil health and crop productivity.

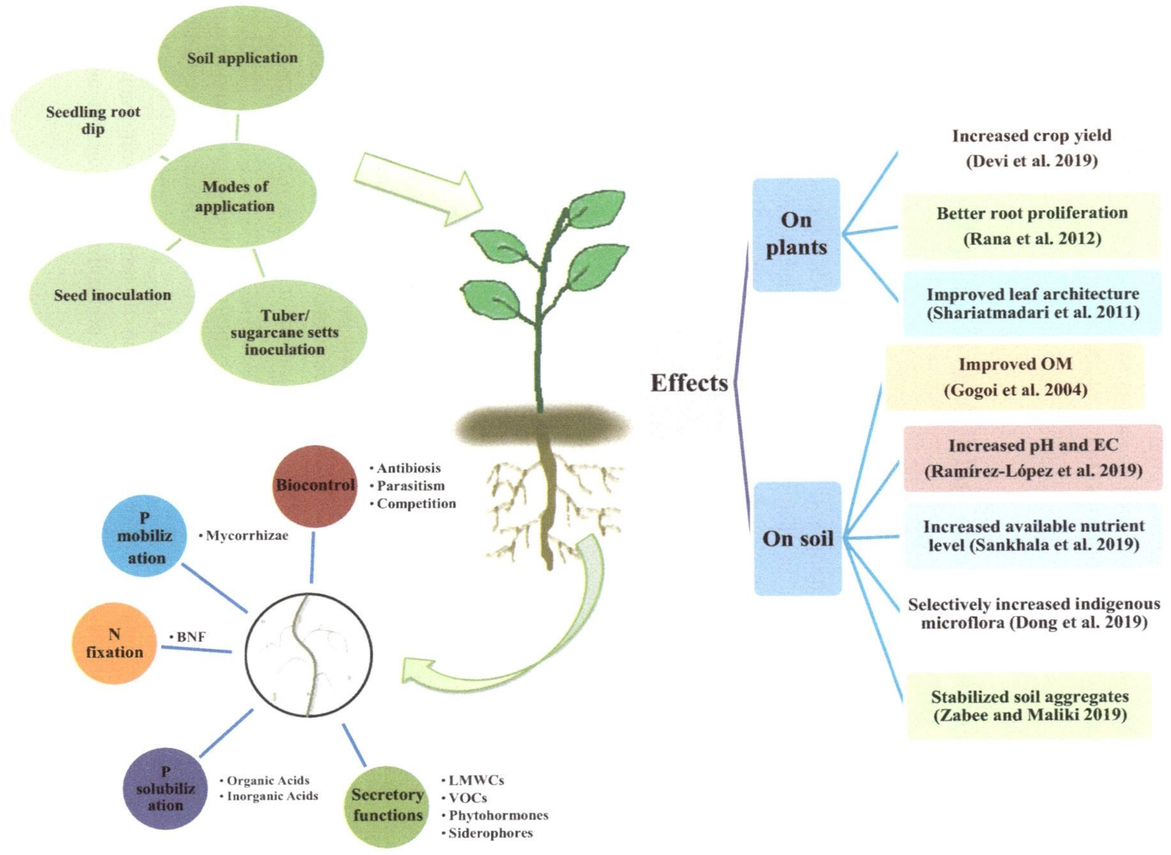


**Fig. 18.1** Different methods of application of bacterial inoculants with their growth-promoting effects

**Table 18.1** Plant growth-promoting potential of bacterial inoculants in various crops

| Bacterial strain | Crop | Growth conditions | Reported PGP role | References |
|---|---|---|---|---|
| Cereal crops | | | | |
| *Acetobacter pasteurianus*, *Stenotrophomonas* | Wheat | Laboratory | Enhanced plant growth and nutrient content | Majeed et al. (2015) |
| *Pseudomonas* | Rice | Laboratory | Increase in growth attributes | Sen and Chandrasekhar (2014) |
| *Bacillus sphaericus*, *Rhizobium* | Rice | Laboratory | Increased seedling emergence, vigor, root length, root surface area and volume | Mia et al. (2012) |
| *Bacillus* sp., *Providencia* sp., *Brevundimonas* sp. | Wheat | Greenhouse | Enhanced plant biometric parameters, N, P, and micronutrient content | Rana et al. (2012) |
| *Bacillus* spp. and nanozeolite | Maize | Greenhouse | • Increased growth parameters, including plant height, leaf area, number of leaves chlorophyll content<br>• Increased total protein content<br>• Higher soil physicochemical parameters.<br>• Threefold increase in soil enzyme activities | Khati et al. (2018) |
| *Agrobacterium* sp. | Maize | Greenhouse | Increase in growth attributes | Yousefi (2018) |
| *Pseudomonas alcaligenes*, *Bacillus Polymyxa*, *Mycobacterium phlei* | Maize | Greenhouse | Increased plant growth and NPK uptake | Egamberdiyeva (2007) |
| *Pseudomonas*, *Bacillus*, *Azospirillum*, *Azotobacter* | Maize, wheat | Greenhouse | Increased growth attributes | Karnwal (2012) |
| *Klebsiella* sp., *Bacillus pumilus*, *Acinetobacter* sp. | Maize | Greenhouse | Increased dry biomass, total N content and bacterial colonization in non-rhizosphere, rhizosphere, and endosphere of maize roots | Kuan et al. (2016) |
| *Azotobacter* sp., *Streptomyces badius* | Wheat | Field | Enhanced yield attributes | Gangwar et al. (2018) |

(continued)

**Table 18.1** (continued)

| Bacterial strain | Crop | Growth conditions | Reported PGP role | References |
|---|---|---|---|---|
| *Serratia* sp., *Bacillus subtilis* | Wheat | Field | • Increased yield, no. of tillers, grain no. per spike, 1000-grain weight and biomass<br>• Increased soil properties in the terms of available N, P and potassium, microbial biomass carbon, soil enzyme activities and population of phosphate-solubilizing bacteria | Sood et al. (2018) |
| *Bacillus*, *Azospirillum* | Wheat | Field | Enhanced plant nutrient content | Turan and Sahin (2012) |
| *Azotobacter* | Wheat | Field | Increase in growth and yield attributes | Singh et al. (2013) |
| *Pseudomonas putida*, *P. fluorescens*, *Azospirillum lipoferum* | Rice | Field | • Improved plant growth and yield<br>• Higher chlorophyll content<br>• Enhanced iron and zinc content | Sharma et al. (2014) |
| *Enterobacter* sp., *Bacillus* sp. | Rice | Field | Increase in growth and yield attributes | Saengsanga (2018) |
| *Pseudomonas* sp., *Azospirillum* sp. | Maize | Field | Increased plant height, seed weight, number of seeds per ear, leaf area and shoot dry weight | Gholami et al. (2009) |
| Leguminous crops | | | | |
| *Sinorhizobium meliloti*, *Delftia* sp. | Alfalfa (*Medicago sativa*) | Laboratory | Increased shoot dry weight and nodulation rate | Morel et al. (2011) |
| *Bacillus*, *Pseudomonas*, *Rhizobium* | Lentil | Laboratory | Enhanced seedling germination, vigour, root and shoot length, fresh and dry weight | Kaur and Khanna (2014) |
| *Rhizobium* and arbuscular mycorrhiza | Pigeon pea | Greenhouse | Enhanced chlorophyll, nitrogen and phosphorus contents | Bhattacharjee and Sharma (2012) |
| *Pseudomonas fluorescens*, *P. putida* | Pea (*Pisum sativum*) | Greenhouse | Increased growth attributes | Zahir et al. (2008) |

(continued)

**Table 18.1** (continued)

| Bacterial strain | Crop | Growth conditions | Reported PGP role | References |
|---|---|---|---|---|
| *Pseudomonas* sp., *Rhizobium leguminosarum*-FB1 | Rajmash (*Phaseolus vulgaris* L.) | Greenhouse | Increased root and shoot dry weight, nodulation, nutrient uptake, pod yield and nutrient content of pods | Mishra et al. (2014) |
| *Escherichia coli*, *Pseudomonas fluorescens*, *Burkholderia* sp. | Chickpea (*Cicer arietinum* L.) | Greenhouse and field | Increased plant height, no. of leaves/plant, pod bearing branches, pods/plant, nodules/plant and 100 seed weight | Dasgupta et al. (2015) |
| *Variovorax paradoxus* | Pea (*Pisum sativum*) | Greenhouse | Increased root and shoot biomass, stomatal conductance, N, P, K, Ca, and Mg uptake | Jiang et al. (2012) |
| Horticultural crops | | | | |
| *Pseudomonas putida*, *Enterobacter cloacae*, *Serratia marcescens*, *P. fluorescens*, *Bacillus* spp. | Cucumber, pepper, tomato | Greenhouse | Increased growth | Kidoglu et al. (2008) |
| *Pseudomonas putida*, *Azotobacter chroococcum* and *Azospirillum lipoferum* | Tomato | Greenhouse | Increased lycopene, antioxidant activity, and potassium contents | Ordookhani et al. (2010) |
| *Bacillus subtilis*, *B. megaterium*, *Acinetobacter baumannii*, *Pantoea agglomerans* | Tomato, cucumber | Greenhouse | • Increased fruit weight, width, length, number per plant, plant length, total soluble solid and dry matter<br>• Enhanced plant mineral content and N, P, Mg, Ca, Na, K, Cu, Mn, Fe, and Zn contents in fruit | Dursun et al. (2010) |
| *Pseudomonas*, *Azotobacter*, *Azosprillum* | Tomato | Greenhouse | • Increased plant biomass and height<br>• Increased NPK and Ca and Mg content | Shahram (2012) |
| *Bacillus cereus*, *Achromobacter xylosoxidans* | Potato | Greenhouse | Increased vegetative growth parameters, photosynthetic pigments and NPK concentrations | Dawwama et al. (2013) |

(continued)

**Table 18.1** (continued)

| Bacterial strain | Crop | Growth conditions | Reported PGP role | References |
|---|---|---|---|---|
| *Pseudomonas aeruginosa*, *Stenotrophomonas rhizophilia* | Potato | Greenhouse | Increased fruit yields, pomological traits, and chemical contents | Dashti et al. (2014) |
| *Pseudomonas stutzeri*, *Bacillus subtilis*, *Stenotrophomonas maltophilia*, *B. amyloliquefaciens* | Cucumber | Greenhouse | Higher levels of germination, seedling vigour, growth and N content in root and shoot tissues | Islam et al. (2015) |
| *Azospirillum* sp., *Agrobacterium* sp., *Pseudomonas* sp., *Enterobacter* sp., *Rhizobium* | Potato | Greenhouse | Increased growth and nitrogen uptake | Naqqash et al. (2016) |
| *Bacillus amyloliquefaciens*, *B. pumilus* | Tomato | Greenhouse | Increased plant height, shoot dry weight and N and P uptake | Fan et al. (2017) |
| *Agrobacterium rubi*, *Burkholderia gladii*, *Pseudomonas putida*, *Bacillus subtilus*, *B. megaterium* | Mint (*Mentha piperita* L.) | Field and greenhouse | Enhanced rooting performance, root length and dry matter Content of root, cuttings of mint | Kaymak et al. (2008) |
| *Pseudomonas fluorescens* | Tomato | Field and greenhouse | Enhanced growth and total fruit yield | Ahirwar et al. (2015) |
| *Bacillus* sp. | Raspberry | Field | Increase the yield, growth and nutrition | Orhan et al. (2006) |
| *Klebsiella* sp., *Erwinia* sp., *Azospirillum brasilense*, *Bacillus sphaericus* | Sweet potato | Field | Enhanced root dry weight and increased the concentrations of N, P, and K in shoots and root | Yasmin et al. (2007) |
| Fodder crops | | | | |
| *A. chroococcum*, *Azospirillum brasilense*, *A. lipoferum*, *P. fluorescens* | Fodder maize | Field | Increased silage fodder yield | Hamidi (2006) |
| *Azotobacter brasilense*, *P. fluorescens* | Sorghum | Field | Increased green fodder yield, dry matter yield, crude protein and ash contents | Chattha et al. (2017) |
| *Azotobacter vinelandii*, *Pantoea agglomerans*, *P. putida* | *Onobrychis sativa* | Field | Increased germination, plant growth, and nutrient uptake | Delshadi et al. (2017) |

(continued)

**Table 18.1** (continued)

| Bacterial strain | Crop | Growth conditions | Reported PGP role | References |
|---|---|---|---|---|
| *Bacillus*, *Acinetobacter*, *Enterobacter* | Alfalfa (*Medicago sativa* L.) | Field | Increased water, chlorophyll, NPK contents, plant height, leaf-to-stem ratio and fresh and dry weight | Daur et al. (2018) |

### 18.6.1 Enhancement of Soil Physicochemical Properties

Application of biofertilizers is considered a promising agriculture management practice. Addition of microorganisms in the form of bacteria, mycorrhizal fungi, cyanobacteria, and microalgae to the soil can alter organic matter content and improve fertility (Wu et al. 2005). Inoculation of a single type of biofertilizer can serve more than one purpose. For example, besides fixing atmospheric N, cyanobacteria may improve soil structure by adding organic matter to soil. Cyanobacteria can also function as primary colonizers by their ability to photosynthesize (Svircev et al. 2019) and can aid the establishment of microflora in soils.

Biofertilizers are supplementary resources generally used in combination with chemical fertilizers. Ramírez-López et al. (2019) combined the application of a nitrogen chemical fertilizer with photosynthetic and N-fixing microbial consortium in soil sown with wheat (*Triticum aestivum* L.) in different proportions. They observed an increase in pH, phosphate content, electrical conductivity, and nitrogenase activity on application of biofertilizers. However, biofertilizers resulted in a decrease in ammonium and nitrate content as compared to 100% chemical fertilizer treatment. Zabee and Maliki (2019) studied the effect of *Glomus mosseae* (mycorrhiza fungus), *Ascophyllum nodosum* (alga), *Saccharomyces cerevisiae* (yeast), and different levels of chemical fertilizers on soil aggregate stability in a potato crop. The highest aggregate stability was observed with the combined application of mycorrhiza and yeast.

The combination of two different types of biofertilizers can have improved effects on crop as compared to a single type of biofertilizer inoculation. A field experiment was conducted by Das and Singh (2014) to study the effect of different types of organic manures on soil properties of a mung bean crop. Application of farmyard manure, cereal and legume compost (sole and combined) with and without a consortium composed of *Rhizobium*, *Azotobacter*, *Pseudomonas*, and *Trichoderma* was performed. Treatments containing the consortium showed increased pH, organic carbon, and available NPK as compared to the treatments containing only organic manures without consortium inoculation. In another study, inoculation with biofertilizers, *Azotobacter*, and *Azospirillum* increased the organic carbon content of soil as compared to control (Gogoi et al. 2004).

Akbar et al. (2019) compared the effect of PGPRs on two different soil types. Seed bio-priming of wheat seeds with different strains of *Bacillus megaterium*,

*Pseudomonas fluorescens*, and *Bacillus subtilis* in combination with different doses of NPK was carried out to observe the effect on soil properties in sandy loam and silty loam soil types. The researchers stated that the effect on soil properties was more prominent in silt loam as compared to the sandy loam soil type.

### 18.6.2 Alterations in Soil Microbial Diversity

As discussed in the previous sections, maintenance of microbial biodiversity or species richness in agro-ecosystems is pivotal. Diversity serves different purposes including nutrient cycling, detoxification of recalcitrant compounds, and disease suppression. The soil microbial community structure is complex and dynamic. It is a direct indicator of soil fertility and productivity (Kalia and Gupta 2004). The soil ecosystem exhibits profound changes from anthropogenic activities through critical alterations in the microflora that result in the loss of soil fertility. Fertilizers and pesticides rank first among the factors responsible for the erosion of species diversity in soils (Kalia and Gosal 2011). Therefore, the ecological function of the soil, which is dependent on the diversity and composition of soil microbial communities, may vary such that these variations in communities can have deteriorating effects on soil productivity, thereby leading to yield loss (Mazzola and Manici 2012). The microbial biodiversity of a particular soil biome depends on its indigenous microflora, abiotic conditions (such as temperature, pH, sunlight, moisture, and aeration), cultivation practices (tillage and fertilization), crop cultivated, and augmentation with plant or agricultural biostimulants. These plant biostimulants are microorganisms and substances such as humic acids, fulvic acids, amino acids, protein hydrolysates, and seaweed extracts that enhance plant growth, nutrient uptake, and tolerance to abiotic stress (Calvo et al. 2014). Regulation of soil microbial communities is prerequisite and can be achieved by inoculation of beneficial microbes to renovate microecology and boost crop yields.

Biofertilizer application over a time period of 1 year can improve the diversity indices in inoculated treatment as compared to untreated controls in rhizospheric soil (Dong et al. 2019). Biofertilizer inoculation also changed the relative abundance of bacterial groups: a decrease in *Gemmatimonadetes* and *Anaerolinea, Cryobacterium*, and *Methanobacterium* during flowering/fruiting and vegetative stages, respectively, whereas increase occurred in *Chloroflexi* and *Anaeromyxobacter, Bdellovibrio*, and *Sphingomonas* during flowering, fruiting/root growth, and vegetative stages, respectively, and a decrease in the relative abundance of *Fusarium*. It has also been documented by various researchers that establishment and maintenance of PGPR populations is necessary for the exhibition of growth-promoting actions on plants (Cattelan et al. 1996; Hatzinger and Alexander 1994; Holl and Chanway 1992). Kang et al. (2013) tested this hypothesis by inoculating strains of *Bacillus pumilus* and *Pseudomonas chlororaphis* in *Vicia faba* soil. The results revealed that the growth-promoting effects of both strains were highly correlated with the occurrence and establishment of *P. chlororaphis* and bacterial community regulation by *B. pumilus*. However, no significant effect on diversity and composition of soil bacterial communities may also occur, as reported by a study in calcareous soil on application of microbial inoculants, *Bacillus megaterium* and *Bacillus mucilaginous* (Zhao et al. 2019).

### *18.6.3 Nutrient Availability in the Rhizo-/Endosphere*

PGPRs exhibit potential for enhancing nutrient availability to plants either by lowering the ethylene levels in the plant or by enhancing the production of phytohormones such as indole-3-acetic acid (IAA) (Simranjit et al. 2019). Addition of biofertilizers can significantly affect the soil NPK status after harvest, as has been observed in a banana crop (Gogoi et al. 2004). The researchers observed increase from 267.21 kg $ha^{-1}$ to 369.43 kg $ha^{-1}$ for N, from 17.68 kg $ha^{-1}$ to 30.30 kg $ha^{-1}$ for P, and from 115.18 kg $ha^{-1}$ to 242.42 kg $ha^{-1}$ for K content in soil after inoculation. Gosal et al. (2012) observed improvement in available soil N content on soil application of *Azotobacter* in sugarcane.

Application of biofertilizers in combination with chemical fertilizers, vermicompost, and castor cake can increase the available soil NPK after harvest in sodic soil sown with green gram (Selvarasu et al. 2019). Similarly, application of biofertilizers (*Rhizobium* and PSB) increased the N and P status of the soil after harvest in a chickpea-fodder maize cropping system (Jat and Ahlawat 2008). Sourcing N from farm yard manure and poultry manure combined with biofertilizers [consortia of *Azospirillum*, phosphate-solubilizing bacteria (PSB), and potassium-solubilizing bacteria (KSB)], *Trichoderma viride*, and neem oil increased postharvest NPK of soil in muskmelon (*Cucumis melo* L.) cv. GMM3 (Sankhala et al. 2019). Simranjit et al. (2019) studied the effect of fungal–bacterial biofilm biofertilizers containing combinations of *Anabaena*, *Trichoderma*, and *Azotobacter* on soil nutrient availability for cucumber grown in field conditions. The results indicated an increase in available P and N, as well as in concentrations of micronutrients Zn, Cu, Mn, and Fe, in the soil.

## 18.7 Enhancement of Growth and Yield for Sustainable Crop Productivity

### *18.7.1 Leguminous Crops*

Legumes are the prime source of N in the human diet. Their production in arid and semiarid regions is limited by inadequate rainfall and water scarcity; hence, the spring-sown plants when exposed to heat stress and drought conditions in their late vegetative and reproductive stages exhibit significant yield losses (Amirnia et al.

2019). Mechanistically, the PGPRs are known to increase stress tolerance in plants via production of phytohormones and improvement of plant–water relationships. Use of such biological stress mitigators should be optimized for better uptake of nutrients, crop productivity, and tolerance to abiotic and biotic constraints.

Nutritionally, legume crops exhibit high phosphorus demands. Phosphorus is required for plant growth and is needed for pod filling and improvement in the grain yield (Sharma et al. 2019). The phosphatic fertilizers are very costly and offer very low nutrient use efficiency (NUE). Therefore, the use of biofertilizers (N fixers and P solubilizers) for enhancing plant growth and yield is lucrative and a mandatory technique for leguminous crops (Thiyagarajan et al. 2003). The rhizosphere of leguminous crops presents a unique ecological niche exhibiting elaborate plant–microbe interactions in which nodule formation is an important aspect of the metabolic activities (Prakash et al. 2013). Biological nitrogen fixation depends on P availability in soil; therefore, efficient root uptake of elemental P is expected to stimulate the symbiotic functions in legumes besides regulating the N nutrition of the plant (Bargaz et al. 2018). Further, balanced input of N fertilizers also affects nodulation and nutrient use pattern in pulses (Thiyagarajan et al. 2003). *Rhizobium leguminosarum* inoculation improved germination percentage and yield in *Pisum sativum* (Sofi et al. 2019). Use of PSB and *Aspergillus awamori* as adjuncts with P fertilization increased nodule number, leaf area index, plant height, and grain and straw yield in mung bean (*Vigna radiata* L.) (Venkatarao et al. 2017).

Synergistic effects of two or more types of microbial inoculants have been studied in various legumes. Sharma et al. (2019) studied the synergistic effects of *Rhizobium*, PSB, AM fungi and *Azotobacter* inoculums with different doses of phosphorus fertilizers on nutrient uptake and yield of chickpea (*Cicer arietinum* L.) under rainfed conditions. The highest yield (56% higher than the control) and NPK uptake was observed in treatment having all the bioinoculants combined with 75% of the recommended dose of phosphate fertilizers. Coinoculation of pea plant with two cyanobacterial strains (*Nostoc endophytum* and *Oscillatoria angustissima*) proved more effective than a single inoculation (Osman et al. 2010). In another experiment on coinoculation using N-fixing bacteria (*Azotobacter*) with mycorrhizal fungi (*Glomus intraradices*), seed yield was improved as compared to a single inoculation in lentil (Amirnia et al. 2019). Similarly, triple inoculation of *Bradyrhizobium japonicum*, *Bradyrhizobium elkanii*, and *Streptomyces griseoflavus* on soybean improved plant growth, nodulation, nutrient uptake, and productivity (Htwe et al. 2019).

Biofilm inoculants are emerging as a new type of biofertilizers. Biofilms can occur in the soil as bacterial, fungal, and fungal–bacterial biofilms (Seneviratne et al. 2008). Bacterial and fungal biofilms are formed on abiotic surfaces in the soil, whereas for bacterial–fungal biofilms, fungi provide a biotic surface for the bacteria to adhere. Biofilm biofertilizers have been used to increase plant growth resulting in significantly improved nutrient acquisition. Use of cyanobacterial and biofilm inoculants including different combinations of *Anabaena torulosa, Bacillus subtilis, Mesorhizobium ciceri*, *Anabaena laxa*, and *Trichoderma viride* as a fungal–bacterial biofilm in chickpea resulted in significant increase in yield (Bidyarani et al. 2016).

The same type of microbial inoculants can exhibit varied effects on the host plant. *Lupinus termis* seeds treated with *Cylindrospermum muscicola* showed significant increase in shoot length, leaf area, and shoot mass whereas a nonsignificant effect was observed with *Anabaena oryzae* treatment as compared to the control (Haroun and Hussein 2003).

### 18.7.2 Cereal Crops and Bacteria

Microbe-based bioinoculants are extensively used as plant growth-promoting agents in cereal crops, especially wheat and rice. Rhizospheric bacteria have a positive effect on the growth attributes of wheat by the application of PGPRs (Khalid et al. 2004). Inoculation with four PGPR isolates in two cultivars of wheat under axenic conditions resulted in 17.3% and 13.5% increase in root length and dry weight, respectively, with 37.7% increased shoot height and 36.3% increased dry weight of wheat seedlings. Seed inoculation with peat-based selected PGPR strains also displayed stimulatory effects on wheat grain yields in pot studies as well as field experiments (Khalid et al. 2004). Application of actinobacteria and phosphate-solubilizing bacteria (PSB) resulted in a positive effect on wheat growth (Rudresh et al. 2005). Rosas et al. (2009) studied the promotion effect of *Pseudomonas aurantiaca* SR1 on maize and wheat in the field. Both crops when inoculated with the SR1 strain exhibited significant promoting effect in growth parameters with higher yields obtained using lower fertilization doses than conventionally applied. Similarly, increase in the yield attributed to traits such as tiller numbers (Salamone et al. 2012) and number of leaves (Ashrafuzzaman et al. 2009) have also been reported with PGPR inoculation of rice plants. In another field study conducted by Gholami et al. (2009), increased dry mass of root was observed in maize on inoculation with *Pseudomonas putida*, *P. fluorescens*, and *Azospirillum lipoferum*. Rana et al. (2012) reported maximum dry root mass for PGPR-treated wheat plants. Yousefi and Barzegar (2014) showed that inoculation of wheat plants with microbial cultures increased the growth characteristics of wheat. The Actinobacteria significantly enhanced tiller numbers, panicle numbers, filled grain numbers and weight, stover yield, grain yield, total dry matter, root length, volume, and dry weight in a rice crop (Gopalakrishnan et al. 2013). Golinska et al. (2015) reported that endophytic *Streptomyces* enhanced plant growth by nutrient mobilization and secondary metabolite production.

### 18.7.3 Horticultural Crops

The benefits of PGPR application have been evaluated in various horticultural crops including vegetable crops across different agroecological regions. These inoculations have been found to be effective in reducing the chemical inputs as well as improving yields (Zaidi et al. 2015). Dong et al. (2019) have compared the effects of growth-promoting biofertilizers (*Burkholderia* and *Rhizobium* consortia) and disease biocontrol biofertilizers (*Actinomyces*, *Bacillus*, and *Aspergillus* consortia) on yield attributes of *Panax ginseng*. Growth-promoting biofertilizers increased the root growth and yield by 16.5% and 17%, respectively, and disease biocontrol biofertilizers improved root growth and yield by 15.7% and 19.1%, respectively, as compared to uninoculated controls. Lily Devi et al. (2019) studied the combined effect of biofertilizers and biocontrol inoculation on growth and yield in brinjal (*Solanum melongena* L.) under polyhouse conditions during off season. In this study, a biofertilizer (mixture of *Azotobacter* and PSB) and biocontrol agents (*Pseudomonas fluorescens* and *Trichoderma*) were used in addition to farmyard manure. The combined inoculation of biofertilizers and biocontrol agents with farmyard manure showed significant increase in plant height, number of fruits, and fruit yield per plant as compared to biofertilizers used alone with farmyard manure.

Microbial inoculation of *Bacillus megaterium* and *B. mucilaginous* increased the shoot biomass, root biomass, and yield of chili pepper by 14.1%, 8.4%, and 28.5%, respectively, as compared to control (Zhao et al. 2019). Seed coating of cabbage (*Brassica oleracea*) with different PGPRs, viz. *Bacillus megaterium*, *Pantoea agglomerans*, and *Bacillus subtilis* increased fresh and dry weight, seedling height, and stem diameter in the inoculated plants (Turan et al. 2014). Seed coating with *Bacillus subtilis* and *Pseudomonas fluorescens* increased fresh and dry plant weight in radish (*Raphanus sativus*) under salt stress conditions (Mohamed and Gomaa 2012). Inoculation with *Burkholderia tropica* increased fruit number and weight in two varieties of tomato (*Lycopersicum esculentum* cv. "superman" Seminis) plants (Bernabeu et al. 2015).

The effect of *Azotobacter chroococcum* and *Azospirillum brasiliense* inoculation was studied on growth and yield of tomato (*Lycopersicum esculentum* Mill.) (Ramakrishnan and Selvakumar 2012). *Azotobacter* inoculation increased the plant height, yield per plant, average fruit weight per plant, and fruit content, and *Azospirillum* inoculation increased the number of leaves and fruits per plant. Dual inoculation of both inoculants increased all the traits as compared to single inoculation. In another study, Bumandalai and Tserennadmid (2019) investigated the effect of the microalga *Chlorella vulgaris* on the germination of tomato and cucumber seeds. Surface-sterilized seeds were placed in petri dishes containing microalgal cultures 3, 6, 9, and 12 days old. In the tomato seedlings, increase in shoot and root length was observed for 3-, 6-, and 9-day cultures but decreased root and shoot length was observed for 12-day-old cultures as compared to control. However, in the cucumber seedlings, increase in root and shoot length was observed in all the cultures as compared to control. Shariatmadari et al. (2011) studied the effect of *Anabaena vaginicola*, *Nostoc* sp., and *Nodularia harveyana* on germination and plant growth in squash (*Cucurbita maxima*), tomato (*Solanum lycopersicum*), and cucumber (*Cucumis sativus*) plants. Inoculation with different cyanobacterial extracts increased the seedling height, root length, plant height, leaf number, and root, shoot, leaf, and stem fresh and dry weight in all the plants to different extents compared to controls.

Biofertilizers have also been used with varied levels of chemical fertilizers to reduce the dosage of chemical fertilizers. Effect of different P levels in PSB and vesicular-arbuscular mycorrhizal (VAM) fungi inoculation (single and dual) on quality parameters and yield of garlic (*Allium sativum* L.) was studied (Meena et al. 2019). The AM fungal inoculation improved both yield and quality characteristics in garlic. Ramandeep et al. (2018) have studied the effect of biofertilizers (*Azotobacter* and PSB) and different levels of NPK fertilizers on growth of potato (*Solanum tuberosum* L.) cv. Kufri Pukhraj and Kufri Jyoti. Application of biofertilizers along with 80% of the recommended NPK fertilizer combined with 1% urea improved the number of leaves per plant, plant height, tuber weight, and tuber yield in both test varieties.

Mohammadi et al. (2019) compared the effect of *Azotobacter* and *Azospirillum* inoculation with different levels of N and P fertilizers on growth of baby corn (*Zea mays* L.). Application of 100 kg N $ha^{-1}$ and 75 kg P $ha^{-1}$ combined with *Azospirillum* inoculation resulted in significantly higher growth as compared to application of lower rates of N and P and *Azotobacter* inoculation. Similarly, a combination of farmyard manure, chemical NPK fertilizers, and microbial consortium (*Azospirillum* and PSB) resulted in maximum root yield in radish (*Raphanus sativus* L.) crop (Dash et al. 2019). Abdel-Razzak and El-Sharkawy (2013) combined biofertilizer application (mixture of *Azotobacter*, *Azospirillum*, and *Klebsiella*) with potassium humate spray as a source of humic acid to achieve improvement in garlic bulb quality, quantity, and storability.

### 18.7.4 Fodder Crops

Fodder crops are crops cultivated primarily for animal feed. These crops may be classified as either temporary or permanent with the former being cultivated and harvested like any other crop while the latter relate to land used permanently (for 5 years or more) for herbaceous forage crops, either cultivated or growing wild, and may include some parts of forest land if it is used for grazing. Green fodders are a staple feed for dairy animals (FAO 2011). Dairy animals producing up to 5–7 L milk per day can be maintained by feeding green fodder exclusively. Inclusion of green fodder in the diet of dairy animals decreases the amount of concentrate feeding and thus improves profit. Therefore, for economical and sustainable dairy farming, fodder production round the year is essential.

Mishra et al. (2008) have studied the effect of inoculation of *Azospirillum brasilense* and mycorrhizal fungal consortia with *Glomus intraradices* inoculum on yield and quality of forage in guinea grass (*Panicum maximum* Jacq.) at different intervals of cuttings. The inoculants were seen to possess synergistic effects leading to enhanced fodder quality and production. Gangwar et al. (2015) isolated 24 isolates from root nodules of berseem plants and evaluated the isolates for plant growth-promoting traits. Isolate B-2 was observed to be high IAA producing ($50.8 \pm 0.90$ μg/ml). Field evaluation of a few of these isolates showed that significant improvement occurred in symbiotic and growth parameters of the berseem plants on seed inoculation.

## 18.8 Innovative Technologies for Improving Soil Health and Crop Productivity

Nanotechnology is emerging as the sixth revolutionary technology (Abobatta 2018) in this era. The nano-scale materials studied have diverse origins, biological, physical, or chemical, with at least one dimension in the range of less than 100 nm. Nanotechnology enables the use of materials and equipment that are capable of exploiting physical and chemical properties of a substance at molecular levels, thus facilitating its use in various disciplines from medicine to agriculture (Fakruddin et al. 2012).

The major arena for nano-agricultural interventions has been in sectors such as food production and postharvest food handling, although nanotechnology-enabled products or devices are also becoming popularized in crop production and protection aspects. In crop production, the main challenge is the low nutrient use efficiency of conventional chemical fertilizers. These fertilizers also exhibit bioaccumulation of toxic trace elements. Further, the loss of applied chemical fertilizers from agricultural fields via leaching and runoff processes can have severe environmental consequences such as eutrophication and contamination of groundwater sources. The N fertilizers also exhibit vulnerability for immediate loss from volatilization. Other nutrients such as P, Fe, Cu, and Zn applied to soils for fertilization transform to bound form quickly, rendering them unavailable for plant uptake and therefore necessitating their repeated application. In recent decades, scientists have emphasized the use of biofertilizers over chemical fertilizers to maintain ecological balance. However, certain limitations of biofertilizers including shorter shelf life, possibility of remaining ineffective under fluctuating environmental conditions, and the requirement of longer incubation durations for establishment in the inoculated niche hinder their wider applicability.

Development of nanofertilizers may possibly help researchers to overcome the limitation of conventional fertilizers (Kalia and Kaur 2019a, b; Kalia et al. 2019, 2020) besides holding promise to improve the efficacy of biofertilizers (Kalia and Kaur 2019c). The nanofertilizers exhibit improved nutrient use efficiencies and can make available nutrients to plants through a slow or controlled release phenomena. Simultaneously, there are possibilities that use of these nanofertilizers containing nano-scale particles at higher concentrations may exhibit toxicity to plants through uptake or release of toxic ions following dissolution of nanoparticles in soil. Therefore, under such circumstances, nanofertilizers may negatively affect plant growth, germination, biomass, and root and leaf growth from the increased production of reactive oxygen species (ROS), may exhibit negative effects on plant–water relationships, and alter plant metabolic pathways and photosynthesis (Kalia and Kaur 2019c; Kumar et al. 2019). At such point, the development of nanobiofertilizers will act as a boon for modern-day agriculture (Kalia and Kaur 2019c). The science of nanotechnology offers nanobiofertilizer formulations in which biofertilizer can be coated in nano-scale polymer matrices, a process known as nano-encapsulation (Kumari et al. 2019), or co-embedding of nano-scale particles with biofertilizer in a polymer matrix for co-delivery to the crop plants (Mahakham et al. 2017; Kalia and Kaur 2019c).

Nanofertilizers have the potential to increase NUE threefold and increase the ability of plants to tolerate environmental stress. These fertilizer formulations may improve crop productivity by promoting metabolic pathways such as nitrogen metabolism, protein synthesis, and the antioxidant system (Iqbal et al. 2019). Nanobiofertilizers offer improved NUE through targeted delivery and controlled release of their biological agent (microbial inoculant) in response to environmental triggers and biological demands. Application of nanobiofertilizer may help in optimizing photosynthesis, nutrient absorption, photosynthate accumulation, and nutrient translocation (Kumari et al. 2019). Mardalipour et al. (2014) studied the effect of nanobiofertilizer on growth parameters of spring wheat. Application of this formulation, which contained *Azotobacter*, *Pseudomonas*, Fe, Zn, and Mn, increased spike length, spike number in square meters, seed number in square meters, and seed weight to physiological maturity. In another study, Farnia and Omidi (2015) observed an increase in grain yield and harvest index in a maize (*Zea mays* L.) crop under water-stressed conditions.

The field of nanotechnology has experienced a sudden increase in research and application in the agriculture sector during the past few years, as demonstrated by the rapidly increasing number of publications and patents (Kah et al. 2019). A number of bottlenecks such as low funding, slow adoption of innovations by agricultural research organizations, negligence under government policies, lack of awareness regarding use of nanofertilizers, and ethical issues regarding their use have risen to obstruct the extensive use of nanobiofertilizers. Moreover, strict safety assessment should be implemented for validation of the permissible and safe limits of crop-specific nanoparticle doses. An understanding of the biodegradability, biotransformation, bioaccumulation of nanobiofertilizers in plants as well as animals and humans must be assessed to gain comprehensive knowledge of the toxicity.

## 18.9 Conclusion

Bacterial inoculants may help in increasing crop yields and, therefore, possess prodigious potential to supplement or even substitute for synthetic agricultural chemicals. Regardless of the enormous capability of bacterial inoculants, the existing literature suggests that positive results can be obtained under greenhouse conditions. However, generally these inoculants fail to perform well under field conditions. The major obstruction for large-scale implementation of bioinoculant technology is the variability in field performance that may result from the cumulative effect of dilution or migration of inoculated microbes in the soil–microbe–plant continuum on interaction with the soil components, native vanguard microbial communities, and prevailing environmental conditions. Therefore, the prerequisites for a field-performing bioinoculant include developing an improved process of selecting suitable strains, identifying the appropriate technique(s) of inoculant application, and understanding the interactions involved between inoculated and native microorganisms. Overall, a holistic approach should be established to enhance agricultural production in a sustainable manner with minimal ill effects.

## References


Abdel-Razzak HS, El-Sharkawy GA (2013) Effect of biofertilizer and humic acid applications on growth, yield, quality and storability of two garlic (*Allium sativum* L.) cultivars. Asian J Crop Sci 5:48–64

Abobatta WF (2018) Nanotechnology application in agriculture. Acta Sci Agric 2:99–102. https://doi.org/10.1007/BF01568729

Afzal Z, Khan S, Shomaila S, Shaheen S (2019) Plant beneficial endophytic bacteria: mechanisms, diversity, host range and genetic determinants. Microbiol Res 221:36–49

Ahemad M, Kibret M (2014) Mechanisms and applications of plant growth promoting rhizobacteria: current perspective. J King Saud Uni-Sci 26:1–20

Ahirwar NK, Gupta G, Singh V, Rawlley RK, Ramana S (2015) Influence on growth and fruit yield of tomato (*Lycopersicon esculentum* Mill.) plants by inoculation with *Pseudomonas fluorescence* (SS5): possible role of plant growth promotion. Int J Curr Microbiol App Sci 4(2):720–730

Ahmad M, Pataczek L, Hilger TH, Zahir ZA, Hussain A, Rasche F, Schafleitner R, Solberg SØ (2018) Perspectives of microbial inoculation for sustainable development and environmental management. Front Microbiol 9:2992. https://doi.org/10.3389/fmicb.2018.02992

Akbar M, Aslam N, Khalil T et al (2019) Effects of seed priming with plant growth- promoting rhizobacteria on wheat yield and soil properties under contrasting soils. J Plant Nutr 42:2080–2091. https://doi.org/10.1080/01904167.2019.1655041

Akhgar R, Arzanlou M, Bakker PAHM, Hamidpour M (2014) Characterization of 1-aminocyclopropane-1-carboxylate (ACC) deaminase-containing *Pseudomonas* sp. in the rhizosphere of salt-stressed canola. Pedosphere 24:161–468

Alori ET, Babalola OO (2018) Microbial inoculants for improving crop quality and human health in Africa. Front Microbiol 9:2213. https://doi.org/10.3389/fmicb.2018.02213

Alori ET, Fawole OB (2017) Microbial inoculants-assisted phytoremediation for sustainable soil management. In: Ansari AA, Gill SS, Lanza GR, Newman L (eds) Phytoremediation: management of environmental contaminants, Switzerland. Springer International Publishing, Berlin

Alori ET, Glick BR, Babalola OO (2017) Microbial phosphorus solubilization and its potential for use in sustainable agriculture. Front Microbiol 8:971. https://doi.org/10.3389/fmicb.2017.00971

Amirnia R, Ghiyasi M, Siavash Moghaddam S et al (2019) Nitrogen-fixing soil bacteria plus mycorrhizal fungi improve seed yield and quality traits of lentil (*Lens culinaris* Medik). J Soil Sci Plant Nutr. 19:592. https://doi.org/10.1007/s42729-019-00058-3

Antoun H, Prevost D (2005) Ecology and plant growth promoting rhizobacteria. In: Siddiqui ZA (ed) PGPR: biocontrol and biofertilization. Springer, Berlin, pp 1–38

Ashrafuzzaman M, Hossen FA, Ismail MR, Hoque MA, Islam MZ, Shahidullah SM, Meon S (2009) Efficiency of plant growth promoting rhizobacteria (PGPR) for the enhancement of rice growth. Afr J Biotechnol 8:1247–1252

Avis TJ, Gravel V, Antoun H, Tweddell RJ (2008) Multifaceted beneficial effects of rhizosphere microorganisms on plant health and productivity. Soil Biol Biochem 40:1733–1740

Azziz G, Bajsa N, Haghjou T, Taulé C, Valverde A, Igual J et al (2012) Abundance, diversity and prospecting of culturable phosphate solubilizing bacteria on soils under crop–pasture rotations in a no-tillage regime in Uruguay. Appl Soil Ecol 61:320–326

Bais HP, Weir TL, Perry LG, Gilroy S, Vivanco JM (2006) The role of root exudates in rhizosphere interactions with plants and other organisms. Annu Rev Plant Biol 57:233–266

Bargaz A, Lyamlouli K, Chtouki M et al (2018) Soil microbial resources for improving fertilizers efficiency in an integrated plant nutrient management system. Front Microbiol 9:1606. https://doi.org/10.3389/fmicb.2018.01606

Barriuso J, Ramos Solano B, Fray RG, Cámara M, Hartmann A, Gutiérrez Mañero FJ (2008) Transgenic tomato plants alter quorum sensing in plant growthpromoting rhizobacteria. Plant Biotechnol J 6:442–452

Becker R, Rokem JS, Ilangumaran G, Lamont J, Praslickova D, Ricci E, Subramanian S, Smith DL (2018) Plant growth-promoting rhizobacteria: context, mechanisms of action, and roadmap to commercialization of biostimulants for sustainable agriculture. Front Plant Sci 9:1473

Beneduzi A, Ambrosini A, Passaglia LMP (2012) Plant growth-promoting rhizobacteria: their potential as antagonists and biocontrol agents. Genet Mol Biol 35:1044–1051

Berendsen RL, Pieterse CMJ, Bakker PAHM (2012) The rhizosphere microbiome and plant health. Trends Plant Sci 17:478–486

Bernabeu PR, Pistorio M, Torres-Tejerizo G et al (2015) Colonization and plant growth-promotion of tomato by *Burkholderia tropica*. Sci Hortic (Amsterdam) 191:113–120. https://doi.org/10.1016/j.scienta.2015.05.014

Bhattacharjee S, Sharma SD (2012) Effect of Dual Inoculation of Arbuscular Mycorrhiza and *Rhizobium* on the Chlorophyll, Nitrogen and Phosphorus Contents of Pigeon Pea (*Cajanus cajan* L.). Adv Microbiol 2:561–564

Bhattacharyya PN, Jha DK (2012) Plant growth-promoting rhizobacteria (PGPR): emergence in agriculture. World J Microbiol Biotechnol 28:1327–1350

Bidyarani N, Prasanna R, Babu S et al (2016) Enhancement of plant growth and yields in chickpea (*Cicer arietinum* L.) through novel cyanobacterial and biofilmed inoculants. Microbiol Res 188–189:97–105. https://doi.org/10.1016/j.micres.2016.04.005

Bulgarelli D, Rott M, Schlaeppi K, van Themaat EVL, Ahmadinejad N, Assenza F, Rauf P, Huettel B, Reinhardt R, Schmelzer E et al (2012) Revealing structure and assembly cues for *Arabidopsis* root-inhabiting bacterial microbiota. Nature 488:91–95

Bumandalai O, Tserennadmid R (2019) Effect of *Chlorella vulgaris* as a biofertilizer on germination of tomato and cucumber seeds. Int J Aquat Biol 7:95–99

Calvo P, Nelson L, Kloepper JW (2014) Agricultural uses of plant biostimulants. Plant Soil 383:3–41

Camerini S, Senatore B, Lonardo E, Imperlini E, Bianco C, Moschetti G et al (2008) Introduction of a novel pathway for IAA biosynthesis to rhizobia alters vetch root nodule development. Arch Microbiol 190:67–77

Cattelan AJ, Hartel PG, Fuhrmann JJ (1996) Screening for plant growth-promoting rhizobacteria to promote early soybean growth. Soil Biol Biochem 3:1670–1680

Chaparro JM, Badri DV, Bakker MG, Sugiyama A, Manter DK, Vivanco JM (2013) Root exudation of phytochemicals in *Arabidopsis* follows specific patterns that are developmentally programmed and correlate with soil microbial functions. PLoS One 8:e55731

Chattha MB, Iqbal A, Chattha MU, Hassan MU, Khan I, Ashraf I, Faisal M, Usman M (2017) PGPR inoculated-seed increases the productivity of forage sorghum under fertilized conditions. J Basic Appl Sci 13:150–153

Chebotar V, Malfanova N, Shcherbakov A, Ahtemova G, Borisov AY, Lugtenberg B, Tikhonovich I (2015) Endophytic bacteria in microbial preparations that improve plant development. Appl Biochem Microbiol 51:271–277

Crowley DE (2006) Microbial siderophores in the plant rhizosphere. In: Barton LL, Abadía J (eds) Iron nutrition in plants and rhizospheric microorganisms. Springer, Netherlands, pp 169–198

Damam M, Kaloori K, Gaddam B, Kausar R (2016) Plant growth promoting substances (phytohormones) produced by rhizobacterial strains isolated from the rhizosphere of medicinal plants. Int J Pharm Sci Rev 37:130–136

Das I, Singh A (2014) Effect of PGPR and organic manures on soil properties of organically cultivated mung bean. Bioscan 9:27–29

Dasgupta D, Ghati A, Sarkar A, Sengupta C, Paul G (2015) Application of plant growth promoting rhizobacteria (PGPR) isolated from the rhizosphere of *Sesbania bispinosa* on the growth of chickpea (*Cicer arietinum* L.). Int J Curr Microbiol App Sci 4(5):1033–1042

Dash SK, Pathak M, Tripathy L, Barik S (2019) Studies on effect of integrated nutrient management on growth and yield attributes in radish (*Raphanus sativus* L.) and its residual effect in coriander (*Coriandrum sativum* L.) in radish-coriander cropping sequence. J Pharmacogn Phytochem 8:319–322

Dashti NH, Montasser MS, Ali NYA, Cherian M (2014) Influence of plant growth promoting rhizobacteria on fruit yield, pomological characteristics and chemical contents in cucumber mosaic virus-infected tomato plants. Kuwait J Sci 41:205–220

Daur I, Saad MM, Eida AA, Ahmad S, Shah ZH, Ihsan MZ, Muhammad Y, Sohrab SS, Heribert H (2018) Boosting *Alfalfa (Medicago sativa* L.) production with rhizobacteria from various plants in Saudi Arabia. Front Microbiol 9:477. https://doi.org/10.3389/fmicb.2018.00477

Dawwama GE, Elbeltagy A, Emara HM, Abbas IH, Hassan MM (2013) Beneficial effect of plant growth promoting bacteria isolated from the roots of potato plant. Ann Agric Sci 58:195–201

Delshadi S, Ebrahimi M, Shirmohammadi E (2017) Influence of plant-growth-promoting bacteria on germination, growth and nutrients uptake of Onobrychis sativa L. under drought stress. J Plant Interact 12:200–208

Dong L, Li Y, Xu J et al (2019) Biofertilizers regulate the soil microbial community and enhance *Panax ginseng* yields. Chinese Med (United Kingdom) 14:1–14. https://doi.org/10.1186/s13020-019-0241-1

Dursun A, EKİNCİ M, Dönmez MF (2010) Effects of foliar application of plant growth promoting bacteriumon chemical contents, yield and growth of tomato (*Lycopersicon esculentum*) and cucumber (*Cucumis sativus*). Pak J Bot 42(5):3349–3356

Egamberdiyeva D (2007) The effect of plant growth promoting bacteria on growth and nutrient uptake of maize in two different soils. Appl Soil Ecol 36:184–189

Enebe MC, Babalola OO (2018) The influence of plant growth-promoting rhizobacteria in plant tolerance to abiotic stress: a survival strategy. Appl Microbiol Biotechnol 102(18):7821–7835. https://doi.org/10.1007/s00253-018-9214-z

Fakruddin M, Hossain Z, Afroz H (2012) Prospects and applications of nanobiotechnology: a medical perspective. J Nanobiotechnol 10:1–8. https://doi.org/10.1186/1477-3155-10-31

Fan X, Zhang S, Mo X, Li Y, Fu Y, Liu Z (2017) Effect of PGPR and N source on plant growth and N, P uptake by tomato grown in calcareous soils. Pedosphere 27:1027–1036

FAO (2011) http://www.fao.org/es/faodef/fdef11e.htm

Farnia A, Omidi MM (2015) Effect of nano-zinc chelate and nano-biofertilizer on yield and yield components of maize (*Zea mays* L.), under water stress condition. Indian J Nat Sci 5:976–997

Fernando WGD, Ramarathnam R, Krishnamoorthy AS, Savchuk SC (2005) Identification and use of potential bacterial organic antifungal volatiles in biocontrol. Soil Biol Biochem 37:955–964

Gangwar M, Dhaliya S, Kaur S (2015) Potential of *Rhiobium* species to enhance growth and symbiosis in Berseem (*Trifolium alexandrinum* L.). Ind J Ecol 42(1):174–178

Gangwar M, Pandove G, Brar S, Sekhon S, Kaur S, Kumar R (2018) Integrated nutrient management in wheat by use of *Azotobacter* sp. and *Streptomyces badius*. Int J Agric Innovat Res 6 (4):2319–1473

Garbeva P, van JA V, van JD E (2004) Microbial diversity in soil: selection of microbial populations by plant and soil type and implications for disease suppressiveness. Annu Rev Phytopathol 42:243–270

Gholami A, Shahsavani S, Nezarat S (2009) The effect of plant growth promoting rhizobacteria (PGPR) on germination, seedling growth and yield of maize. Int J Biol Life Sci 1:35–40

Gilchrist M, Greko C, Wallinga D, Beran G, Riley D, Thorne P (2007) The potential role of concentrated animal feeding operations in infectious disease epidemics and antibiotics resistance. Environ Health Perspect 115:313–316

Gogoi D, Kotoky U, Hazarika S (2004) Effect of biofertilizers on productivity and soil characteristics in banana. Indian J Hortic 61:354–356

Golinska P, Wypij M, Agarkar G, Rathod D, Dahm H, Rai M (2015) Endophytic actinobacteria of medicinal plants: diversity and bioactivity. Antonie Van Leeuwenhoek 108:267–289

Gopalakrishnan S, Vadlamudi S, Apparla S, Bandikinda P, Vijayabharathi R, Bhimineni RK, Rupela O (2013) Evaluation of *Streptomyces* spp. for their plant growthpromotion traits in rice. Can J Microbiol 59:534–539

Gosal SK, Kalia A, Uppal SK et al (2012) Assessing the benefits of *Azotobacter* bacterization in sugarcane: a field appraisal. Sugar Tech 14(1):61–67. https://doi.org/10.1007/s12355-011-0131-z

Goswami D, Thakker JN, Dhandhukia PC (2016) Portraying mechanics of plant growth promoting rhizobacteria (PGPR): a review. Cogent Food Agric 2:1–19

Gouda S, Kerry RG, Dasc G, Paramithiotisd S, Shine HS, Patrac JK (2018) Revitalization of plant growth promoting rhizobacteria for sustainable development in agriculture. Microbiol Res 206:131–140

Hamidi A (2006) The Effects of application of plant growth promoting rhizobacteria (PGPR) on the yield of fodder maize (*Zea Mays* L.). Tarbiat Modarres University, Agriculture, Faculty Agronomy, Crops Ecology (PhD thesis)

Haroun SA, Hussein M (2003) The promotive effect of algal biofertilizers on growth, protein pattern and some metabolic activities of *Lupinus termis* plants grown in siliceous soil. Asian J Plant Sci 2:944–951

Hassanein WA, Awny NM, El-Mougith AA, Salah El-Dien SH (2009) The antagonistic activities of some metabolites produced by Pseudomonas aeruginosa Sha. J Appl Sci Res 5:404–414

Hatzinger PB, Alexander M (1994) Relationship between the number of bacteria added to soil or seeds and their abundance and distribution in the rhizosphere of alfalfa. Plant Soil 158:211–222. https://doi.org/10.1007/BF00009496

Ho YN, Mathew DC, Huang CC (2017) Plant-microbe ecology: interactions of plants and symbiotic microbial communities. In: Plant ecology-traditional approaches to recent trends. InTec, Rijeka, pp 93–119

Holl FB, Chanway CP (1992) Rhizosphere colonization and seedling growth promotion of lodgepole pine by *Bacillus polymyxa*. Can J Microbiol 38:303–308. https://doi.org/10.1139/m92-050

Htwe AZ, Moh SM, Moe K, Yamakawa T (2019) Biofertiliser production for agronomic application and evaluation of its symbiotic effectiveness in soybeans. Agronomy 9:162. https://doi.org/10.3390/agronomy9040162

Igiehon NO, Babalola OO (2018) Below-ground-above-ground plant-microbial Interactions: focusing on Soybean, Rhizobacteria and Mycorrhizal Fungi. Open Microbiol J 12:261–279

Iqbal M, Umar S, Mahmooduzzafar (2019) Nano-fertilization to enhance nutrient use efficiency and productivity of crop plants. In: Husen A, Iqbal M (eds) Nanomaterials and plant potential. Springer, Cham, pp 473–505

Islam S, Akanda AM, Prova A, Islam MT, Hossain MDM (2015) Isolation and identification of plant growth promoting rhizobacteria from cucumber rhizosphere and their effect on plant growth promotion and disease suppression. Front Microbiol 6:1360. https://doi.org/10.3389/fmicb.2015.01360

Jat RS, Ahlawat IPS (2008) Direct and residual effect of vermicompost, biofertilizers and phosphorus on soil nutrient dynamics and productivity of chickpea-fodder maize sequence. J Sustain Agric 28:41–54. https://doi.org/10.1300/J064v28n01

Jha CK, Saraf M (2015) Plant growth promoting rhizobacteria (PGPR): a review. J Agric Res Dev 5:108–119

Jiang F, Chen L, Andrey AB, Shaposhnikov AI, Gong G, Xu M, Wolfram H, Dieter W, William JD, Ian CD (2012) Application of plant growth promoting rhizobacteria (PGPR) isolated from the rhizosphere of *Sesbania bispinosa* on the growth of chickpea (*Cicer arietinum* L.). J Exp Bot 63:6421–6430

Kah M, Tufenkji N, White JC (2019) Nano-enabled strategies to enhance crop nutrition and protection. Nat Nanotechnol 14:532–540. https://doi.org/10.1038/s41565-019-0439-5

Kai M, Haustein M, Molina F, Petri A, Scholz B, Piechulla B (2009) Bacterial volatiles and their action potential. Appl Microbiol Biotechnol 81:1001–1012

Kalia A, Gosal SK (2011) Effect of pesticide application on soil microorganisms. Arch Agron Soil Sci 57:569–596. https://doi.org/10.1080/03650341003787582

Kalia A, Gupta RP (2004) Disruption of soil food web by pesticides. Indian J Ecol 31:85–92

Kalia A, Gupta R (2005) Conservation and utilization of microbial diversity. NBA Sci Bull 1:1–40

Kalia A, Kaur H (2019a) Nanofertilizers: an innovation towards new generation fertilizers for improved nutrient use efficacy (NUE) and environmental sustainability. In: Bhoop B, Katare O, Souto E (eds) Emerging trends in nanobiomedicine. Taylor & Francis (CRC), Boca Raton, FL, pp 45–61

Kalia A, Kaur H (2019b) Agri-applications of nano-scale micronutrients: prospects for plant growth promotion. In: Raliya R (ed) Nanoscale engineering in agricultural management. Taylor & Francis, CRC, Boca Raton, pp 81–105

Kalia A, Kaur H (2019c) Nano-biofertilizers: harnessing dual benefits of nano-nutrient and bio-fertilizers for enhanced nutrient use efficiency and sustainable productivity. In: Pudake RN, Chauhan N, Kole C (eds) Nanoscience for sustainable agriculture. Springer, Cham, pp 51–73

Kalia A, Sharma SP, Kaur H (2019) Nanoscale fertilizers: harnessing boons for enhanced nutrient use efficiency and crop productivity. In: Abd-Elsalam KA, Prasad R (eds) Nanobiotechnology applications in plant protection, Nanotechnology in the life sciences. Springer Nature, Switzerland, pp 191–208

Kalia A, Sharma SP, Kaur H (2020) Novel nanocomposite-based controlled-release fertilizer and pesticide formulations: prospects and challenges. In: Abd-Elsalam KA (ed) Multifunctional hybrid nanomaterials for sustainable agri-food and ecosystems. Elsevier Inc., Cambridge, MA, pp 99–134. https://doi.org/10.1016/B978-0-12-821354-4.00005-4

Kang SM, Joo GJ, Hamayun M, Na CI, Shin DH, Kim HY, Hong JK, Lee IJ (2009) Gibberellin production and phosphate solubilization by newly isolated strain of *Acinetobacter calcoaceticus* and its effect on plant growth. Biotechnol Lett 31:277–281. https://doi.org/10.1007/s10529-008-9867-2

Kang Y, Shen M, Wang H, Zhao Q (2013) A possible mechanism of action of plant growth-promoting rhizobacteria (PGPR) strain *Bacillus pumilus* WP8 via regulation of soil bacterial community structure. J Gen Appl Microbiol 59:267–277. https://doi.org/10.2323/jgam.59.267

Karnwal A (2012) Screening of plant growth–promoting rhizobacteria from maize (*Zea mays*) and wheat (*Triticum aestivum*). Afr J Food Agric Nutr Dev 12:6170–6185

Kaur S, Khanna V (2014) Screening indole acetic-acid over producing rhizobacteria for improving growth of Lentil under axenic conditions. J Food Legumes 27(2):37–40

Kaur S, Kalia A, Gangwar M (2019) Bioprospecting endophytic actinobacteria of medicinal plants as potential anticancer therapeutic agents. In: Pollutants and protectants: valuation and assessment techniques, pp 196–204

Kaymak HC, Yarali F, Guvenc I, Donmez MF (2008) The effect of inoculation with plant growth rhizobacteria (PGPR) on root formation of mint (*Mentha piperita* L.) cuttings. Afr J Biotechnol 7(24):4479–4483

Khalid A, Arshad M, Zahir ZA (2004) Screening plant growth promoting rhizobacteria for improving growth and yield of wheat. J Appl Microbiol 46:473–480

Khati P, Parul, Bhatt P, Nisha, Kumar R, Sharma A (2018) Effect of nanozeolite and plant growth promoting rhizobacteria on maize. 3 Biotech 8(3):141. https://doi.org/10.1007/s13205-018-1142-1

Kidoglu F, Gül A, Ozaktan H, Tuzel Y (2008) Effect of rhizobacteria on plant growth of different vegetables. Acta Hortic 801:1471–1477

Kothe E, Turnau K (2018) Mycorrhizosphere. Communication: mycorrhizal fungi and endophytic fungus-plant interactions. Lausanne: Frontiers Media. Front Microbiol 9:6–9

Kuan KB, Radziah O, Rahim KA, Shamsuddin ZH (2016) Effectivenes of rhizobacteria containing ACC deaminase for growth promotion of Peas (Pisum sativum) under drought conditions. J Microbiol Biotechnol 18(5):958–963

Kumar P, Dubey RC (2012) Plant growth promoting rhizobacteria for biocontrol of phytopathogens and yield enhancement of Phaseolus vulgaris. J Curr Perspect Appl Microbiol 1:6–38

Kumar A, Gupta K, Dixit S et al (2019) A review on positive and negative impacts of nanotechnology in agriculture. Int J Environ Sci Technol 16:2175–2184. https://doi.org/10.1007/s13762-018-2119-7

Kumari B, Mallick MA, Solanki MK et al (2019) Applying nanotechnology to bacteria: an emerging technology for sustainable agriculture. In: Role of plant growth promoting microorganisms in sustainable agriculture and nanotechnology. Elsevier, Cambridge, pp 121–143

Kundan R, Pant G, Jadon N, Agrawal PK (2015) Plant growth promoting rhizobacteria: mechanism and current prospective. J Fertil Pestic 6:155. https://doi.org/10.4172/2471-2728.1000155

Lakshmanan V, Selvaraj G, Bais HP (2014) Functional soil microbiome: belowground solutions to an aboveground problem. Plant Physiol 166:689–700

Lee SK, Lur HS, Lo KJ, Cheng KC, Chuang CC, Tang SJ et al (2016) Evaluation of the effects of different liquid inoculant formulations on the survival and plantgrowth-promoting efficiency of *Rhodopseudomonas palustris* strain PS3. Appl Microbiol Biotechnol 100:7977–7987

Lily Devi K, Chettri S, Sharma APM et al (2019) Effect of biofertilizers and biocontrol agents on growth and yield in off season brinjal under low cost polyhouse. Curr J Appl Sci Technol 34:1–5. https://doi.org/10.9734/cjast/2019/v34i530143

Liu D, Lian B, Dong H (2012) Isolation of *Paenibacillus* sp. and assessment of its potential for enhancing mineral weathering. J Geom 29:413–421

Liu L, Guo G, Wang Z, Ji H, Mu F, Li X (2014) Auxin in plant growth and stress responses. In: LSP T, Pal S (eds) Phytohormones: a window to metabolism, signaling and biotechnological applications. Springer, New York, pp 1–35

Lugtenberg B, Rozen DE, Kamilova F (2017) Wars between microbes on roots and fruits. F1000Research 6:343. https://doi.org/10.12688/f1000research.10696.1

Lundberg DS, Lebeis SL, Paredes SH, Yourstone S, Gehring J, Malfatti S, Tremblay J, Engelbrektson A, Kunin V, del Rio TG et al (2012) Defining the core *Arabidopsis thaliana* root microbiome. Nature 488:86–90

Mahakham W, Sarmah AK, Maensiri S, Theerakulpisut P (2017) Nanopriming technology for enhancing germination and starch metabolism of aged rice seeds using phytosynthesized silver nanoparticles. Sci Rep 7:1–21. https://doi.org/10.1038/s41598-017-08669-5

Mahdi SS, Hassan GI, Samoon SA et al (2010) Bio-fertilizers in organic agriculture. J Phytol 2:42–54

Majeed A, Abbasi MK, Hameed S, Imran A, Rahim N (2015) Isolation and characterization of plant growth-promoting rhizobacteria from wheat rhizosphere and their effect on plant growth promotion. Front Microbiol 6:198. https://doi.org/10.3389/fmicb.2015.00198

Mallesh SB (2008) Plant growth promoting rhizobacteria, their characterization and mechanisms in the suppression of soil borne pathogens of Coleus and Ashwagandha. M.Sc. thesis. University of Agricultural Sciences, Dharwad

Malusá E, Sas-Paszt L, Ciesielska J (2012) Technologies for beneficial microorganisms inocula used as biofertilizers. Sci World J 2012:491206. https://doi.org/10.1100/2012/491206

Mardalipour M, Zahedi H, Sharghi Y (2014) Evaluation of nano biofertilizer efficiency on agronomic traits of spring wheat at different sowing date. Biol Forum-An Int J 6:349–356

Martínez-Viveros O, Jorquera MA, Crowley DE, Gajardo G, Mora ML (2010) Mechanisms and practical considerations involved in plant growth promotion by rhizobacteria. J Soil Sci Plant Nutr 10:293–319

Mazzola M, Manici LM (2012) Apple replant disease: role of microbial ecology in cause and control. Annu Rev Phytopathol 50:45–65. https://doi.org/10.1146/annurev-phyto-081211-173005

Meena D, Ram RB, Verma RS, Shivran BC (2019) Effect of phosphorus levels and bio-fertilizers on yield and quality parameters of garlic (*Allium sativum* L.) cv. G-282. Pharma Innov J 8:681–683

Mia MAB, Shamsuddin ZH, Mahmood M (2012) Effects of rhizobia and plant growth promoting bacteria inoculation on germination and seedling vigor of lowland rice. Afr J Biotechnol 11 (16):3758–3765

Mimmo T, Hann S, Jaitz L, Cesco S, Gessa C, Puschenreiter M (2011) Time and substrate dependent exudation of carboxylates by *Lupinus albus* L. and *Brassica napus* L. Plant Physiol Biochem 49:1272–1278. https://doi.org/10.1016/j.plaphy.2011.08.012

Mishra S, Sharma S, Vasudevan P (2008) Comparative effect of biofertilizers on fodder production and quality in guinea grass (*Panicum maximum* Jacq.). J Sci Food Agric 88:1667–1673. https://doi.org/10.1002/jsfa.3267

Mishra PK, Bisht SC, Jeevanandan K, Kumar S, Bisht JK, Bhatt JC (2014) Synergistic effect of inoculating plant growth-promoting Pseudomonas spp. and *Rhizobium leguminosarum*-FB1 on growth and nutrient uptake of rajmash (Phaseolus vulgaris L.). Arch Agron Soil Sci 60:799–815

Moe LA (2013) Amino acids in the rhizosphere: from plants to microbes. Am J Bot 100:1692–1705

Mohamed HI, Gomaa EZ (2012) Effect of plant growth promoting *Bacillus subtilis* and *Pseudomonas fluorescens* on growth and pigment composition of radish plants (*Raphanus sativus*) under NaCl stress. Photosynthetica 50:263–272. https://doi.org/10.1007/s11099-012-0032-8

Mohammadi NK, Hekmat AW, Ghosh G, Atif A (2019) Effect of biofertilizers with different levels of nitrogen and phosphorus on growth and growth attributes of baby corn (*Zea mays* L.). Int J Chem Stud 7:2300–2305

Montesinos E (2003) Plant-associated microorganisms: a view from the scope of microbiology. Int Microbiol 6:221–223

Morel MA, Ubalde M, Braña V, Castro-Sowinski S (2011) *Delftia* sp. JD2: a potential Cr (VI)-reducing agent with plant growth-promoting activity. Arch Microbiol 193:63–68

Nair DN, Padmavathy S (2014) Impact of endophytic microorganisms on plants, environment and humans. Sci World J, 1–11. 250693. https://doi.org/10.1155/2014/250693

Naqqash T, Hameed S, Imran A, Hanif MK, Majeed J, Elsas JD (2016) Differential response of potato toward inoculation with taxonomically diverse plant growth promoting rhizobacteria. Front Plant Sci 7:144. https://doi.org/10.3389/fpls.2016.00144

Nelson LM (2004) Plant growth promoting rhizobacteria (PGPR): prospects for new inoculants. Crop Manag 3:301–305

Nivya RM (2015) A study on plant growth promoting activity of the Endophytic bacteria isolated from the root nodules of *Mimosa pudica* plant. Int J Innov Res Sci Eng Technol 4:6959–6968

Oldroyd GED, Murray JD, Poole PS, Downie JA (2011) The rules of engagement in the legume-rhizobial symbiosis. Annu Rev Genet 45:119–144

Ordookhani K, Khavazi K, Moezzi A, Rejali F (2010) Influence of PGPR and AMF on antioxidant activity, lycopene and potassium contents in tomato. Afr J Agric Res 5(10):1108–1116

Orhan E, Esitken A, Ercisli S, Turan M, Sahin F (2006) Effects of plant growth promoting rhizobacteria (PGPR) on yield, growth and nutrient contents in organically growing raspberry. Sci Hortic 111:38–43

Osman MEH, El-Sheekh MM, El-Naggar AH, Gheda SF (2010) Effect of two species of cyanobacteria as biofertilizers on some metabolic activities, growth, and yield of pea plant. Biol Fertil Soils 46:861–875. https://doi.org/10.1007/s00374-010-0491-7

Oteino N, Lally RD, Kiwanuka S, Lloyd A, Ryan D, Germaine KJ, Dowling DN (2015) Plant growth promotion induced by phosphate solubilizing endophytic *Pseudomonas* isolates. Front Microbiol 6:745

Parmar P, Sindhu SS (2013) Potassium solubilization by rhizosphere bacteria: influence of nutritional and environmental conditions. J Microbiol Res 3(1):25–31

Peiffer JA, Spor A, Koren O, Jin Z, Tringe SG, Dangl JL et al (2013) Diversity and heritability of the maize rhizosphere microbiome under field conditions. Proc Natl Acad Sci U S A 110:6548–6553

Piechulla B, Pott MB (2003) Plant scents-mediator of inter- and intraorganismic communication. Planta 217:687–689

Prakash J, Yadav J, Nath K, Kumar A (2013) Effect of indigenous *Mesorhizobium* spp. and plant growth promoting rhizobacteria on yields and nutrients uptake of chickpea (*Cicer arietinum* L.) under sustainable agriculture. Ecol Eng 51:282–286. https://doi.org/10.1016/j.ecoleng.2012.12.022

Prashar P, Kapoor N, Sachdeva S (2013) Rhizosphere: its structure, bacterial diversity and significance. Rev Environ Sci Biotechnol, 1–18

Ramakrishnan K, Selvakumar G (2012) Effect of biofertilizers on enhancement of growth and yield on tomato (*Lycopersicum esculentum* Mill.). Int J Res Bot 2:20–23

Ramandeep SS, Kumari S, Singh SK (2018) Impact of bio-fertilizers and fertilizers on potato (*Solanum tuberosum* L.) cv. Kufri Pukhraj and Kufri Jyoti cultivation. Int J Chem Stud 6:29–31

Ramette A, Frapolli M, defado G, Moenne-Loccoz Y (2003) Phylogeny of HCN synthase-encoding hcnBC genes in biocontrol fluorescent Pseudomonad and its relationship with host plant species and HCN synthesis ability. Amer Phytopathol Soc 16:525–535

Ramírez-López C, Esparza-García FJ, Ferrera-Cerrato R et al (2019) Short-term effects of a photosynthetic microbial consortium and nitrogen fertilization on soil chemical properties, growth, and yield of wheat under greenhouse conditions. J Appl Phycol 31:3617. https://doi.org/10.1007/s10811-019-01861-2

Rana A, Saharan B, Nain L, Prasanna R, Shivay YS (2012) Enhancing micronutrient uptake and yield of wheat through bacterial PGPR consortia. Soil Sci Plant Nutr 58:573–582

Reddy CA, Saravanan RS (2013) Polymicrobial multi-functional approachfor enhancement of crop productivity. Adv Appl Microbiol 82:53–113

Rosas SB, Avanzin G, Carlier E, Pasluosta C, Pastor N, Rovera M (2009) Root colonization and growth promotion of wheat and maize by *Pseudomonas aurantiaca* SR1. Soil Biol Biochem 41:1802–1806

Rosenblueth M, Martínez-Romero E (2006) Bacterial endophytes and their interactions with hosts. Mol Plant-Microbe Interact 19:827–837

Rudresh DL, Shivaprakash MK, Prasad RD (2005) Tricalcium phosphate solubilizing abilities of *Trichoderma* spp. in relation to P uptake and growth and yield parameters of chickpea (*Cicer arietinum* L.). Can J Microbiol 51:217–222. https://doi.org/10.1139/w04-127

Saengsanga T (2018) Isolation and characterization of indigenous plant growth-promoting rhizobacteria and their effects on growth at the early stage of Thai Jasmine Rice (*Oryza sativa* L. KDML105). Arab J Sci Eng 43:3359–3369

Salamone IEG, Hynes RK, Nelson LM (2012) Role of cytokinins in plant growth promotion by rhizosphere bacteria. In: Siddiqui ZA (ed) PGPR: biocontrol and biofertilization. Springer, Dordrecht, pp 173–195

Sankhala G, Verma P, Nandre B et al (2019) Effect of organic nutrient management on growth and flowering of muskmelon (*Cucumis melo* L.) cv. GMM3. Int J Farm Sci 7:362–363. https://doi.org/10.5958/2250-0499.2019.00054.5

Santos ML, Berlitz DL, Wiest SLF, Schünemann R, Knaak N, Fiuza LM (2018) Benefits associated with the interaction of endophytic bacteria and plants. Braz Arch Biol Technol 61:1–11

Selvarasu K, Avudaithai S, Somasundaram S, Sundar M (2019) Effect of INM on chemical properties of soil, nutrient uptake and yield of greengram in sodic soil. Int J Chem Stud 7:2053–2055

Sen S, Chandrasekhar CN (2014) Effect of PGPR on growth promotion of rice (*Oryza sativa* L.) under salt stress. Asian J Plant Sci Res 4(5):62–67

Seneviratne G, Kecskés M, Kennedy I (2008) Biofilmed biofertilisers: novel inoculants for efficient nutrient use in plants. In: Kennedy IR, ATMA C, Kecskés ML, Rose MT (eds) Proceedings of a project (SMCN/2002/073) workshop 'Efficient nutrient use in rice production in Vietnam achieved using inoculant biofertilizers'. Hanoi, Vietnam, pp 126–130

Setiawati TC, Mutmainnah L (2016) Solubilization of potassium containing mineral by microorganisms from sugarcane rhizosphere. Agri Sci Procedia 9:108–117

Shahram S (2012) Effects of PGPR on growth and nutrients uptake of tomato. Int J Adv Eng Technol 27(2):1–31

Shaikh SS, Wani SJ, Sayyed RZ (2018) Impact of Interactions between rhizosphere and rhizobacteria: a review. J Bacteriol Mycol 5(1):1058

Shariatmadari Z, Riahi H, Shokravi S (2011) Study of soil blue-green algae and their effect on seed germination and plant growth of vegetable crops. Rostaniha 12:101–110

Sharma A, Johri BN (2003) Growth promoting influence of siderophore-producing *Pseudomonas* strains GRP3A and PRS9 in maize (*Zea mays* L.) under iron limiting conditions. Microbiol Res 158:243–248

Sharma A, Johri BN, Sharma AK, Glick BR (2013) Plant growth-promoting bacterium *Pseudomonas* sp. strain GRP3 influences iron acquisition in mung bean (*Vigna radiata* L. Wilzeck). Soil Biol Biochem 35:887–894

Sharma A, Shankhdhar D, Sharma A, Shankhdhar SC (2014) Growth promotion of the rice genotypes by pgprs isolated from rice rhizosphere. J Soil Sci Plant Nutr 14(2):505–517

Sharma V, Sharma S, Sharma S, Kumar V (2019) Synergistic effect of bio-inoculants on yield, nodulation and nutrient uptake of chickpea (*Cicer arietinum* L.) under rainfed conditions. J Plant Nutr 42:374–383. https://doi.org/10.1080/01904167.2018.1555850

Simranjit K, Kanchan A, Prasanna R et al (2019) Microbial inoculants as plant growth stimulating and soil nutrient availability enhancing options for cucumber under protected cultivation. World J Microbiol Biotechnol 35:1–14. https://doi.org/10.1007/s11274-019-2623-z

Singh RP, Jha PN (2015) Molecular identification and characterization of rhizospheric bacteria for plant growth promoting ability. Int J Curr Biotechnol 3:12–18

Singh N, Chaudhary F, Patel B (2013) Effectiveness of *Azotobacter* bio-inoculant for wheat grown under dryland condition. Acad Environ Biol 34:927–932

Singh M, Kumar A, Singh R, Pandey KD (2017) Endophytic bacteria: a new source of bioactive compounds. 3 Biotech 7:315–328

Sofi KA, Chesti MH, Peer QJA, Aziz T (2019) Effect of inoculations of *Rhizobium leguminosarum* on germination and yield of field pea (*Pisum sativum*) under temperate conditions of Kashmir. J Pharmacogn Phytochem 8:819–821

Sood GAC, Kaushal RB, Chauhan AA, Gupta SA (2018) Indigenous plant-growth-promoting rhizobacteria and chemical fertilisers: impact on wheat (*Triticum aestivum*) productivity and soil properties in North Western Himalayan region. Crop Pasture Sci 69:460–468

Spence C, Bais H (2013) Probiotics for plants: rhizospheric microbiome and plant fitness. In: de Bruijn FJ (ed) Molecular microbial ecology of the rhizosphere: vols 1 and 2. John Wiley & Sons, Hoboken, NJ, pp 713–721

Spence C, Alff E, Johnson C, Ramos C, Donofrio N, Sundaresan V, Bais H (2014) Natural rice rhizospheric microbes suppress rice blast infections. BMC Plant Biol 14:130

Srutek M, Urbn J (2008) Organic farming. In: Encyclopedia of ecology. Academic, Oxford, pp 2582–2587

Steer J, Harris JA (2000) Shifts in the microbial community in rhizosphere and non-rhizosphere soils during the growth of Agrostis stolonifera. Soil Biol Biochem 32:869–878

Sureshbabu K, Amaresan N, Kumar K (2016) Amazing multiple function properties of plant growth promoting rhizobacteria in the rhizosphere soil. Int J Curr Microbiol Appl Sci 5:661–683

Svircev Z, Dulic T, Obreht I et al (2019) Cyanobacteria and loess—an underestimated interaction. Plant Soil 439:293–308

Teixeira LCRS, Peixoto RS, Cury JC, Sul WJ, Pellizari VH, Tiedje J, Rosado AS (2010) Bacterial diversity in rhizosphere soil from Antarctic vascular plants of Admiralty Bay, maritime Antarctica. ISME J 4:989–1001

Thiyagarajan T, Backiyavathy M, Savithri P (2003) Nutrient management for pulses: a review. Agric Rev 24:40–48

Turan M, Ekinci M, Yildirim E et al (2014) Plant growth-promoting rhizobacteria improved growth, nutrient, and hormone content of cabbage (*Brassica oleracea*) seedlings. Turk J Agric For 38:327–333. https://doi.org/10.3906/tar-1308-62

Turan MG, Sahin MF (2012) Effects of plant-growth-promoting rhizobacteria on yield, growth, and some physiological characteristics of wheat and barley plants. Commun Soil Sci Plant Anal 43:1658–1673. https://doi.org/10.1080/00103624.2012.681739

Venkatarao CV, Naga SR, Yadav BL et al (2017) Effect of phosphorus and biofertilizers on growth and yield of mungbean [*Vigna radiata* (L.) Wilczek]. Int J Curr Microbiol Appl Sci 6:3992–3997. https://doi.org/10.20546/ijcmas.2017.607.413

Verhagen BW, Glazebrook J, Zhu T, Chang HS, van Loon LC et al (2004) The transcriptome of rhizobacteria-induced systemic resistance in Arabidopsis. Mol Plant-Microbe Interact 17:895–908

Vurukonda SSKP, Giovanardi D, Stefani E (2018) Plant growth promoting and biocontrol activity of Streptomyces spp. as endophytes. Int J Mol Sci 19:952. https://doi.org/10.3390/ijms19040952

Wheatley RE (2002) The consequences of volatile organic compound mediated bacterial and fungal interactions. Antonie Van Leeuwenhoek 81:357–364

Wu SC, Cao ZH, Li ZG, Cheung KC, Wong MH (2005) Effects of biofertilizers containing N-fixers, P and K solubilizers and AM fungi on maize growth: a greenhouse trial. Geoderma 125:155–166

Yang Y, Wang N, Guo X, Zhang Y, Ye B (2017) Comparative analysis of bacterial community structure in the rhizosphere of maize by high-throughput pyrosequencing. PLoS One 12(5): e0178425

Yasmin F, Radziah O, Sijam K, Saad MS (2007) Effect of PGPR inoculation on growth and yield of sweetpotato. J Biol Sci 7(2):421–424

Yousefi NMH (2018) Capability of plant growth-promoting rhizobacteria (PGPR) for producing indole acetic acid (IAA) under extreme conditions. European J Biol Res 8:174–182

Yousefi AA, Barzegar AR (2014) Effect of *Azotobacter* and *Pseudomonas* bacteria inoculation on wheat yield under field condition. Int J Agric Crop Sci 7(9):616

Yuan M, He H, Xiao L, Zhong T, Liu H, Li S, Deng P, Ye Z, Jing Y (2014) Enhancement of Cd phytoextraction by two *Amaranthus* species with endophytic Rahnella sp. JN27. Chemosphere 103:99–104

Zabee MA, Maliki SA (2019) Interactions between biofertilizers and chemical fertilizers affected soil biological properties and potato yield. Euphrates J Agric Sci 11

Zahir ZA, Munir A, Asghar HN, Shaharoona B, Arshad M (2008) Multiple impacts of the plant growth-promoting rhizobacterium *Variovorax paradoxus* 5C-2 on nutrient and ABA relations of *Pisum sativum* drought conditions. J Microbiol Biotechnol 18:982–987

Zaidi A, Khan MS, Ahemad M, Oves M, Wani PA (2009) Recent advances in plant growth promotion by phosphate-solubilizing microbes. In: Khan M, Zaidi A, Musarrat J (eds) Microbial strategies for crop improvement. Springer, Berlin, pp 23–50

Zaidi A, Ahmad E, Khan MS et al (2015) Role of plant growth promoting rhizobacteria in sustainable production of vegetables: current perspective. Sci Hortic (Amsterdam) 193:231–239. https://doi.org/10.1016/j.scienta.2015.07.020

Zhang Q, Jacquelinne JA, Nitza GI, María LM, Sergio R, Michael JS, Milko AJ (2019) Endophytic bacterial communities associated with roots and leaves of plants growing in chilean extreme environments. Sci Rep 9:4950–4962

Zhao Y, Zhang M, Yang W, Di HJ, Ma L, Liu W, Li B (2019) Effects of microbial inoculants on phosphorus and potassium availability, bacterial community composition, and chili pepper growth in a calcareous soil: a greenhouse study. J Soils Sediments 19(10):3597–3607
Zinniel DK, Lambrecht P, Harris NB, Feng Z, Kuczmarski D, Higley P, Ishimaru CA, Arunakumari A, Barletta RG, Vidaver AK (2002) Isolation and characterization of endophytic colonizing bacteria from agronomic crops and prairie plants. Appl Environ Microbiol 68:2198–2208

# Chapter 19
# Role of Rhizomicrobiome in Maintaining Soil Fertility and Crop Production

**Maddur Puttaswamy Raghavendra and Aralakuppe Narayana Santhoshkannada**

**Abstract** Microorganisms are involved in several major biological events associated with almost all the living organisms. Agriculture, in particular, which involves and revolves around the cultivated crops with special reference to its growth and yield demands association of microorganisms. As the world of microbiome associated with several animals especially human beings started highlighting the need of association of microorganisms for health condition, the interest is also started pointing towards the role of microbiome/rhizomicrobiome which is associated with plant roots and thereby promotes plant growth. Rhizosphere is a hot spot of microbial diversity; its distribution is guided by root exudate pattern and the soil type. There is a complex interaction between soil, plant and microbes which decides the growth of the plants. Especially rhizobiome is considered as an alternate genome of the plants, which guides several physiological and ecological activities of the plants. In recent times, rhizosphere engineering, application of synthetic microbiome, agent-based modelling of the rhizobiome, metagenomic studies of the different agricultural soil, understanding of the complex interaction between microorganisms in biome, role of plants in developing these biomes, role of individual or group of microbes associated with rhizobiome in plant growth promotion and finally documenting the different rhizobiome of the agricultural crops are gaining momentum. These new areas of research are providing more insights into understanding of microorganisms associated with plants in general and its roots in particular. The bioformulations of these microorganisms which support plant growth would be the good solution for future green revolution strategies.

**Keywords** Rhizobiome · Growth promotion · Sustainable agriculture

---

M. P. Raghavendra (✉)
Maharani's Science College for Women, Mysuru, Karnataka, India

A. N. Santhoshkannada
Maharaja's Government Pre-University College, Mysuru, Karnataka, India



B. Giri, A. Varma (eds.), *Soil Health*, Soil Biology 59,
https://doi.org/10.1007/978-3-030-44364-1_19

## 19.1 Introduction

To achieve self-sufficiency in the food production for rapidly growing population, there has been extensive use of chemical fertilizers and pesticides that in turn resulted in increased environmental and health hazards. Microorganisms associated with plants internally and externally (plant microbiome and rhizobiome) have also been disturbed, which resulted in the loss of beneficial interaction between two components (plant–microbe) of the ecosystem associated with primary productivity. On the basis of research experimentations, it is tempted to state that crop yield can be increased with the use of indigenous microbiome without environmental and health hazardous, especially rhizosphere microflora, which is known to increase the efficiency of water absorption and mineral nutrition from the soil for plant usage. The rhizosphere microbes are known as hot spot for plant root interactions and also considered as second genome of the plants (Berendsen et al. 2012).

The microbiome is a total aggregate of microbes in a community (Orozco-Mosqueda et al. 2018). Those, the total number of microbes present in plant rhizosphere can be termed as rhizomicrobiome and other parts of the plant can be termed as plant microbiome. One may define rhizosphere in their own ways, like it is a root surface plus attached soil, total roots plus soil or soil cores containing some roots (Badri et al. 2009; Doornbos et al. 2011). Balandreau and Knowles (1978) definition was considered to define rhizosphere as a soil that is attached to as affected by the roots or the vicinity of the soil which is under the influence of the root exudates.

The root-associated complexes are divided into three important types, such as the endosphere (root interior), rhizoplane (root surface), and rhizosphere (soil close to the root surface), each of which possesses distinct microbiome. The ectorhizosphere and rhizoplane comprises root surface where microbes resides. The endorhizosphere includes microbes present inside root tissues, including rhizodermis, cortex and stele, which is more specific term than endosphere (Berg et al. 2014), which also consists of aboveground part of the plants. The root associated microbes are the constituents of both rhizoplane and endorhizospheres.

The microbiome composition in controlled environment varies with soil source and genotype. Under field conditions, the factors such as geographical location, cultivation practice, namely organic versus conventional were different contributors for various microbiomes (Edwards et al. 2015). According to Berendsen et al. (2012) the totality of rhizosphere community includes the microbes and genetic elements present. The specific microbial community with specific functions associated with plants like rhizosphere, endosphere and phyllosphere (Berg et al. 2016) is even though fixed for a given period of growth it is dynamic in response to root exudation pattern and circadian rhythm. In the soil root microbiome assembly, the rhizoplane plays an important role in multistep microbiomes complex model.

## 19.2 Microbiome of the Plants

Plant microbiome determines the health, growth, fitness and consequent plant productivity (Lakshmanan et al. 2014). The microbiome induces different functional expression in plants such as primary productivity and soil health and fertility. Plant microbiome studies help in understanding the plant–microbe interactions as well as the genetic and functional capacity of the host and vital aspects of metabolism and physiology can be understood by studying the microbiome structure and functions of specific plants (Rout and Southworth 2013). Even though a root association with microbiome is currently a forefront research field through which understanding and unravelling of importance of microbiome has begun (Geldner and Salt 2014).

The plant–microbiome interaction requires connecting link that mediates between them, such as root exudates and soil parameters. They help plants in repelling or attracting rhizobiome, which initiates significant effect on plants (Bais et al. 2006). All types of root exudates are not involved in plant nutrition and growth. Some exudates act as signal molecules mediating interactions in the rhizobiome. The different rhizome exudates possess different compounds such as monosaccharides—fructose, mannose, glucose; five carbon sugars—arabinose; disaccharides—maltose, oligocaccharides; amino acids—aspartate, asparagines, glutamine, arginine and cysteine; organic acids—ascorbic, acetic, benzoic, ferulic, oxalic and malic acids; phenolic compounds—coumarin and high molecular weight compounds such as flavonoids, enzymes, fatty acids, auxin, gibberellins, nucleotides, tannins, steroids, terpenoids, alkaloids, polyacetylenes, and vitamins (Gunina and Kuzyakov 2015; Hayat et al. 2017). Along with these, the vital carbon cost compounds continuously secreted from root exudates were carbon fixed by photosynthesis (like phytoanticipins and phytoalexins) (Bamji and Corbitt 2017).

On the other hand, microbes produce secondary metabolites that can alter plant metabolism and signalling, which help in receiving nutrients in addition with antibiotics (Brazelton et al. 2008; Kim et al. 2011). Even though host-symbiotic organisms were studied widely, factors that contributes to host–microbiome interactions, and evolution and ecology of the symbioses were poorly understood. The host-associated microbial communities play basic roles in plants or animals nutrition. The fundamental key for the evolution of land plants and underlining fundamental ecosystem processes are root-associated microorganisms (Fitzpatricka et al. 2018).

## 19.3 Rhizobiome of the Agricultural Fields

Even though the role and importance of rhizosphere microbiome for plant growth is known, there are numerous microbes that exist in agricultural fields, but remaining unknown. The study of these microbes will enhance the knowledge on plant growth and health (Mendes et al. 2013). For example, methanogens are commonly present in every spatial compartment of rice fields, which make the rice fields as one of the major sources of global methane emissions. The detailed study of this may improve the potential networks involved in cycling of methane and also provides information on other phyla associated with root microbiomes of rice and other plants (Edwards et al. 2015). They observed that the rice field microbial diversity has significant influences on the field site. The α-diversity measurement of the field rhizospheres indicates that cultivation site significantly impacts microbial diversity.

It is obvious that as the plant traits varies the root microbial diversity also varies, i.e. traits show variation even between rhizosphere and endosphere compartments. Endosphere diversity was positively associated with increasing root hair density, while rhizosphere diversity was positively associated with host productivity and negatively associated with root length (Fitzpatricka et al. 2018).

To understand the microbial diversity, rice (Edwards et al. 2015), wheat (Ofek-Lalzar et al. 2014), maize (Peiffer et al. 2013), soybean (Mendes et al. 2014) and *Arabidopsis* (Bulgarelli et al. 2012) rhizobiomes are already characterized and documented. These studies revealed the presence of microorganisms in rhizobiome which are subset of the bulk soil microbiome. Microorganisms derived from the bulk soil are filtered by the plants to create the rhizobiome with specific structure and function. The composition of the microbes is also influenced majorly by the type of the soil and environmental conditions along with plant genotype. Among the several factors influencing, it is the plant genotype and root exudate pattern which majorly decide the rhizosphere structure and composition (Haney et al. 2015). These rhizobiome microflora are coordinating with the plants in natural ecosystems from time immemorial, hence they might have coevolved understanding the needs of each other for ecosystem fitness (Philippot et al. 2013).

The enrichment of respective microbes in rhizosphere can be done by altering the composition of root exudates and inducing release of more amounts of favourable exudates by means of reprogramming microbes of plants (Prikyrl et al. 1985; Bulgarelli et al. 2013). Thus, for the establishment of microbial communities in the rhizosphere and for competitive niche exclusion antimicrobials and secondary metabolites are vital factors. Those for achieving coordinated communication between microbes, plants, and their environmental signals, competitiveness for establishment and dominance of communities are highly essential.

The independent studies of rice rhizosphere indicated that bacterial community is largely occupied by Proteobacteria (mainly Alpha, Beta and Delta proteobacteria classes), *Acidobacteria, Actinobacteria* and *Chloroflexi* phyla. The bacterial community composition in the rice rhizosphere is distinct from that in the rhizosphere of other crops such as maize, soybean, potato, *Populus* and *Arabidopsis* based on the selected studies.

The rice is usually cultivated in flooded paddy soils compared to other crops. The oxic zones were created around the rice roots (e.g. rhizosphere) that are surrounded by anoxic bulk soil, which is a result of secretion of oxygen through rice root aerenchymatous tissues (Yuan et al. 2016; Zhao et al. 2018). The oxic–anoxic interface created due to this will favour the growth of aerobic, anaerobic or facultative anaerobic microbes (Revsbech et al. 1999; Li and Wang 2013). The main biogenic sources of methane ($CH_4$) release was due to methanogens inhabiting the rice rhizosphere at its anoxic interface accounts to nearly 5–19% of global methane (Bao et al. 2016). The higher demand for food and increasing global climate change resulted in major challenges in rice agriculture to obtain high yield with limited cultivation area and with alleviated methane emissions in paddy soils. Indeed, plant productivity is greatly influenced by microbial diversity (Conrad 2007), the crop health promotion and sustainable productivity of paddy ecosystems can be significantly increased by comprehensive understanding and exploitation of the overall microbial community (i.e. microbiome) associated with rice roots. However, studies of the microbiomes residing in the rice root related compartments, i.e. endosphere, rhizoplane and rhizosphere were still lacking.

Indeed, many rhizosphere sampling strategies because of low resolution are likely not reproducible enough. The maize and *Brachypodium* studies indicates that different root types harbour different microbial communities (Kawasaki et al. 2016; Yu et al. 2018), and this suggests that microbiome varies with different tissues of the same plant. In single-root types, longitudinal separation of developmentally diverse root zones such as root tip, elongation zone, differential zone, and maturation zone will be beneficial to comprehensively detect the development-dependent microbial gradients of host–microbe interactions. In addition, rhizosphere and endosphere microflora isolation by laser capture microdissection in combination with generation sequencing would be important tool to systemically target the direct interaction between plants and microbes at the cellular level.

## 19.4 Beneficial Aspects of the Rhizosphere Microflora

As it is observed with human gut microflora, plants and their microbiomes are interdependent, making microbes as superorganisms (Berendsen et al. 2012) and hence plants depend on rhizosphere microbes for specific functions and characters. The rhizomicrobiome of plants were significant in:

1. Plant nutrition, like nitrogen fixation or phosphate solubilization
2. Biological control, by competing with pathogens or by inducing resistance in plants
3. Plant growth and stress release by phytohormone production or degradation (Martinez-Viveros et al. 2010; Berendsen et al. 2012; Gaiero et al. 2013)

In brief, microorganisms associated with plants as microbiome and rhizomicrobiome will impact several parameters of the plants either positively or negatively (Fig. 19.1).

It is observed that several microorganisms associated with plants are not fully understood whether it is culturable or nonculturable. Keeping its importance in plant growth promotion or diseases, it is imperative to evaluate the microflora in detail to document all the microbes present in different agricultural fields in different growth conditions and under diurnal alterations. These studies will provide more insights

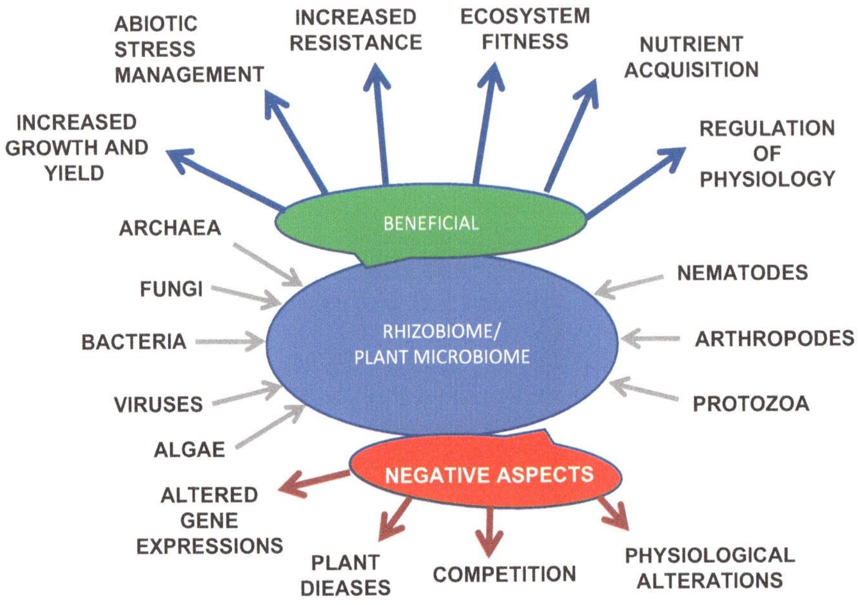


**Fig. 19.1** Role of microbiome/rhizobiome on plants

into the structure and functions of the rhizomicrobiomes and through which a detailed interaction within or between plant and microbes can be easily recorded. Even though little information are available with model plant systems such as *Arabidopsis* (Lundberg et al. 2012; Bulgarelli et al. 2012; Schlaeppi et al. 2014; Bai et al. 2015), it is important to understand how rhizobiomes are assembled into a specific structure with specific composition in other cultivated crops. It helps in understanding the principle underlying the beneficial interaction between the micro-organisms and plants in enhancing the plant growth.

## 19.5 Signalling Between Species

Rhizosphere microbes mediate root–soil, microbe–microbe and soil–microbe interactions (Berendsen et al. 2012). It is reported that interactions between and within species of bacteria are through complex chemical lexicon. Quorum sensing and the array of chemicals serving as a signalling in the process is thrust area of research for plant biologists. The mechanisms of enhancing and inhibiting quorum sensing were recognized in both prokaryotes and eukaryotes, which indicates that quorum sensing is subjected to regulation both within bacterial interactions as well as bacteria–eukaryotic associations (Henke and Bassler 2004).

Even though much attention was paid to molecular aspects of cell–cell communication, the ecological reasons for why bacteria produce signalling molecules that results in intraspecific and interspecific signals demands a thorough research (Crespi 2001), but in recent times volumes of information is available on how this interaction takes place and its impact on disease or growth progression in plants.

Especially, the study on quorum quenching between prokaryotic bacteria and eukaryotes await attention, it shows that different chemical communications are involved in interfering this mechanism. It is helping the biotechnology industry too especially in developing significant strategies for controlling bacterial infections, e.g. the quorum quenching study of eukaryotic sea weed *Delisea pulchra* with prokaryote halogenated furanones revealed inhibition of *Serratia liquefaciens* and also reported to inhibit quorum sensing dependent biofilm formation in *Pseudomonas aeruginosa* (Ren et al. 2002). Even the *Bacillus* aiiA gene encoding for the AHL lactonase is associated with quorum sensing which is reported to mediate resistance in transgenic potato and tobacco plants to *Erwinia carotovora*, thus resulting in control of the virulence factors associated with plant pathogens (Dong et al. 2001). Thus it is observed that AHL-dependent quorum sensing behaviour such as virulence and biofilm formation is a heterologous expression of aiiD gene in *Ps. aeruginosa* (Lin et al. 2003).

The autoinducer signals for the different interspecies interactions has clear signalling effect involving LuxS and AI-2, which is associated with control of gene regulation in different bacteria (Surette et al. 1999; Xavier and Bassler 2003). The biofilm formation in dental plaque communities like *Streptococcus gordonii* and *Porphyromonas gingivalis* requires AI-2. If both species do not have LuxS, the biofilm is not produced on glass coat between *P. gingivalis* and *S. gordonii*. The introduction of LuxS into either species again initiates the mixed biofilm formation (McNab et al. 2003).

This interesting finding shows that *P. aeruginosa* does not contain the LuxS gene and therefore not able to produce AI-2. But AI-2 is produced from the nonpathogenic microflora of host in sputum of cystic fibrosis. That results in composition of complex microbial community of pathogens *Ps. aeruginosa* and non-virulent bacteria in cystic fibrosis lungs. As a result sputum collected from patients of cystic fibrosis showed high level of AI-2, which in turn induces the production of virulence factors in *Ps. aeruginaosa* like elastase, exoenzyme T, rhamnolipid and phenazin (Duan et al. 2003).

AI-2 detection is not exclusive to bacterial "eavesdropping" because AHL signals will be intercepted by *Salmonella enterica*. Thus the Gram-negative bacteria produce LuxI enzyme which induces response to AHLs, resulting in expression of rck operon and other genes that protect *S. enterica* from defences of host in intestine (Ahmer 2004).

These findings are also observed with plant pathogens. The AHL based signalling is used to control bacteria to bacteria exchange of plasmid containing gene required for virulence of plant pathogen *Agrobacterium tumefaciens* (Piper et al. 1993). The expression of plasmid-encoded genes occurs during infection, where bacteria deliver the plasmid to plant resulting in formation of tumours. The tumours producing molecule opine is required for nourishment, increase of gene exchange between bacteria and infective capacity in population (Fuqua and Winans 1996).

Cis-11-methyl-2-dodeceenoic acid a novel a, b, unsaturated fatty acid produced from *Xanthomonas campestris* is used as a signal molecule in regulation of polysaccharide and extra cellular enzymes production, which are main virulence factors. The molecules are structurally related to farnesoic acid (produced by *Candida albicans* that inhibits filaments growth) and are diffusible signalling factor (DSF) (Oh et al. 2001). The diffusible signal factor of *Xanthomonas campestris* inhibits *Candida albicans* growth, increase farnesoic acid concentration and also important component of pathogenicity (Wang et al. 2004). The biological significance and molecular mechanism of this cross domain signalling still requires further study.

The proteomic study to understand the influence of these diffusible signals on gene regulation shows that over 150 protein expression in legume *Medicago truncatula* is regulated by AHL signals produced by *Sinorhizobium meliloti* and *Ps. aeruginosa*. The AI-2 production was regulated by AHLs by affecting the secretions of plant compounds and also by stimulating AHL quorum sensing reporter strains. These will encourage AHL signalling between the bacteria which produce AHL than AI-2 producers. Hence, to understand the effect of these plant metabolites on the bacteria, with which *M. truncatula* naturally associated requires further understanding (Mathesius et al. 2003).

The cooperative signal does not simply involve quorum sensing chemicals for interactions, but entails requirement of other cues and chemical manipulations. These signals may be useful or involved in conflicts within or between species. Thus, it is important to study why and how bacteria react with chemical substances produced by other bacteria (Keller and Surette 2006).

## 19.6 Factors Associated with Development of Rhizobiome

Plant species and its genotype, soil and environmental conditions play a major role in shaping the plant microbiome/rhizomicrobiome; these variables were hard to disentangle from each other in natural systems. For example, *Populus* in its different soil condition shows different root associated bacteria as well as endophytic and ectomycorhizal fungi. The particular *Populus* genotype is not associated with particular single taxon or consortium of microbes. It is observed that the fungi and bacteria are represented from different orders such as Rhizobiales, Cytophagales, Chitinophagales and Burkholderiales and fungi are represented by arbuscular mycorrhizal, endophytic and ectomycorrhizal fungi (Bonito et al. 2019).

The experiment on *Populus* rhizobiome decouples plant genotype under the influence of soil property impact structure of fungal and bacterial communities in soil. The 16S and ITS datasets sequences were assessed for co-association patterns of fungal and bacterial taxa (Bonito et al. 2019). They showed that the impact was at the community level rather than at the species level. It was supported by their further observation that 20 fungi are commonly abundant in all the test samples irrespective of the plant genotypes. Instead it was observed that the geographic location, soil texture and phosphorous level show positive correlation with networks of fungal and bacterial communities. This indicates that microbial taxa across landscape were not evenly distributed and the composition of plant microbiome is largely influenced by edaphic and local environmental factors.

It is also evident from the other works also, which reports the difference between endosphere and rhizosphere communities of *Populus deltoides* due to distinct nature of seasonal, geographical, and habitat differences. Similar results were also reported from *Arabidopsis*, agave and pine (Lundberg et al. 2012; Talbot et al. 2014; Fonseca-García et al. 2016).

On the other hand, it was observed that the structure of bacteria and fungi within the plant rhizomicrobiome is influenced by host species and its genotype (Bonito et al. 2014). The aboveground tissues (leaves, stems) and belowground root tissues is reported to have varied influence on microbiome (Bálint et al. 2013; Cregger et al. 2018). But the influence of soil properties on plant genotype directly is yet to be proved and it seems it will have indirect influence through soil microflora.

## 19.7 Biofilm Formation for Better Communication

The relationship for biofilm formation especially between prokaryotes and eukaryotes was first discovered based on the quorum sensing communication between marine bacteria and green seaweed *Enteromorpha* (Joint et al. 2002). The *Enteromorpha* zoospores attach to marine bacterial biofilms (*Vibrio anguillarum*) and initiate its growth. However, if the conditions are not ideal for seaweed, they simply detach and continue its planktonic lifestyle. Based on the observations it has been hypothesized that autoinducers produced by bacteria are recognized by *Enteromorpha*.

The AHLs (3-oxo-$C_{10}$-HSL, $C_6$-HSL and 3-hydroxy-$C_6$-HSL) is used in regulating quorum sensing in a Gram-negative bacteria *Vibrio anguillarum*. The mutants lacking the synthases for production of quorum sensing signals were generated to understand the importance of these signals in biofilm formation. The AHL signals in seaweed play a crucial role, which in its absence fails the *Enteromorpha* zoospores to adhere and it may be due to unpredictable phenotype created by quorum sensing mutation in *Vibrio anguillarum*. After transferring genes necessary for synthesizing three AHLs to *E. coli*, authors observed binding of zoospores to *E. coli* biofilm. Finally, they also proved that if biofilm lack AHL synthases, the attachment can be restored by adding synthetic 3-oxo-$C_{10}$-HSL, 3-hydroxy-$C_6$-HSL and $C_6$HSL.

Biofilm formation is also observed in several plant-associated microorganisms whether it is beneficial or harmful. Table 19.1 provides brief information on the application of biofilm formation on the host or the bacteria in combination or individually.

**Table 19.1** Biofilm formation for beneficial/harmful micobial interaction with plants

| Biofilm forming species | Plant host | Active component | Uses | References |
|---|---|---|---|---|
| *Rhizobium leguminosarum* | *Trifoliuum, Pisum, Vicia, Phaseolus* | Lipopolysaccharides | Affects adhesive forces among bacteria. Alters the cell surface hydrophobicity required for infection and nodulation | Priefer (1989); Cava et al. (1989); García and Brom (1997); Rolfe et al. (1996). |
| *Xylella fastidiosa* | Citrus, alfalfa, coffee | Rhamnose O-antigen | Cell aggregation | Clifford et al. (2013). |
| *Azospirillum brasilense* | Grass | Outer membrane proteins (67-kDa Outer membrane lectin) | Promotes bacterial flocculation, autoaggregation. | Burdman et al. (1999). |
| *Ensifer meliloti* | Leguminous plants | Galactoglucan (exopolysaccharides II) | Acts as molecular glue, that initiates and maintains contact between cells | González et al. (1996). |
| *Rhizobium leguminosarium* | *Vicia sativa*, Pea | Cellulose microfibrils | Agglutinating activity | Ausmees et al. (1999); Laus et al. (2005); Williams et al. (2008). |
| *Azospirillum brasilense* | | Arabinose-rich extracellular polysaccharides | Aggregative activity | Bogino et al. (2013) |
| *Bradyrhizobium japonicum* | Glycine max | O-antigen Exopolysaccharide (pentasaccharide units) | Enhance adhesion to plastic support. Biofilm formation on both inert and biotic surfaces. As role in initial attachment of rhizobia to root epidermal cells | Lee et al. (2010); Minamisawa (1989); Poveda et al. (1997); Pérez-Giménez et al. (2009). |
| *Ensifer meliloti* | Medicago sativa | Succinoglycan Galactoglucan | Required for biofilm formation and effective symbiosis | Reuber and Walker (1993); Cheng and Walker (1998); Her et al. (1990). |

(continued)

**Table 19.1** (continued)

| Biofilm forming species | Plant host | Active component | Uses | References |
|---|---|---|---|---|
| *Agrobacterium tumefaciens* | Ubiquitous plant pathogen | Succinoglycan | Reduces the attachment and biofilm formation. | Cangelosi et al. (1987); Tomlinson et al. (2010). |
| *Xylella fastidiosa* | Plant pathogen | Putative Fastidian gum | Bacterial pathogenicity. Cell attachment and overall biofilm formation | Roper et al. (2007). |
| *Pantoea stewarii* | Plant pathogen | Stewartan | Essential for adhesion and maturation of biofilm structure. Colonization and efficient dissemination through xylem. | Nimtz et al. (1996); Koutsoudis et al. (2006). |
| *Erwinia amylovora* | Plant pathogen | Amylovoran, Levan | Biofilm formation. Virulence factor helps in biofilm formation. | Nimtz et al. (1996); Koczan et al. (2009). |
| *Pseudomonas brassicacearum* J12 | Tomato | 2,4-diacetylphloroglucinol (2,4-DAPG) hydrogencyanide (HCN) | Antimicrobial | Zhou et al. (2012). |
| *Bacillus amyloliquefaciens* SQR9 | Cucumber | Citric acid | Biofilm formation | Zhang et al. (2014). |
| *Bacillus subtilis* N11 | Banana | Fumaric acid | Biofilm formation | Zhang et al. (2014). |
| *Bacillus subtilis* FB 17 | *Arabidopsis thaliana* | L-malic acid | Colonization and biofilm formation | Rudrappa et al. (2008). |
| *Paenibacillus polymyxa* SQR 21 | Watermelon | Malic acid and citric acid | Colonization and biofilm formation | Ling et al. (2011). |
| *X. campestris* *X. axonopodis* | Plant associated pathogen | Xanthan gum | Microcolony formation. Structured biofilm formation for abiotic surfaces. | Jansson et al. (1975); Rigano et al. (2007a, b); Torres et al. (2007). |

(continued)

**Table 19.1** (continued)

| Biofilm forming species | Plant host | Active component | Uses | References |
|---|---|---|---|---|
| *Ralstonia solanacearum* | Solanum plant | Acidic exopolysaccharides (composed by *N*-acetylgalactosamine and amino sugars (bacillosamine, galactosaminuronic acid) | Major virulence factor | Cook and Sequeira (1991); Chapman and Kao (1998) |
| *Mesorhizobium tianshanense* | *Glycyrrhiza uralensis* | Exopolysaccharide | Biofilm formation and involve in successful establishment of symbiosis | Wang et al. (2008) |

## 19.8 Unravelling Secrets of Interaction: Transcriptomic Analysis of Rhizosphere Microflora

The rhizosphere is a "hot spot" of microbial colonization and hence Yi et al. (2017) suggested that transcriptomic analysis of the soil or endophytic microorganisms with special reference to root exudation pattern will provide actual gene regulation pattern in these organisms associated with plant growth promotion.

Presently, there are several reports available on the transcriptomic studies of rhizosphere and endosphere of land plants. As revealed by Yu et al. (2018) plant–microbe interactions have a significant correlation with genetic factors of rhizosphere species and genotype of host plants. Even though advancement in metagenomic analysis shows only 5% of bacteria were cultured, yet other bacteria and their functions were not known (Mendes et al. 2013). On the other hand, the bacterial quorum sensing signal demonstrates induction of gene expression in legume like *Medicago truncatula* (Mathesius et al. 2003).

The metatranscriptomic analysis with evaluation of rRNA transcripts reveals the active community. Like for, oat, wheat and pea, the rhizosphere and bulk soil compartments show active prokaryotic diversity with each plant species. The slightly less diverse at higher taxonomic, i.e. the genus level was found in oat rhizosphere soil. The number of microbial taxa was considered approximately belongs to 300–400 genera per gram of soil showing high richness. But with regard to composition of community at genus level, all rhizosphere soils were different from the bulk soil (Turner et al. 2013). This indicates that in the rhizosphere soil specific bacteria were active and plants too plays a significant role on the composition of the microbiome in soil rhizosphere.

The functional characterization of communities can also be done by parallel metagenome analysis (Ofek-Lalzar et al. 2014). The individual functional groups represent by their relative abundance of gene fragments, where the functional profiles of root and bulk soils were different significantly. Interestingly, there was enrichment in genes linked to motility, chemotaxis, lipopolysaccharide biosynthesis and plant cell wall polysaccharide degradation enzymes in root associated communities. This suggests a strong functional and metabolic adaptation of microbes to the root compartments.

Even application of xenobiotics to agricultural field is known to influence the microbial interaction in different way and it is usually unnatural. It is observed in commercial agriculture that use of specific agrochemicals potentially affects population and interactions of rhizosphere microflora, which was probably due to alteration of gene expression in these organisms, e.g. changes in bacterial gene expression within the rhizosphere of corn tolerant to glyphosate and soybean in response to long-term glyphosate treatment was observed. In both, the rhizosphere bacterial metatranscriptomes were dominated by transcripts of RNA and carbohydrate metabolism. This finally determines long-term glyphosate use will affect rhizosphere bacterial activities and potentially helps in shift of bacterial community composition favouring bacteria more tolerant to glyphosate (Newman et al. 2016) leading to artificial evolution surrounding the plant root zones.

Once it is proven that the rhizobiome is subjected to alteration and it is not static, several recent advanced methods started engineering the rhizosphere microflora. The rhizosphere modification processes through changes in plant structure and physiology can be done using tools like clustered regulatory interspaced short palindromic repeats (CRISPR)-Cas. The studies of how plants control microbiome assembly and function in the rhizosphere by systematic alterations in plant architecture or physiology that helps in mechanistic studies is also systematically worked out (Nemudryi et al. 2014).

Similarly, the modification of root exudation processes by genetic engineering and gene-editing helps in disruption of mutualistic interactions between microorganisms which will aid in nutrient capture and plants that provide carbon sources for microbial growth (Ahkami et al. 2017). The specific families of compounds are produced in the rhizosphere by modification of root exudation processes. The study of plant selection for functionally distinct microbial taxa was also tried by altering host physiology through transcription factors (Su et al. 2015). The functional and taxonomic shifts in the rhizosphere could help further in understanding the selection of specific microbial groups.

For present day sustainable agricultural gains, the interactions in the rhizosphere of all partners (plant roots, soil and microbes) of the tripartite interactions can be manipulated or engineered to shift direction in favour of plants (Ryan et al. 2009; Zhang et al. 2015; Dessaux et al. 2016).

## 19.9 Cooperative Interactions Among Rhizosphere Microbes for Plant Growth

Today there is an explosion of research related to interactions between the cells, i.e. bacteria can communicate, cooperate and alter their behaviour, according to changes that occur in social environment. In nature, cooperation between the cells is widespread phenomenon, ranging from microbial populations to multicellular organisms (Celiker and Gore 2013).

Cooperation between genetically identical microbes is often natural (Riley and Gordon 1999; Griffin et al. 2004; West et al. 2006), whereas in genetically heterogenous social groups, organisms proficient at cooperation will frequently interact (Strassmann et al. 2000; Diggle et al. 2007; Vos and Velicer 2009; Liu et al. 2014). Therefore, the evolutionary reasons of divergence in social fitness between natural isolates that cooperate each other is often unclear (Vos and Velicer 2009; Buttery et al. 2009; Strassmann and Queller 2011; Ostrowski et al. 2015).

Earlier, the three types of interactions such as (i) Cooperation between PGPR and *Rhizobium* for effect $N_2$-fixation; (ii) Microbial antagonism for biocontrol of plant pathogen; and (iii) Interactions between rhizosphere microbes and AM fungi to establish a functional mycorrhizosphere were generally considered because of their relevance in sustainable agroecosystem development (Barea et al. 2005).

Further, this has led to understanding of group-to-group and community-level microbial interaction influencing soil nutrient cycling and other plant growth promotion activity by applying evolutionary biology to microbiome experiments. The components of complex microbiome study and its impact on plant host biology provides significant knowledge on unknown concepts of host–microbe biology of crop production, interaction and protection (Garcia and Kao-Kniffin 2018). The increasing interest in host–microbe biology, the development of mixtures of new and re-established terms in recent years, describes the complex associations of organisms that are heritable (Rosenberg and Zilber-Rosenberg 2016).

As mentioned earlier, soil microbial communities are largely influenced and affected by environment parameters, but the microbial interactions dominate in microscale environments such as rhizosphere and plant microbiome communities (Cordero and Datta 2016). It has been reported that rather microbes in bulk soil, the interaction of root-associated microbes are more complex, which may be due to diversity and density of microbial cells in rhizosphere and also their predominant positive interactions (>80%), it reveals the greater potential for mutualistic associations in rhizosphere (Shi et al. 2016). The indirect mediations and interactions of plant roots for plant beneficial and pathogenic microbes through root exudates signals, benefits plant themselves. The soil-borne pathogen attack of crops stimulates recruitment of antagonistic bacteria by roots, which is considered as a better way of defence strategy against infectious pathogens (Berendsen et al. 2012).

Plant rhizospheres encompass a dynamic zone of interactions between microorganisms and their respective plant hosts. During past few years' research helped in understanding complex interactions influencing different aspects of plant growth, development and evolution. Earlier, in root zone the plant–microbial interactions were focused on single microbial species or strains on a plant host effects. But presently, the studies on understanding complex microbiome effects on the biology of plants are unravelling. Further better understanding of how rhizosphere interactions influence plant growth development and its implementation by novel framework research methodologies need to be experimented (Garcia and Kao-Kniffin 2018).

For example, in a split root system increase of cucumber root secretions like citric acid and fumaric acid due to infection of wilt disease pathogen *Fusarium oxysporum* f.sp. *cucumerinum* helped to enhance its root colonization and to recruit plant beneficial *B. amyloliquefaciens* SQR9 (Liu et al. 2014). The barley roots infected with *Pythium ultimum* induced the expression of 2,4-diacetylphloroglucinol (DAPG) by biosynthesis gene phlA used for biocontrol of bacterium *Pseudomonas fluorescens* CHA0 in another side of roots. The induction of DAPG production of CHA0 in vitro at very low concentration was due to secretion of vanillic acid, fumaric acid and p-coumaric acid (Jousset et al. 2011).

Some *Pseudomonas* strains increase the number of nodules and reduces acetylene in soyabean plants when incubated with *B. japonicum*. It has been demonstrated that gus-A marked rhizobacteria used in bacterial colonization of roots (Chebotar et al. 2001). Manero et al. (2003) showed that enhancement of nodule formation takes place by metabolites other than phytohormones, such as siderophores, phytoalexins and flavonoids, but this hypothesis is subjected for verification (Lucas-Garcia et al. 2004).

The increase of nodulation and nitrogen fixation in alfalfa plants occurs by inoculation of phosphate-solubilizing bacteria (PBS), which helps in increasing the phosphate content of plant tissues (Toro et al. 1998). This shows that phosphate dependent processes were responsible for increase of nodulation and nitrogen fixation by nitrogen fixing bacteria (Barea et al. 2005).

In another study, the increasing nodulation of clover plants growing in soil rich in Cd, in presence of PGPR was observed (Vivas et al. 2005). PGPR present in Cd rich soil was reported to reduce the Cd concentrations and its uptake by plants and as well as *Rhizobia*, thus preventing Cd toxicity and increasing nodulation. Along with this, PGPR also increased production of β-glucosidase, phosphatase, dehydrogenase and was found to increase auxin production around PGPR inoculated roots.

Similar results are observed in the direct antagonism and induction of plant resistance involved in effective multiple biocontrol methods. For example rhizobacteria from genus *Pseudomonas* spp. produces several antimicrobial compounds which inhibit other bacteria and fungi (Hass and Keel 2003). Thomashow and Weller (1988) itself demonstrated clearly that bacteria produce antibiotic that suppress other plant pathogens in an ecosystem. The direct correlation between the phenazine antibiotic production by fluorescent *Pseudomonas* spp. and its effectiveness against all diseases of wheat makes it good biocontrol agent. Another key factor of this organism is its ability of compete and production of siderophores such as pyoverdine and pyochelin. It is this ability for siderophore-mediated competition for iron that helps in biocontrol of *Fusarium* and *Pythium* in soils (Duijff et al. 1994; Raaijmakers et al. 1995).

Compared to individual strains, in most of cases selected strains of bacteria used as a consortium in field trails are supporting the significant gain in yield from the plants. Considering this the present day research is focussed on multipartite interactions involving array of microorganisms viz., bacteria, fungi and other living organisms which holistically support the plant growth. For example the co-inoculation of *Aspergillus niger*, *Salmonella* serovars such as *typhimurium* resulted in increased plant growth which is significant compared to individual performances in the field. It was observed that the synergistic interaction between these two organisms supported one another to support the plant growth. In this interaction *A. niger* supported the growth of the bacteria, on the other hand the biofilm produced by the bacteria protected the fungus from antifungal agents such as cycloheximide. Similar result was also observed in the legume plant *Amorpha canescens* when inoculated with arbuscular mycorrhizal fungi in association with *Rhizobium* bacteria which supported the greater plant biomass production than the individual treatments (Larimer et al. 2014).

Another result reveals that the legume seedlings when associated with both AMF and nitrogen fixing rhizobia show 15-fold higher productivity than only with either of them. Hence Larimer et al. (2010) is of the opinion that these mutualistic studies in microbes provide more appropriate information than meta-analysis. The synergism serves as an indication that the cumulative impact of multipartite interactions overtakes effects of single microbial species. Is observed in case of inoculation of *Cicer arietinum* with AMF *Glomus intraradices* and PGPRs *Pseudomonas alcaligenes* and *Bacillus pumilus* significantly reduced the impact of *M. incognita* and *M. phaseolina* compared to the individual performances (Akhtar and Siddiqui 2008). Hence present day research is involved in increasing the effect of multipartite interactions on plant health through manipulation of microbial communities in the rhizosphere.

## 19.10 Natural Selection Favours Defectors in Simple Environments

The natural selection acting on whole groups of organisms in addition to individuals is called multilevel selection as proposed by Darwin (Wilson and Wilson 2008). The knowledge of plant and their associated microbiomes are very little than compared to their study on animals. The study on types of selective pressures that dictate microbial density-dependent rhizosphere processes such as nutrient cycling has a significant application in plant nutrition (Garcia and Kao-Kniffin 2018).

It is observed that cellular cooperation involves production of public goods (i.e. exopolysaccharides, extracellular enzymes, antibiotics, siderophores and quorum sensing molecules) which benefit other cells in the population (West et al. 2007). Cooperative action of cells are cost providing one, which range from a small metabolic burden to the death of individual cells, like self-destructive cooperation, where cooperative individuals die while formation of public goods (Ackermann et al. 2008). The cooperation evolves, where cooperating cells assort that was achieved by various mechanism like spatial heterogeneity of population structure, reciprocity, multilevel selection and other ecological factors (Nowak 2006).

The different prevalence of cooperation within a population depend on frequency and size of the disturbances such as ecological disturbances that temporarily alter the environment. The interaction of cooperating and cheating cells is affected by limiting resources and efficiency of competing species. The side effects on the cooperative population dynamics was seen when unrelated environmental conditions, for example hypermutability in response to stress which leads to evolution of final traits (Celiker and Gore 2013).

The primary functions of some antibiotic compounds released at low concentrations by some microbes are for communication rather than inhibition or exclusion of competitors (Aminov 2009). The antimicrobial compounds because of range of functions were key in establishing microbial communities in the rhizosphere. A wide array of Antibiotic Resistance Genes (ARG) are reported in soil microbiomes, which serves as a potential candidate for horizontal ARG transfer, that has important implications in agriculture as well as human health due to presence of several human pathogenic bacteria in soil (Cytryn 2013).

## 19.11 Multilevel Selection

The evolution of social traits was affected by metapopulation dynamics (Damore and Gore 2011, 2012). The selection can favour subpopulations with higher fitness when different groups of spatially separated subpopulations compete with one another. Thus the cooperative traits evolution theory suggests that selection can drive the evolution of cooperation (Nowak 2006).

It was also observed that the cooperator frequency decreases at the level of subpopulation. At the same time the frequency cooperators increase when observed at the level of global population, due to higher productivity of groups with higher cooperator frequency. A modelling study showed that demographic fluctuations lead to large variance in the composition of subpopulations, which further promotes the evolution of cooperation in this system (Cremer 2012). The population structure traits affect by concurrent evolution of cooperators (Powers et al. 2011).

It proves that the evolution process now a day is driven by several unnatural driving forces. Evolving microbes due to these multilevel selection and cooperation with other microbes may lead to development of microbiomes which may or may not support the plant growth. It is alarming issue that need to be addressed through thorough scientific research on factors associated with this problem.

## 19.12 Intraspecific Interactions Between Microorganisms

The selective force that promotes cooperation during intraspecific interaction in species that lack complex cognitive abilities is kin selection. This is a general principle of altruistic acts that directs towards relative production of important kind of reproductive compensation. The intraspecific communication has relatedness between bacteria, low cost of signal production (e.g. AI-2-dependent signalling) and high benefits for bacteria to behave in coordinated manner, e.g. coordinate regulation of virulence genes or in fruiting body formation in *Myxococcus*.

The signal production is costly or benefits of coordinated behaviour are limited, when the communication is less common when multiple strains of the same species of bacteria are mixed. Experimental study of pathogenic bacterium *Ps. aeruginosa* showed higher levels of cooperation as a result of higher relatedness. Similarly, experiments with strains of *Myxococcus xanthus* showed most pairings, when competed against one another in all possible pair-wise combinations showed at least one competitor as strong antagonism towards its partner. It shows that bacteria can perceive the presence of different strains, which change in overall relatedness that have profound effects on growth of population and its survival.

The extra cellular enzymes (that degrade macromolecules) from bacteria present in environment regulate their expression by use of quorum sensing. The enzymes and their products will diffuse away from cell to cell initiating catabolic production, when there are sufficient numbers of cells to scavenge all the products making it more beneficial. This more clearly indicates that rather than in single isolated cells, microcolony is more efficient and coordinate expression of the enzymes using intraspecies signalling systems would be optimal. In this respect it is important to note that quorum sensing pathways connected with virulence determinants are mostly extracellular (Keller and Surette 2006).

The sum of plant response to the environment and to the microbiome (including endophytes and pathogens) is called plant phenotype and on the other hand microbiome also responds to the environment and interacts with each other (Hardoim et al. 2015). The similarities between gut and plant rhizosphere microbiomes is that both are open systems, with a gradient of oxygen, water and pH resulting in a large number and diversity of microorganisms due to the different existing conditions. Likewise there is difference in microbiome compositions; therefore, the similarities related to nutrient acquisition, immune system modulation and protection against infection were there (Mendes and Raaijmakers 2015). The functional diversity is considered to be the key factor in microbiome diversity whereas it is lost in the present agricultural fields due to more human interventions and aggressive agricultural practices (Berg et al. 2015).

The wide ranges of bioactive natural products are produced by actinomycetes. The compounds produced in each interaction (*Streptomyces coelicolor* with actinomycetes) were unique with differential response in each case. It is reported that 227 compounds are differentially produced in interactions and half of these were known metabolites such as actinorhodins, prodigines, acyl-desferrioxamines and coelichelins (Traxler et al. 2013).

The production of specific fungal secondary metabolites due to fungal–bacterial interactions, which not only is a diffusible compound, but also act in communication along with contribution from other physical interactions. The intimate physical interaction between *Aspergillus nidulans* and actinomycetes *Streptomyces rapamycinius* leads to the activation of fungal secondary metabolite genes related to production of aromatic polyketides (Schroeckh et al. 2009). It is because of the alterations of fungal histone acetylation through Saga/Ada complex that triggered induction of PKS cluster of actinomycetes. It signifies that bacteria can trigger alterations of histone acetylation in fungi (Nutzmann et al. 2011).

The symbiotic relationships of remarkable complex of inter-kingdom interaction between bacterial genus *Burkholderia* and *Rhizopus*, genus of pathogen that causes rice seedling blight also stands as a better example. The phytotoxin "rhizoxin" compound is responsible for rice seedling blight produced by endosymbiotic bacteria *Burkholderia* species (Partida-Martinez and Hertweck 2005). It was observed in this case that exopolysaccharides (EPS) produced by *Burkholderia rhizoxinica* played a key role in the interaction. Further experiment on mutant showed that EPS production did not affect the endosymbiotic interactions with *Rhizopus microspores* (Uzum et al. 2015). But still the complex symbiont–pathogen–plant interaction and communication mechanisms were poorly understood. It is also observed that nacyloxins, an antibiotic compound, was produced by *Burkholderia gladioli* only in co-culture with *R. microspores* (Ross et al. 2014).

The analysis of multicellular cooperative behaviour of microbes in the light of evolutionary biology needs careful consideration of the unique aspects of the genetics and ecology of microbes. Ultimately to have great benefits in microbial and evolutionary fields enhancement of cooperative interdisciplinary research of both microbiologists and evolutionary biologists is highlighted (Dunny et al. 2008). The new approaches for managing plant microbiomes for health and sustainability are to identify and understand the functions of keystone microbial species in plant microbiomes (Bonito et al. 2019).

## 19.13 Division of Labour

In present day, the fundamental open issue in biology, relevant to plant and human health, metabolic engineering and environmental sustainability is based on understanding how microbes assemble into communities. The possible mechanism of microbial interactions is through cross-feeding, i.e. the exchange of small molecules. These exchanges may allow different microbes to specialize in distinct tasks and evolve division of labour (Thommes et al. 2019).

The supply of some essential metabolites relying on neighbouring organisms and microbes face a trade-off between being metabolically independent organisms. This balance of conflicting strategies is an important implication in microbiome research and synthetic ecology affecting microbial community structure and dynamics. At present, the emergence of obligate mutualism is coupled with sharp metabolic network differentiation and exchange of different chemicals (Thommes et al. 2019).

The obligate mutualism is specifically displayed by construction (or evolution) of artificial microbial consortia which shows interdependencies. In the interactions of synthetic communities, one strain is unable to synthesize essential metabolite (like amino acid) that is supplied via overproduction or leakage by another strain (Shou et al. 2007; Harcombe 2010; Wintermute and Silver 2010; Kerner et al. 2012; Mee et al. 2014; Bernstein et al. 2012; Henry et al. 2016; Hoek et al. 2016). This ensures that, in order to grow, the two strains require each other's presence (Pacheco et al. 2018). The involvement of complex metabolic mutualism beyond single amino acid exchanges was given due to complexity of metabolism, its evolutionary history and possibly due to naturally evolved cross-feeding strategies (Medlock et al. 2018). These information are useful in novel strategies such as bioproduction, design of bioformulations and engineering of consortia for specific metabolic pathways.

In microbial applications, the significant drop in cell performance was result of metabolic burdens and hence division of labour through the natural and synthetic microbial consortia will address this problem. Today success of bioformulations in field condition depends on understanding different aspects of synergy in plant microbiome and rhizomicrobiome. It is also important to understand the strains used as inoculums, its nutritional requirement and divergence, cross-feeding or co-metabolic activity and mutualistic evolutionary growth history to develop the stable consortia. Along with these, the metabolic behaviour of these strains with a single or multiple substrates need to be modelled through metabolic modelling and $^{13}$C-metabolic flux analysis which will throw light on the metabolism of the substrates in mixed culture and co-metabolic activity of the strains used in consortia. Computational approach to understand performance of consortia in group or individually will also provide a concrete information (Roell et al. 2019).

## 19.14 Concluding Remarks

To provide the insights into the structure of the particular microbial community, it is necessary for attempting to define the interaction, rather than with more generic descriptors. In broader terms, to understand the progress in quorum sensing and to develop bacteria as better model organisms in ecology and evolution, it is important to link branches between microbiology and field of ecology and evolution (Keller and Surette 2006). To unravel the rhizobiome species composition the modern molecular techniques are used, even though understanding interactions were in progress, the thorough investigation is yet to be made on interplay between the rhizobiome root exudates and other factors involved in maintenance of plant health. To improve the plant health, optimization of growth and productivity of host plants and microbial diversity of rhizobiome and modulatory techniques need to be clearly understood (Olanrewaju et al. 2019). Quorum sensing, under many environmental conditions, from soil and water (where quorum sensing genes influence nutrients cycling) to animal hosts (where quorum sensing regulated genes determine pathogen virulence) moderates bacterial metabolism. How we might modify bacterial behaviour for environmental gains can be realized by understanding the ecology of quorum sensing that yield vital clues (Mund et al. 2017).

Moreover, microbiome will be shaped by environmental factors such as geography and soil type. The health and fitness of the whole system is determined by relationships between the root-associated microbiome, architectural variations and functional switches within the root system. Yu et al. (2018) suggested that the detailed information on the important factors guiding plant–microbe interactions is the fundamental agricultural importance and supports significantly even plant breeding through crop improvement. Microbiome engineering associated with modulating microbial communities in different living organisms is an upcoming branch of research (Mendes et al. 2013). By altering the ecological processes such as modulation in community diversity, changing structure of microbial interaction networks and by altering the evolutionary processes the microbiome engineering can be achieved. It leads to extinction of few microbial species in microbiome, horizontal gene transfer and mutations that can restructure microbial genomes (Mueller and Sachs 2015). The complex interactions involving multiple players will replace focus by single microbial isolate effects. The recent popularity of examining synthetic communities comprised multiple microbial strains that help to advance microbiome science further, which would be beneficial to move beyond cultivation-dependent methods. The reduction of diversity of complex microbiomes associated with a plant trait could enable by applying selective filters more to top-down and bottom-up approaches that comprised cultivation-dependent and -independent multiplayer interaction studies. The complexity of the rhizosphere studies could ultimately help in developing a better understanding of how rhizosphere microbiomes influence plant growth, development and fitness (Garcia and Kao-Kniffin 2018).

The molecular approaches, particularly for specific plant genotypes and mutants, and major research coupling of reduction list will unravel how plants shape their known microbiome and clarify causal relationships in complex root communities. The nature of how plants recognize and filter the root communities should be known in detail using well-defined plant mutants instead of different cultivar or species (Reinhold-Hurek et al. 2015). Several approaches can also be used in order to define which microorganisms should be inoculated. The first approach is to know core microbiome associated with plant health or to understand the function of rhizobiomes by sequencing techniques, which can be followed by experiments in tissue culture technique or green house experiments. The microbiome with a specific plant growth promoting activity individually and in consortia can then be applied in field as bioinoculants for sustainable agriculture.

## References

Ackermann M, Stecher B, Freed NE, Songhet P, Hardt WD, Doebeli M (2008) Self-destructive cooperation mediated by phenotypic noise. Nature 454(7207):987

Ahkami AH, White RA, Handakumbura PP, Jansson C (2017) Rhizosphere engineering: enhancing sustainable plant ecosystem productivity. Rhizosphere 3:233–243

Ahmer BM (2004) Cell-to-cell signalling in *Escherichia coli* and *Salmonella enterica*. Mol Microbiol 52:933–945

Akhtar MS, Siddiqui ZA (2008) *Glomus intraradices*, *Pseudomonas alcaligenes* and *Bacillus pumilus*: effect agents for the control of root-rot disease complex of chickpea (*Cicer arietinum* L.). J Gen Plant Pathol 74(1):53–60

Aminov RI (2009) The role of antibiotics and antibiotic resistance in nature. Environ Microbiol 11 (12):2970–2988

Ausmees N, Jonsson H, Höglund S, Ljunggren H, Lindberg M (1999) Structural and putative regulatory genes involved in cellulose synthesis in *Rhizobium leguminosarum* bv. *trifolii*. Microbiology 145:1253–1262

Badri DV, Quintana N, El Kassis EG, Kim HK, Choi YH et al (2009) An ABC transporter mutation alters root exudation of phytochemicals that provoke an overhaul of natural soilmicrobiota. Plant Physiol 151:2006–2017

Bai Y, Muller DB, Srinivas G, Garrido-Oter R, Potthoff E, Rott M, Dombrowski N, Münch PC, Spaepen S, Remus-Emsermann M et al (2015) Functional overlap of the Arabidopsis leaf and root microbiota. Nature 528:364–369

Bais HP, Weir TL, Perry LG, Gilroy S, Vivanco JM (2006) The role of root exudates in rhizosphere interactions with plants and other organisms. Annu Rev Plant Biol 57:233–266

Balandreau J, Knowles R (1978) The rhizosphere. In: Dommergues YR, Krupa SV (eds) Interactions between non-pathogenic soil microorganisms and plants. Elsevier, Amsterdam, pp 243–268

Bálint M, Tiffin P, Hallström B, O'Hara RB, Olson MS, Fankhauser JD et al (2013) Host genotype shapes the foliar fungal microbiome of balsam poplar (*Populus balsamifera*). PLoS One 8:53987

Bamji SF, Corbitt C (2017) Glyceollins: soybean phytoalexins that exhibit a wide range of health-promoting effects. J Funct Foods 34:98–105

Bao Q, Huang Y, Wang F et al (2016) Effect of nitrogen fertilizer and/or rice straw amendment on methanogenic archaeal communities and methane production from a rice paddy soil. Appl Microbiol Biotechnol 100:5989–5998

Barea JM, Pozo MJ, Azcon R, Azcon-Aguilar C (2005) Microbial co-operation in the rhizosphere. J Exp Bot 56(417):1761–1778

Berendsen RL, Pieterse CM, Bakker PA (2012) The rhizosphere microbiome and plant health. Trends Plant Sci 17(8):478–486

Berg G, Grube M, Schloter M, Smalla K (2014) Unravelling the plant microbiome: looking back and future perspectives. Front Microbiol 5:148

Berg G, Krause R, Mendes R (2015) Cross-kingdom similarities in microbiome ecology and biocontrol of pathogens. Front Microbiol 6:1311

Berg G, Rybakova D, Grube M, Koberl M (2016) The plant microbiome explored: implications for experimental botany. J Exp Bot 67:995–1002

Bernstein HC, Paulson SD, Carlson RP (2012) Synthetic *Escherichia coli* consortia engineered for syntrophy demonstrate enhanced biomass productivity. J Biotechnol 157:159–166

Bogino PC, Oliva MDLM, Sorroche FG, Giordano W (2013) The role of bacterial biofilms and surface components in plant-bacterial associations. Int J Mol Sci 14(8):15838–15859

Bonito G, Reynolds H, Robeson MSII, Nelson J, Hodkinson BP, Tuskan G et al (2014) Plant host and soil origin influence fungal and bacterial assemblages in the roots of woody plants. Mol Ecol 23:3356–3370

Bonito G, Benucci GMN, Hameed K, Weighill D, Jones P, Chen K-H, Jacobson D, Schadt C, Vilgalys R (2019) Fungal-bacterial networks in the *Populus* Rhizobiome are impacted by soil properties and host genotype. Front Microbiol 10:481–502

Brazelton JN, Pfeufer E, Sweat TA, Gardener BB, Coenen C (2008) 2,4-Diacetylphloroglucinol alters plant root development. Mol Plant Microbe Interact 21(10):1349–1358

Bulgarelli D, Rott M, Schlaeppi K, Ver Loren van Themaat E, Ahmadinejad N et al (2012) Revealing structure and assembly cues for *Arabidopsis* root-inhabiting bacterial microbiota. Nature 488:91–95

Bulgarelli D, Schlaeppi Spaepen S, Ver L, van Themaat E, Schulze-Lefert P (2013) Structure and functions of the bacterial microbiota of plants. Annu Rev Plant Biol 64:807–838

Burdman S, Jurkevitch E, Schwartsburd B, Okon Y (1999) Involvement of outer membrane proteins in the aggregation of *Azospirillum brasilense*. Microbiology 145:1145–1152

Buttery NJ, Rozen DE, Wolf JB, Thompson CRL (2009) Quantification of social behavior in *D. discoideum* reveals complex fixed and facultative strategies. Curr Biol 19:1373–1377

Cangelosi GA, Hung L, Puvanesarajah V, Stacey G, Ozga DA, Leigh JA, Nester EW (1987) Common loci for *Agrobacterium tumefaciens* and *Rhizobium meliloti* exopolysaccharide synthesis and their roles in plant interactions. J Bacteriol 169:2086–2091

Cava JR, Elias PM, Turowski DA, Noel KD (1989) *Rhizobium leguminosarum* CFN42 genetic region encoding lipopolysaccharide structures essential for complete nodule development on bean plants. J Bacteriol 171:8–15

Celiker H, Gore J (2013) Cellular cooperation: insights from microbes. Trends Cell Biol 23:9–15

Chapman MR, Kao CC (1998) EpsR modulates production of extracellular polysaccharides in the bacterial wilt pathogen *Ralstonia* (*Pseudomonas*) *solanacearum*. J Bacteriol 180:27–34

Chebotar VK, Asis CA, Akao S (2001) Production of growth promoting substances and high colonization ability of rhizobacteria enhance the nitrogen fixation of soybean when inoculated with *Bradyrhizobium japonicum*. Biol Fertil Soils 34:427–432

Cheng HP, Walker GC (1998) Succinoglycan is required for initiation and elongation of infection threads during nodulation of alfalfa by *Rhizobium meliloti*. J Bacteriol 180:5183–5191

Clifford JC, Rapicavoli JN, Roper MC (2013) A rhamnose-rich O-antigen mediates adhesion, virulence and host colonization for the xylem-limited phytopathogen, *Xylella fastidiosa*. Mol Plant Microbe Interact 26:676–685

Conrad R (2007) Microbial ecology of methanogens and methanotrophs. Adv Agron 96:1–63

Cook D, Sequeira L (1991) Genetic and biochemical characterization of a gene cluster from *Pseudomonas solanacearum* required for extracellular polysaccharide production and for virulence. J Bacteriol 173:1654–1662

Cordero OX, Datta MS (2016) Microbial interactions and community assembly at microscales. Curr Opin Microbiol 31:227–234

Cregger MA, Veach AM, Yang ZK, Crouch MJ, Vilgalys R, Tuskan GA et al (2018) The *Populus* holobiont: dissecting the effects of plant niches and genotype on the microbiome. Microbiome 6:31

Cremer J (2012) Growth dynamics and the evolution of cooperation in microbial populations. Sci Rep 2:281

Crespi BJ (2001) The evolution of social behavior in microorganisms. Trends Ecol Evol 16:178–183

Cytryn E (2013) The soil resistome: the anthropogenic, the native, and the unknown. Soil Biol Biochem 63:18–23

Damore JA, Gore J (2011) A slowly evolving host moves first in symbiotic interactions. Evolution 65:2391–2398

Damore JA, Gore J (2012) Understanding microbial cooperation. J Theor Biol 299:31–41

Dessaux Y, Grandclement C, Faure D (2016) Engineering the rhizosphere. Trends Plant Sci 21:266–278

Diggle SP, Griffin AS, Campbell GS, West SA (2007) Cooperation and conflict in quorum-sensing bacterial populations. Nature 450:411–414

Dong YH, Wang YH, Xu J, Zhang HB, Zhang XF, Zhang LH (2001) Quenching quorum-sensing-dependent bacterial infection by an N-acyl homoserine lactonase. Nature 411:813–817

Doornbos RF, Geraats BP, Kuramae EE, Van Loon LC, Bakker PA (2011) Effects of jasmonic acid, ethylene, and salicylic acid signaling on the rhizosphere bacterial community of *Arabidopsis thaliana*. Mol Plant Microbe Interact 24:395–407

Duan K, Dammel C, Stein J, Rabin H, Surette MG (2003) Modulation of *Pseudomonas aeruginosa* gene expression by host microflora through interspecies communication. Mol Microbiol 50:1477–1491

Duijff BJ, Bakker PAHM, Schippers B (1994) Suppression of fusarium wilt of carnation by *Pseudomonas putida* WCS358 at different levels of disease incidence and iron availability. Biocontrol Sci Technol 4:279–288

Dunny GM, Brickman TJ, Dworkin M (2008) Multicellular behaviour in bacteria: communication, cooperation, competition and cheating. BioEssays 30:296–298

Edwards J, Johnson C, Santos-Medellín C, Lurie E, Podishetty NK, Bhatnagar S, Eisen JA, Sundaresan V (2015) Structure, variation, and assembly of the root-associated microbiomes of rice. Proc Natl Acad Sci USA 112:911–920

Fitzpatricka CR, Copeland J, Wang PW, Guttmanc DS, Kotanena PM, Johnson MTJ (2018) Assembly and ecological function of the root microbiome across angiosperm plant species. Proc Natl Acad Sci USA 22:1157–1165

Fonseca-García C, Coleman-Derr D, Garrido E, Visel A, Tringe SG, Partida-Martínez LP (2016) The cacti microbiome: interplay between habitat-filtering and host-specificity. Front Microbiol 7:150

Fuqua C, Winans SC (1996) Localization of OccR-activated and TraR-activated promoters that express two ABC-type permeases and the traR gene of Ti plasmid pTiR10. Mol Microbiol 20:1199–1210

Gaiero JR, McCall CA, Thompson KA, Day NJ, Best AS, Dunfield KE (2013) Inside the root microbiome: bacterial root endophytes and plant growth promotion. Am J Bot 100:1738–1750

García DLSA, Brom S (1997) Characterization of two plasmid-borne lps beta loci of *Rhizobium etli* required for lipopolysaccharide synthesis and for optimal interaction with plants. Mol Plant Microbe Interact 10:891–902

Garcia J, Kao-Kniffin J (2018) Microbial group dynamics in plant rhizospheres and their implications on nutrient cycling. Front Microbiol 9:1516

Geldner N, Salt DE (2014) Focus on roots. Plant Physiol 166:453–454

González JE, Reuhs BL, Walker GC (1996) Low molecular weight EPS II of *Rhizobium meliloti* allows nodule invasion in *Medicago sativa*. Proc Natl Acad Sci USA 93:8636–8641

Griffin AS, West SA, Buckling A (2004) Cooperation and competition in pathogenic bacteria. Nature 430:1024–1027

Gunina A, Kuzyakov Y (2015) Sugars in soil and sweets for microorganisms: review of origin, content, composition and fate. Soil Biol Biochem 90:87–100

Haney CH, Samuel BS, Bush J, Ausubel FM (2015) Associations with rhizosphere bacteria can confer an adaptive advantage to plants. Nat Plants 1:15051

Harcombe WR (2010) Novel cooperation experimentally evolved between species. Evolution 64:2166–2172

Hardoim PR, van Overbeek LS, Berg G et al (2015) The hidden world within plants: ecological and evolutionary considerations for defining functioning of microbial endophytes. Microbiol Mol Biol Rev 79:293–320

Hass D, Keel C (2003) Regulation of antibiotic production in root colonizing *Pseudomonas* spp. and relevance for biological control of plant disease. Annu Rev Phytopathol 41:117–153

Hayat S, Faraz A, Faizan M (2017) Root exudates: composition and impact on plant–microbe interaction. In: Ahmand I, Hussain FM (eds) Biofilms in plant and soil health. Wiley, Hoboken, NJ, pp 179–193

Henke JM, Bassler BL (2004) Bacterial social engagements. Trends Cell Biol 14(11):648–659

Henry CS, Bernstein HC, Weisenhorn P, Taylor RC, Lee JY, Zucker J, Song HS (2016) Microbial community metabolic modeling: a community data-driven network reconstruction. J Cell Physiol 231:2339–2345

Her GR, Glazebroock J, Walker GC, Reinhold VN (1990) Structural studies of a novel exopolysaccharide produced by a mutant of *Rhizobium meliloti* strain Rm1021. Carbohydr Res 198:305–312

Hoek TA, Axelrod K, Biancalani T, Yurtsev EA, Liu J, Gore J (2016) Resource availability modulates the cooperative and competitive nature of a microbial cross-feeding mutualism. PLoS Biol 15(6):e1002606

Jansson PE, Kenne L, Lindberg B (1975) Structure of the extracellular polysaccharide from *Xanthomonas campestris*. Carbohydr Res 45:274–282

Joint I, Tait K, Callow ME, Callow JA, Milton D, Williams P, Cámara M (2002) Cell-to-cell communication across the prokaryote-eukaryote boundary. Science 8:1207

Jousset A, Rochat L, Lanoue A, Bonkowski M, Keel C, Scheu S (2011) Plants respond to pathogen infection by enhancing the antifungal gene expression of root-associated bacteria. Mol Plant Microbe Interact 24:352–358

Kawasaki A, Donn S, Ryan PR, Mathesius U, Devilla R, Jones A et al (2016) Microbiome and exudates of the root and rhizosphere of *Brachypodium distachyon*, a model for wheat. PLoS One 11:0164533

Keller L, Surette MG (2006) Communication in bacteria: an ecological and evolutionary perspective. Nat Rev 4:249–258

Kerner A, Park J, Williams A, Lin XN (2012) A programmable *Escherichia coli* consortium via tunable symbiosis. PLoS One 7:34032

Kim YC, Leveau J, McSpadden Gardener BB, Pierson EA, Pierson LS III, Ryu C (2011) The multifactorial basis for plant health promotion by plant-associated bacteria. Appl Environ Microbiol 77:1548–1555

Koczan JM, McGrath MJ, Zhao Y, Sundin GW (2009) Contribution of *Erwinia amylovora* exopolysaccharides amylovoran and levan to biofilm formation: implications in pathogenicity. Phytopathology 99:1237–1244

Koutsoudis MD, Tsaltas D, Minogue TD, von Bodman SB (2006) Quorum-sensing regulation governs bacterial adhesion, biofilm development, and host colonization in *Pantoea stewartii* subspecies stewartii. Proc Natl Acad Sci USA 103:5983–5988

Lakshmanan V, Selvaraj G, Bais HP (2014) Functional soil microbiome: belowground solutions to an above ground problem. Plant Physiol 166:689–700

Larimer AL, Bever JD, Clay K (2010) The interactive effects of plant microbial symbionts: a review and meta-analysis. Symbiosis 51:139–148

Larimer AL, Clay K, Bever JD (2014) Synergism and context dependency of interactions between arbuscular mycorrhizal fungi and rhizobia with a prairie legume. Ecology 95(4):1045–1054

Laus MC, van Brussel AA, Kijne JW (2005) Role of cellulose fibrils and exopolysaccharides of *Rhizobium leguminosarum* in attachment to and infection of *Vicia sativa* root hairs. Mol Plant-Microbe Interact 18:533–538

Lee YW, Jeong SY, In YH, Kim KY, So JS, Chang WS (2010) Lack of O-polysaccharide enhances biofilm formation by *Bradyrhizobium japonicum*. Lett Appl Microbiol 50:452–456

Li Y, Wang X (2013) Root-induced changes in radial oxygen loss, rhizosphere oxygen profile, and nitrification of two rice cultivars in Chinese red soil regions. Plant Soil 365:115–126

Lin YH, Xu JL, Hu J, Wang LH, Ong SL, Leadbetter JR, Zhng LH (2003) Acyl-homoserine lactone acylase from *Ralstonia* strain XJ12B represents a novel and potent class of quorum quenching enzymes. Mol Microbiol 47:849–860

Ling N, Raza W, Ma J, Huang Q, Shen Q (2011) Identification and role of organic acids in watermelon root exudates for recruiting *Paenibacillus polymyxa* SQR-21 in the rhizosphere. Eur J Soil Biol 47:374–379

Liu Y, Fire AZ, Boyd S, Olshen RA (2014) Estimating clonality. Proc Natl Acad Sci USA 108:10823–10830

Lucas-Garcia JA, Probanza A, Ramos B, Colón-Flores JJ, Gutierrez-Manero FJ (2004) Effects of plant growth promoting rhizobateria (PGPRs) on the biological nitrogen fixation, nodulation and growth of *Lupinus albus* cv. *multolupa*. Eng Life Sci 4:71–77

Lundberg DS, Lebeis SL, Paredes SH, Yourstone S, Gehring J, Malfatti S, Tremblay J, Engelbrektson A, Kunin V, del Rio TG et al (2012) Defining the core *Arabidopsis thaliana* root microbiome. Nature 488:86–90

Manero FJ, Probanza A, Ramos B, Flores JJ, Garcia-Lucas JA (2003) Effects of culture filtrates of rhizobacteria isolated from wild lupin on germination, growth, and biological nitrogen fixation of lupin seedlings. J Plant Nutr 26:1101–1115

Martinez-Viveros O, Jorquera MA, Crowley DE, Gajardo G, Mora ML (2010) Mechanisms and practical considerations involved in plant growth promotion by rhizobacteria. J Soil Sci Plant Nutr 10:293–319
Mathesius U, Mulders S, Gao M, Teplitski M, Caetano-Anolles G, Rolfe BG, Bauer WD (2003) Extensive and specific responses of a eukaryote to bacterial quorum-sensing signals. Proc Natl Acad Sci USA 100:1444–1449
McNab R, Ford SK, El-Sabaenv A, Barbieri B, Cook GS, Lamont RJ (2003) LuxS-based signaling in Streptococcus gordonii: autoinducer 2 controls carbohydrate metabolism and biofilm formation with *Porphyromonas gingivalis*. J Bacteriol 185:274–284
Medlock GL, Carey MA, McDuffie DG, Mundy MB, Giallourou N, Swann JR, Kolling GL, Papin JA (2018) Inferring metabolic mechanisms of interaction within a defined gut microbiota. Cell Syst 7:245–257
Mee MT, Collins JJ, Church GM, Wang HH (2014) Syntrophic exchange in synthetic microbial communities. Proc Natl Acad Sci USA 111:2149–2156
Mendes R, Raaijmakers JM (2015) Cross-kingdom similarities in microbiome functions. ISME J 9:1905–1907
Mendes R, Garbeva P, Raaijmakers JM (2013) The rhizosphere microbiome: significance of plant beneficial, plant pathogenic, and human pathogenic microorganisms. FEMS Microbiol Rev 37:634–663
Mendes LW, Kuramae EE, Navarrete AA, van Veen JA, Tsai SM (2014) Taxonomical and functional microbial community selection in soybean rhizosphere. ISME J 8:1577–1587
Minamisawa K (1989) Comparison of extracellular polysaccharide composition, rhizobiotoxine production, and hydrogenase phenotype among various strains of *Bradyrhizobium japonicum*. Plant Cell Physiol 30:877–884
Mueller UG, Sachs JL (2015) Engineering microbiomes to improve plant and animal health. Trends Microbiol 23:606–617
Mund A, Diggle SP, Harrison F (2017) The fitness of *Pseudomonas aeruginosa* quorum sensing signal cheats is influenced by the diffusivity of the environment. mBio 8:e00353-17
Nemudryi AA, Valetdinova KR, Medvedev SP, Zakian SM (2014) TALEN and CRISPR/Cas genome editing systems: tools of discovery. Acta Nat 6:19–40
Newman MM, Lorenz N, Hoilett N, Lee NR, Dick RP, Liles MR, Ramsier C, Kloepper JW (2016) Changes in rhizosphere bacterial gene expression following glyphosate treatment. Sci Total Environ 553:32–41
Nimtz M, Mort A, Wray V, Domke T, Zhang Y, Coplin DL, Geider K (1996) Structure of stewartan, the capsular exopolysaccharide from the corn pathogen *Erwinia stewartii*. Carbohydr Res 288:189–201
Nowak MA (2006) Five rules for the evolution of cooperation. Science 314:1560–1563
Nutzmann HW, Reyes-Dominguez Y, Scherlach K et al (2011) Bacteria-induced natural product formation in the fungus *Aspergillus nidulans* requires Saga/Ada-mediated histoneacetylation. Proc Natl Acad Sci USA 108:14282–14287
Ofek-Lalzar M, Sela N, Goldman-Voronov M, Green SJ, Hadar Y, Minz D (2014) Niche and host-associated functional signatures of the root surface microbiome. Nat Commun 5:4950
Oh KB, Miyazawa H, Naito T, Matsuoka H (2001) Purification and characterization of an autoregulatory substance capable of regulating the morphological transition in *Candida albicans*. Proc Natl Acad Sci USA 98:4664–4668
Olanrewaju OS, Ayangbenro AS, Glick BR, Babalola OO (2019) Plant health: feedback effect of root exudates-rhizobiome interactions. Appl Microbiol Biotechnol 103:1155–1166
Orozco-Mosqueda MDC, Rocha-Granados MDC, Glick BR, Santoyo G (2018) Microbiome engineering to improve biocontrol and plant growth-promoting mechanisms. Microbiol Res 208:25–31
Ostrowski EA, Shen Y, Tian X, Sucgang R, Jiang H, Qu J, Katoh-Kurasawa M et al (2015) Genomic signatures of cooperation and conflict in the social amoeba. Curr Biol 25:1661–1665

Pacheco AR, Moel M, Segre D (2018) Costless metabolic secretions as drivers of interspecies interactions in microbial ecosystems. Nat Commun 10:103–115

Partida-Martinez LP, Hertweck C (2005) Pathogenic fungus harbours endosymbiotic bacteria for toxin production. Nature 437:884–888

Peiffer JA, Spor A, Koren O, Jin Z, Tringe SG, Dangl JL, Buckler ES, Ley RE (2013) Diversity and heritability of the maize rhizosphere microbiome under field conditions. Proc Natl Acad Sci USA 110:6548–6553

Pérez-Giménez J, Mongiardini EJ, Althabegoiti MJ, Covelli J, Quelas JI, López-García SL, Lodeiro AR (2009) Soybean lectin enhances biofilm formation by *Bradyrhizobium japonicum* in the absence of plants. Int J Microbiol 2009:719367–719373

Philippot L, Raaijmakers JM, Lemanceau P, van der Putten WH (2013) Going back to the roots: the microbial ecology of the rhizosphere. Nat Rev Microbiol 11:789–799

Piper KR, Beck VBS, Farrand SK (1993) Conjugation factor of *Agrobacterium tumefaciens* regulates Ti plasmid transfer by autoinduction. Nature 362:448–450

Poveda A, Santamaria M, Bernabe M, Prieto A, Briux M, Corzo J, Jimenez-Barbero J (1997) Studies on the structure and the solution conformation of an acidic extracellular polysaccharide isolated from *Bradyrhizobium*. Carbohydr Res 304:209–217

Powers ST, Penn AS, Watson RA (2011) The concurrent evolution of cooperation and the population structures that support it. Evolution 65:1527–1543

Priefer UB (1989) Genes involved in lipopolysaccharide production and symbiosis are clustered on the chromosome of *Rhizobium leguminosarum* biovar *viciae* VF39. J Bacteriol 171:6161–6168

Prikyrl Z, Vancura V, Wurst M (1985) Auxin formation by rhizosphere bacteria as a factor of root growth. Biol Plant 27:159–163

Raaijmakers JM, Leeman M, Van Oorschot MMP, Van der Sluis I, Schippers B, Bakker PAHM (1995) Dose–response relationships in biological control of fusarium wilt of radish by *Pseudomonas* spp. Phytopathology 85:1075–1081

Reinhold-Hurek B, Bunger W, Burbano CS, Sabale M, Hurek T (2015) Shaping their microbiome: global hotspots for microbial activity. Annu Rev Phytopathol 53:403–424

Ren D, Sims JJ, Wood TK (2002) Inhibition of biofilm formation and swarming of *Bacillus subtilis* by (5Z)-4-bromo-5-(bromomethylene)-3-butyl-2(5H)- furanone. Lett Appl Microbiol 34:293–299

Reuber TL, Walker GC (1993) Biosynthesis of succinoglycan, a symbiotically important exopolysaccharide of *Rhizobium meliloti*. Cell 74:269–280

Revsbech NP, Pedersen O, Reichardt W et al (1999) Microsensor analysis of oxygen and pH in the rice rhizosphere under field and laboratory conditions. Biol Fertil Soils 29:379–385

Rigano LA, Payette C, Brouillard G, Marano MR, Abramowicz L, Torres PS, Yun M, Castagnaro AP, Oirdi ME, Dufour V et al (2007a) Bacterial cyclic beta-(1,2)-glucan acts in systemic suppression of plant immune responses. Plant Cell 19:2077–2089

Rigano LA, Siciliano F, Enrique R, Sendín L, Filippone P, Torres PS, Qüesta J, Dow JM, Castagnaro AP, Vojnov AA et al (2007b) Biofilm formation, epiphytic fitness and canker development in *Xanthomonas axonopodis* pv. *citri*. Mol Plant Microbes Interact 20:1222–1230

Riley MA, Gordon DM (1999) The ecological role of bacteriocins in bacterial competition. Trends Microbiol 7:129–133

Roell GW, Jian Zha J, Carrl RR, Koffas MA, Fong SS, Tang YJ (2019) Engineering microbial consortia by division of labor. Microb Cell Fact 18:35–46

Rolfe BG, Carlson RW, Ridge RW, Dazzo RW, Mateos FB, Pankhurst CE (1996) Defective infection and nodulation of clovers by exopolysaccharide mutants of *Rhizobium leguminosarum* bv. *trifolii*. Aust J Plant Physiol 23:285–303

Roper MC, Greve LC, Labavitch JM, Kirkpatrick BC (2007) Detection and visualization of an exopolysaccharide produced by *Xylella fastidiosa in vitro* and *in planta*. Appl Environ Microbiol 73:7252–7258

Rosenberg E, Zilber-Rosenberg I (2016) Microbes drive evolution of animals and plants: the hologenome concept. mBio 7:01395-15

Ross C, Opel V, Scherlach K, Hertweck C (2014) Biosynthesis of antifungal and antibacterial polyketides by *Burkholderiagladioli* in coculture with *Rhizopus microsporus*. Mycoses 57:48–55

Rout ME, Southworth D (2013) The root microbiome influences scales from molecules to ecosystems: the unseen majority. Am J Bot 100(9):1689–1691

Rudrappa T, Czymmek KJ, Pare PW, Bais HP (2008) Root secreted malic acid recruits beneficial soil bacteria. Plant Physiol 148:1547–1556

Ryan PR, Dessaux Y, Thomashow LS, Weller DM (2009) Rhizosphere engineering and management for sustainable agriculture. Plant Soil 321:363–383

Schlaeppi K, Dombrowski N, Oter RG, Ver Loren van Themaat E, Schulze-Lefert P (2014) Quantitative divergence of the bacterial root microbiota in *Arabidopsis thaliana* relatives. Proc Natl Acad Sci USA 111:585–592

Schroeckh V, Scherlach K, Nutzmann HW et al (2009) Intimate bacterial–fungal interaction triggers biosynthesis of archetypal polyketides in *Aspergillus nidulans*. Proc Natl Acad Sci USA 106:14558–14563

Shi S, Nuccio EE, Shi ZJ, He Z, Zhou J, Firestone MK (2016) The interconnected rhizosphere: high network complexity dominates rhizosphere assemblages. Ecol Lett 19:926–936

Shou W, Ram S, Vilar J (2007) Synthetic cooperation in engineered yeast populations. Proc Natl Acad Sci USA 104:1877–1882

Strassmann JE, Queller DC (2011) Evolution of cooperation and control of cheating in a social microbe. Proc Natl Acad Sci USA 108:10855–10862

Strassmann JE, Zhu Y, Queller DC (2000) Altruism and social cheating in the social amoeba *Dictyostelium discoideum*. Nature 408:965–967

Su J, Hu C, Yan X, Jin Y, Chen Z, Guan Q et al (2015) Expression of barley SUSIBA2 transcription factor yields high-starch low-methane rice. Nature 523:602–606

Surette MG et al (1999) Quorum sensing in *Escherichia coli*, Salmonella typhimurium, and Vibrio harveyi: a new family of genes responsible for autoinducer production. Proc Natl Acad Sci USA 96:1639–1644

Talbot JM, Bruns TD, Taylor JW, Smith DP, Branco S, Glassman SI et al (2014) Endemism and functional convergence across the north American soil mycobiome. Proc Natl Acad Sci USA 111:6341–6346

Thomashow LS, Weller DM (1988) Role of a phenazine antibiotic from *Pseudomonas fluorescens* in biological control of *Gaeumannomyces graminis* var. *tritici*. J Bacteriol 170:3499–3508

Thommes M, Wang T, Zhao Q, Paschalidis IC, Segrè D (2019) Designing metabolic division of labor in microbial communities. mSystems 4:00263-18

Tomlinson AD, Ramey-Hartung B, Day TW, Merritt PM, Fuqua C (2010) *Agrobacterium tumefaciens* ExoR represses succinoglycan biosynthesis and is required for biofilm formation and motility. Microbiology 156:2670–2681

Toro M, Azcon R, Barea JM (1998) The use of isotopic dilution techniques to evaluate the interactive effects of rhizobium genotype, mycorrhizal fungi, phosphate-solubizing rhizobacteria and rock phosphate on nitrogen and phosphorus acquisition by *Medicago sativa*. New Phytol 138:265–273

Torres PS, Malamud F, Rigano LA, Russo DM, Marano MR, Castagnaro AP, Zorreguieta A, Bouarab K, Dow JM, Vojnov AA (2007) Controlled synthesis of the DSF cell-cell signal is required for biofilm formation and virulence in *Xanthomonas campestris*. Environ Microbiol 9:2101–2109

Traxler MF, Watrous JD, Alexandrov T, Dorrestein PC, Kolter R (2013) Interspecies interactions stimulate diversification of the *Streptomyces coelicolor* secreted metabolome. MBio 4(4): e00459–e00413

Turner TR, James EK, Poole PS (2013) The plant microbiome. Genome Biol 14(6):209

Uzum Z, Silipo A, Lackner G, De Felice A, Molinaro A, Hertweck C (2015) Structure, genetics and function of an exopolysaccharide produced by a bacterium living within fungal hyphae. Chembiochem 16:387–392

Vivas A, Barea JM, Azcon R (2005) Interactive effect of Brevibacillus brevis and Glomus mosseae, both isolated from Cd-contaminated soil, on plant growth, physiological mycorrhizal fungal characteristics and soil enzymatic activities in Cd polluted soil. Environ Pollut 134:257–266

Vos M, Velicer GJ (2009) Social conflict in centimeter and global-scale populations of the bacterium *Myxococcus xanthus*. Curr Biol 19:1763–1767

Wang LH, He Y, Gao Y, Wu JE, Dong YH, He C, Wang SX, Weng LX, Xu JL, Tay L, Fang RX, Zhang LH (2004) A bacterial cell-cell communication signal with cross-kingdom structural analogues. Mol Microbiol 51:903–912

Wang P, Zhong Z, Zhou J, Cai T, Zhu J (2008) Exopolysaccharide biosynthesis is important for *Mesorhizobium tianshanense*: plant host interaction. Arch Microbiol 189:525–530

West SA, Griffin AS, Gardner A, Diggle SP (2006) Social evolution theory for microorganisms. Nat Rev Microbiol 4:597–607

West SA, Diggle SP, Buckling A, Gardner A, Griffin AS (2007) The social lives of microbes. Annu Rev Ecol Evol Syst 38:53–77

Williams A, Wilkinson A, Krehenbrink M, Russo DM, Zorreguieta A, Downie JA (2008) Glucomannan-mediated attachment of *Rhizobium leguminosarum* to pea root hairs is required for competitive nodule infection. J Bacteriol 190:4706–4715

Wilson DS, Wilson EO (2008) Evolution "for the good of the group". Am Sci 96:380–389

Wintermute EH, Silver PA (2010) Emergent cooperation in microbial metabolism. Mol Syst Biol 6:407

Xavier KB, Bassler BL (2003) LuxS quorum sensing: more than just a numbers game. Curr Opin Microbiol 6:191–197

Yi Y, de Jong A, Fremzel E, Kuiperrs OP (2017) Comparative transcriptomics of *Bacillus mycoides* strains in response to potato-root exudates reveals different genetic adaptation of endophytic and soil isolates. Front Microbiol 8:1487

Yu P, Wang C, Baldauf JA, Tai H, Gutjahr C, Hochholdinger F et al (2018) Root type and soil phosphate determine the taxonomic landscape of colonizing fungi and the transcriptome of field-grown maize roots. New Phytol 217:1240–1253

Yuan H, Zhu Z, Liu S et al (2016) Microbial utilization of rice root exudates: 13C labeling and PLFA composition. Biol Fertil Soils 52:615–627

Zhang N, Wang D, Liu Y, Li S, Shen Q, Zhang R (2014) Effects of different plant root exudates and their organic acid components on chemotaxis, biofilm formation and colonization by beneficial rhizosphere-associated bacterial strains. Plant Soil 374(1–2):689–700

Zhang Y, Ruyter-Spira C, Bouwmeester HJ (2015) Engineering the plant rhizosphere. Curr Opin Biotechnol 32:136–142

Zhao Z, Gunina A, Ge T, Kuzyakov Y (2018) Carbon and nitrogen availability in paddy soil affects rice photosynthate allocation, microbial community composition, and priming: combining continuous $^{13}C$ labeling with PLFA analysis. Geophys Res Abs 20. https://doi.org/10.1007/s11104-018-3873-5

Zhou T, Chen D, Li C, Sun Q, Li L, Liu F, Shen B (2012) Isolation and characterization of *Pseudomonas brassicacearum* J12 as an antagonist against *Ralstonia solanacearum* and identification of its antimicrobial components. Microbiol Res 167(7):388–394

The manufacturer's authorised representative in the EU is Springer Nature Customer Service Centre GmbH, Europaplatz 3, 69115 Heidelberg, Germany. If you have any concerns regarding our products, please contact ProductSafety@springernature.com

Printed and bound by CPI Group (UK) Ltd, Croydon, CR0 4YY
30/09/2026
02225648-0016